Synthetic Chemistry of Amides

Synthetic Chemistry of Amides

Editor: Caleb Murphy

NY RESEARCH PRESS

New York

Published by NY Research Press
118-35 Queens Blvd., Suite 400,
Forest Hills, NY 11375, USA
www.nyresearchpress.com

Synthetic Chemistry of Amides
Edited by Caleb Murphy

International Standard Book Number: 978-1-64725-454-4 (Hardback)

Cataloging-in-Publication Data

Synthetic chemistry of amides / edited by Caleb Murphy.
p. cm.
Includes bibliographical references and index.
ISBN 978-1-64725-454-4
1. Amides--Synthesis. 2. Organic compounds--Synthesis. 3. Chemistry. I. Murphy, Caleb.
QD341.A7 S96 2023
547.04--dc23

Contents

Preface

Over the recent decade, advancements and applications have progressed exponentially. This has led to the increased interest in this field and projects are being conducted to enhance knowledge. The main objective of this book is to present some of the critical challenges and provide insights into possible solutions. This book will answer the varied questions that arise in the field and also provide an increased scope for furthering studies.

Amides are functional groups in which a carbonyl carbon atom is linked by a single bond to a nitrogen atom, and either a hydrogen or a carbon atom. The general formula of amide is RC (=O)NR′R″, where R, R′, and R″ represent organic groups or hydrogen atoms. Some common examples of amides are benazamide, acetamide, nylon, paracetamol and dimethylformamide. The amide bond is found in many naturally occurring and synthetic compounds. There are several areas of chemistry such as biochemistry, organic synthesis, polymers and drug discovery, wherein the amide groups are used as the fundamental functional group. Amide groups can participate in a wide variety of transformations synthetically and are particularly versatile when exposed to electrochemical conditions. This book aims to shed light on the synthetic chemistry of amides. It presents researches and studies performed by experts across the globe. A number of latest researches have been included to keep the readers up-to-date with the global concepts in this area of study.

I hope that this book, with its visionary approach, will be a valuable addition and will promote interest among readers. Each of the authors has provided their extraordinary competence in their specific fields by providing different perspectives as they come from diverse nations and regions. I thank them for their contributions.

Editor

1

Synthetic Methods of Phosphonopeptides

Jiaxi Xu

State Key Laboratory of Chemical Resource Engineering, Department of Organic Chemistry, College of Chemistry, Beijing University of Chemical Technology, Beijing 100029, China; jxxu@mail.buct.edu.cn

Abstract: Phosphonopeptides are phosphorus analogues of peptides and have been widely applied as enzyme inhibitors and antigens to induce catalytic antibodies. Phosphonopeptides generally contain one aminoalkylphosphonic acid residue and include phosphonopeptides with C-terminal aminoalkylphosphonic acids and phosphonopeptides with a phosphonamidate bond. The phosphonamidate bond in the phosphonopeptides is generally formed via phosphonylation with phosphonochloridates, condensation with coupling reagents and enzymes, and phosphinylation followed by oxidation. Pseudo four-component condensation reaction of amides, aldehydes, alkyl dichlorophosphites, and amino/peptide esters is an alternative, convergent, and efficient strategy for synthesis of phosphonopeptides through simultaneous construction of aminoalkylphosphonic acids and formation of the phosphonamidate bond. This review focuses on the synthetic methods of phosphonopeptides containing a phosphonamidate bond.

Keywords: phosphonamidate; phosphonopeptide; β-phosphonopeptide; γ-phosphonopeptide; peptide

---

1. Introduction

Phosphonopeptides are phosphorus analogues of peptides. They generally contain one aminoalkylphosphonic acid residue and include phosphonopeptides with C-terminal aminoalkylphosphonic acids and phosphonopeptides containing a phosphonamidate bond [1,2]. Phosphonopeptides have been used as antibacterial agents [3]. They have been widely applied as enzyme inhibitors [4–8] and antigens for inducing catalytic antibodies [9–12] due to their tetrahedral structural feature. Phosphonopeptides containing C-terminal aminoalkylphosphonic acids have been prepared via coupling of *N*-protected amino acyl chlorides with aminoalkylphosphonic acids [13], condensation of *N*-protected amino acids or peptides and aminoalkylphosphonic acids with coupling reagents [14,15], aminolysis of *N*-chloroacetyl aminoalkylphosphonic acids [16], and the Mannich-type reactions of *N*-protected amino amides or aminoalkanesulfonamides, aldehydes, and phosphorus trichloride followed by hydrolysis [17,18]. Synthesis of phosphonopeptides with C-terminal aminoalkylphosphonic acids was reviewed recently [19]. This review focuses on the synthetic methods of phosphonopeptides, including α-, β-, γ-, and δ-phosphonopeptides (Figure 1), with a phosphonamidate bond, especially focuses on the synthetic strategies for the formation of the phosphonamidate bond, excluding the modification of phosphonopeptides.

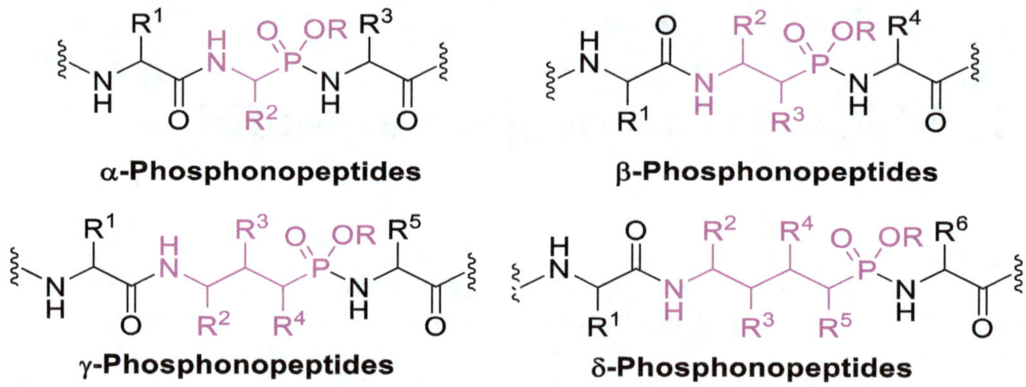

Figure 1. Different classes of phosphonopeptides.

2. Synthesis of Phosphonopeptides via Phosphonochloridates

Phosphonylation of amino/peptide esters with alkyl *N*-protected aminoalkylphosphonochloridates is a general and widely applied method for the synthesis of phosphonopeptides containing a phosphonamidate bond. The phosphonochloridates are usually prepared via chlorination of the corresponding dialkyl phosphonates with phosphorus pentachloride [13,20] or phosphorus oxychloride [21], chlorination of phosphonic monoesters with thionyl chloride [22,23] or oxalyl chloride [24,25], and chlorination of alkyl trimethylsilyl phosphonates [26] or alkyl phosphinates [27] with carbon tetrachloride. Phosphonobromidates are seldom applied as intermediates in the synthesis of phosphonopeptides and generated via bromination of alkyl phosphinates with bromine [28]. Each of the above-mentioned methods will be presented as following.

2.1. Chlorination of Dialkyl Phosphonates with Phosphorus Pentachloride

After aminomethylphosphonic acid was isolated from numerous organisms and animal and human organs [29,30], to understand the biological significance of this new class of compounds, phosphonodipeptide was synthesized in 1973. Diisopropyl *N*-phthalyl(Phth)aminomethylphosphonate was prepared from phthalylaminomethyl chloride and triisopropyl phosphite and further treated with phosphorus pentachloride to give the corresponding isopropyl *N*-phthalylaminomethylphosphonochloridate, which reacted with ethyl glycinate in the presence of triethylamine to give rise to protected phosphonopeptide (Scheme 1) [13]. This is the first chemical synthesis of α-phosphonopeptide with a phosphonamidate linkage.

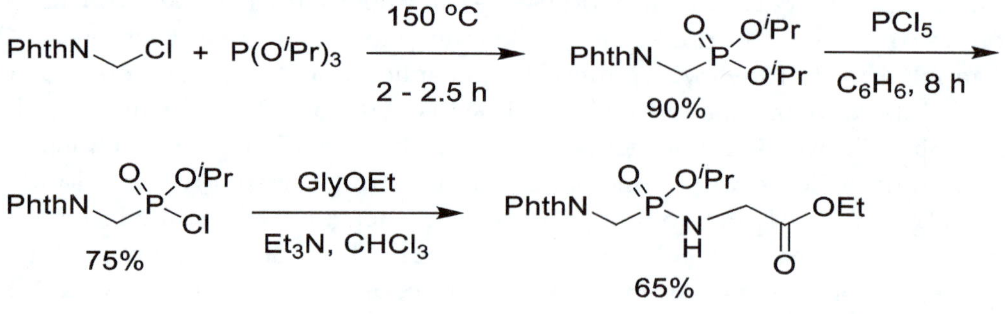

Scheme 1. First synthesis of phosphonopeptide.

Similarly, *N*-phthalylaminomethylphosphonochloridate was prepared and further coupled with dipeptide esters to afford phosphonotripeptides. After hydrazinolysis and acetylation with acetic anhydride and acylation with acyl chlorides or *N*-benzyloxycarbonyl(Cbz)-protected dipeptides, the phosphonotripeptides were transformed into *N*-acetyl phosphonotripeptides, *N*-acyl phosphonotripeptides, and phosphonopentapeptides after hydrogenolysis, respectively (Scheme 2). As inhibitors of enkephalinase and angiotensin-converting enzyme (ACE), these phosphonopeptides exhibited good inhibitory potency against enkephalinase with several of the analogs having K_i values in the submicromolar range as contrasted to micromolar or higher toward ACE [20]. Another series of phosphonopeptides were synthesized as potential inhibitors of ACE [21]. The phosphonamide bond in the phosphonopeptides is generally stable under weak acidic and basic conditions.

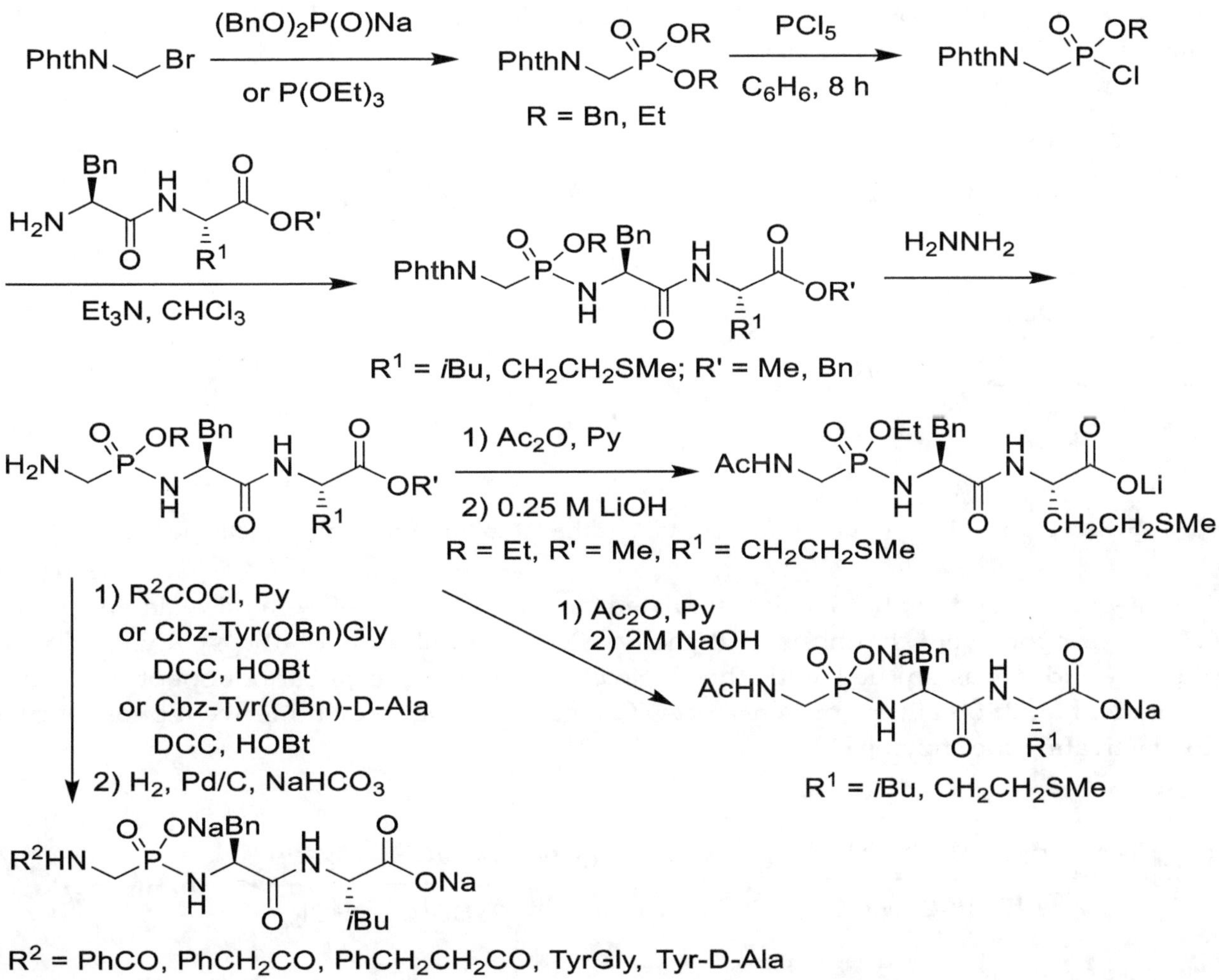

Scheme 2. Synthesis of phosphonopeptides as inhibitors of enkephalinase and angiotensin-converting enzyme.

2.2. Chlorination of Dialkyl Phosphonates with Phosphorus Oxychloride

Besides phosphorus pentachloride, phosphorus oxychloride was also applied in the conversion of dialkyl phosphonates into phosphonochloridates. Phosphonopeptide was synthesized from ethyl glycylglycinate hydroloride and ethyl phosphonochloridate derived from direct chlorination of diethyl phosphonate with phosphorus oxychloride (Scheme 3) [21].

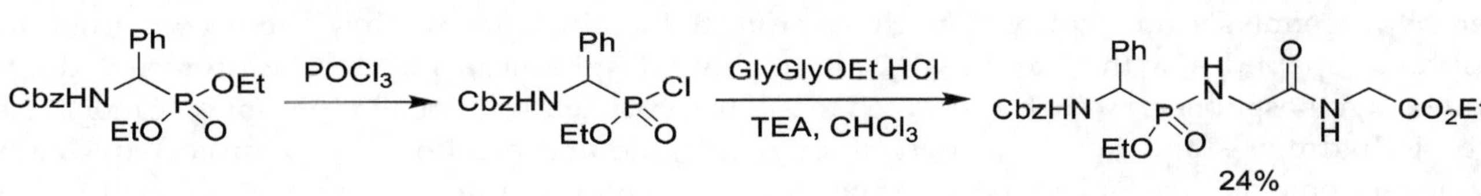

Scheme 3. Synthesis of phosphonopeptide from phosphonochloridate generated by chlorination of diethyl phosphonate with phosphorus oxychloride.

2.3. Chlorination of Alkyl Phosphonic Acid Monoesters with Thionyl Chloride

To search for competitive inhibitors of D-alanine:D-alanine ligase, phosphonodipeptide was prepared from methyl alaninate and *N*-Cbz-protected methyl 1-aminoethylphosphonochloridate, which was prepared from *N*-Cbz-protected diphenyl 1-aminoethylphosphonate via transesterification, selective hydrolysis, and chlorination with thionyl chloride (Scheme 4). The phosphonodipeptide was the competitive inhibitor of D-alanine:D-alanine ligase, with K_i close to 10^{-6} M [22].

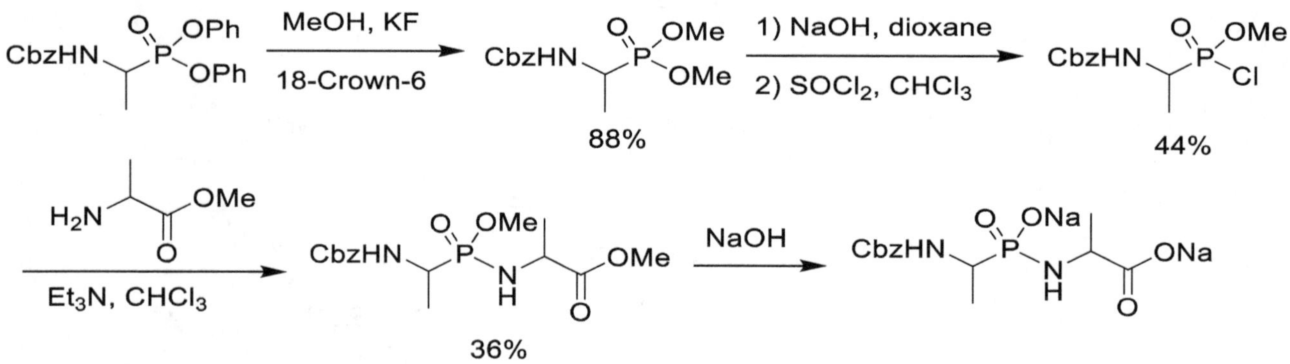

Scheme 4. Synthesis of phosphonopeptide as the inhibitor of D-alanine:D-alanine ligase.

Some α-phosphonodipeptides which simulated the transition state of D-alanyl-D-alanine synthetase (EC 6.3.2.4) reaction were synthesized (Scheme 5). The inhibition of the synthetase by these phosphonopeptides was studied with the *S. faecalis* enzyme. The phosphonopeptides exhibited time-dependent inhibition in the presence of ATP, suggesting that they underwent phosphorylation prior to inactivating the enzyme [23].

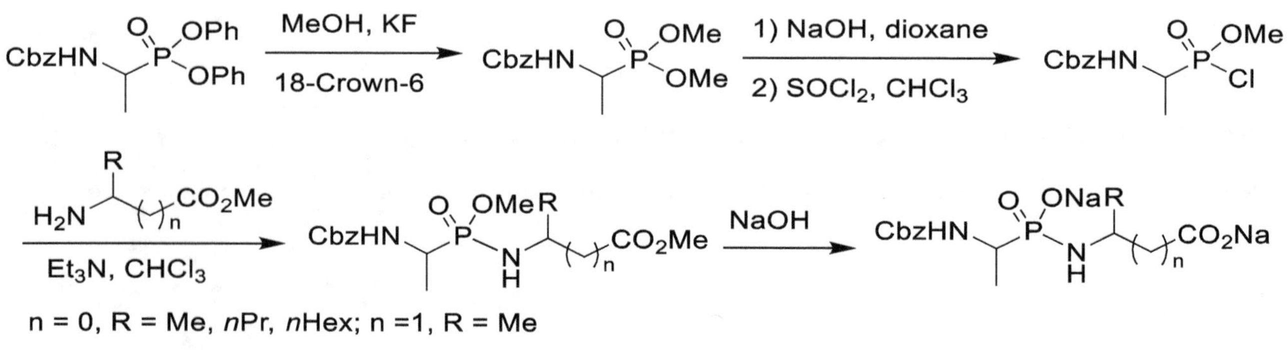

Scheme 5. Synthesis of phosphonopeptides from aminoalkylphosphintes.

N-Cbz-protected diethyl 1-aminoalkylphosphonates were prepared via three component condensation reaction of benzyl carbamate, aldehydes, and diethyl phosphite in acetyl chloride. After hydrogenolysis and coupled with *N*-protected amino esters, they were converted into phosphonodipeptides with C-terminal 1-aminoalkylphosphonic acids. *N*-Cbz-protected diethyl 1-aminoalkylphosphonates were transformed to *N*-Cbz-protected 1-aminoalkylphosphonic monoesters via basic hydrolysis and treatment with thionyl chloride and alcohol. *N*-Cbz-protected 1-amino alkylphosphonic monoesters were treated with thionyl chloride followed by reactions with amino esters or 1-aminoalkylphosphonates, affording phosphonopeptides composing of one or two 1-aminoalkylphosphonic acid residue(s) (Scheme 6) [31].

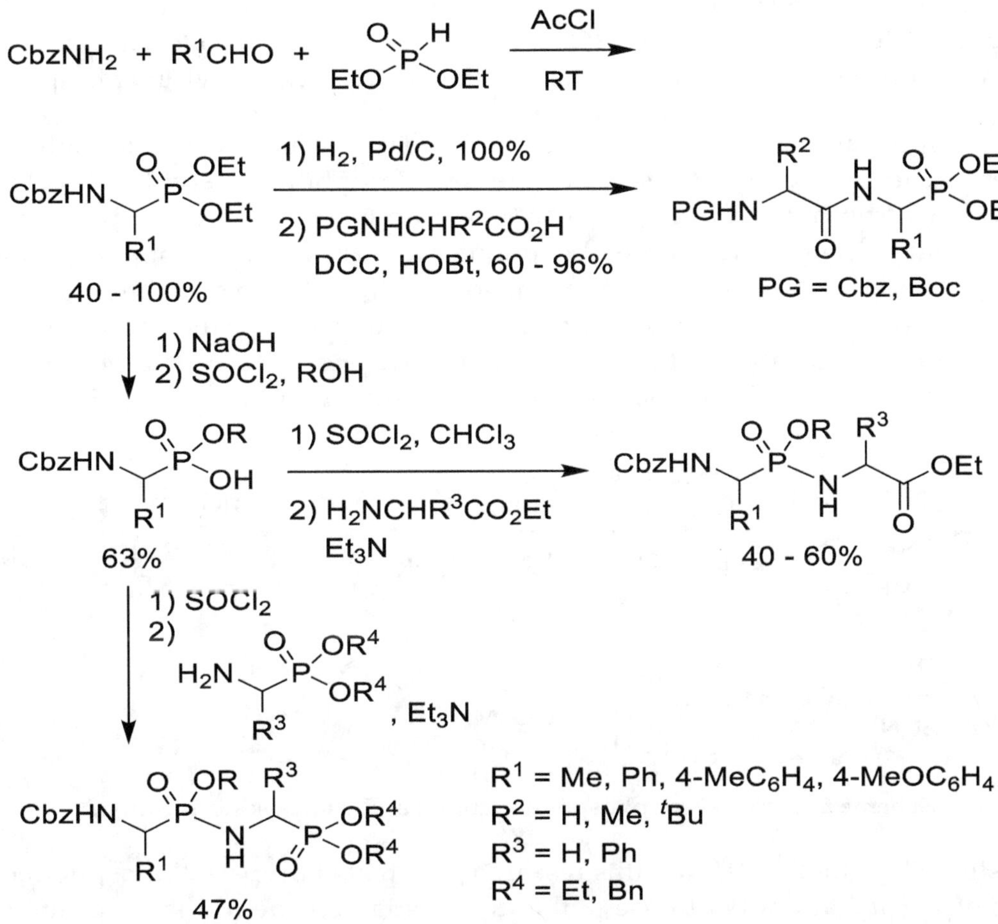

Scheme 6. Synthesis of phosphonopeptides composing of one or two 1-aminoalkylphosphonic acid residue(s).

To prepare novel inhibitors of VanX, a Zn(II) metalloenzyme that was required for high-level vancomycin resistance in bacteria, *N*-[(1-aminoethyl)hydroxyphosphinyl]-D-alanine was synthesized via coupling of *N*-Cbz protected methyl 1-aminoethylphosphonochloridate with methyl D-alaninate followed by basic hydrolysis and hydrogenolysis (Scheme 7). Bioassay results indicated that phosphonodipeptide was shown to be a partial competitive inhibitor of VanX with a *K*i of 36 μM [7,32].

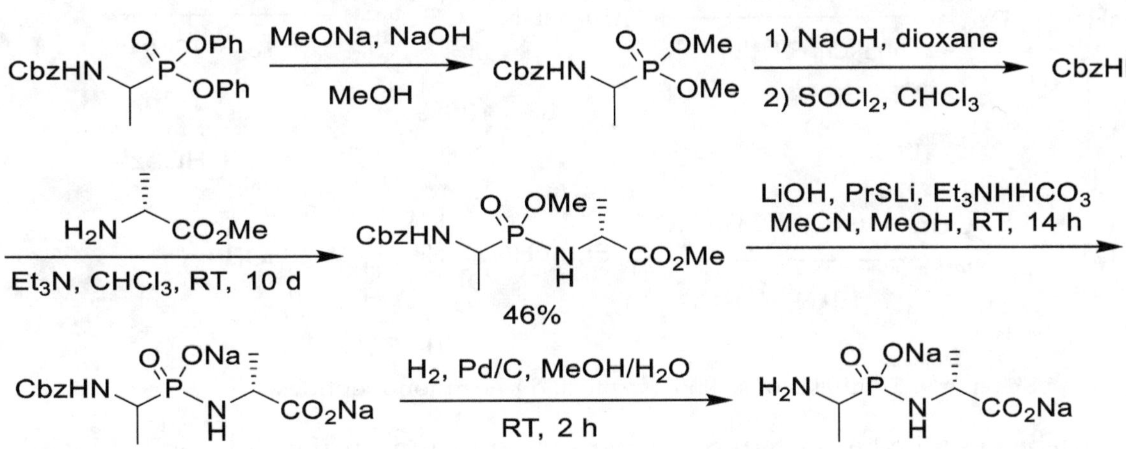

Scheme 7. Synthesis of phosphonopeptide inhibitors of VanX.

Phosphonopeptides $CbzNHCH_2P(O)(OEt)$-L-ProOBn, $CbzNHCH_2P(O)(OEt)$-D-Pro, $CbzNHCH_2P$ (O)(OEt)-L-thioProOMe, and $H_2NCH_2P(O)(OEt)$-L-Pro were prepared following similar method for investigation on the relative catalytic efficiency of β-lactamase catalyzed acyl and phosphyl transfer [33].

O-Phenyl phosphonamidates have been designed to bind covalently by nucleophilic substitution to the serine residue in the active site of serine proteases. Phosphonodipeptide with *O*-phenyl on the phosphorus atom was synthesized from phenyl *N*-Cbz 1-amino(phenyl)methylphosphonochloridate and methyl L-valinate. The stability of the phosphonamidates as a model of phosphonopeptides in aqueous solutions and their selectivity in the reaction against alcohols vs. thiols proved that they constituted a class of potential inhibitors of serine proteases and valuable tools to study the mechanism of inhibition. The transesterification of the phosphonopeptide phenyl ester with MeOH in the presence of Et_3N/KF gave the corresponding phosphonopeptide methyl ester (Scheme 8) [34].

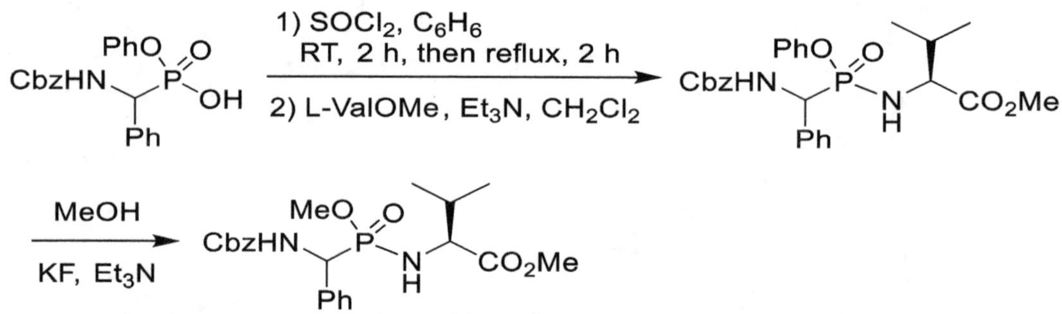

Scheme 8. Synthesis of phosphonopeptide and its transesterification.

By using similar synthetic method, this research group also prepared phosphonodipeptides as new inhibitors of leucine aminopeptidase [35] and investigated the influence of the nature of the *N*-terminal functional group on their hydrolysis [36

Besides Cbz protecting group, Fmoc group was also used in the synthesis of phosphonopeptides. Benzyl hydrogen α-(9-fluorenylmethoxycarbonyl(Fmoc)amino)alkylphosphonates were obtained by the sodium metaperiodate oxidation of the corresponding phosphinates and converted to the phosphonochloridates, which were coupled with N^{ε}-protected lysine benzyl ester to afford phosphonodipeptides containing a lysine residue (Scheme 9) [37].

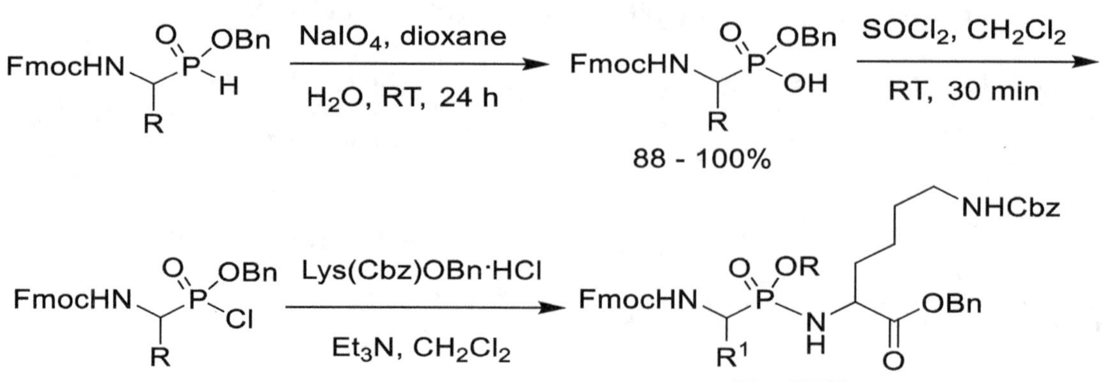

Scheme 9. Synthesis of *N*-Fmoc protected phosphonopeptides.

Cramer and Klebe described a procedure for the synthesis and purification of functionalized phosphonopeptides that were able to generate inhibitors for the metalloprotease thermolysin for use in biophysiological experiments. The method utilized an allyl ester/allyloxycarbonyl(Aloc) protection strategy and showed advantage of a fast and effective solid-phase purification step. They first prepared diallyl *N*-Cbz aminomethylphosphonate and transformed it into allyl *N*-Cbz aminomethylphosphonochloridate, which was further reacted with dipeptide allyl esters or amino amide derivatives. After deprotection under the catalysis of $Pd(PPh_3)_4$ followed by treatment with LiOH, phosphonopeptide lithium salts were obtained (Scheme 10). By using the strategy, they synthesized a series of highly polar phosphonopeptide inhibitors with amino- and hydroxy-functionalized side chains in excellent purity [38].

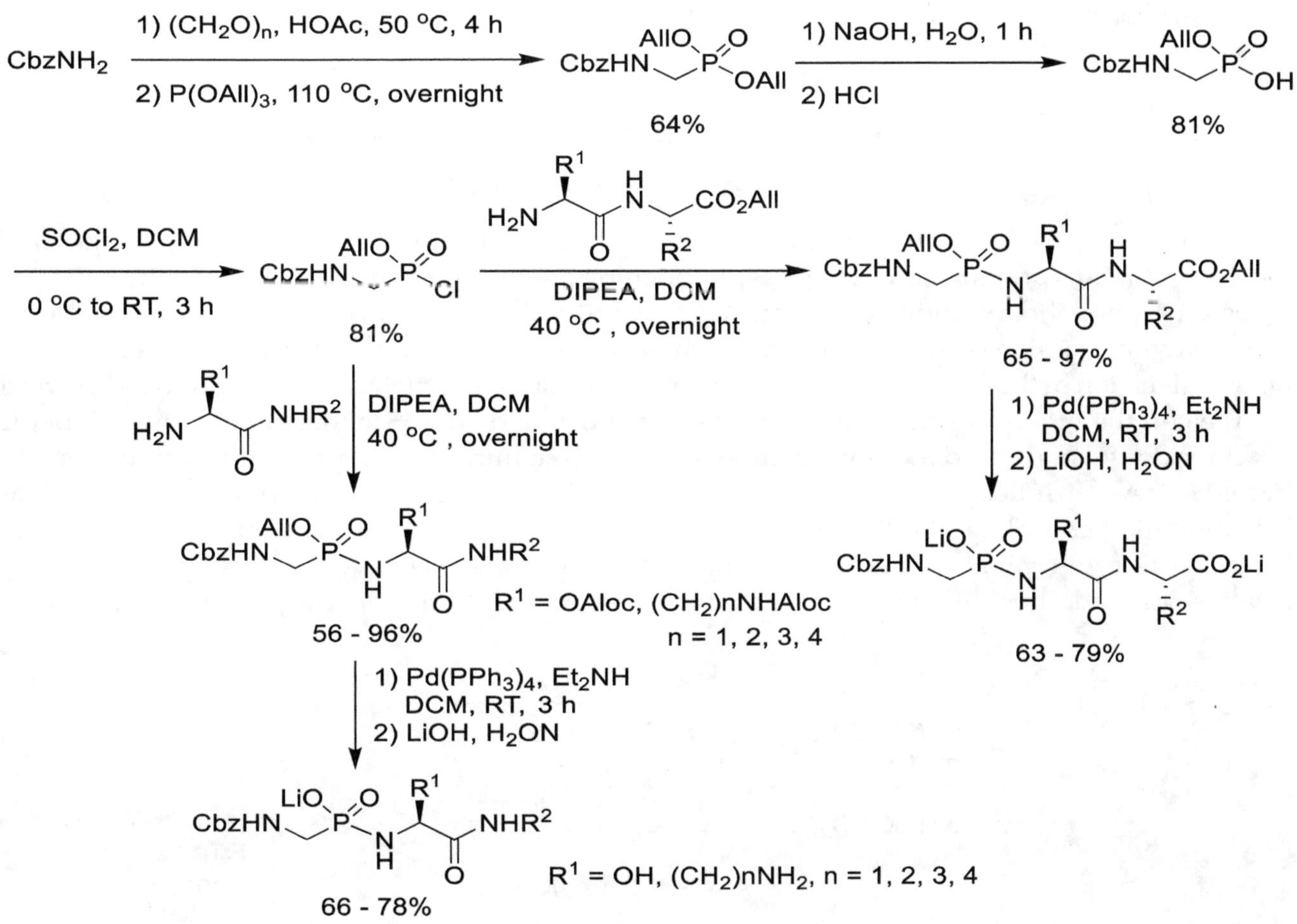

Scheme 10. Synthesis of phosphonopeptides from diallyl *N*-Cbz aminomethylphosphonate.

To search for effective and stable inhibitors of cytosolic leucine aminopeptidase, a β-phosphonopeptide containing aromatic *N*-terminal amino group was designed and synthesized. Diethyl 2-nitrophenylphosphonate was prepared and converted to the corresponding phosphonochloridate, which was coupled with methyl glycinate hydrochloride followed by reduction and basic hydrolysis, affording β-phosphonopeptide. The decrease in basicity of the terminal amino moiety of the β-phosphonopeptide resulted in satisfactory improvement of hydrolytic stability of the P–N bond. However, it did not exhibit inhibition activity up to millimolar concentration in enzymic assays towards leucine aminopeptidase possibly because diminishing the basic character of the terminal amino group resulted in a change of its affinity towards the zinc ions in the aminopeptidase. On the other hand, the decrease of the inhibition activity might also be ascribed to the steric constrain induced by the planar phenyl ring (Scheme 11) [39].

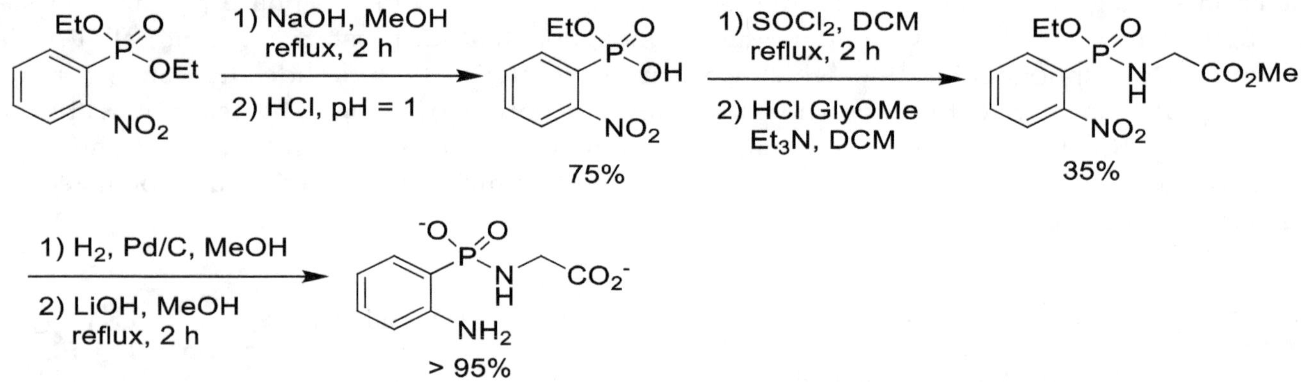

Scheme 11. Synthesis of β-phosphonopeptide containing aromatic *N*-terminal amino group.

2.4. Chlorination of Alkyl Phosphonic Acid Monoesters with Oxalyl Chloride

Moroder's research group tested different chlorination conditions and found that oxalyl chloride-mediated preparation of phosphonochloridates allowed the improvement of the synthesis of phosphonopeptides, in comparison with thionyl chloride. A catalytic amount of *N,N*-dimethylformamide (DMF) promoted the chloridation. Particularly if AgCN was used as catalyst or upon conversion of the corresponding phosphonochloridates into their 7-aza-1-hydroxybenzotriazole (HOAt) ester, the yields of phosphonopeptides even reached 90% (Scheme 12) [24]. When peptide hydrochlorides were utilized as amino components triethylamine or diisoproylethylamine was added as base. However, triphosgene was an inefficient chloridating reagent. Furthermore, the Mukaiyama procedure failed completely in the present case. The mixed anhydrides of aminoalkylphosphonic acid monoesters with pivaloyl chloride generated, but affording exclusively to the *N*-pivaloylpeptide derivatives instead phosphonopeptides, despite the steric bulk of the *tert*-butyl group [24].

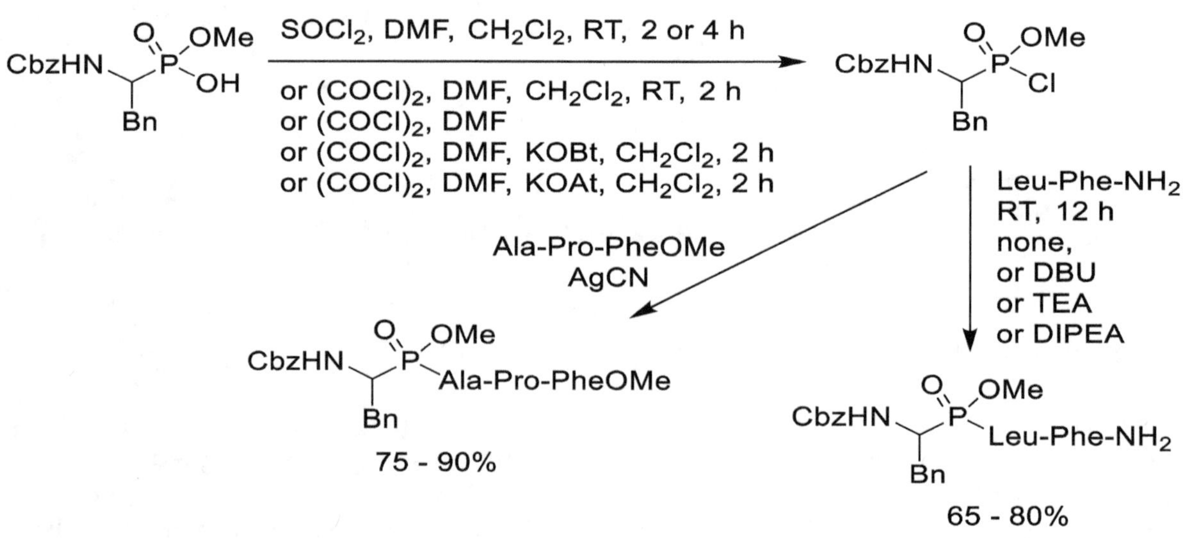

Scheme 12. Synthesis of phosphonopeptides under mild conditions.

Various methods have been explored for the convergent synthesis of phosphonopeptides via phosphonochloridates. PCl_5 is an effective reagent for the conversion of diethyl and diisopropyl phosphonates to the corresponding phosphonochloridates in the absence of complex functionality. Subsequent reaction with amino and peptide esters generated the phosphonopeptides in good yields. However, a milder method using oxalyl chloride to generate the phosphonochloridates from phosphonic monoesters was required in the presence of more complex functionality. Aminolysis of the dimethyl phosphonates or basic hydrolysis of diethyl phosphonates generated the requisite phosphonic monoesters for the preparation of phosphonochloridates [40].

New muramyl dipeptide (MDP) analogs related to LK 423 as potential immunomodulators was synthesized by coupling of methyl *N*-Fmoc 1-aminoethylphosphonochloridate and dibenzyl D-glutamate toluenesulfonate, followed by basic deprotection and coupling with 5-phthalimidopentanoic acid. The dipeptide part of the lead compound was modified by introducing a phosphonamidate bond instead of the amide bond between L-alanine and D-glutamic acid (Scheme 13) [25].

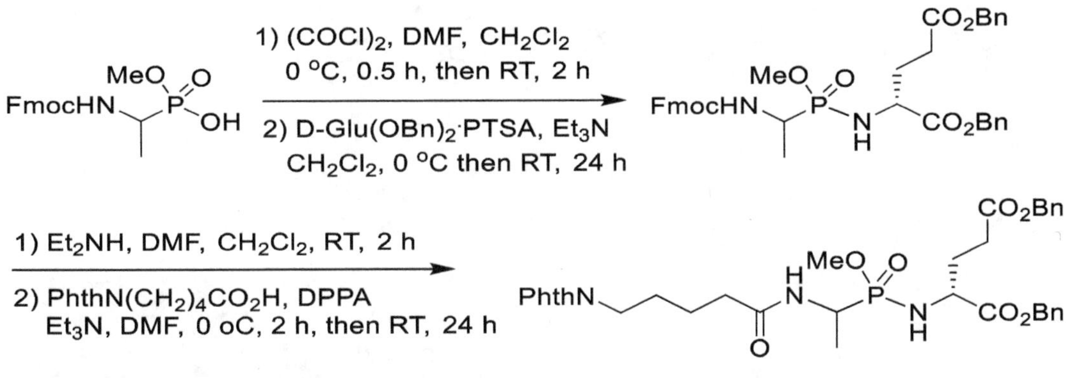

Scheme 13. Synthesis of phosphonopeptides as potential immunomodulators.

The synthesis of different muramyl dipeptide analogue LK 415 derivatives as potential immunomodulators was reported. The D-alanine and D-isoglutamine of LK 415 were replaced by their phosphonic analogues L,D-phosphonoalanine [$H_2NCHMeP(O)(OMe)_2$] and $H_2NCH[CH_2CH_2P(O)(OEt)_2]COCH_2Ph$, yielding the LK 415 analogues *N*-[2-[2-[(1-adamantylcarbonyl) amino]ethoxy]acetyl]-NHCHMeP(O)(OMe)-D-Glu($OEt)_2$ and *N*-[2-[2-[(1adamantylcarbonyl)amino] ethoxy]acetyl]-L-Ala-NHCH[$CH_2CH_2P(O)(OEt)_2$]$COCH_2Ph$, respectively [41].

To prepare phosphorus analogues of γ-glutamyl peptide, *N*-Cbz L-glutamic acid was first transformed to the corresponding dimethyl phosphonate. After aminolysis and chlorination it was transformed to the corresponding phosphonochloridate, which was reacted with diethyl glutamate, affording γ-phosphonodipeptide in 64% yield. After hydrogenolysis, *N*-terminal free γ-phosphonopeptide was obtained in 87% yield. However, the coupling of the phosphonochloridate with diethyl 2-hydroxyglutarate generated γ-phosphonodepsipeptide in only 6.7% yield, indicating the strategy was not suitable for the synthesis of γ-phosphonodepsipeptide (Scheme 14) [42].

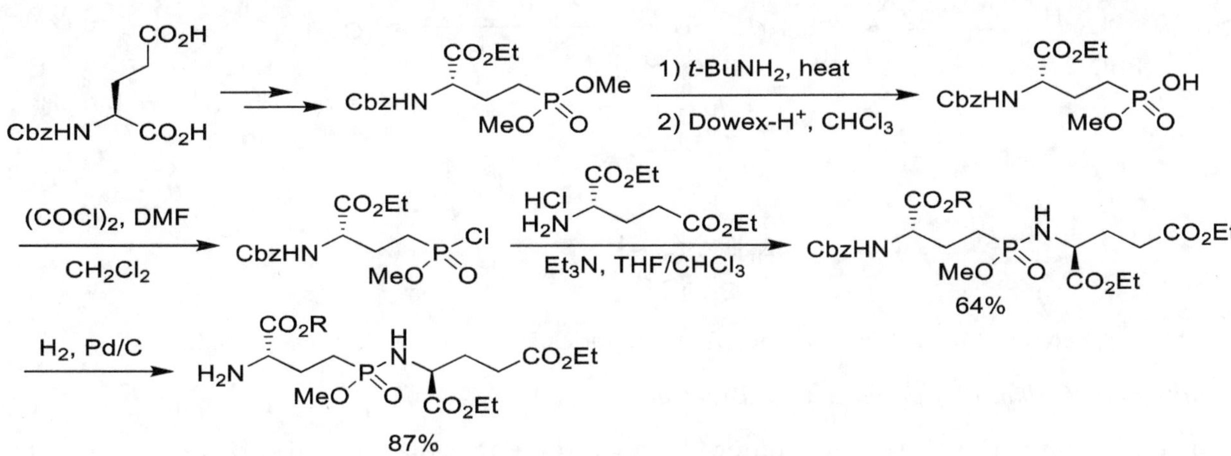

Scheme 14. Synthesis of γ-phosphonopeptides.

2.5. *Chlorination of Alkyl Trimethylsilyl Phosphonites or Alkyl Phosphinates with Carbon Tetrachloride*

To minimize the side reactions (formation of oxazaphospholines) of *N*-Cbz protected 1-aminoalkylphosphonochloridates, a convenient method for the synthesis of phosphonopeptides was described. *N*-Cbz protected 2-aminoalkylphosphinates were converted to their trimethylsilyl phosphonite tautomers with the treatment with bis(trimethylsilyl)acetamide. They were oxidized with CCl_4 to generate the corresponding phosphonochloridates as intermediates, which reacted with amino acid esters to give rise to the desired phosphonopeptides. The oxidative activation was carried out in the presence of the amine nucleophiles so that stoichiometric formation of the phosphonochloridates were avoided and side reactions were minimized (Scheme 15) [26].

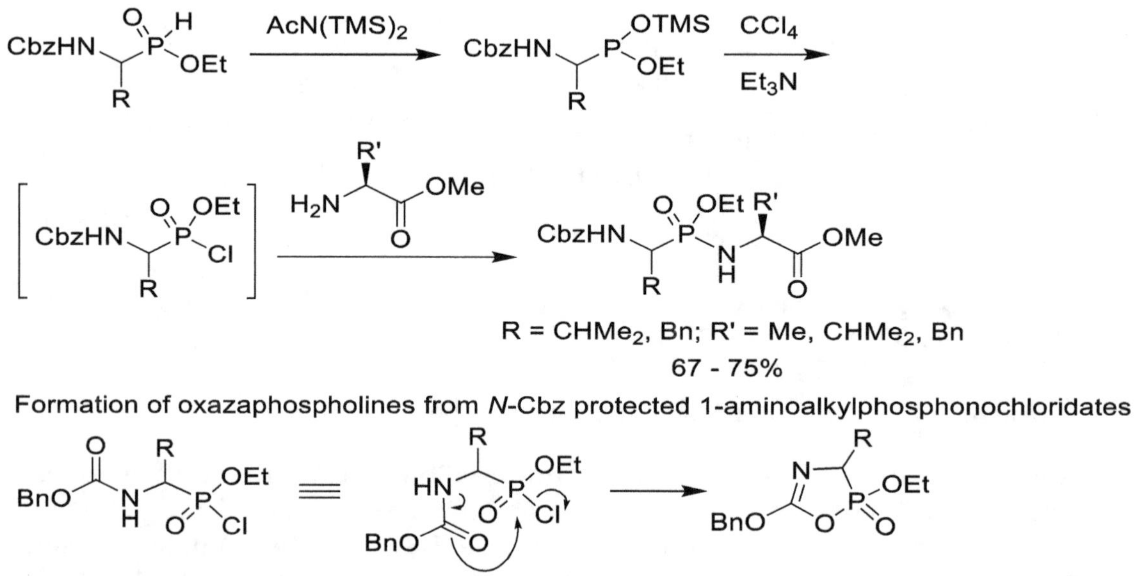

Scheme 15. Synthesis of phosphonopeptides via chlorination of aminoalkylphosphonites.

Phosphonylation of amino acid esters with *N*-Cbz 1-aminoalkylphosphonochloridates was a general method for the preparation of phosphonopeptides containing a phosphonamidate bond. *N*-Cbz 1-aminoalkylphosphonochloridates were prepared from both *N*-Cbz 1-aminoalkylphosphinates with carbon tetrachloride in the presence of trimethylamine through the Atherton–Todd reaction and the corresponding *N*-Cbz 1-aminoalkylphosphonic acid monoesters with thionyl chloride (Scheme 16) [27].

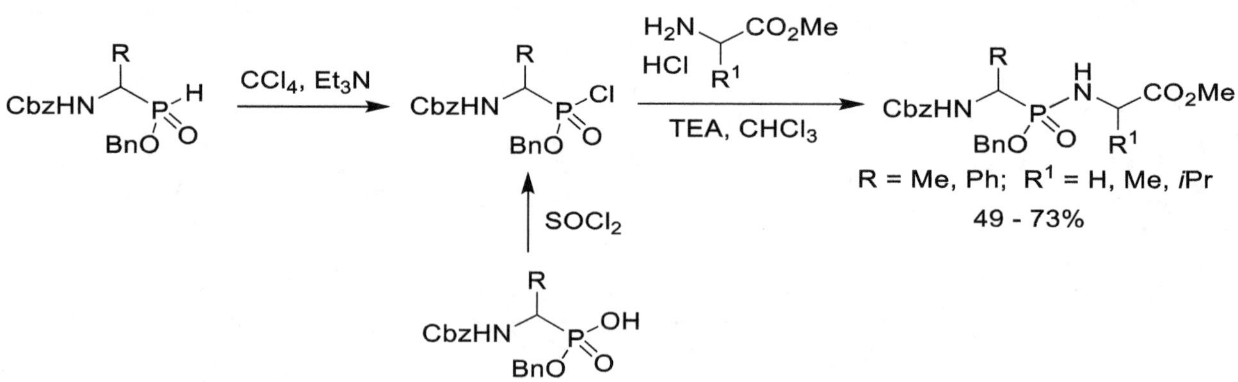

Scheme 16. Synthesis of phosphonopeptides from aminoalkylphosphinates.

2.6. Bromination of Alkyl Phosphinates with Bromine

Yao and Yuan developed a general and efficient one-pot procedure for converting 1,1-diethoxyalkylphosphinates into phosphonates or phosphonamides by the use of bromine. For α-aminoalkylphosphinates, the transformation could be realized without the protection of the amino group. Enantiopure phosphinate reacted stereospecifically with bromine and subsequently couples with nucleophile to generate the corresponding optically active derivatives with retention of configuration at the phosphorus center (Scheme 17) [28]. The developed method was applied in the synthesis of phosphonopeptides.

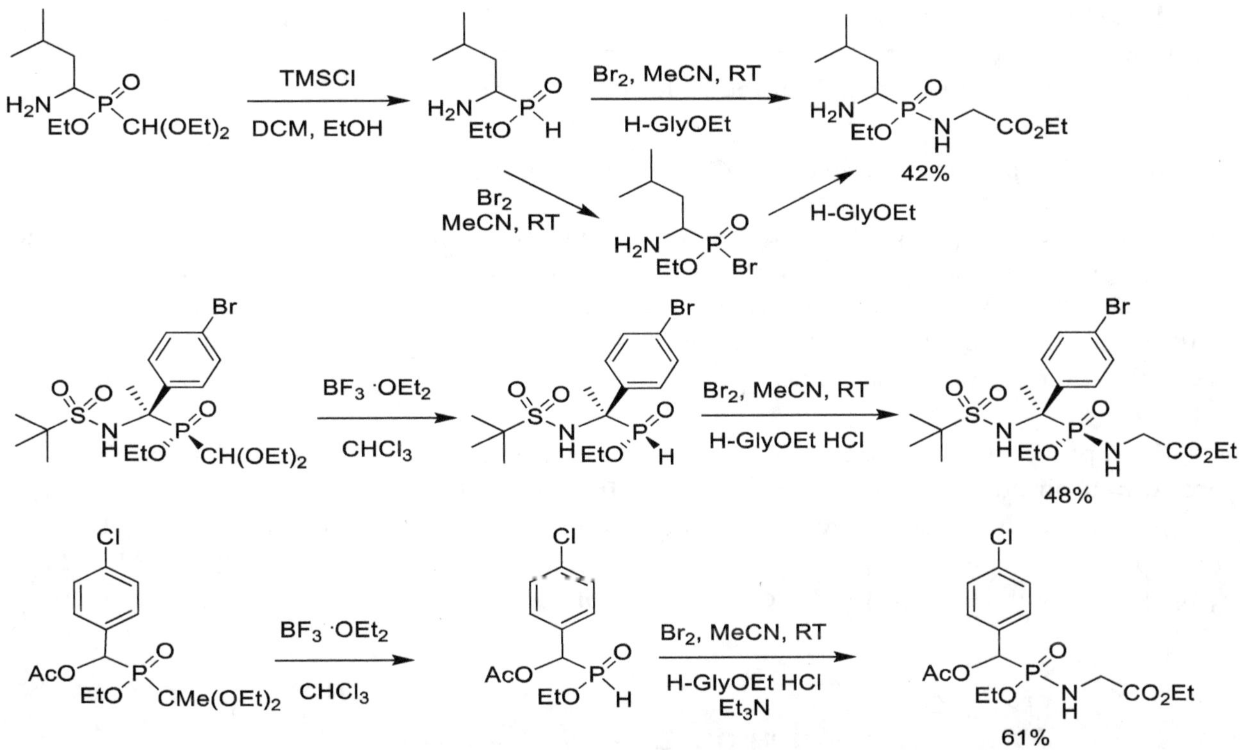

Scheme 17. Synthesis of phosphonopeptides via phosphonobromidates.

A new method for the synthesis of *tert*-butoxycarbonyl(Boc)-protected phosphonopeptides was developed with diethyl 1-azidoalkylphosphonates as starting materials. After partial hydrolysis and treatment with oxalyl chloride under the catalysis of DMF, diethyl 1-azidoalkylphosphonates were transformed to ethyl 1-azidoalkylphosphonochloridates, which were coupled with amino ester hydrochlorides to afford azidophosphonodipeptides. They were further converted *N*-Boc protected phosphonodipeptides under hydrogenolysis in the presence of Boc_2O in ethyl acetate (Scheme 18) [43].

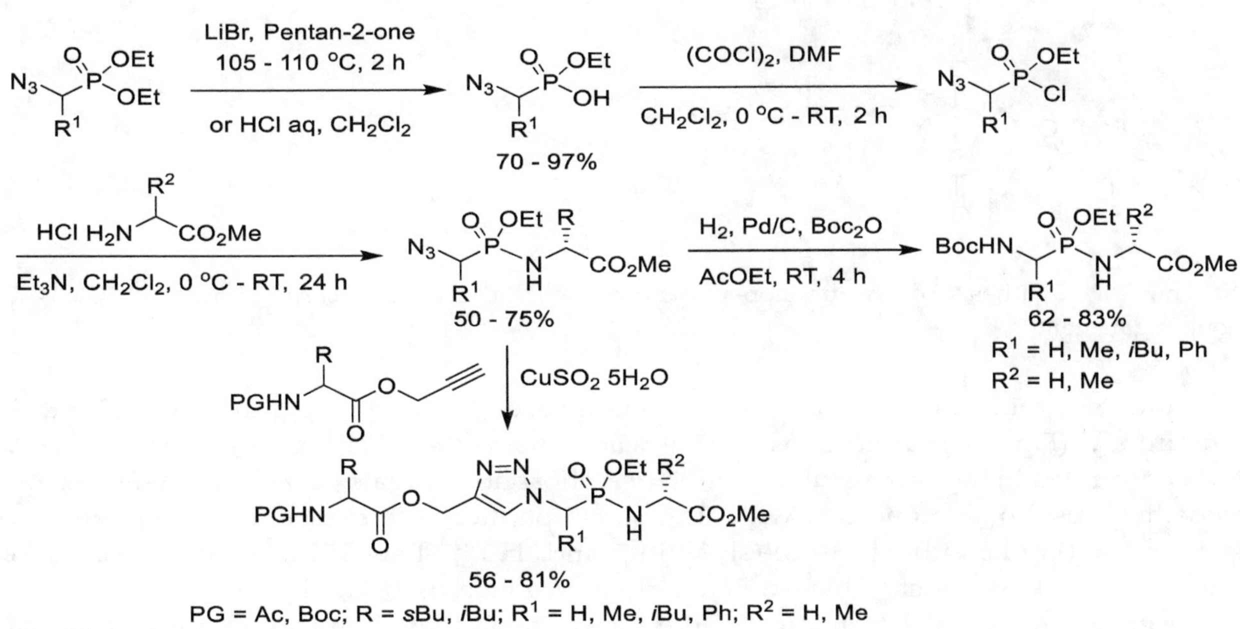

Scheme 18. Synthesis of phosphonopeptides and triazole-containing phosphonopeptides via 1-azidoa lkylphosphonochloridates.

The click reaction of azidophosphonodipeptides and *N*-protected leucine/isoleucine propargyl esters with different protecting group (PG) under the catalysis of copper sulfate pentahydrate, affording triazole-containing phosphonopeptides (Scheme 18) [44].

3. Synthesis of Phosphonopeptides with Coupling Reagents

Condensation of *N*-protected amino acids or peptides with amino/peptide esters is a general method for peptide synthesis. The method has also been applied in the synthesis of phosphonopeptides with carbodiimides, *N,N′*-dicyclohexylcarbodiimide (DCC) and *N,N′*-diisopropylcarbodiimide (DIC), diphenylphosphoryl azide (DPPA), and benzotriazole-1-yl-oxytripyrrolidinophosphonium hexafluorophosphate (PyBop) as coupling reagents, respectively.

The diketopiperazine-derived diphosphonic monoethyl ester was condensed with ethyl glycinate to afford diketopiperazine-derived phosphonodipeptide. Strict selectivity was observed upon enzyme-catalyzed hydrolysis. Enzyme α-chymotrypsin catalyzed the hydrolysis of the diketopiperazine ring and the ethyl ester group to give the free acids $HO_2CCH_2NHCH_2P(O)R^1NHCH_2CO_2H$. While phosphodiesterase I catalyzed the hydrolysis of only the ethyl ester group (Scheme 19) [45].

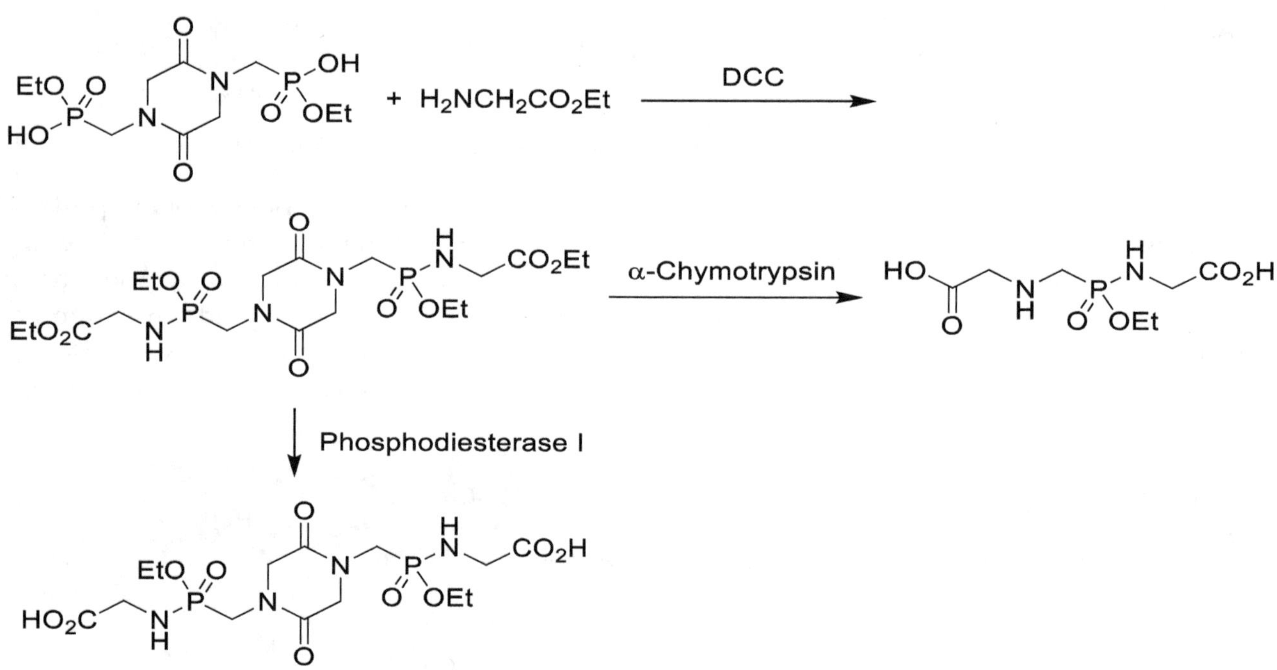

Scheme 19. Synthesis of phosphonopeptides with *N,N′*-dicyclohexylcarbodiimide (DCC) as a coupling reagent.

Kitamura’s group developed a sulfonamide-based protecting group, (9*H*-fluoren-9-yl) methanesulfonyl (Fms). It was used in a similar way to the well-established Fmoc protecting group. It was demonstrated in the successful formation of a phosphonamidate between an *N*-Fms-protected α-aminoethylphosphonic monoester with amino or dipeptide esters, including (*S*)-phenylalanine *tert*-butyl ester (H-Phe-O*t*Bu), H-Pro-Gly-O*t*Bu, and H-Phe-Phe-O*t*Bu, without formation of oxazaphospholine byproduct, which was a serious problem associated with the Fmoc and Cbz protecting groups during the formation of the phosphonamidate bond in both coupling reagent method [46] and phosphonochloridate method [24] (Scheme 20).

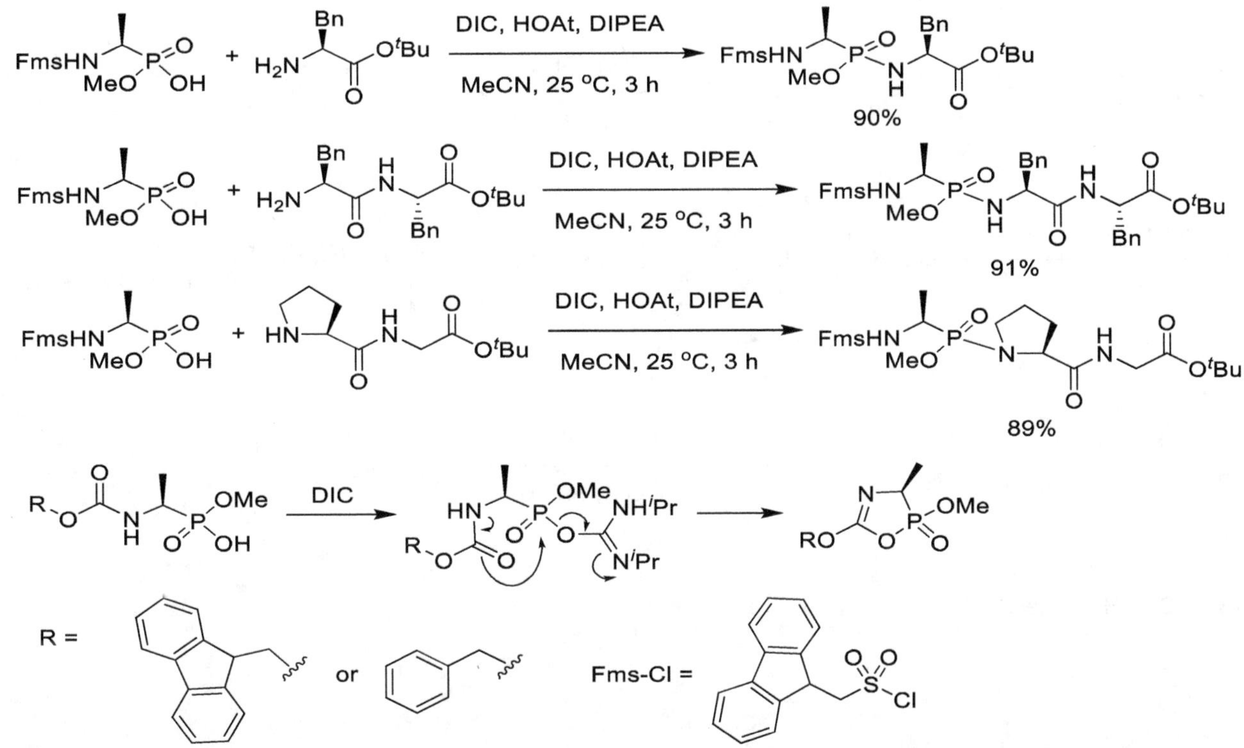

Scheme 20. Synthesis of phosphonopeptides with *N,N′*-diisopropylcarbodiimide (DIC) as a coupling reagent.

MDP phosphorus analogs related to LK 423 as potential immunomodulators were synthesized by coupling of methyl 1-(*N*-benzyloxycarbonyl)aminoethylphosphonate and methyl D-isoglutamate hydrochloride with DPPA as coupling reagent, followed by hydrogenolysis and coupling with 2-(2-phthalimidoethoxy)acetic acid. The dipeptide part of the lead compound was modified by introducing a phosphonamidate isostere instead of the amide bond between L-alanine and D-isoglutamine (Scheme 21) [25].

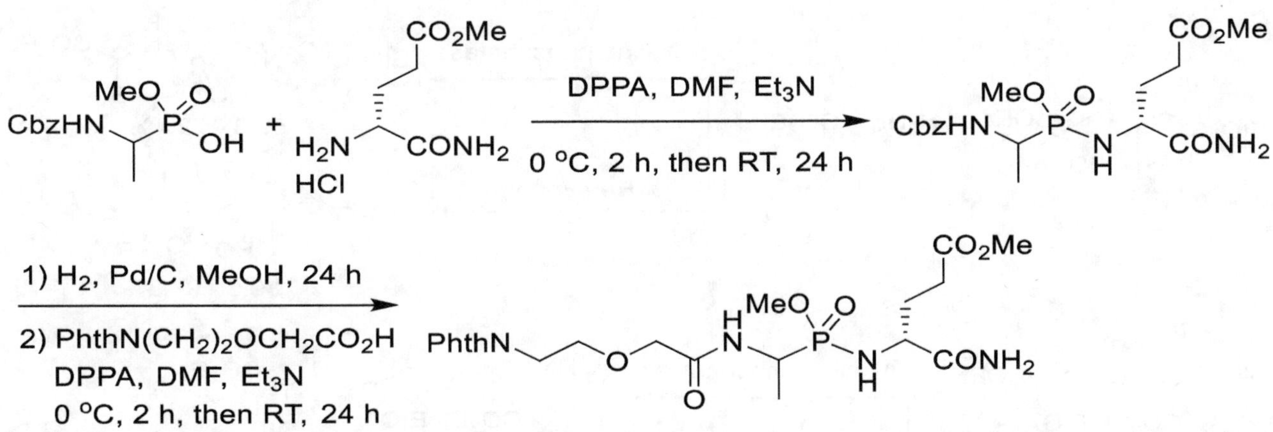

Scheme 21. Synthesis of phosphonopeptides with diphenylphosphoryl azide (DPPA) as coupling reagent.

To investigate the ligand–protein binding thermodynamics of the hydration waters of ligand-thermolysin complexes and hydrophobic binding, various thermolysin phosphonopeptide inhibitors were synthesized via the PyBop coupling of *N*-Cbz-protected aminomethylphosphonic methyl monoester with dipeptide esters or amino amides. The thermodynamic study provided important understanding on the water role in ligand–protein binding (Scheme 22) [47–49].

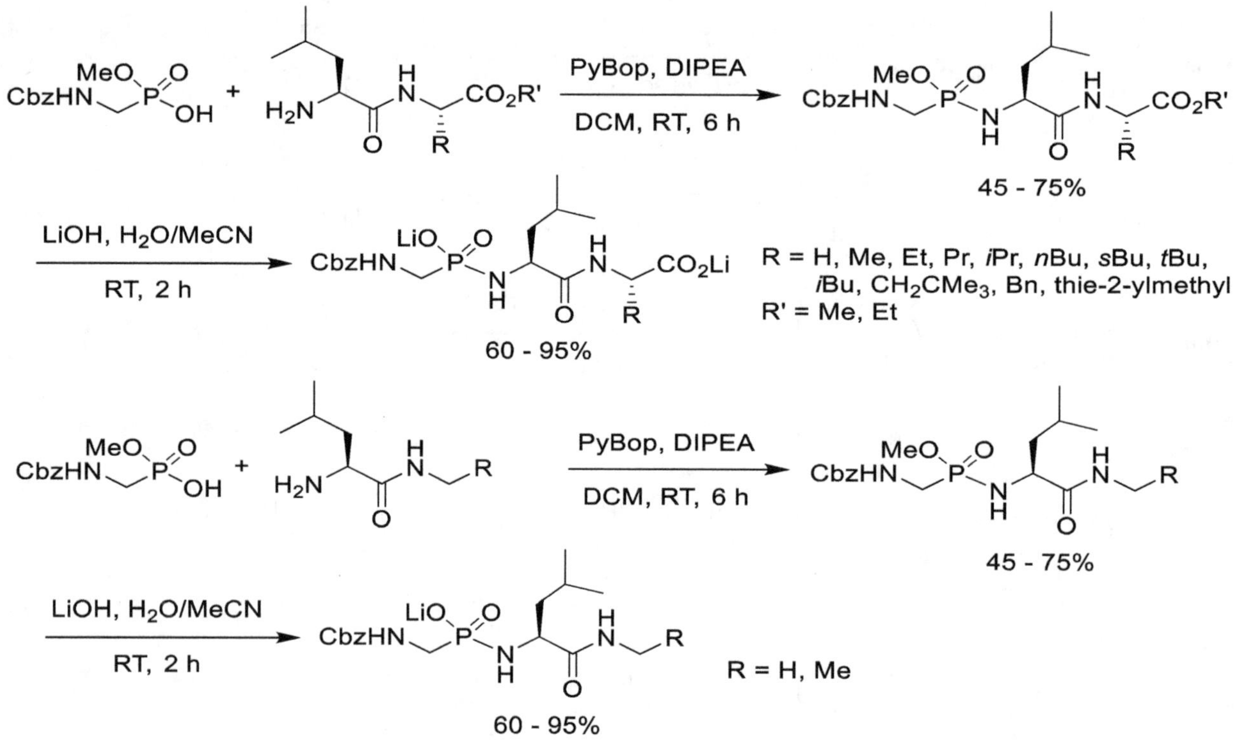

Scheme 22. Synthesis of phosphonopeptides with benzotriazole-1-yl-oxytripyrrolidinophosphonium hexafluorophosphate (PyBop) as a coupling reagent.

4. Synthesis of Phosphonopeptides Catalyzed by Enzyme

Natchev investigated the enzyme-catalyzed synthesis of phosphonopeptides. The alkaline phosphatase-catalyzed reaction of 1-aminomethylphosphonate and amino acid esters gave rise to the corresponding phosphonodipeptide esters, while the bee venom-catalyzed reaction of 1-aminomethylphosphonate and amino acid esters generated free phosphonodipeptides (Scheme 23) [50].

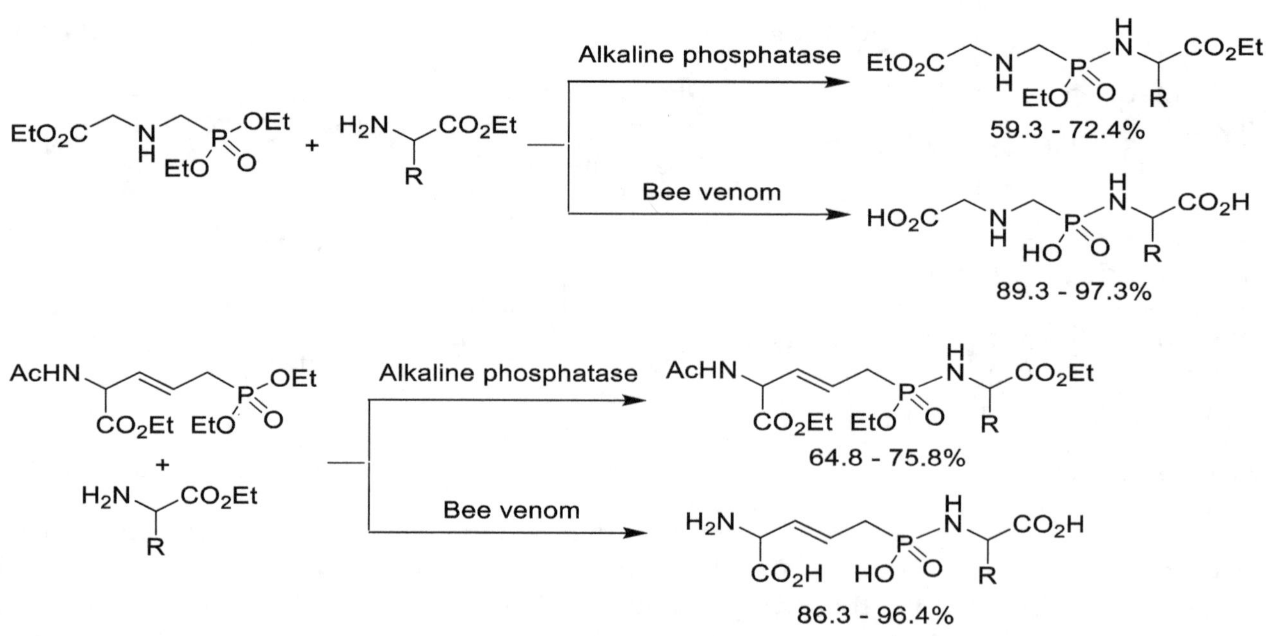

Scheme 23. Synthesis of α- and δ-phosphonopeptides.

Alkaline mesintericopeptidase selectively hydrolyzed *N*-acetyl protection group to afford *N*-terminal free phosphonodipeptide, which was further reacted with phosphonate under the catalysis of phosphodiesterase I to give protected phosphonotripeptide. After deprotection under the catalysis of bee venom, free phosphonotripeptide was obtained (Scheme 24) [50].

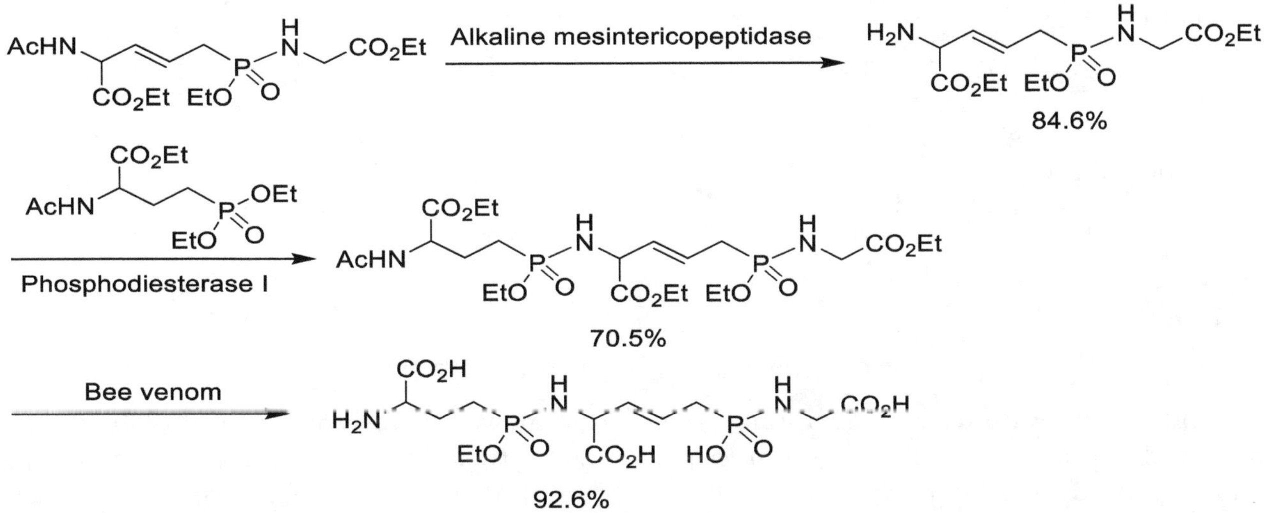

Scheme 24. Synthesis of phosphonopeptides composing of γ- and δ-aminoalkylphosphonic acid residues.

The alkaline phosphatase-catalyzed reaction of cyclic dipeptide with two molecules of 1-aminomethylphosphonate generated the corresponding phosphonotetrapeptide ester with a cyclic dipeptide moiety in the center. The bee venom-catalyzed hydrolysis of the protected phosphonotetrapeptide realized synthesis of free phosphonopeptide (Scheme 25) [50].

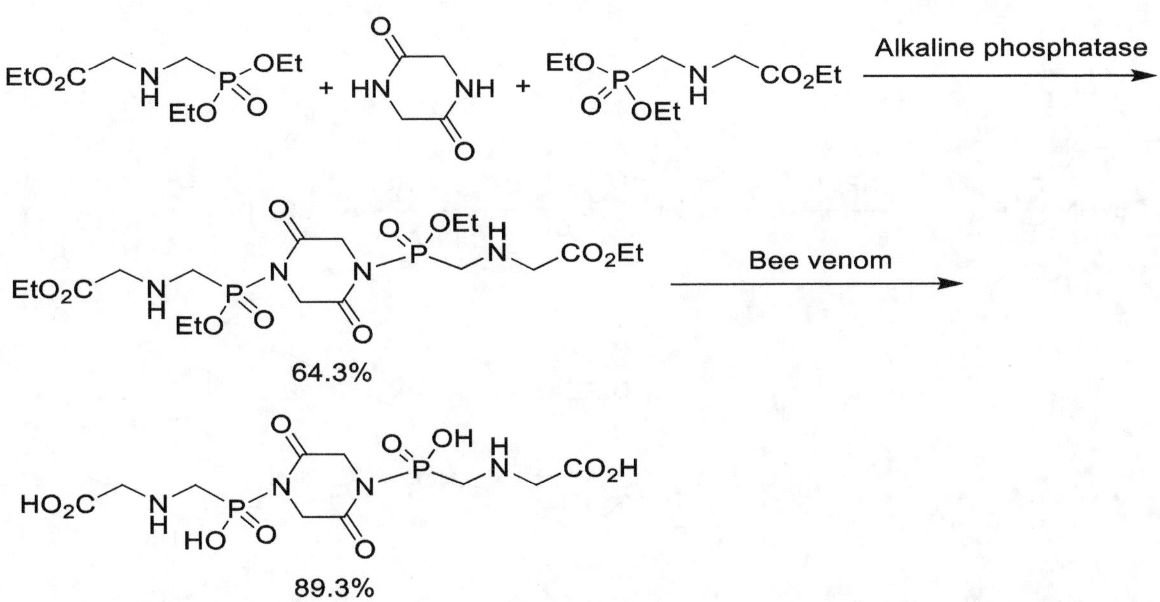

Scheme 25. Synthesis of phosphonopeptides containing a cyclodipeptide core.

5. Synthesis of Phosphonopeptides via Phosphonochloridite Followed by Oxidation

Hammer's group developed a new strategy to synthesize phosphonopeptides via coupling of phosphonochloridites and amino esters and subsequent sulfur oxidation. They used *N*-Boc protected 1-aminoalkylphosphinate as starting material and converted it into the corresponding phosphonochloridite with dichlorotriphenylphosphorane. The phosphonochloridite was further coupled with ethyl glycinate hydrochloride followed by sulfurization with sulfur, affording phosphonothiopeptide in one-pot activation-coupling-oxidation procedure. They mentioned that the phosphonochloridites were more active species than the corresponding phosphonochloridates (Scheme 26) [51].

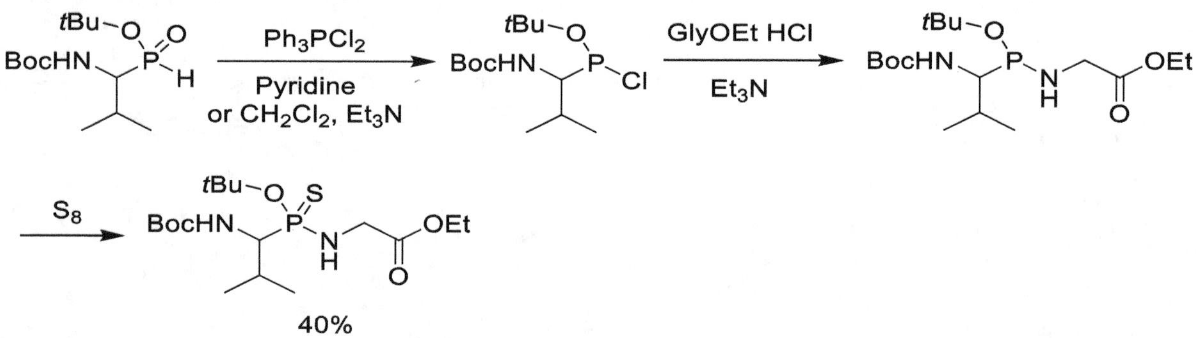

Scheme 26. Synthesis of thiophosphonopeptides.

Rushing and Hammer further applied the new strategy in the synthesis of *N*-Cbz protected phosphonodipeptides and phosphonothiopeptides using their P(III) one-pot activation-coupling-oxidation procedure. After treatment with dichlorophosphorane, *N*-Cbz protected 1-amino (cyclohexyl)methylphosphinate was transformed to the P(III) intermediate, *N*-Cbz protected 1-amino(cyclohexyl)methylphosphonochloridite, which was reacted with ᴅ-tryptophan derivatives in the presence of diisopropylethylamine (DIPEA) followed by oxidation with tertial butyl peroxide and sulfur, affording phosphonopeptides and phosphonothiopeptides, respectively. It was found that base DIPEA could cyclize the *N*-Cbz 1-aminomethylphosphonochloridite into 5-(benzyloxy)-2,3-dihydro-1,4,2-oxazaphosphole. However, it was further reacted with ᴅ-tryptophan derivatives to generate the same intermediate phosphinamides in the reaction system (Scheme 27) [52].

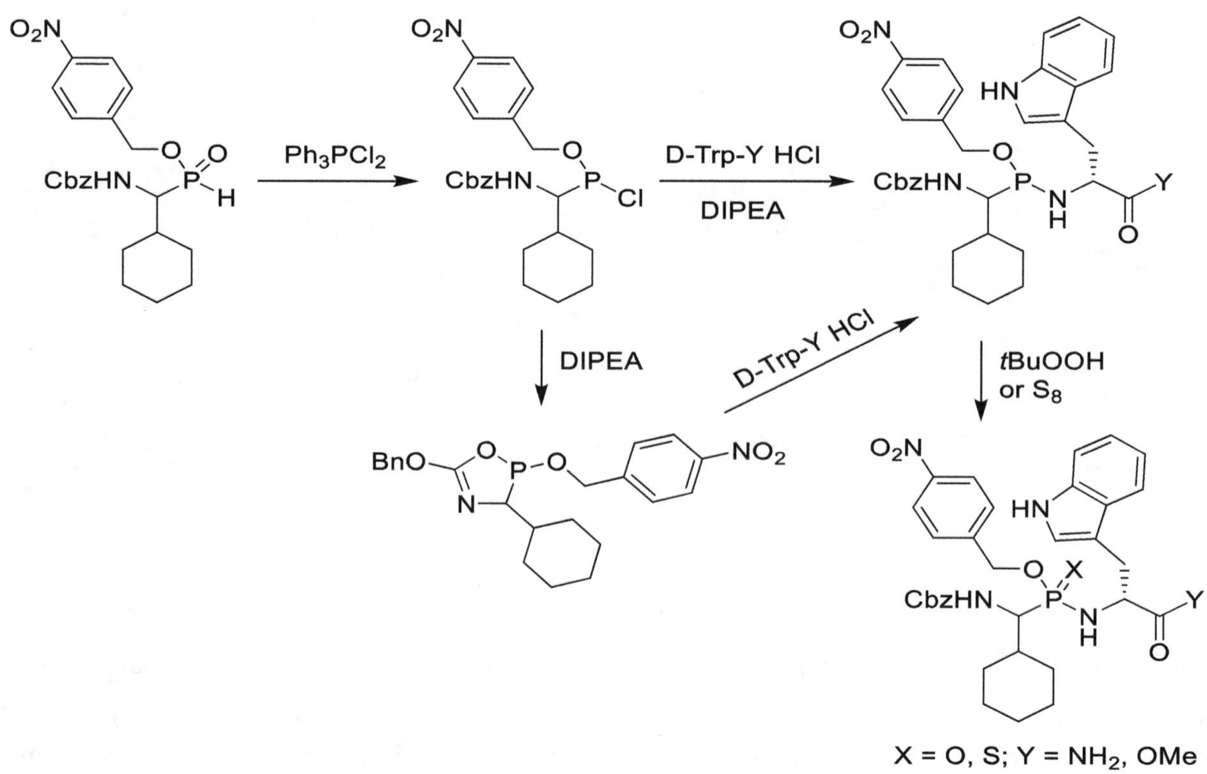

Scheme 27. Synthesis of phosphonopeptides via phosphonochliridites.

6. Synthesis of Phosphonopeptides via Pseudo Four-Component Condensation Reaction

Previously the Mannich-type reaction of benzyl carbamate, aldehydes, and trialkyl phosphites in acetyl chloride gave rise to *N*-Cbz-1-aminoalkylphosphonates [31]. We further developed the method as a pseudo-four-component condensation reactions for synthesis of *N*-Cbz 1-aminoalkylphosphonic acid derivatives, including phosphonamidates [53], mixed esters [54,55], and phosphonodepsipeptides [56–59]. The four-component condensation was applied as a direct method for the preparation of phosphonopeptides in construction of 1-aminoalkylphosphonic acids simultaneously formed the phosphonamidate bond. Using this method, phosphonopeptides were prepared in acceptable yields directly from simple and commercially available chemicals in one-pot reactions of benzyl carbamate, aldehydes, and methyl dichlorophosphite, followed by aminolysis with amino acid esters (Scheme 28) [60].

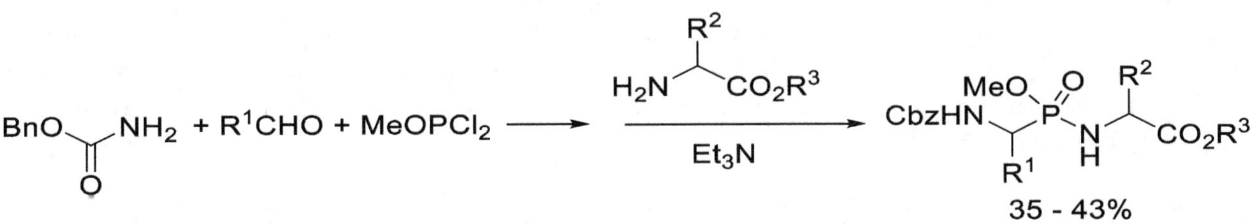

Scheme 28. Synthesis of phosphonopeptides via pseudo four-component condensation.

The pseudo four-component condensation reaction mechanism is proposed as following. Benzyl carbamate first nucleophilically attack aldehydes followed by proton transfer to generate α-amino alcohols **B**, which undergo a nucleophilic substitution with methyl dichlorophosphite on the phosphorus atom, yielding chlorophosphites **C**. Chlorophosphites **C** undergo an elimination to give rise imines **D** and chlorophosphonous acid **E** in the presence of HCl. Chlorophosphonous acid **E** can exist as **E** and H-phosphonic chloride **F**. However, chlorophosphonous acid **E** is efficient species to nucleophilically add to the C=N bond or protonated C=N bond of imines **D** to produce the phophonochloridates as intermediates. After aminolysis with amino esters, the desired phosphonopeptides were obtained (Scheme 29) [54,57].

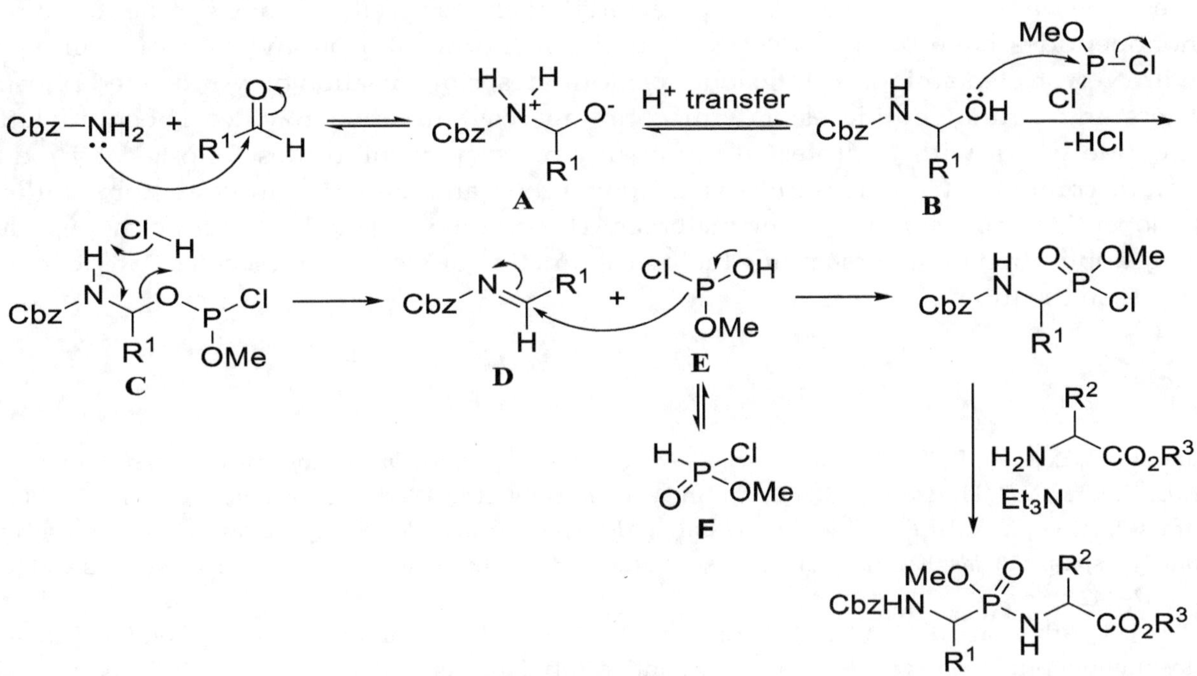

Scheme 29. Mechanism for the synthesis of phosphonopeptides via pseudo four-component condensation.

The pseudo four-component condensation method is an efficient and convergent strategy for the synthesis of phosphonopeptides from simple and commercially available chemicals, such as, benzyl carbamate, aldehydes, methyl dichlorophosphite, and amino acid esters. Compared with other strategies, the strategy combines the construction of α-aminoalkylphosphonic acids and the formation of the phosphonamidate bond in a one-pot mode. However, other reported methods construct α-aminoalkylphosphonic acid derivatives first and then form the phosphonamidate bond.

7. Synthesis by Nucleophilic Addition

Gololobov and Nesterova synthesized cyclic phosphonodipeptides via the annulation of *N*-benzylidenemethylamine and alkyl *N*-(dialkoxyphosphanyl)-*N*-alkylglycinates, which were prepared from alkyl *N*-(dichlorophosphanyl)-*N*-alkylglycinates with alcohols R^2OH. The annulation was a sequence of P-nucleophilic addition and intramolecular aminolysis accompanying an N-nucleophilic addition followed by intramolecular aminolysis to yield imidazolidin-4-ones as byproducts (Scheme 30) [61].

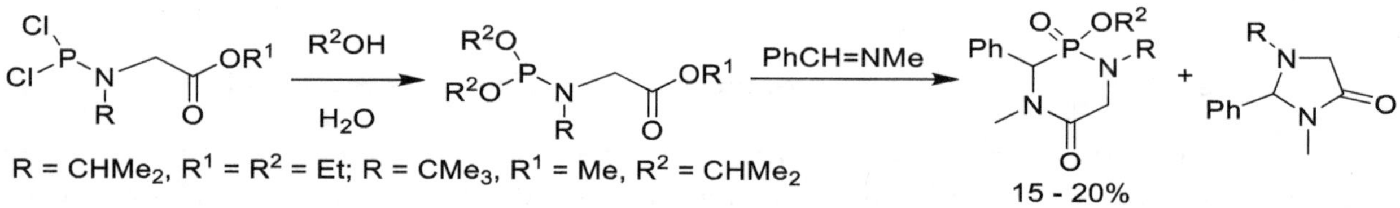

Scheme 30. Synthesis of cyclophosphonopeptides via nucleophilic addition.

8. Conclusions

Phosphonopeptides are a class of phosphorus analogues of peptides containing a tetrahedral phosphonamidate bond, which can mimic the transition state of amide hydrolysis. Thus, phosphonopeptides have been widely applied as enzyme inhibitors and antigens for catalytic antibodies. They have been utilized as antibacterial agents as well. Various synthetic methods of phosphonopeptides have been developed, mainly including phosphonylation of amino/peptide esters with *N*-protected aminoalkylphosphonochloridates, condensation of *N*-protected phosphonic monoesters and amino/peptide ester with coupling reagents and enzyme, phosphinylation of amino/peptide esters with *N*-protected aminoalkylphosphonochloridites followed by oxidation. Pseudo four-component condensation is a convergent and efficient strategy for synthesis of phosphonopeptides from simple starting materials. The synthetic methods of phosphonopeptides will show wide application in understanding biological function of biomacromolecules and development of medicines in the future.

References

1. Kafarski, P.; Lejczak, B. *Synthesis of Phosphono- and Phosphinopeptides in Aminophosphonic and Aminophosphinic Acids*; Kukhar, V.P., Hudson, H.R., Eds.; John Wiley & Sons Ltd.: West Sussex, UK, 2000; pp. 173–203.
2. Kafarski, P.; Lejczak, B. *The Biological Activity of Phosphono- and Phosphinopeptides in Aminophosphonic and Aminophosphinic Acids*; Kukhar, V.P., Hudson, H.R., Eds.; John Wiley & Sons Ltd.: West Sussex, UK, 2000; pp. 407–442.
3. Allen, J.G.; Atherton, F.R.; Hall, M.J.; Hassall, C.H.; Holmes, S.W.; Lambert, R.W.; Nisbet, L.J.; Ringrose, P.S. Phosphonopeptides, a new class of synthetic antibacterial agents. *Nature* **1978**, *272*, 56–58. [CrossRef]
4. Bartlett, P.A.; Marlowe, C.K. Evaluation of intrinsic binding energy from a hydrogen bonding group in an enzyme inhibitor. *Science* **1987**, *235*, 569–571. [CrossRef]

5. Krimmer, S.G.; Cramer, J.; Betz, M.; Fridh, V.; Karlsson, R.; Heine, A.; Klebe, G. Rational design of thermodynamic and kinetic binding profiles by optimizing surface water networks coating protein-bound ligands. *J. Med. Chem.* **2016**, *59*, 10530–10548. [CrossRef]
6. Grembecka, J.; Mucha, A.; Cierpicki, T.; Kafarski, P. The most potent organophosphorus inhibitors of leucine aminopeptidase. Structure-based design, chemistry, and activity. *J. Med. Chem.* **2003**, *46*, 2641–2655. [CrossRef]
7. Yang, K.W.; Cheng, X.; Zhao, C.; Liu, C.C.; Jia, C.; Feng, L.; Xiao, J.M.; Zhou, L.S.; Gao, H.Z.; Yang, X.; et al. Synthesis and activity study of phosphonamidate dipeptides as potential inhibitors of VanX. *Bioorg. Med. Chem. Lett.* **2011**, *21*, 7224–7227. [CrossRef]
8. Demange, L.; Moutiez, M.; Dugave, C. Synthesis and evaluation of Glyψ(PO_2R-N)Pro-containing pseudopeptides as novel inhibitors of the human cyclophilin hCyp-18. *J. Med. Chem.* **2002**, *45*, 3928–3933. [CrossRef]
9. Schultz, P.G.; Lerner, R.A. Antibody catalysis of difficult chemical transformations. *Acc. Chem. Res.* **1993**, *26*, 392–395. [CrossRef]
10. Jacobsen, J.R.; Schultz, P.G. Antibody catalysis of peptide-bond formation. *Proc. Nat. Acad. Sci. USA* **1994**, *91*, 5888–5892. [CrossRef]
11. Xie, G.Y.; Chen, J.H.; Lin, S.S.; Jin, S. The synthesis of a novel phosphorus containing antigen. *Chin. Chem. Lett.* **2001**, *12*, 497–498.
12. Xie, G.Y.; Chen, L.M.; Chen, J.H. Synthesis of a novel antigen containing phosphorus. *Chem. J. Chin. Univ.* **2003**, *24*, 1037–1039.
13. Hariharan, M.; Chaberek, S.; Martell, A.E. Synthesis of phosphonopeptide derivatives. *Synth. Commun.* **1973**, *3*, 375–379. [CrossRef]
14. Jackson, D.S.; Fraser, S.A.; Ni, L.-M.; Kam, C.-M.; Winkler, U.; Johnson, D.A.; Froelich, C.J.; Hudig, D.; Powers, J.C. Synthesis and evaluation of diphenyl phosphonate esters as inhibitors of the trypsin-like granzymes A and K and mast cell tryptase. *J. Med. Chem.* **1998**, *41*, 2289–2301. [CrossRef]
15. Cervera-Villanueva, J.M., Jr.; Viveros-Ceballos, J.L.; Linzaga-Elizalde, I.; Ordonez, M. Practical synthesis of novel phosphonopeptides containing AibP. *J. Peptide Sci.* **2016**, *22*, 70–75. [CrossRef]
16. Kudzin, Z.H.; Depczynski, R.; Kudzin, M.H.; Drabowicz, J. 1-(*N*-chloroacetylamino)-alkylphosphonic acids—synthetic precursors of phosphonopeptides. *Amino Acids* **2008**, *34*, 163–168. [CrossRef]
17. Sun, B.Y.; Xu, J.X. Convenient synthesis of phosphonodipeptides containing *C*-terminal α-aminoalkylphosphonic acids. *J. Pept. Sci.* **2015**, *21*, 615–619. [CrossRef]
18. Sun, B.Y.; Xu, J.X. Convenient and facile synthesis of hybrid sulfonophosphonodipeptides. *Synthesis* **2015**, *47*, 1479–1487.
19. Kafarski, P. Phosphonopeptides containing free phosphonic groups: Recent advances. *RSC Adv.* **2020**, *10*, 25898–25910. [CrossRef]
20. Elliott, R.L.; Marks, N.; Berg, M.J.; Portoghese, P.S. Synthesis and biological evaluation of phosphonamidate peptide inhibitors of enkephalinase and angiotensin-converting enzyme. *J. Med. Chem.* **1985**, *28*, 1208–1216. [CrossRef]
21. Xu, J.X.; Xia, C.F.; Yu, L.; Zhou, Q.Z. Synthesis of phosphonopeptides containing 1-aminoalkylphosphonic acid. *Phosphorus Sulfur Silicon Relat. Elem.* **1999**, *152*, 35–44. [CrossRef]
22. Vo Quang, Y.; Gravey, A.M.; Simonneau, R.; Vo Quang, L.; Lacoste, A.M.; Le Goffic, F. Towards new inhibitors of D-alanine:D-alanine ligase: The synthesis of 3-aminobutenylphosphonic and aminophosphonamidic acids. *Tetrahedron Lett.* **1987**, *28*, 6167–6170. [CrossRef]
23. Lacoste, A.M.; Chollet-Gravey, A.M.; Vo Quang, L.; Vo Quang, Y.; Le Goffic, F. Time-dependent inhibition of Streptococcus faecalis D-alanine:D-alanine ligase by α-aminophosphonamidic acids. *Eur. J. Med. Chem.* **1991**, *26*, 255–260. [CrossRef]
24. Musiol, H.-J.; Grams, F.; Rudolph-Boehner, S.; Moroder, L. The chemistry of phosphapeptides: Formation of functionalized phosphonochloridates under mild conditions and their reaction with alcohols and amines. On the synthesis of phosphonamidate peptides. *J. Org. Chem.* **1994**, *59*, 6144–6146. [CrossRef]
25. Gobec, S.; Urleb, U. Synthesis of new lipophilic phosphonate and phosphonamidate analogs of *N*-acetylmuramyl-D-isoglutamine related to LK 423. *Molecules* **2002**, *7*, 394–404. [CrossRef]
26. Sampson, N.S.; Bartlett, P.A. Synthesis of phosphonic acid derivatives by oxidative activation of phosphinate esters. *J. Org. Chem.* **1988**, *53*, 4500–4503. [CrossRef]

27. Mucha, A.; Kafarski, P.; Plenat, F.; Cristau, H.-J. The preparation of phosphono peptides containing a phosphonamidate bond. *Tetrahedron* **1994**, *50*, 12743–12754. [CrossRef]
28. Yao, Q.L.; Yuan, C.Y. Stereospecific transformation of protected P–H group into P–O or P–N group in one-pot reaction. *J. Org. Chem.* **2012**, *77*, 10985–10990. [CrossRef]
29. Horiguchi, M.; Kandatsu, M. Isolation of 2-aminoethane phosphonic acid from *Rumen Protozoa*. *Nature* **1959**, *184*, 901–902. [CrossRef]
30. Kittredge, J.S.; Roberts, E.; Simonsen, D.G. The occurrence of free 2-aminoethylphosphonic acid in the sea anemone, *Anthopleura elegantissima*. *Biochemistry* **1962**, *1*, 624–628. [CrossRef]
31. Yuan, C.Y.; Wang, G. H Studies on organophosphorus compounds. 65. A facile synthetic route to phosphonopeptides. *Phosphorus Sulfur Silicon Relat. Elem.* **1992**, *71*, 207–212. [CrossRef]
32. Yang, K.-W.; Brandt, J.J.; Chatwood, L.L.; Crowder, M.W. Phosphonamidate and phosphothioate dipeptides as potential inhibitors of VanX. *Bioorg. Med. Chem. Lett.* **2000**, *10*, 1085–1087. [CrossRef]
33. Slater, M.J.; Laws, A.P.; Page, M.I. The relative catalytic efficiency of β-lactamase catalyzed acyl and phosphyl transfer. *Bioorg. Chem.* **2001**, *29*, 77–95. [CrossRef]
34. Mucha, A.; Kafarski, P. Transesterification of monophenyl phosphonamidates-chemical modeling of serine protease inhibition. *Tetrahedron* **2002**, *58*, 5855–5863. [CrossRef]
35. Grembecka, J.; Mucha, A.; Cierpicki, T.; Kafarski, P. Structure-based design and synthesis of dipeptide analogs as new inhibitors of leucine aminopeptidase. *Phosphorus Sulfur Silicon Relat. Elem.* **2002**, *177*, 1739–1743. [CrossRef]
36. Mucha, A.; Grembecka, J.; Cierpicki, T.; Kafarski, P. Hydrolysis of the phosphonamidate bond in phosphono dipeptide analogs—the influence of the nature of the *N*-terminal functional group. *Eur. J. Org. Chem.* **2003**, 4797–4803. [CrossRef]
37. Dumy, P.; Escale, R.; Girard, J.P.; Parello, J.; Vidal, J.P. A convenient synthetic approach to new α-(9-fluorenylmethoxycarbonylamino)alkylphosphonic acid derivatives. *Synthesis* **1992**, 1226–1228. [CrossRef]
38. Cramer, J.; Klebe, G. An allyl protection and improved purification strategy enables the synthesis of functionalized phosphonamidate peptides. *Synthesis* **2017**, *49*, 1857–1866.
39. Mucha, A.; Kunert, A.; Grembecka, J.; Pawelczak, M.; Kafarski, P. A phosphonamidate containing aromatic *N*-terminal amino group as inhibitor of leucine aminopeptidase-design, synthesis and stability. *Eur. J. Med. Chem.* **2006**, *41*, 768–772. [CrossRef]
40. Malachowski, W.P.; Coward, J.K. The chemistry of phosphapeptides: Formation of functionalized phosphonochloridates under mild conditions and their reaction with alcohols and amines. *J. Org. Chem.* **1994**, *59*, 7616–7624. [CrossRef]
41. Urleb, U.; Gobec, S. Synthesis of new adamantane substituted acyclic MDP analogs related to LK 415. *Acta Pharm.* **2000**, *50*, 173–184.
42. Malachowski, W.P.; Coward, J.K. The chemistry of phosphapeptides: Investigations on the synthesis of phosphonamidate, phosphonate, and phosphinate analogs of glutamyl-γ-glutamate. *J. Org. Chem.* **1994**, *59*, 7625–7634. [CrossRef]
43. Sikora, D.; Nonas, T.; Gajda, T. *O*-Ethyl 1-azidoalkylphosphonic acids-versatile reagents for the synthesis of protected phosphonamidate peptides. *Tetrahedron* **2001**, *57*, 1619–1625. [CrossRef]
44. Artyushin, O.I.; Sharova, E.V.; Yarkevich, A.N.; Genkina, G.K.; Vinogradova, N.V.; Brel, V.K. Design of phosphonate analogs of short peptides by "Click" chemistry. *Russ. Chem. Bull.* **2015**, *64*, 2172–2177. [CrossRef]
45. Natchev, I. Glycine phospha C-peptides. *Izvestiya po Khimiya* **1988**, *21*, 40–45.
46. Ishibashi, Y.; Miyata, K.; Kitamura, M. (9*H*-Fluoren-9-yl)methanesulfonyl (Fms): An amino protecting group complementary to Fmoc. *Eur. J. Org. Chem.* **2010**, 4201–4204. [CrossRef]
47. Nasief, N.N.; Tan, H.W.; Kong, J.; Hangauer, D. Water mediated ligand functional group cooperativity: The contribution of a methyl group to binding affinity is enhanced by a COO^- group through changes in the structure and thermodynamics of the hydration waters of ligand-thermolysin complexes. *J. Med. Chem.* **2012**, *55*, 8283–8302. [CrossRef]
48. Nasief, N.N.; Hangauer, D. Influence of neighboring groups on the thermodynamics of hydrophobic binding: An added complex facet to the hydrophobic effect. *J. Med. Chem.* **2014**, *57*, 2315–2333. [CrossRef]

49. Nasief, N.N.; Hangauer, D. Additivity or cooperativity: Which model can predict the influence of simultaneous incorporation of two or more functionalities in a ligand molecule? *Eur. J. Med. Chem.* **2015**, *90*, 897–915. [CrossRef]
50. Natchev, I.A. Organophosphorus analogs and derivatives of the natural L-aminocarboxylic acid and peptides. VII. Enzyme synthesis of phospha-C peptides. *Tetrahedron* **1991**, *47*, 1239–1248. [CrossRef]
51. De Fatima Fernandez, M.; Vlaar, C.P.; Fan, H.; Liu, Y.-H.; Fronczek, F.R.; Hammer, R.P. Synthesis of phosphonate and thiophosphonate esters and amides from hydrogen-phosphinates by a novel one-pot activation-coupling-oxidation procedure. *J. Org. Chem.* **1996**, *60*, 7390–7391. [CrossRef]
52. Rushing, S.D.; Hammer, R.P. Synthesis of phosphonamide and thiophosphonamide dipeptides. *J. Am. Chem. Soc.* **2001**, *123*, 4861–4862. [CrossRef]
53. Xu, J.X.; Fu, N.Y. A facile synthesis of *N*-protected 1-aminoalkylphosphonamidate derivatives. *Synth. Commun.* **2000**, *30*, 4137–4145. [CrossRef]
54. Xu, J.X.; Fu, N.Y. A novel and convenient method for synthesizing unsymmetrical *N*-benzyloxycarbonyl-protected 1-amino-1-aryl-alkylphosphonate mixed diesters. *J. Chem. Soc. Perkin. Trans. 1* **2001**, 1223–1226. [CrossRef]
55. Xu, J.X.; Wei, M. A convenient method for the synthesis of N-protected 1-aminoalkyl-phosphonate mixed monothioesters and dithioesters. *Synth. Commun.* **2001**, *31*, 1489–1497. [CrossRef]
56. Xu, J.X. Convergent synthesis of phosphonopeptides and phosphonodepsipeptides. *Sci. Sin. Chim.* **2013**, *43*, 995–1004. [CrossRef]
57. Xu, J.X. Convergent synthesis of phosphonopeptides via phospha-Mannich reactions: Rationale on reactivity and mechanistic insights. *Phosphorus Sulfur Silicon Relat. Elem.* **2019**, *194*, 487–492. [CrossRef]
58. Xu, J.X.; Gao, Y.H. Straightforward synthesis of depsiphosphonopeptides via Mannich-type multiple component condensation. *Synthesis* **2006**, 783–788. [CrossRef]
59. Liu, H.; Cai, S.Z.; Xu, J.X. Asymmetric synthesis of *N*-protected chiral 1-aminoalkylphosphonic acids and synthesis of side-chain functionalized depsiphosphonopeptides. *J. Peptide Sci.* **2006**, *12*, 337–340. [CrossRef]
60. Fu, N.Y.; Zhang, Q.H.; Duan, L.F.; Xu, J.X. Facile synthesis of phosphonamidate and phosphonate-linked phosphonopeptides. *J. Peptide Sci.* **2006**, *12*, 303–309. [CrossRef]
61. Gololobov, Y.G.; Nesterova, L.I. *N*-Hydrophosphoryl derivatives of *N*-alkylaminoacetic acid in the synthesis of 1,4,2-diazaphosphorinan-5-ones. *Zhurnal Obshchei Khimii* **1982**, *52*, 2465–2467.

2

Straightforward Synthetic Protocol to Bio-Based Unsaturated Poly(ester amide)s from Itaconic Acid with Thixotropic Behavior

Lazaros Papadopoulos [1], Marcel Kluge [2,3], Dimitrios N. Bikiaris [1] and Tobias Robert [2,*]

1 Laboratory of Polymer Chemistry and Technology, Department of Chemistry, Aristotle University of Thessaloniki, GR-541 24 Thessaloniki, Greece; lazaros.geo.papadopoulos@gmail.com (L.P.); dbic@chem.auth.gr (D.N.B.)

2 Fraunhofer Institute for Wood Research—Wilhelm-Klauditz-Institut WKI, Bienroder Weg 54E, 38108 Braunschweig, Germany; marcel.kluge@wki.fraunhofer.de

3 Institute for Chemistry of Renewable Resources, Department of Chemistry, University of Natural Resources and Life Sciences Vienna, Konrad-Lorenz-Straße 24, 3430 Tulln, Austria

* Correspondence: tobias.robert@wki.fraunhofer.de

Abstract: In the field of polymer chemistry, tremendous efforts have been made over the last decade to replace petrochemical monomers with building blocks from renewable resources. In this respect, itaconic acid has been used as an alternative to acrylic acid or maleic acid in unsaturated polyesters for thermal or UV-curing applications. However, examples of poly(ester amide)s from itaconic acid are scarce. Under standard polycondensation reactions, the presence of free amines leads to aza-Michael addition reactions at the α,β-unsaturated double bond of the itaconic acid and isomerization reactions to mesaconic acid. Both reactions make the resulting materials useless as UV-curing polymer resins. To avoid these undesired side reactions, we herein report the use of preformed, well-defined diols containing internal amide bonds. The resulting unsaturated poly(ester amide) resins were analyzed before and after UV-induced crosslinking. Viscosity measurements revealed a strong thixotropic behavior induced by the amide groups, which is usually not detected in structurally similar polyester resins.

Keywords: bio-based polymer; poly(ester amide); itaconic acid; thixotropy; photo-DSC

---

1. Introduction

Over the last years, polymers from renewable resources have attracted considerable attention from both industry and academia in an effort to limit the environmental impact arising from the excessive use of polymeric materials derived from fossil resources [1–4]. However, in order to be able to compete with these established materials, the bio-based counterparts have to be either economically competitive or exhibit unprecedented material properties [5,6]. A viable way to achieve the latter is the utilization of bio-based monomeric building blocks, which are not economically accessible from petrochemical pathways. These can be used to produce novel polymeric structures that in turn can lead to materials with new properties [7–9].

In this context, itaconic acid (IA) has lately drawn considerable attention as it exhibits a promising structure and is available at competitive prices (<2 €/kg) [10]. This monomer is already produced on industrial scale (>80.000 t/a) via the fermentation of sugars with the fungi *aspergillus terreus* [11–13]. Another interesting feature of IA is its versatility. Since it possesses two carboxylic groups and an *exo*-double bond, it is a good substrate for different chemical transformations. This versatility also applies for the production of polymers. Two main strategies emerge: first, the utilization of the unsaturated double bonds for the radical polymerization of homo-polymers and co-polymers [14–18] and second, the synthesis of unsaturated polyesters [19–25]. Those materials find a number of applications in printing inks, shape memory polymers, and coatings [26–30], while they can also be exploited for post-polymerization reactions [31–34].

Despite the increased research activities in itaconic acid-based materials described above, poly(ester amide)s from itaconic acid have so far received little attention. The main reason for this is the fact that the α,β-unsaturated double bond is prone to a nucleophilic attack of the diamine, leading to the formation of pyrrolidone rings [35,36]. While this effect can be exploited for the synthesis of bio-based polyamides and poly(ester amide)s [37,38], it also results in the complete conversion of the unsaturated double bonds and thus the inability to produce polymers susceptible to radical curing reactions. Recently, we reported two synthetic strategies in order to circumvent these undesired side reactions [39]. The first method involved the synthesis of an itaconic acid-based oligoester and an oligo(ester amide), both of which were synthesized independently. In the next step, these two oligomers were combined in a transesterification reaction. The second synthetic pathway involved the in situ formation of a bis-ester amide that was subsequently reacted with IA and further monomers. While both approaches resulted in the successful synthesis of the envisaged structures, the first synthetic strategy involved three synthetic steps, which makes the synthesis quite cumbersome. The second strategy suffered from a base-catalyzed isomerization of the itaconate moiety to the mesaconate, which was due to some free amines still present after the in situ formation of the bis-ester amide.

Another promising and more straightforward approach to this class of unsaturated poly(ester amide)s is the used of well-defined preformed monomers containing amide moieties. In this respect, we recently reported the use of symmetric diamido-α,ω-diols (amido diols) as building blocks for poly(ester amide)s derived from succinic acid [40–42]. By using these monomers, we were able to prevent the formation of succinimide that usually interferes with the synthesis of polyamides or poly(ester amide)s derived from succinic acid. Intrigued by these promising results, we were interested in whether these amido diols could also be used as building blocks for poly(ester amide)s derived from itaconic acid. This would allow for the introduction of amide segments without the occurrence of the undesired aza-Michael additions described above. In this work, two series of unsaturated poly(ester amide)s were synthesized based on itaconic acid and two different preformed amido diols. The properties of the polymer resins as well as UV-cured materials were thoroughly characterized, revealing a thixotropic behavior of the resins and considerable influence of the amide content on the thermal and mechanical properties of the final cured materials.

2. Materials and Methods

2.1. Materials

Dimethyl itaconate (DMI, 98%), dimethyl sebacate (DMS, 98%), 1,4-diaminobutane (≥98%), and 1,6-diaminohexane (≥99%) were purchased from Tokyo Chemicals Industry (TCI), Tokyo, Japan. Glycerol (Gly, ≥98%) and chloroform-d_1 (99.8% D) + 0.03% TMS *v*/*v* were provided by Carl Roth, Karlsruhe, Germany. 2,3-Butanediol (2,3-BD, a mixture of meso-D- and L-form, ≥99%) and 2,6-di-*tert*-butyl-4-methylphenol (BHT, 99%) were bought from Merck, Darmstadt, Germany. 4-Methoxyphenol (MeHQ, 99%) was purchased from Sigma-Aldrich Chemie, Steinheim, Germany. ε-Caprolactone (99%) was obtained from Alfa Aesar, Kandel, Germany. 1,3-Propanediol (1,3-PD, 99.7%) was kindly provided by DuPont Tate & Lyle Bio Products, Loudon, NH, USA. Isosorbide (Is) was obtained from Ecogreen Oleochemicals, Dessau-Roßlau, Germany. FASCAT 4101 catalyst was provided by PMC Group, Mount Laurel, NJ, USA. Solvents were reagent or analytical grade and were purchased from VWR International, Fontenay-sous-Bois, France. All reagents were used without further purification except for isosorbide, which was recrystallized from ethyl acetate.

2.2. Measurements

FTIR spectra were obtained on a Thermo Scientific Nicolet iS5 FTIR (Thermo Fisher Scientific, Waltham, MA, USA) using the ATR technique (32 scans, resolution of 4 cm^{-1}).

NMR experiments were conducted on a Bruker Avance III 400-MHz spectrometer (Bruker, Billerica, MA, USA). ^1H NMR shifts of polymers are reported in ppm (δ) downfield from tetramethylsilane (TMS) and were determined by reference to the residual solvent peak (chlorofrom-d_1, 7.26 ppm for hydrogen atoms).

Size-exclusion chromatography (SEC) was performed on a Malvern Viscotek GPCmax (Malvern Instruments, Malvern, United Kingdom) equipped with triple detection, consisting of a Malvern Dual detector and Schambeck RI2012 refractive index detector. The separation was performed by utilizing two PLgel 5 mm MIXED-C, 300 mm columns from Agilent Technologies (Santa Clara, CA, USA) at 35 °C. Chloroform was used as the eluent at a flow rate of 0.5 mL min^{-1}. Data acquisition and calculations were performed using Viscotek OmniSec software version 5.0. The samples were filtered over a 0.2 mm PTFE filter prior to injection.

Rheological properties of resins were studied on a Bohlin CVO 100 rheometer (Bohlin Instruments, Cirencester, UK) equipped with cone-plate geometry (CP 4°, 40 mm). For the table temperature scan, measurements were done with a shear rate of 100 s^{-1} for 10 s. Five measurements were carried out for each temperature (25 °C, 37.5 °C, 50 °C, 62.5 °C, and 75 °C), and the average value was calculated for each temperature. For the shear thinning tests and to measure time-dependent behavior, each sample was left on the rheometer for five minutes prior to measurement to ensure a temperature equilibrium of 25 °C. Subsequently, a shear rate of 1 s^{-1} was applied for three minutes in order to reach the equilibrium state. Then, the shear rate was increased at 100 s^{-1} for the application of the mechanical load for a duration of two minutes. Finally, a shear rate of 1 s^{-1} was applied again over a prolonged period of time until the system reached an equilibrium state again. The ability of recovery as a measure of thixotropic behavior for each system was evaluated by comparing the viscosity value of the first and the third stage of the measurement. Five samples of each resin were studied, and the average values are given in the results.

Photo-differential scanning calorimatry (Photo-DSC) measurements were conducted on a DSC 3+ (Mettler-Toledo, Greifensee, Switzerland) equipped with a Lightningcure LC8 UV spot light source (Hamamatsu Photonics, Hamamatsu, Japan) at 70% of its intensity. To get the integration of the heat of reaction, two runs were carried out with a short break between the runs to let the resin cool down. The second run was made once the material was fully cured and the baseline was stable. Then, the second curve was subtracted from the first to obtain the curve related to the curing only. Each run is conducted as follows: 30 s at the set temperature (25 °C at atmospheric pressure under air) without the lamp; then, the lamp is turned on for 2.5 min. The break between the runs lasts 30 s.

Dynamic mechanical analysis (DMA) was carried out on a Tritec 2000 DMA (Triton Technology, Loughborough, UK) in dual cantilever bending mode. Measurements were performed on rectangle bars (80 × 10 × 2 mm) which were prepared using a Teflon mold. Irgacure-1173 was added to the resin as photoinitator (5 wt %). Before curing, the mixture was left for 1 h at elevated temperature (50 to 90 °C depending on the viscosity of each sample) to reduce the amount of air trapped in the resin that could lead to cracks. Subsequently, it was poured into the mold and cured under UV light using an Aktiprint mini UN50029 UV dryer (Technigraf, Grävenwiesbach, Germany), equipped with a UV 4/120-2 lamp (254 nm, max. 120W cm^{-1}). The conveyor belt was set at 5 m min^{-1}, and five runs were required to fully cure the specimens. The irradiation energy (UV dose) was measured before and after curing with an average value of 460 mJ cm^{-2}. Samples were studied under nitrogen in a temperature range of −30 to 120 °C (5 K min^{-1}, 1 Hz, maximum displacement 0.05 mm). The exact dimensions of specimens were determined before each experiment, and five test bars were measured for each sample.

Thermogravimetric analysis (TGA) of cured resins was performed on a TGA/DSC 1 (Mettler-Toledo, Greifensee, Switzerland) under nitrogen (35 mL min^{-1}). Samples of ca. 10 mg were heated from 25 to 800 °C with a heating rate of 10 K min^{-1}.

2.3. *Synthesis of Amido Diols*

Two amido diols (AD) with varying alkylene chain lengths were synthesized as described previously [40,42]. In brief, a certain amount of diamine (1,4-diaminobutane or 1,6-diaminohexane) was placed in a three-necked flask equipped with mechanical stirrer and cooler and dissolved in 2-propanol (200 mL mol^{-1}). ε-Caprolactone (2 equivalents) in 2-propanol (50 mL mol^{-1}) was slowly added to the diamine solution at room temperature using a pump (0.6 mL/min). Stirring was continued overnight. The resulting white pasty product was diluted with acetone and then washed several times with the same solvent to remove residual monomers and potentially formed oligomeric by-products by (homo)polymerization of the lactone. It was further purified by recrystallization from methanol and acetone and then dried under reduced pressure, whereby the respective product was obtained as a white powder. Amido diols are named analogously to polyamides of the AABB type, while the first number indicates the number of carbon atoms in the diamine and the second indicates the number of carbon atoms in the lactone. Hence, AD 4.6 was obtained from 1,4-diaminobutane and ε-caprolactone, while AD 6.6 was synthesized from 1,6-diaminohexane.

2.4. *Polymer Synthesis*

Six unsaturated poly(ester amide) (PEA) resins as well as a neat polyester, that served as a reference, were prepared via the transesterification method, where an excess of 30% hydroxyl groups was targeted (dimethyl ester:diol ratio of 1:1.3). As an example, the synthesis of the neat polyester resin, named PE, will be described. For the synthesis of 60 g of resin, 35.57 g of DMI (224 mmol, 0.8 eq), 12.95 g of DMS (56 mmol, 0.2 eq), 14.71 g of 1,3-PD (193 mmol, 0.75 eq), 5.14 g of Is (35 mmol, 0.125 eq), 3.24 g of Gly (35 mmol, 0.125 eq) and 6.33 g of 2,3-BD (70 mmol, 0.25 eq), along with 0.12 g of FASCAT 4101 (0.2 wt %), 24 mg of BHT (0.08 wt %), and 18 mg of MeHQ (0.06 wt %) were charged into a three-necked round-bottom flask equipped with mechanical stirrer, thermometer, condenser and an immersed thermocouple to control the temperature of the heating mantle. The mixture was heated gradually until homogenization, and then the temperature was set at 180 °C for 3 h. Finally, a slight vacuum (20 mbar) was applied for 30 min, and then the system was left to cool down to room temperature. The product was obtained as a yellow-orange resin. For the preparation of the resins containing amide moieties in their backbone, 0.0625, 0.125, and 0.1875 eq of 1,3-PD were substituted by the respective amido diol.

3. Results and Discussion

3.1. *Polymer Synthesis*

One of the main objects of this work was the introduction of amide moieties into the backbone of a bio-based unsaturated polyester resin derived from itaconic acid. As it was already discussed, the presence of free $-NH_2$ moieties leads to the formation of the pyrrolidone lactam ring from itaconic acid, while it also supports the isomerization of itaconic acid to mesaconic acid. For this reason, preformed symmetrical diols that have two internal amide bonds were used for the synthesis of poly(ester amide)s. In order to assure that the amido diol structure would remain intact under the conditions applied for the synthesis, the dimethyl esters of the respective acids were used. Previous studies have shown that the amido diol structure can be lost by an acidolysis-type reaction if free carboxylic acid groups are present during reaction [43–45]. Hence, the dimethyl esters of dicarboxylic acids were utilized to avoid this undesirable side reaction.

Another challenge in the synthesis of poly(ester amide) resins is the formation of hydrogen bonds, which should lead to a high increase in viscosity. Indeed, preliminary studies showed very high viscosity values for the produced resins, rendering them nearly non-processable. In an attempt to counter this effect, a range of bio-based diols were incorporated in the polymer backbone to obtain highly irregular polymer chains, which should reduce the intermolecular interaction. Therefore, we anticipated obtaining poly(ester amide) resins with low viscosities. Even though this strategy was successful to some extent, the phenomenon could only be mildly suppressed, as the viscosity of the resins increased dramatically even with very small amido diol content. So, in another attempt to reduce the viscosity, the reaction time was set at 3.5 h for all resins, as prolonged reaction times can lead to undesired side reactions and an increased molecular weight that would result in higher viscosities of the resins.

Six unsaturated bio-based poly(ester amide) resins containing 5 to 15 mol% of an amido diol (AD 4.6 or 6.6) and a neat polyester, that served as a reference, were successfully synthesized via a transesterification method of DMI, DMS, and a variety of sterically demanding diols (Scheme 1). The distance between amide bonds was varied by the choice of the diamine used for the synthesis of amido diols. All polymers were obtained following the same protocol, considering the above-mentioned limitations. The monomers and composition used for the synthesis of the resins are given in Table 1. PEA labels include the molar fraction of targeted ester-amide sequences in the resin, which corresponds to the molar percentage of the respective amido diol in the feed. For example, 5 mol% of ester amide moieties were aimed for PEA 4.6-05, corresponding to the equal amount of AD 4.6 in the feed. All building blocks used, but the amido diols, are accessible from renewable resources, resulting in PEA resins with very high bio-based content.

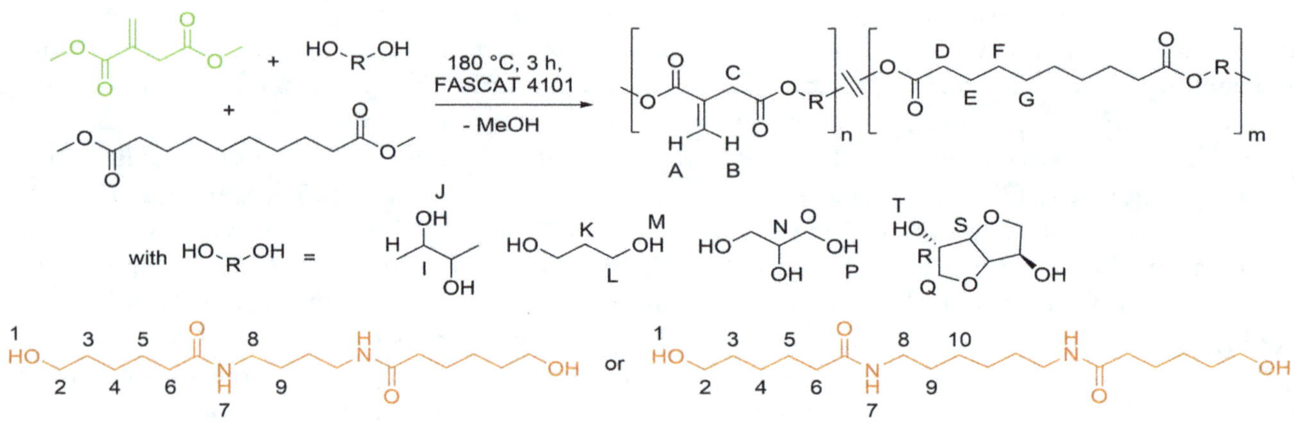

Scheme 1. Synthesis route of the unsaturated polyester and poly(ester amide) resins obtained in the course of this study.

Table 1. Composition of the unsaturated neat polyester and poly(ester amide) resins synthesized. DMI: dimethyl itaconate, DMS: dimethyl sebacate.

Sample	DMI (eq)	DMS (eq)	1,3 PD (eq)	2,3 BD (eq)	Is (eq)	Gly (eq)	Amido Diol (eq)	Amido Diol (mol%)
PE	0.8	0.2	0.75	0.25	0.125	0.125	-	-
PEA 4.6-05	0.8	0.2	0.6875	0.25	0.125	0.125	0.0625	5
PEA 4.6-10	0.8	0.2	0.625	0.25	0.125	0.125	0.125	10
PEA 4.6-15	0.8	0.2	0.5625	0.25	0.125	0.125	0.1875	15
PEA 6.6-05	0.8	0.2	0.688	0.25	0.125	0.125	0.0625	5
PEA 6.6-10	0.8	0.2	0.625	0.25	0.125	0.125	0.125	10
PEA 6.6-15	0.8	0.2	0.5625	0.25	0.125	0.125	0.1875	15

3.2. Structural Characterization

The ATR-FTIR spectra of synthesized resins are presented in Figure 1. In the spectra of all polymers, the presence of ester bonds is confirmed by the signal at 1720 cm^{-1}, which can be assigned to the C=O stretching vibrations. Peaks at 1640 cm^{-1} and 810 cm^{-1}, corresponding to the C–C stretch and C–C deformation vibrations of the double bond of itaconic acid, confirm that the double bond did not undergo undesired side reactions during synthesis. A broad peak around 3500 cm^{-1} can be found in the spectra due to the OH end groups of polymer chains, since diols were used in excess. After the incorporation of an amido diol into the polyester backbone, new signals associated with amide groups can be found in the spectra. The intensity of the signal at 1640 cm^{-1} increased, since the peak of the double bond of itaconic acid and those associated with the C=O (Amide I) stretching vibrations of the amido diol are overlapping. In addition, a new absorption peak can be found at 1540 cm^{-1} that is attributed to the N–H (Amide II) bending vibrations of amide bonds. Furthermore, a new signal is observed at 3300 cm^{-1}, which is related to hydrogen-bonded N–H groups. By increasing the amido diol content, the intensity of the peaks associated with amide moieties also increased.

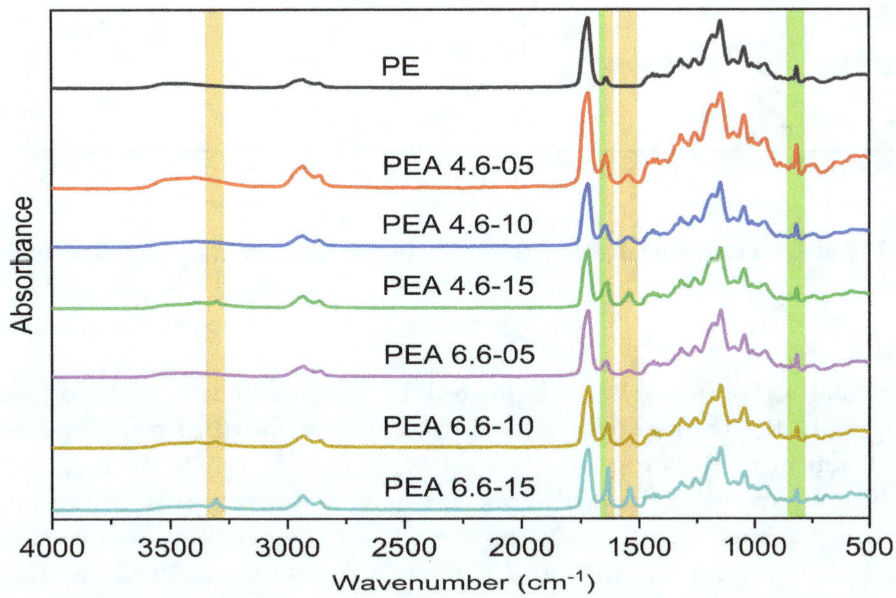

Figure 1. FTIR spectra of synthesized resins. C–C double bond vibrations are highlighted in green, vibrations derived from the amide group are in orange.

The chemical structure of the synthesized resins was also investigated by means of ^{1}H NMR spectroscopy. All the spectra of the resins show the signals of the neat polyester resin (PE). The poly(ester amide)s show additional peaks that can be assigned to the ester-amide sequences. In addition, by increasing the amido diol content, the respective peaks increase in intensity. Detailed analysis of the spectra can be found in the Supplementary Materials.

Both FTIR and NMR spectra support the successful incorporation of the amido diol building block into the polymer backbone. In addition, the results suggest that aforementioned side reactions, such as ring formation or isomerization, can be suppressed when the preformed amido diols are used. Pyrrolidone lactam ring formation was completely avoided, while the isomerization of itaconic acid was below 3%, which is similar to what is observed when standard polyesters are synthesized from itaconic acid. These results confirm that the copolymerization of amido diols is a viable synthetic strategy for the insertion of amide moieties into unsaturated polyester resins based on itaconic acid.

As discussed earlier, the reaction time for the synthesis of the resins in question was limited to obtain PEAs with a viscosity suitable for processing. However, this practice resulted in some methyl ester end groups remaining in the polymer chain, as shown in the ^{1}H NMR spectra (Figure 2). Therefore, rather low molecular weights are assumed for the resins produced. Nevertheless, we consider this practice mandatory, because the increased viscosity makes these resins very difficult to manipulate for the characterizations needed as well as the envisaged UV-curing applications, and a longer reaction would have rendered them non-processable.

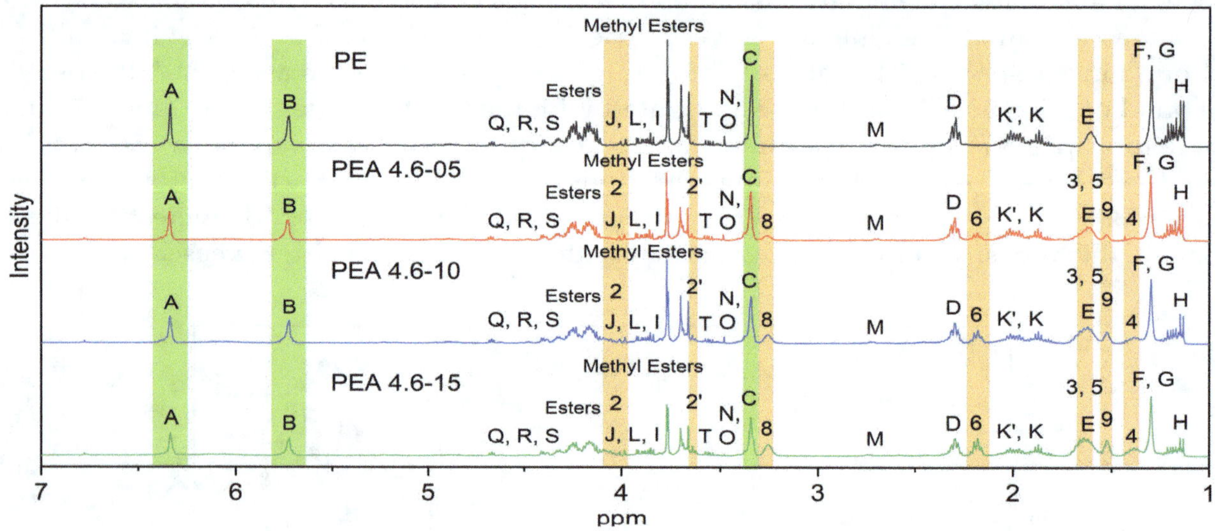

Figure 2. ^{1}H NMR spectra of the neat polyester resin and PEAs containing different amounts of amido diol 4.6.

3.3. SEC Measurements

SEC results are shown in Figure 3. As expected for the products of polycondensation reactions, the molecular weights of the resins are obtained in gaussian distribution. The absence of any high molecular weight fraction in the elugrams, along with the ^{1}H NMR data, shows that undesired side reactions leading to cross-linking of the polymers were successfully prevented. The molecular weight of the produced resins is rather low, as a result of the synthetic procedure, in the range of 300 to 5000 g/mol. However, the M_n, M_w, and Đ were not calculated, because the calibration of the measurements was limited to 575 g/mol. Nevertheless, they are still suitable for applications in the coatings field. The resins derived from AD 6.6 present larger retention volume peaks compared to those based on AD 4.6. This indicates that a higher molecular weight is to be expected when the amide groups are separated by a longer alkylene chain.

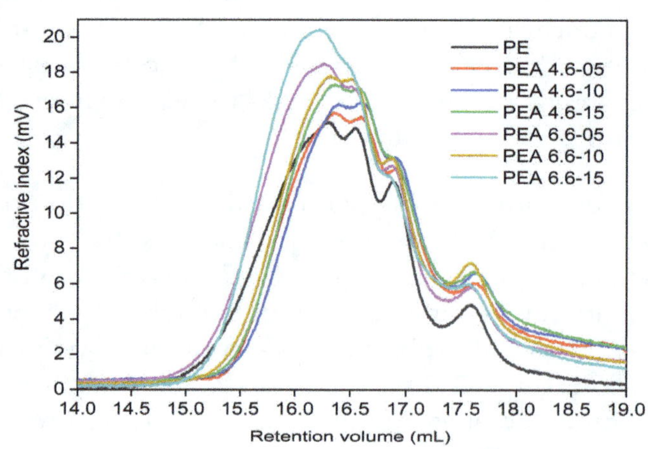

Figure 3. SEC elugrams of the resins synthesized in this study.

3.4. Viscosity

As anticipated, the incorporation of the amido diol into the polyester backbone led to increased viscosities. However, as a somewhat unexpected feature, the resins also became opaque and almost turned into a waxy solid when allowed to cool overnight in the storage container. At this state, the resins did not flow any more unless an external force was applied. This phenomenon became more pronounced with increasing amido diol content, suggesting that it was derived from intermolecular hydrogen bonds. Similar claims were made in other works where polyesters containing amide moieties were examined, and it was found via temperature-dependent IR spectroscopy that as the temperature increased, the amide-ester hydrogen bonds were destroyed [46,47].

For the first set of measurements, viscosity values were obtained at different temperatures under a constant shear rate of 100 s^{-1}. As can be seen in Figure 4, an increase of the temperature resulted in an exponential decrease of the viscosity. In addition to the temperature dependency, we were intrigued to investigate whether the resins also exhibit a non-Newtonian behavior—thus, if the viscosity can be influenced by a difference in shear stress and whether there is a time-dependent behavior. This behavior was studied by monitoring the viscosity as a function of the shear rate. In detail, this was done by first applying a low shear rate of 1 s^{-1} until a constant viscosity was obtained. Then, a high shear rate of 100 s^{-1} was applied for 2 min to break down the interactions between the polymer chains. In a last phase, the shear rate was reduced again to 1 s^{-1}. For both transitions of the shear rates, the effect on the viscosity was examined, as well as a time dependency of the latter, which would indicate a thixotropic behavior. In addition, the ability of the resins to recover to the initial state was examined [48–51]. A strong shear thinning effect could be observed for all resins with amido diols incorporated in the polymer chain (Figure 5). Even at low amido diol contents of 5 mol%, this effect could be observed with AD 6.6 resulting in a more pronounced influence on the shear thinning behavior (Figure 5b). With increasing ester-amide content, the initial viscosity increased to a point where PEA 6.6–15 could not be measured anymore. However, for PEA 4.6–15, the viscosity is reduced from 1127 Pa·s at 1 s^{-1} to 30 Pa·s at 100 s^{-1}, which corresponds to 2.6% of the initial viscosity (Table 2). This shear thinning effect also showed a time-dependent behavior, indicating a thixotropic behavior of the resins. For the neat polyester, no change in viscosity could be observed. The recovery of the initial viscosity was measured in the third phase of the rheology experiment. Here, the thixotropic behavior was more easily observed, as it took several minutes to reach constant viscosities. In addition, it became apparent that resins containing amido diol 6.6 exhibited very high recovery rates that were significantly higher than those of the corresponding resins with amido diol 4.6 as a building block (Table 2). This difference is especially pronounced for the resins with 10 mol% amido diol, where the resin based on AD 4.6 recovers only 69% of the initial viscosity in comparison to 95% in the case of that derived from AD 6.6. These results suggest that in the case of the former amido diol, the initial arrangement of the resin is not achieved within the measurement time, resulting in a lower amount of polymer chain interaction via hydrogen bonds. Most probably, this behavior can be explained by the different structure of the amido diols, since the more flexible structure is the one with higher recovery rates. The increased length of the spacer unit between the two amide bonds (1,6-diaminohexane and 1,4-diaminobutane, respectively) gives the chain an extra flexibility that contributes to the faster recreation of the hydrogen bonds after their structure is destroyed by the high shear rate.

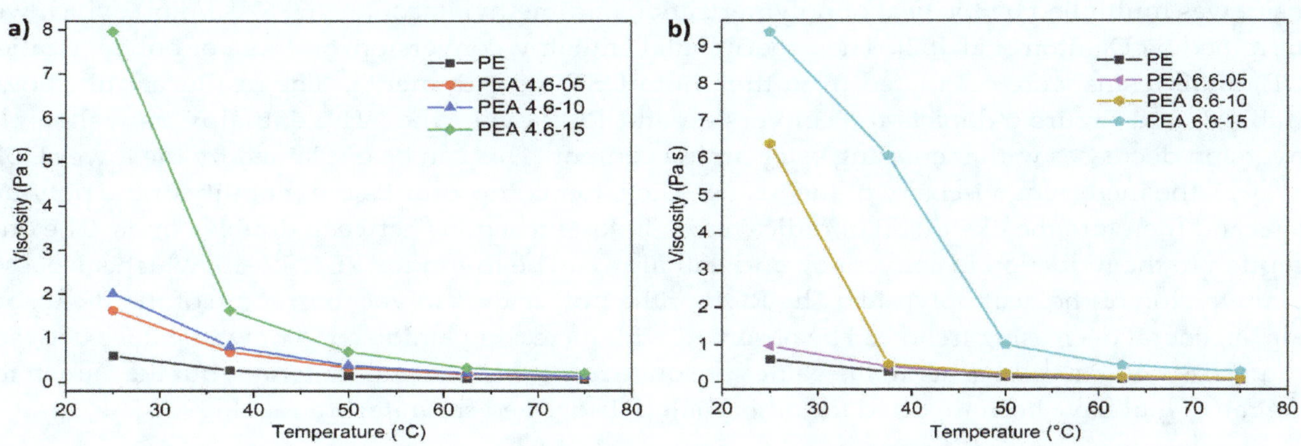

Figure 4. Temperature dependency of the viscosity of resins synthesized in the course of this study with (**a**) amido diol 4.6 and (**b**) amido diol 6.6.

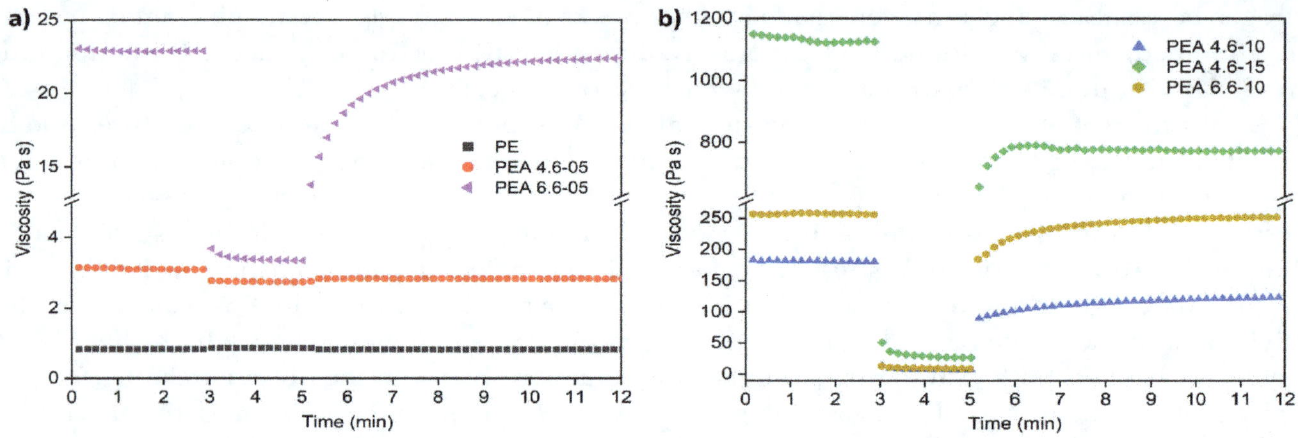

Figure 5. Viscosity experiment conducted to study the thixotropic behavior of the resins containing (**a**) 5 mol% amido diol and (**b**) 10 mol% and higher amounts of amido diol. Conditions: 3 min at 1 s^{-1}, then 2 min at 100 s^{-1}, finally 7 min at 1 s^{-1}.

Table 2. Viscosity values of the polymers at different shear rates and percentage of the initial viscosity recovered at the end of the experiment.

Sample	Starting Viscosity (Pa·s at 1 s^{-1})	Viscosity at High Shear Rate (Pa·s at 100 s^{-1})	% of Starting Viscosity at High Shear Rate	Viscosity at Recovery (Pa·s at 1 s^{-1})	% of Recovered Viscosity
PE	0.84	0.84	-	0.84	-
PEA 4.6-05	3.1	2.75	88.7	2.83	91
PEA 4.6-10	181	7	3.8	123	69
PEA 4.6-15	1127	30	2.6	773	68
PEA 6.6-05	22.9	3.37	14.7	22.3	97
PEA 6.6-10	258	10	3.8	251	95
PEA 6.6-15			Unable to measure at 25 °C		

3.5. UV-Curing Behavior

Photo-DSC measurements were conducted to evaluate the reactivity of the resins toward UV-curing. The double-bond density (DBD) of the resin was calculated from the composition of the resins. In order to obtain the theoretical enthalpy of the UV-induced radical polymerization reaction, the double-bond density was multiplied by the heat of polymerization of dimethyl itaconate (60.67 KJ/mol), which was determined by Dainton et al. [52]. The experimental enthalpy, conversion, and rate of polymerization (ROP) of all resins were calculated from the photo-DSC measurements. The results are all shown in Table 3, and Figure 6 depicts the conversion and ROP over time. The enthalpy, as well as the conversion decreases with increasing ester-amide content. This can be explained by the lower DBD, as well as the increasing viscosity of the resins. The latter compromises the mobility of the polymer chains and therefore the likelihood of radical cross-linking reactions between double bonds. The only exception to the reduction in conversion and enthalpy can be found for PEA 4.6-10, which surpasses the conversion of the neat polyester. The ROP of the polyester is lower than some of the PEAs, but again, a general decreasing trend can be observed with increasing amide content with values between 6.3 and 10.8 × 10^{-3} s^{-1}. In general, the values are quite low for UV-curing materials, but they are in the same range that have been reported for other materials derived from itaconic acid [39,53].

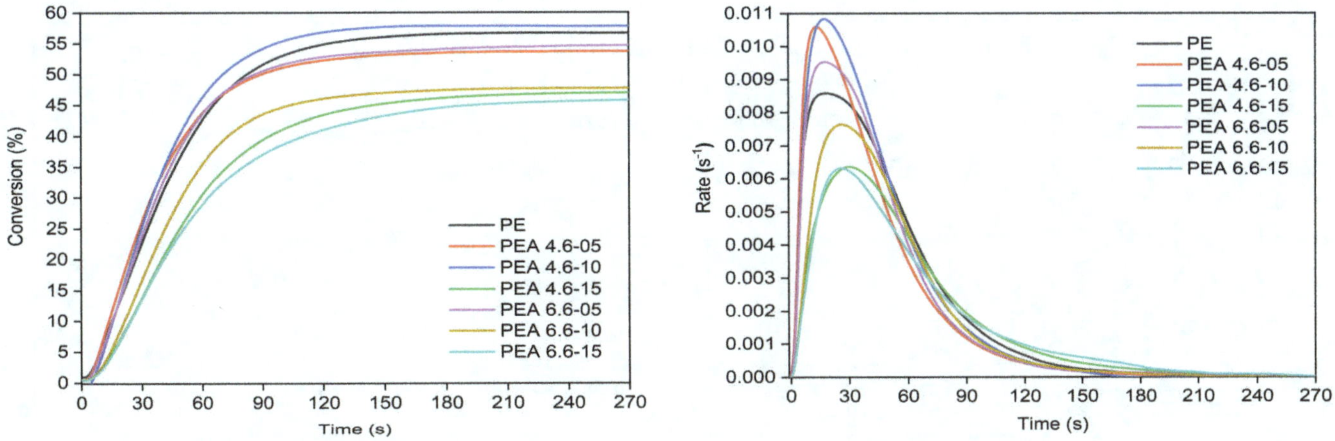

(a) (b)

Figure 6. Photo-DSC of the PEAs: Conversion (**a**) and rate of polymerization (**b**).

Table 3. Double-bond density (DBD), theoretical and experimental enthalpy of curing, conversion, and maximum rate of polymerization (ROP) of the resins at 25 °C.

Sample	DBD (mmol/g)	ΔH_{theo} (J/g)	ΔH (J/g)	C (%)	ROP (1000 s^{-1})
PE	3.7	224.5	127	57	8.6
PEA 4.6-05	3.4	206.3	111	54	10.6
PEA 4.6-10	3.2	194.1	112	58	10.8
PEA 4.6-15	3.0	182.0	85	47	6.4
PEA 6.6-05	3.4	206.3	113	55	9.5
PEA 6.6-10	3.2	194.1	92	48	7.7
PEA 6.6-15	3.0	182.0	83.	46	6.3

3.6. DMA Analysis

The PEAs were also cast into a Teflon mold and cured under UV light. The resulting bars (80 × 10 × 2 mm) were analyzed by means of DMA measurements. Figure 7 shows the storage modulus (E′) as a function of the temperature for the crosslinked polymers. The values for all materials at 25 °C are between 11.74 and 2.03 GPa and decrease with increasing temperature, reaching the range of 1.58–0.14 GPa in the rubbery state. The storage modulus E′ is a measurement of material stiffness and can be used as providing the information regarding the molecular weight of a polymer and its crosslink density [54]. By introducing the long aliphatic chain of the amido diols into the structure of the polyesters, the spacing between the double bonds increases, leading to a decrease of the crosslinking density in the cured polyesters. The phenomenon gets more pronounced by the introduction of a higher amido diol content into the resins, as well as by the use of AD 6.6, in which the amide bonds are separated by a longer alkylene chain compared to AD 4.6. This statement is supported by the crosslinking density values (ν_e) i.e., the molar number of elastically effective network chains per cubic centimeter of a sample, which was calculated from the formula: $\nu_e = E'/3RT$, where E′ is the storage modulus, R is the gas constant, and T is the Kelvin temperature, which are given in Table 4.

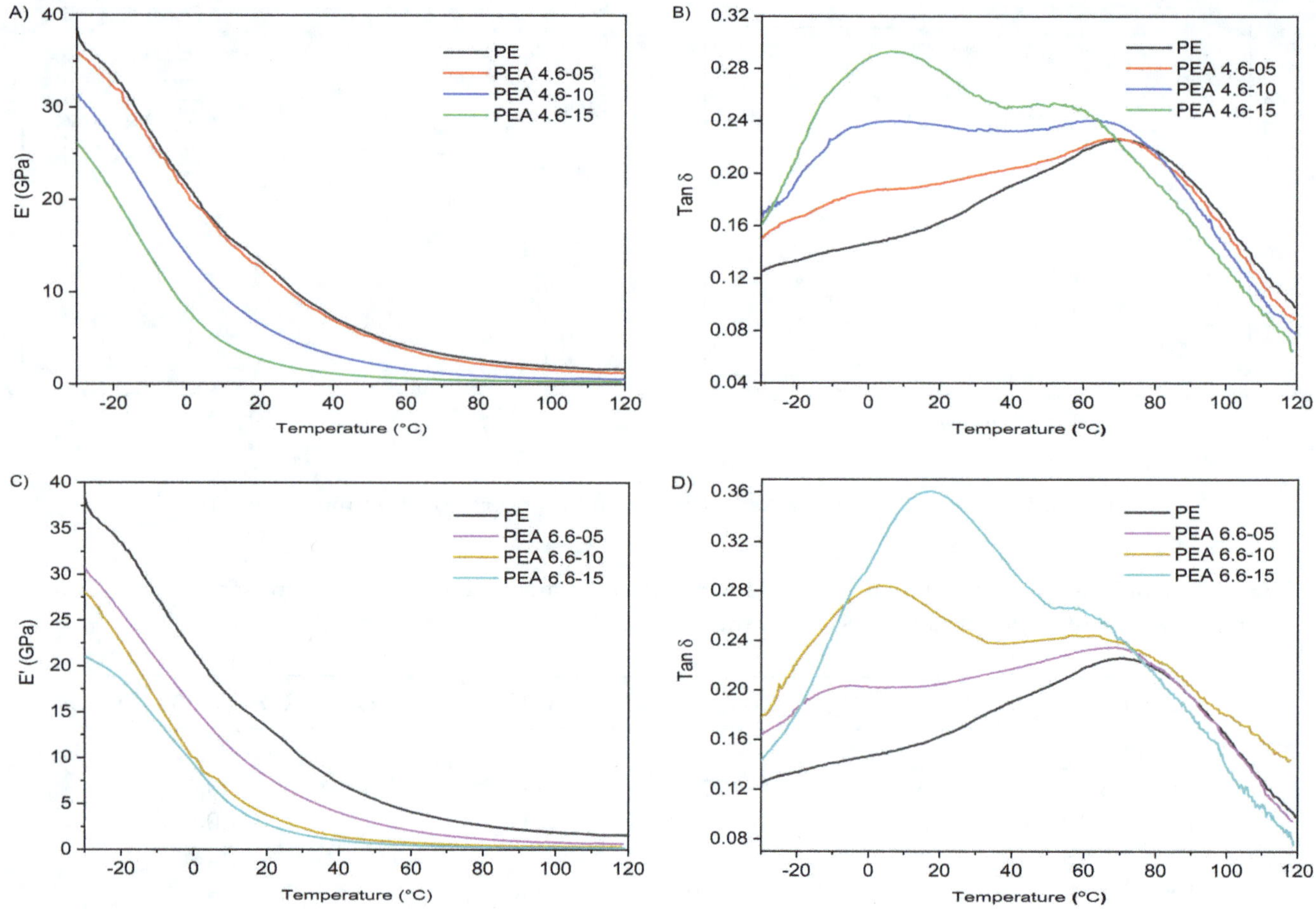

Figure 7. Storage modulus (E') and tan δ values of the PEAs after UV curing. PEAs with amido diol 4.6 (**A**,**B**) and PEAs with amido diol 6.6 (**C**,**D**).

Tan δ graphs are also presented in Figure 7. The tan δ peak corresponds to the T_g of the materials, while its broadening occurs when a less homogenous network is created [23,55]. In all cases, a very broad tan δ peak can be observed. The tan δ peak of the cured neat polyester resin is found at 71 °C, and by increasing the amido diol content, a proportional decrease is observed. Simultaneously, the incorporation of the amido diol into the resins resulted in a second peak emerging in the tan δ graphs at temperatures below 0 °C. This peak is shifted to higher values and gains in intensity with increasing AD content in the resins. This is an indication that possibly the crosslinked and non-crosslinked regions of the polymer are affected in different ways from the addition of the amido diols. As the distance between the double bonds is increasing, a more sparsely crosslinked region is created, and this is reflected on the transfer of the peak from 71 °C to lower values. In the same time, movement restrictions of the non-crosslinked polymer chains get lifted, and they can interact more with each other, resulting in more hydrogen bonds being formed and thus the second peak shifting to higher temperatures.

Table 4. DMA results of the cured resins.

Sample	E' at 25 °C (GPa)	E' at 115 °C (GPa)	Crosslink Density ($10^6 \times mol \times m^{-3}$)	Tan δ Peaks (°C)	
PE	11.74	1.58	0.163	-	71
PEA 4.6-05	11.02	1.19	0.123	−0.4	69
PEA 4.6-10	5.41	0.51	0.053	3.6	63
PEA 4.6-15	2.06	0.22	0.023	6.8	55
PEA 6.6-05	6.70	0.63	0.065	−9	68
PEA 6.6-10	3.01	0.28	0.029	4	61
PEA 6.6-15	2.03	0.14	0.0145	17	58

3.7. Thermal Degradation

The TGA results of the cured samples are given in Figure 8. The thermal degradation of the crosslinked resins occurs in three main steps as shown by the first derivative of the of the TGA curve (dTG graph). In the first step, a mass loss of 10%–15% is observed by 250 °C, which is assumed to correspond to non-reacted oligomers that have not been incorporated to the crosslinked network. However, the introduction of an amido diol seems to enhance the thermal stability of the material, since an increase of 5 to 16°C was found in the $T_{d,10\%}$ data that are shown in Table 5. In the second step of degradation, polyester segments decompose in the range of 350–400 °C, which is very common among polyester resins [55,56]. The incorporation of the amido diol leads to an extra peak in the dTG graph, which originally appeared as a shoulder peak for resin PE and as the amido diol mol% increases, that peak gets separated from the main polyester degradation peak and two clear peaks can be observed around 350 °C and 400 °C. Although it is unclear why this splitting occurs, it might be associated with the fact that the crosslinking density becomes sparser by introducing higher amido diol mol% in the polymer chains. In the third and final step, the amide segments are degrading, as it is shown by the third peak of the dTG graph, around 450 °C, the intensity of which increased along with the amido diol content of the resins. Lastly, the resins containing amide moieties presented higher residues at 800 °C than the PE resin, while resins containing amido diol 4.6 present slightly higher values for both $T_{d,10\%}$ and $R_{800°C}$, suggesting better thermal stability.

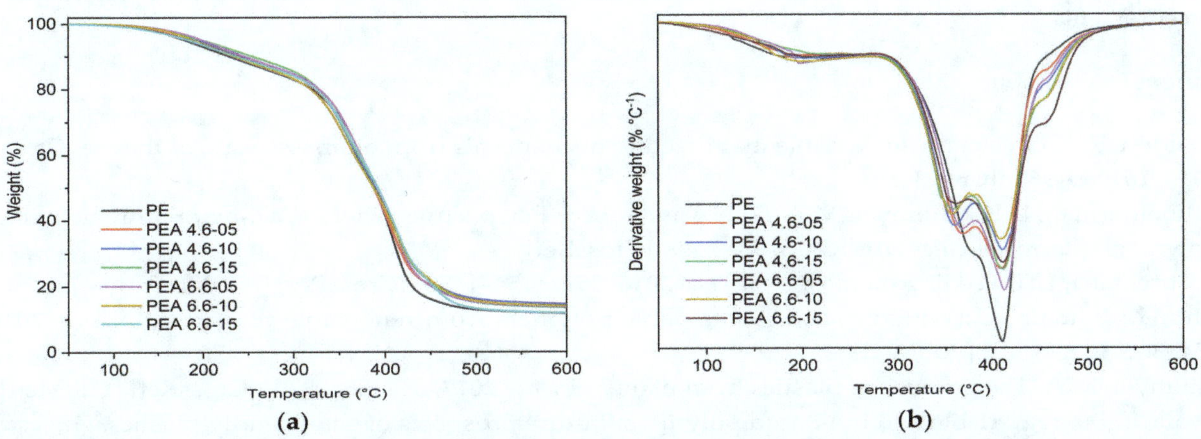

Figure 8. TGA thermograph (**a**) and dTG (**b**) of the cured resins.

Table 5. TGA results of the cured resins.

Sample	$T_{d,10\%}$ (°C)	$R_{800\,°C}$ (%)
PE	230	11.2
PEA 4.6-05	242	13.3
PEA 4.6-10	245	13.8
PEA 4.6-15	255	12.8
PEA 6.6-05	246	13.1
PEA 6.6-10	235	13.3
PEA 6.6-15	237	11.3

4. Conclusions

In this work, we report the synthesis of bio-based unsaturated poly(ester amide)s, trying to overcome aza-Michael additions and isomerization reactions that usually take place when itaconic acid and primary diamines are reacted under polycondensation conditions. Our synthetic strategy involved the use of preformed symmetrical amido diols with two internal amide bonds. Evaluation of structural composition verified the successful synthesis of the desired poly(ester amide)s. In addition, both undesired side reactions were effectively suppressed even at amido diol contents of up to 15 mol%.

The introduction of amide bonds in the polymer structure had a huge impact on the physicochemical properties of the resins e.g., the viscosity and reactivity toward UV-induced crosslinking. The latter was affected, as the conversion of the double bonds along with the rate of polymerization were lowered at higher amido diol content. However, at lower amounts of the amido diol building blocks incorporated into the PEA resins, the curing behavior was not significantly affected. For PEA with 10 mol% of the amido diol 4.6, the conversion and ROP was even higher than for the neat PE. In addition, all PEA resins exhibited an intriguing thixotropic behavior. This feature is very interesting, and it could be achieved even with amido diol loadings as low as 5 mol%. Furthermore, despite the lower conversion of some PEA resins, all samples were still suitable for UV-curing applications. The cured test specimen obtained in this study were examined on their thermomechanical properties by means of DMA, with a considerable impact of the amido diol on the properties of the samples. Therefore, the PEAs described in this study could be suitable as additives for certain UV-curing application where thixotropic behavior of the resins is desired.

Author Contributions: L.P. and M.K.: Conceptualization, methodology, investigation, writing—original draft preparation; D.N.B.: Conceptualization, supervision, writing—review and editing; T.R.: Conceptualization, methodology, formal analysis, supervision, writing—review and editing. All authors have read and agreed to the published version of this manuscript.

Acknowledgments: The authors would like to thank K.U. Loos und A.J.J. Woortman for the performance of the SEC measurements.

References

1. Sheldon, R.A. Green and sustainable manufacture of chemicals from biomass: State of the art. *Green Chem.* **2014**, *16*, 950–963. [CrossRef]
2. Schneiderman, D.K.; Hillmyer, M.A. 50th Anniversary Perspective: There Is a Great Future in Sustainable Polymers. *Macromolecules* **2017**, *50*, 3733–3749. [CrossRef]
3. Bornscheuer, U.T. Feeding on plastic. *Science* **2016**, *351*, 1154–1155. [CrossRef] [PubMed]
4. Zhu, Y.; Romain, C.; Williams, C.K. Sustainable polymers from renewable resources. *Nature* **2016**, *540*, 354–362. [CrossRef] [PubMed]
5. Hillmyer, M.A. The promise of plastics from plants. *Science* **2017**, *358*, 868–870. [CrossRef] [PubMed]
6. Iwata, T. Biodegradable and bio-based polymers: Future prospects of eco-friendly plastics. *Angew. Chem. Int. Ed.* **2015**, *54*, 3210–3215. [CrossRef] [PubMed]
7. Williams, C.K.; Hillmyer, M.A. Polymers from renewable resources: A perspective for a special issue of polymer reviews. *Polym. Rev.* **2008**, *48*, 1–10. [CrossRef]
8. Niaounakis, M. *Biopolymers: Processing and Products*; William Andrew: Oxford, UK; Waltham, MA, USA, 2014.
9. Niaounakis, M. *Biopolymers: Reuse, Recycling, and Disposal*; William Andrew: Oxford, UK; Waltham, MA, USA, 2013.
10. Werpy, T.; Petersen, G. *Top Value Added Chemicals from Biomass: Volume I—Results of Screening for Potential Candidates from Sugars and Synthesis Gas*; U.S. Department of Energy: Washington, DC, USA, 2004; pp. 1–69. [CrossRef]
11. Willke, T.; Vorlop, K.D. Biotechnological production of itaconic acid. *Appl. Microbiol. Biotechnol.* **2001**, *56*, 289–295. [CrossRef]
12. Krull, S.; Lünsmann, M.; Prüße, U.; Kuenz, A. Ustilago Rabenhorstiana—An Alternative Natural Itaconic Acid Producer. *Fermentation* **2020**, *6*, 4. [CrossRef]
13. Saha, B.C.; Kennedy, G.J. Phosphate limitation alleviates the inhibitory effect of manganese on itaconic acid production by Aspergillus terreus. *Biocatal. Agric. Biotechnol.* **2019**, *18*, 101016. [CrossRef]
14. Chakraborty, S.; Ju, L.; Galuska, A.A.; Moore, R.B.; Turner, S.R. Suspension polymerization of itaconic acid diesters. *J. Appl. Polym. Sci.* **2018**, *135*, 46417. [CrossRef]
15. Lopez-Carrasquero, F.; de Ilarduya, A.M.; Cardenas, M.; Carrillo, M.; Arnal, M.L.; Laredo, E.; Torres, C.; Mendez, B.; Muller, A.J. New comb-like poly(n-alkyl itaconate)s with crystalizable side chains. *Polymer* **2003**, *44*, 4969–4979. [CrossRef]

16. Retuert, J.; Pedram, M.Y.; Martínez, F.; Jeria, M. Soluble Itaconic Acid–Ethylene Glycol Polyesters. *Bull. Chem. Soc. Jpn.* **1993**, *66*, 1707–1708. [CrossRef]
17. Arrighi, V.; McEwen, I.J.; Holmes, P.F. Dielectric Relaxations in Poly(di-n-alkyl itaconate)s. *Macromolecules* **2004**, *37*, 6210–6218. [CrossRef]
18. Inciarte, H.; Orozco, M.; Fuenmayor, M.; López-Carrasquero, F.; Oliva, H. Comb-like copolymers of n-alkyl monoitaconates and styrene. *e-Polymers* **2006**, *6*. [CrossRef]
19. Robert, T.; Friebel, S. Itaconic acid-a versatile building block for renewable polyesters with enhanced functionality. *Green Chem.* **2016**, *18*, 2922–2934. [CrossRef]
20. Pellis, A.; Hanson, P.A.; Comerford, J.W.; Clark, J.H.; Farmer, T.J. Enzymatic synthesis of unsaturated polyesters: Functionalization and reversibility of the aza-Michael addition of pendants. *Polym. Chem.* **2019**, *10*, 843–851. [CrossRef]
21. Li, P.; Ma, S.; Dai, J.; Liu, X.; Jiang, Y.; Wang, S.; Wei, J.; Chen, J.; Zhu, J. Itaconic Acid as a Green Alternative to Acrylic Acid for Producing a Soybean Oil-Based Thermoset: Synthesis and Properties. *ACS Sustain. Chem. Eng.* **2017**, *5*, 1228–1236. [CrossRef]
22. Dai, J.Y.; Ma, S.Q.; Teng, N.; Dai, X.Y.; Shen, X.B.; Wang, S.; Liu, X.Q.; Zhu, J. 2,5-Furandicarboxylic Acid- and Itaconic Acid-Derived Fully Biobased Unsaturated Polyesters and Their Cross-Linked Networks. *Ind. Eng. Chem. Res.* **2017**, *56*, 2650–2657. [CrossRef]
23. Brännström, S.; Malmström, E.; Johansson, M. Biobased UV-curable coatings based on itaconic acid. *J. Coat. Technol. Res.* **2017**, *14*, 851–861. [CrossRef]
24. Schoon, I.; Kluge, M.; Eschig, S.; Robert, T. Catalyst influence on undesired side reactions in the polycondensation of fully bio-based polyester itaconates. *Polymers* **2017**, *9*, 693. [CrossRef] [PubMed]
25. Mehta, L.B.; Wadgaonkar, K.K.; Jagtap, R.N. Synthesis and characterization of high bio-based content unsaturated polyester resin for wood coating from itaconic acid: Effect of various reactive diluents as an alternative to styrene. *J. Dispers. Sci. Technol.* **2019**, *40*, 756–765. [CrossRef]
26. Patil, D.M.; Phalak, G.A.; Mhaske, S.T. Design and synthesis of bio-based UV curable PU acrylate resin from itaconic acid for coating applications. *Des. Monomers Polym.* **2017**, *20*, 269–282. [CrossRef] [PubMed]
27. Dai, J.; Ma, S.; Wu, Y.; Han, L.; Zhang, L.; Zhu, J.; Liu, X. Polyesters derived from itaconic acid for the properties and bio-based content enhancement of soybean oil-based thermosets. *Green Chem.* **2015**, *17*, 2383–2392. [CrossRef]
28. Goerz, O.; Ritter, H. Polymers with shape memory effect from renewable resources: Crosslinking of polyesters based on isosorbide, itaconic acid and succinic acid. *Polym. Int.* **2013**, *62*, 709–712. [CrossRef]
29. Guo, B.; Chen, Y.; Lei, Y.; Zhang, L.; Zhou, W.Y.; Rabie, A.B.M.; Zhao, J. Biobased Poly(propylene sebacate) as Shape Memory Polymer with Tunable Switching Temperature for Potential Biomedical Applications. *Biomacromolecules* **2011**, *12*, 1312–1321. [CrossRef]
30. Robert, T.; Eschig, S.; Biemans, T.; Scheifler, F. Bio-based polyester itaconates as binder resins for UV-curing offset printing inks. *J. Coat. Technol. Res.* **2019**, *16*, 689–697. [CrossRef]
31. Farmer, T.J.; Clark, J.H.; Macquarrie, D.J.; Ogunjobi, J.K.; Castle, R.L. Post-polymerisation modification of bio-derived unsaturated polyester resins via Michael additions of 1,3-dicarbonyls. *Polym. Chem.* **2016**, *7*, 1650–1658. [CrossRef]
32. Winkler, M.; Lacerda, T.M.; Mack, F.; Meier, M.A.R. Renewable polymers from itaconic acid by polycondensation and ring-opening-metathesis polymerization. *Macromolecules* **2015**, *48*, 1398–1403. [CrossRef]
33. Guarneri, A.; Cutifani, V.; Cespugli, M.; Pellis, A.; Vassallo, R.; Asaro, F.; Ebert, C.; Gardossi, L. Functionalization of Enzymatically Synthesized Rigid Poly(itaconate)s via Post-Polymerization Aza-Michael Addition of Primary Amines. *Adv. Synth. Catal.* **2019**, *361*, 2559–2573. [CrossRef]
34. Farmer, T.J.; Macquarrie, D.J.; Comerford, J.W.; Pellis, A.; Clark, J.H. Insights into post-polymerisation modification of bio-based unsaturated itaconate and fumarate polyesters via aza-michael addition: Understanding the effects of C=C isomerisation. *J. Polym. Sci. Part A Polym. Chem.* **2018**, *56*, 1935–1945. [CrossRef] [PubMed]
35. Medway, A.M.; Sperry, J. Heterocycle construction using the biomass-derived building block itaconic acid. *Green Chem.* **2014** *16*, 2084–2101. [CrossRef]
36. Paytash, P.L.; Sparrow, E.; Gathe, J.C. The Reaction of Itaconic Acid with Primary Amines. *J. Am. Chem. Soc.* **1950**, *72*, 1415–1416. [CrossRef]
37. Wang, Z.; Wei, T.; Xue, X.; He, M.; Xue, J.; Song, M.; Wu, S.; Kang, H.; Zhang, L.; Jia, Q. Synthesis of fully

bio-based polyamides with tunable properties by employing itaconic acid. *Polymer* **2014**, *55*, 4846–4856. [CrossRef]

38. Nsengiyumva, O.; Miller, S.A. Synthesis, characterization, and water-degradation of biorenewable polyesters derived from natural camphoric acid. *Green Chem.* **2019**, *21*, 973–978. [CrossRef]
39. Ouhichi, R.; Pérocheau Arnaud, S.; Bougarech, A.; Abid, S.; Abid, M.; Robert, T. First Example of Unsaturated Poly(Ester Amide)s Derived From Itaconic Acid and Their Application as Bio-Based UV-Curing Polymers. *Appl. Sci.* **2020**, *10*, 2163. [CrossRef]
40. Kluge, M.; Bikiaris, D.N.; Robert, T. Enhancing the properties of poly(propylene succinate) by the incorporation of crystallizable symmetrical amido diols. *Eur. Polym. J.* **2019**, *120*, 109195. [CrossRef]
41. Klonos, P.A.; Kluge, M.; Robert, T.; Kyritsis, A.; Bikiaris, D.N. Molecular dynamics, crystallization and hydration study of Poly(Propylene succinate) based Poly(Ester amide)s. *Polymer* **2020**, *186*, 122056. [CrossRef]
42. Kluge, M.; Rennhofer, H.; Lichtenegger, H.C.; Liebner, F.W.; Robert, T. Poly(ester amide)s from poly(alkylene succinate)s and rapid crystallizing amido diols: Synthesis, thermal properties and crystallization behavior. *Eur. Polym. J.* **2020**, *129*, 109622. [CrossRef]
43. Stapert, H.R.; Bouwens, A.-M.; Dijkstra, P.J.; Feijen, J. Environmentally degradable aliphatic poly(ester-amide)s based on short, symmetrical and uniform bisamide-diol blocks, 1. Synthesis and interchange reactions. *Macromol. Chem. Phys.* **1999**, *200*, 1921–1929. [CrossRef]
44. Kotliar, A.M. Interchange Reactions Involving Condensation Polymers. *J. Polym. Sci. Macromol. Rev.* **1981**, *16*, 367–395. [CrossRef]
45. Kricheldorf, H.R.; Denchev, Z. Interchange Reactions in Condensation Polymers and Their Analysis by NMR Spectroscopy. In *Transreactions in Condensation Polymers*; Wiley-VCH Verlag GmbH: Weinheim, Germany, 2007; pp. 1–78.
46. Pillai, C.K.S.; Sandhya, K.Y.; Sudha, J.D.; Saminathan, M. Influence of hydrogen bonding on the generation and stabilization of liquid crystalline polyesters, poly(esteramide)s and polyacrylates. *Pramana* **2003**, *61*, 417–423. [CrossRef]
47. Lips, P. Aliphatic Segmented Poly(Ester amide)s Based on Symmetrical Bisamide-Diols. PhD Thesis, University of Twente, Enschede, The Netherlands, 2005.
48. Shenoy, A. *Thermoplastic Melt Rheology and Processing*; CRC Press: Boca Raton, FL, USA, 1996.
49. Han, C.D. *Rheology and Processing of Polymeric Materials, Volume 1: Polymer Rheology*; Oxford University Press: New York, NY, USA, 2007.
50. Shaw, M.T. *Introduction to Polymer Rheology*; John Wiley & Sons, Inc.: Hoboken, NJ, USA, 2012; Volume 66.
51. Kontopoulou, M. *Applied Polymer Rheology*; John Wiley & Sons, Inc.: Hoboken, NJ, USA, 2011.
52. Dainton, F.S.; Ivin, K.J.; Walmsley, D.A.G. The heats of polymerization of some cyclic and ethylenic compounds. *Trans. Faraday Soc.* **1960**, *56*, 1784–1792. [CrossRef]
53. Pérocheau Arnaud, S.; Andreou, E.; Pereira Köster, L.V.G.; Robert, T. Selective Synthesis of Monoesters of Itaconic Acid with Broad Substrate Scope: Biobased Alternatives to Acrylic Acid? *ACS Sustain. Chem. Eng.* **2020**, *8*, 1583–1590. [CrossRef]
54. Zhang, Y.; Asif, A.; Shi, W. Highly branched polyurethane acrylates and their waterborne UV curing coating. *Prog. Org. Coat.* **2011**, *71*, 295–301. [CrossRef]
55. Panic, V.V.; Seslija, S.I.; Popovic, I.G.; Spasojevic, V.D.; Popovic, A.R.; Nikolic, V.B.; Spasojevic, P.M. Simple One-Pot Synthesis of Fully Biobased Unsaturated Polyester Resins Based on Itaconic Acid. *Biomacromolecules* **2017**, *18*, 3881–3891. [CrossRef]
56. Gubbels, E.; Jasinska-Walc, L.; Noordover, B.A.J.; Koning, C.E. Linear and branched polyester resins based on dimethyl-2,5-furandicarboxylate for coating applications. *Eur. Polym. J.* **2013**, *49*, 3188–3198. [CrossRef]

3

Palladium-Catalyzed Room Temperature Acylative Cross-Coupling of Activated Amides with Trialkylboranes

Weijia Shi and Gang Zou *

School of Chemistry & Molecular Engineering, East China University of Science & Technology, 130 Meilong Rd, Shanghai 200237, China; 10111048@mail.ecust.edu.cn

* Correspondence: zougang@ecust.edu.cn

Abstract: A highly efficient acylative cross-coupling of trialkylboranes with activated amides has been effected at room temperature to give the corresponding alkyl ketones in good to excellent yields by using 1,3-bis(2,6-diisopropyl)phenylimidazolylidene and 3-chloropyridine co-supported palladium chloride, the PEPPSI catalyst, in the presence of K_2CO_3 in methyl *tert*-butyl ether. The scope and limitations of the protocol were investigated, showing good tolerance of acyl, cyano, and ester functional groups in the amide counterpart while halo group competed via the classical Suzuki coupling. The trialkylboranes generated in situ by hydroboration of olefins with BH_3 or 9-BBN performed similarly to those separately prepared, making this protocol more practical.

Keywords: acylative cross-coupling; trialkylborane; amide activation; palladium; *N*-heterocyclic carbene

---

1. Introduction

Amides are unique and ubiquitous substructures in natural and artificial organic functional molecules, because the strong resonance between the carbonyl and amino groups leads to a highly inert and significantly planar linkage. However, the synthetic utility of amides as an acyl source had remained underdeveloped until the seminal publications in 2015 independently from Garg [1], Szostak [2], and ourselves [3], taking advantage of palladium or nickel catalysis for cleavage of electronically or sterically activated amide C-N bond, and formation of new carbon-carbon or carbon-oxygen bonds. In the past three years, many efforts have been made to expand the scope of amide counterparts, developing a variety of activated amides suitable for the acylative cross-coupling, e.g., *N*-acyl imides [4–6], *N*-Boc and *N*-Ts/Ms amides [7–15], *N*-acylsaccharins [16–20], and amides of heteroaromatic compounds [21,22]. Comparably, the carbon-centered nucleophile counterparts are still rather undeveloped, in particular, with respect to alkyl ones, although alkyl ketones have been widely found in biologically important molecules and synthetic building blocks for fine chemicals. In fact, besides the closely related esters [23–25], there are only two reports on the acylative cross-coupling of amides with alkyl reagents effected by using nickel catalysts. Garg and co-workers reported the first alkylation of amide derivatives by nickel-catalyzed acylative coupling with alkyl zinc reagents in 2016 [26]. Early this year, Rueping and co-workers described the other nickel-catalyzed cross-coupling of B-alkyl-9-BBN with amides [27]. After extension of the nucleophile counterparts from arylboronic acids to cost-effective diarylborinic acids and tetraarylboronates in palladium-catalyzed acylative cross-coupling of amides [28,29], we are interested in the reactivity of trialkylboranes, which could be readily prepared from alkyl Grignard reagents or olefins via hydroboration. Herein we report a highly efficient palladium-catalyzed acylative cross-coupling of activated amides with trialkylboranes at room temperature by using 1,3-bis(2,6-diisopropyl)phenylimidazolylidene (IPr) and 3-chloropyridine (3-ClPy) co-supported palladium chloride, the PEPPSI catalyst developed by Organ et al. [30].

2. Results and Discussion

Reaction of 4-methoxy-*N*-methyl-*N*-tosylbenzamide (**1a**) with the commercially available triethylborane (**2a**) (BEt_3, 1M in THF) was chosen as the model to establish an optimal catalyst system for the acyl alkylation of activated amides (Scheme 1).

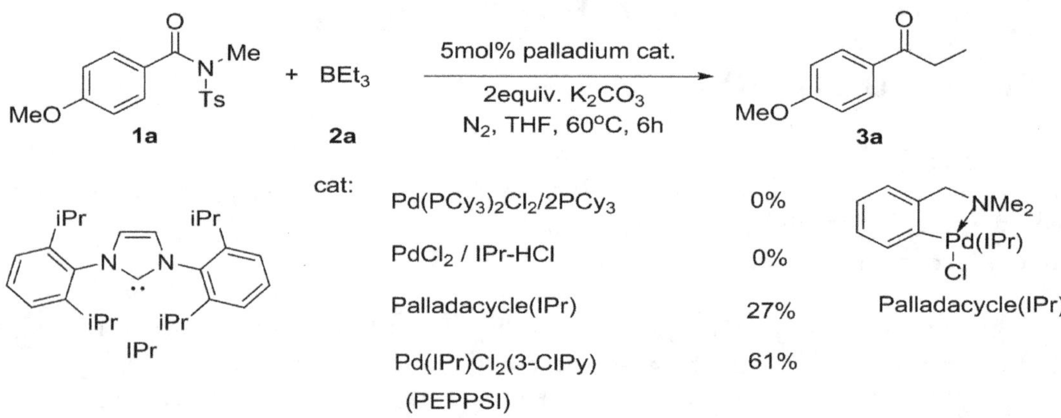

Scheme 1. Screening of palladium catalyst for acyl alkylation of amides with alkylboranes.

The catalyst system $Pd(PCy_3)_2Cl_2/PCy_3$, which we previously established for acylative cross-coupling of amides with aryl boron compounds, was investigated, at first, in the presence of K_2CO_3 as base in THF. Unfortunately, palladium black developed immediately upon heating and no reaction was detected after 6h at 60 °C, indicating the incompatibility of the $Pd(PCy_3)_2Cl_2/PCy_3$ system with the high reducing ability of trialkylboranes. We then turned to the sterically demanding *N*-heterocyclic carbene (NHC) supported palladium catalysts pioneered by Nolan group [31], for example, 1,3-bis(2,6-diisopropylphenyl)imidazolium chloride (IPr-HCl)/palladium chloride (IPr-HCl/$PdCl_2$) [32], IPr supported 2-((dimethylamino)methyl)phenyl palladium chloride (Palladacycle(IPr)) [33], and IPr and 3-chloropyridine co-supported PEPPSI catalyst (IPr)$PdCl_2$(3-ClPy) [30]. Although the IPr-HCl/$PdCl_2$ system also rapidly decomposed to palladium black and lost activity, the palladacycle/IPr and PEPPSI catalysts provided the desired cross-coupling product **3a** in 27% and 61% yields, respectively, from **1a** with 1.0 equiv **2a** in the presence of 2 equiv K_2CO_3. To test if dialkylboranes could be used in the acyl alkylation with amides, we also carried out the model reaction by using diethylborinate ($Et_2B(OMe)$, 1M in THF) under otherwise identical conditions (Scheme 2). Surprisingly, the C-O cross-coupling product, methyl 4-methoxybenzoate, was obtained in 74% yield, while the C-C coupling product **3a** was isolated in low (~5%) yield, implying a much slower transmetalation of alkyl group from boron to palladium than that of alkoxyl group in dialkylborinates.

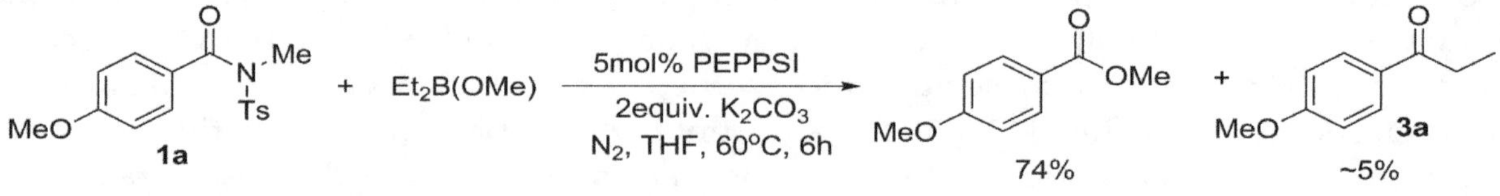

Scheme 2. Unexpected acyl C-O cross-coupling of methyl diethylborinate.

To increase the yield of the desired C-C coupling product in the PEPPSI-catalyzed acyl alkylation of amides, the reaction conditions were optimized with respect to substrate ratio, solvent, base, and reaction temperature, etc. Bases proved to be crucial since no reaction took place in the absence of bases or in the presence of organic base, e.g., NEt_3 and pyridine (Table 1, entries 2, 8 and 9). The stronger inorganic bases, Cs_2CO_3 (25%) and K_3PO_4 (33%), gave the lower yields of **3a** than that of K_2CO_3, albeit the reaction proceeded faster while the weaker ones, e.g., Na_2CO_3, NaOAc, and $NaHCO_3$, resulted in recovering most of amide substrate **1a** (Table 1, entries 3–7). The yield of **3a** could be increased to 80% with 1.5 equiv BEt_3 (Table 1, entry 10). Methyl *tert*-butyl ether (MTBE) appeared to be the choice of solvent, giving **3a** in 90% yield (Table 1, entry 12). The reaction occurred even at room temperature and an excellent yield (98%) for **3a** was obtained in 24 h (Table 1, entry 14). Given the advantages of room temperature organic synthesis [34], we reinvestigated the model reaction in THF or MTBE with K_2CO_3 or K_3PO_4 as the base (Table 1, entries 15–17). The results confirmed the best performance of the combination of MTBE with K_2CO_3.

Table 1. PEPPSI-catalyzed acyl alkylation reaction of amide **1a** with triethylborane. [a]

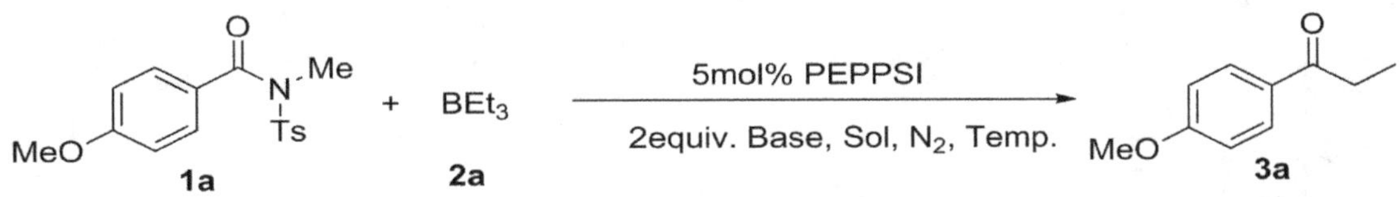

Entry	Equiv of 2a	Solvent	Base [b]	T (°C)	Time (h)	Yield (%) [c]
1	1.1	THF	K_2CO_3	60	5	61
2	1.1	THF	/	60	5	NR [d]
3	1.1	THF	Cs_2CO_3	60	5	25
4	1.1	THF	K_3PO_4	60	5	33
5	1.1	THF	Na_2CO_3	60	8	15
6	1.1	THF	$NaHCO_3$	60	8	Trace [d]
7	1.1	THF	NaOAc	60	8	Trace [d]
8	1.1	THF	Et_3N	60	8	NR [d]
9	1.1	THF	Pyridine	60	8	NR [d]
10	1.5	THF	K_2CO_3	60	5	80
11	1.5	Dioxane	K_2CO_3	60	8	35
12	1.5	MTBE	K_2CO_3	55	8	90
13	1.5	MeCN	K_2CO_3	60	8	17
14	1.5	MTBE	K_2CO_3	rt	24	98
15	1.5	MTBE	K_3PO_4	rt	24	84
16	1.5	THF	K_2CO_3	rt	24	82
17	1.5	THF	K_3PO_4	rt	12	40

[a] Reaction was run at 0.5 mol scale with respect to **1a**; [b] 2 equiv base used but no base used in Entry 2; [c] Isolated yield; [d] Amide **1a** recovered.

Scope and limitations of the PEPPSI-catalyzed room temperature acylative cross-coupling of activated amides with trialkylboranes were briefly explored (Table 2). Influence of the amide structure was investigated at first under the optimized reaction conditions. Similar to tosyl-activated amide **1a**, mesyl (Ms, **1b**) analog also reacted efficiently to give **3a** in 98% yield, while *tert*-butyloxycarbonyl (Boc) activated amide (**1c**) showed comparably lower reactivity (75%) (Table 2, entries 1 and 2). Reaction of benzamides bearing an electron-withdrawing group, acyl (**1e**), cyano (**1f**), or ester (**1g**), at the *para*-position of the benzene ring gave the corresponding ketones (**3b, 3c, or 3d**) in 92%, 70%, or 91% yields, respectively, indicating good functional group compatibility of the PEPPSI-catalyzed acyl alkylation (Table 2, entries 4–6). When 4-chloro-*N*-methyl-*N*-tosylbenzamide (**1h**) was used as the substrate, the desired acylative cross-coupling product 1-(4-chlorophenyl)propanone was not obtained. Instead, double alkylation product 1-(4-ethylphenyl)propanone (**3e**) was isolated in 47% yield, which could be increased to 88% yield with 2.5 equiv BEt_3 (Table 2, entry 7), similar to our previous investigation on the palladium-catalyzed acylative Suzuki coupling of arylboronic acids [3]. A small *ortho*-substituent at the benzene ring appeared to slightly hamper the coupling. In fact, 1-(*o*-tolyl)propan-1-one (**3f**) was isolated in 72% yield from *N*-methyl-*N*-tosyl-2-methylbenzamide (**1i**) (Table 2, entry 8). Alkyl amides (**1j** and **1k**) reacted similarly to give dialkyl ketones **3g** and **3h** in 95% and 98% yields, respectively (Table 2, entries 9 and 10).

Table 2. Scope and limitations of the PEPPSI-catalyzed acyl alkylation reaction of amides with trialkylboranes [a].

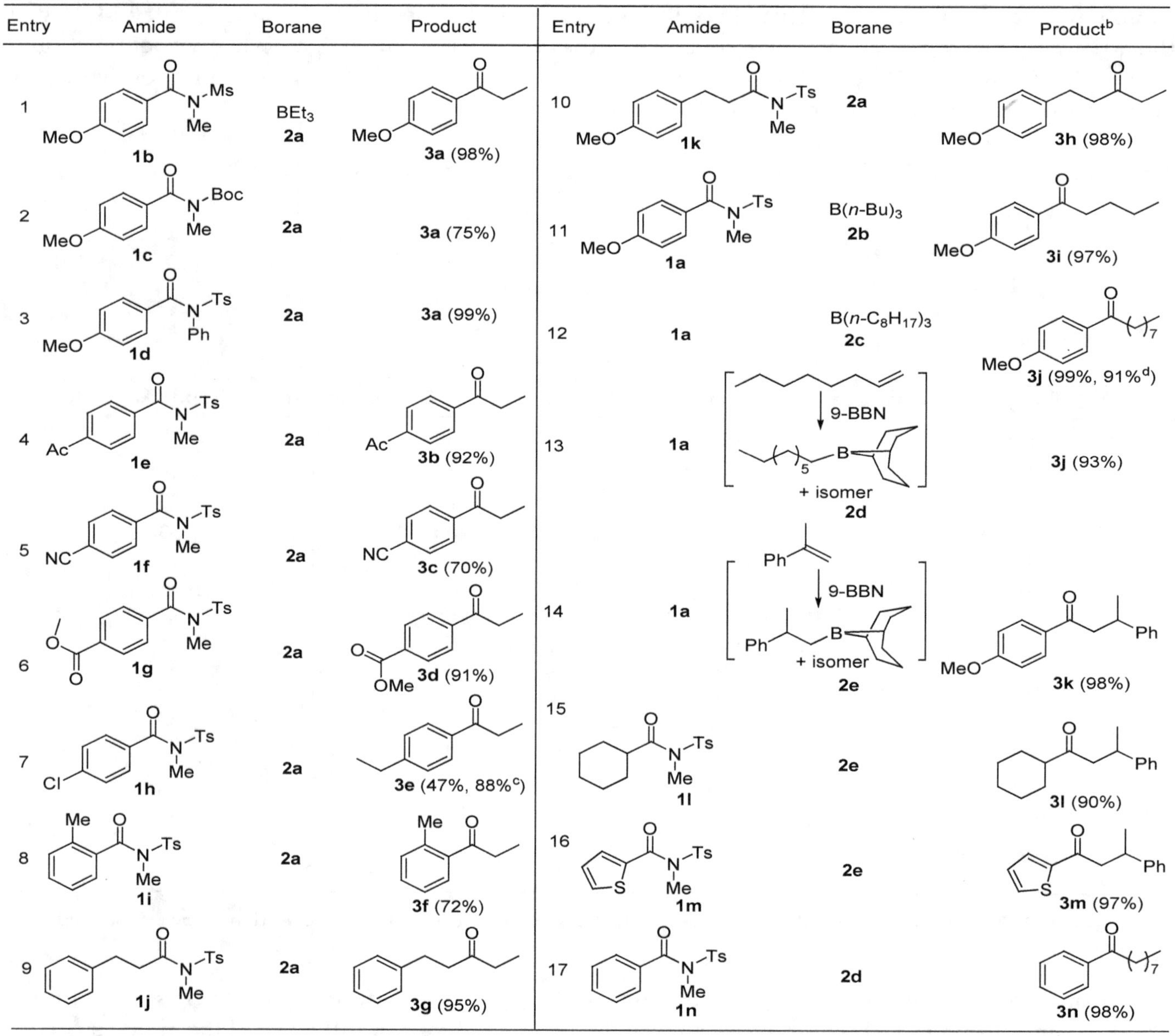

Entry	Amide	Borane	Product[b]
1	1b	BEt_3 2a	3a (98%)
2	1c	2a	3a (75%)
3	1d	2a	3a (99%)
4	1e	2a	3b (92%)
5	1f	2a	3c (70%)
6	1g	2a	3d (91%)
7	1h	2a	3e (47%, 88%[c])
8	1i	2a	3f (72%)
9	1j	2a	3g (95%)
10	1k	2a	3h (98%)
11	1a	$B(n\text{-}Bu)_3$ 2b	3i (97%)
12	1a	$B(n\text{-}C_8H_{17})_3$ 2c	3j (99%, 91%[d])
13	1a	1-octene, 9-BBN → + isomer 2d	3j (93%)
14	1a	α-methylstyrene, 9-BBN → + isomer 2e	3k (98%)
15	1l	2e	3l (90%)
16	1m	2e	3m (97%)
17	1n	2d	3n (98%)

[a] Reaction was run at 0.5 mol scale with respect to amides with 1.5 equiv trialkylboranes; [b] Isolated yield; [c] With 2.5 equiv BEt_3; [d] $B(n\text{-}C_8H_{17})_3$ and branched isomers in situ generated from hydroboration of octene with 1.0 equiv BH_3 in THF at room temperature was used.

Tri(*n*-butyl)borane (**2b**) reacted with **1a** similarly to triethylborane (**2a**) while no reaction of tricyclohexylborane was observed, implying the failure of transmetalation of *secondary*-alkyl group from boron to palladium. Hydroboration of alkenes represents an alternative route to primary alkyl boranes. However, the long-chain primary alkyl boranes prepared by hydroboration of alkenes are generally contaminated by the presence of branched-isomers due to the non-regiospecific addition of B-H to C-C double bonds [35]. Due to the inertness of the *secondary*-alkyl group in alkylboranes, reaction of **1a** with tri(octyl)boranes prepared by hydroboration of 1-octene proceeded comparably to that of tri(*n*-octyl)borane (**2c**) from Grignard reagent [36], giving **3j** in 91% and 99% yields, respectively (Table 2, entry 12). Given the fact that only one primary alkyl group of trialkylboranes is useful in the palladium-catalyzed acylative cross-coupling, it is more practical to use long-chain alkyl boranes prepared by hydroboration of corresponding olefins by dialkylborane, a more stable B-H source, e.g., 9-borabicyclo[3.3.1]nonane (9-BBN). Therefore, we further investigated use of B-octyl-9-BBN formed in situ from octene as the trialkylborane counterpart and obtained **3j** in 93% yield (Table 2, entry 13). β-Phenyl propiophenones, which show intrinsic deactivation of the lowest and thermally populated n, π^*-triplet excited states in aryl alkyl ketones [37], could be readily obtained in 98% yield (**3k**) by the acyl alkylation of amide **1a** with B-(2-phenylpropyl)-9-BBN generated in situ from 9-BBN and α-methylstyrene. Alkyl and heterocyclic analogs, e.g., 1-cyclohexyl-3-phenylbutanone (**3l**, 90%) and 3-phenyl-1-(thiophen-2-yl)butanone (**3m**, 97%) could also be prepared efficiently (Table 2, entries 14–16).

However, when amide **1a** was subjected to the hydroboration solution of *n*-butyl vinyl ether with 9-BBN the C-O coupling product (vide supra) butyl 4-methoxybenzoate, instead of the desired β-butoxy-4-methoxypropiophenone, was obtained in 83% yield, probably due to the facile β-alkoxy elimination of labile B-(2-butoxyethyl)-9-BBN via intramolecular O-B coordination to B-butoxy-9-BBN for the subsequent C-O cross-coupling (Scheme 3).

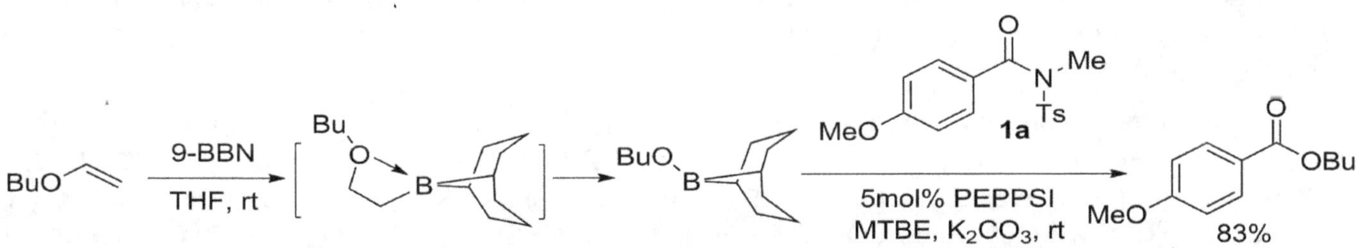

Scheme 3. Formation of borinate and subsequent acyl C-O cross-coupling.

3. Materials and Methods

3.1. General

Chemicals obtained commercially were used as received. Nuclear magnetic resonance (NMR) spectra were recorded on a Bruker DPX-400 spectrometer (Bruker Co., Billerica, MA, USA) using residue of the deuterated solvent or tetramethylsilane (TMS) as the internal standard (Cambridge Isotope Lab. Inc., Tewksbury, MA, USA). For copies of ^{1}H and ^{13}C-NMR spectra of the products, please see Supplementary Materials. All products were isolated by flash chromatography using petroleum ether (Sinopharm Chemical Reagent Co. Ltd., Shanghai, China) (60–90 °C)/ethyl acetate (Sinopharm Chemical Reagent Co. Ltd., Shanghai, China) as eluents. Triethylborane (1 M in THF), tributylborane, 9-BBN (1 M in THF) and BH_3 (1 M in THF) were purchased from J&K chemicals (Beijing, China). Amides [3], PEPPSI catalyst [30], 1,3-bis(2,6-diisopropylphenyl)imidazolium chloride (IPr-HCl) [32], palladacycle (IPr) [33], and tri(*n*-octyl)borane [36] were prepared according to the procedures reported previously. Solvents, methyl *tert*-butyl ether (MTBE), tetrahydrofuran (THF), and dioxane were dried over sodium while acetonitrile was distilled over CaH_2 prior to use.

3.2. General Procedure for the PEPPSI-Catalyzed Cross-Coupling of N-Methyl-N-Tosylamides with Trialkylboranes

A Schlenk tube (20 mL) charged with amide (0.5 mmol), PEPPSI (0.025 mmol, 5 mol%), and K_2CO_3 (2 equiv) was degassed and then refilled with nitrogen, three times. Then, solvent MTBE (6.0 mL) was added via syringe followed by 0.75 mL (1.5 equiv) borane solution (1 M in THF), which was commercially available or was in situ prepared by hydroboration of olefin. The resulted mixture was stirred for 24 h at room temperature under N_2 atmosphere. The reaction was quenched with iced water (5 mL) and extracted with MTBE (2 $\times$ 5 mL). The combined MTBE extracts were dried over anhydrous Na_2SO_4. After filtration, solvents were removed by rotavapor to afford the crude product, which was purified by flash column chromatography on silica gel using petroleum ether/EtOAc as the eluents.

1-(4-Methoxyphenyl)propanone (**3a**) [38]: Colorless oil (80.3 mg, 98%). ^{1}H-NMR (400 MHz, $CDCl_3$) δ(ppm): 7.91 (d, J = 8.8 Hz, 2H, Ar), 6.89 (d, J = 8.8 Hz, 2H, Ar), 3.82 (s, 3H, OMe), 2.91 (q, J = 7.2 Hz,

2H, CH_2), 1.18 (t, *J* = 7.2 Hz, 3H, CH_3). ^{13}C-NMR (100 MHz, $CDCl_3$) δ(ppm): 199.34 (C=O), 163.16 (C^4_{Ar}), 130.06 ($C^{2,6}_{Ar}$), 129.85 (C^1_{Ar}), 113.52 ($C^{3,5}_{Ar}$), 55.29 (OMe), 31.26 (CH_2), 8.29 (CH_3).

1-(4-Acetylphenyl)propanone (**3b**) [39]: White powder (80.9 mg, 92%), mp 69–71 °C. ^{1}H-NMR (400 MHz, $CDCl_3$) δ(ppm): 8.00 (s, 4H, Ar), 3.01 (q, *J* = 7.2 Hz, 2H, CH_2), 2.62 (s, 3H, Ac), 1.21 (t, *J* = 7.2 Hz, 3H, CH_3). ^{13}C-NMR (100 MHz, $CDCl_3$) δ(ppm): 200.13 (C=O), 197.46 (C=O), 139.96 ($C^{1/4}_{Ar}$), 139.90 ($C^{1/4}_{Ar}$), 128.41 (C_{Ar}), 128.07 (C_{Ar}), 32.15 (CH_2), 26.81 (*Me*CO), 7.97 (CH_3).

4-Propionylbenzonitrile (**3c**) [40]: White powder (55.8 mg, 70%), mp 53–55 °C. ^{1}H-NMR (400 MHz, $CDCl_3$) δ(ppm): 7.97 (d, *J* = 8.4 Hz, 2H, Ar), 7.69 (d, *J* = 8.4 Hz, 2H, Ar), 2.95 (q, *J* = 7.2 Hz, 2H, CH_2), 1.15 (t, *J* = 7.2 Hz, 3H, CH_3). ^{13}C-NMR (100 MHz, $CDCl_3$) δ(ppm): 199.16 (C=O), 139.64 (C^4_{Ar}), 132.33 (C_{Ar}), 128.23 (C_{Ar}), 117.84 (CN/C^1_{Ar}), 115.95 (CN/C^1_{Ar}), 32.01 (CH_2), 7.75 (CH_3).

Methyl 4-propionylbenzoate (**3d**) [41]: White crystalline powder, (87.1 mg, 91%), mp 81–83 °C. ^{1}H-NMR (400 MHz, $CDCl_3$) δ(ppm): 8.08 (d, *J* = 8.4 Hz, 2H, Ar), 7.97 (d, *J* = 8.4 Hz, 2H, Ar), 3.92 (s, 3H, OMe), 3.05 (q, *J* = 7.2 Hz, 2H, CH_2), 1.21 (t, *J* = 7.2 Hz, 3H, CH_3). ^{13}C-NMR (100 MHz, $CDCl_3$) δ(ppm): 200.15 (C=O), 166.17 (OC=O), 140.01 ($C^{1/4}_{Ar}$), 133.59 ($C^{1/4}_{Ar}$), 129.72 (C_{Ar}), 127.78 (C_{Ar}), 52.35 (OMe), 32.10 (CH_2), 7.96 (CH_3).

1-(4-Ethylphenyl)propanone (**3e**) [42]: Colorless oil (71.1 mg, 88%, with 2.5 equiv BEt_3). ^{1}H-NMR (400 MHz, $CDCl_3$) δ(ppm): 7.81 (d, *J* = 8.4 Hz, 2H, Ar), 7.19 (d, *J* = 8.4 Hz, 2H, Ar), 2.90 (q, *J* = 7.2 Hz, 2H, $COCH_2$), 2.61 (q, *J* = 7.6 Hz, 2H, $ArCH_2$), 1.17 (t, *J* = 7.6 Hz, 3H, $COCH_2CH_3$), 1.13 (t, *J* = 7.2 Hz, 3H, Ar CH_2CH_3). ^{13}C-NMR (100 MHz, $CDCl_3$) δ(ppm): 200.51 (C=O), 149.71 (C^4_{Ar}), 134.59 (C^1_{Ar}), 128.14 (C_{Ar}), 127.97 (C_{Ar}), 31.60 ($COCH_2$), 28.85 ($ArCH_2$), 15.16 ($ArCH_2CH_3$), 8.27 ($COCH_2CH_3$).

1-(o-Tolyl)propanone (**3f**) [43]: Colorless oil (53.4 mg, 72%). ^{1}H-NMR (400 MHz, $CDCl_3$) δ(ppm): 7.53 (d, *J* = 7.6 Hz, 1H), 7.29–7.25 (m, 1H, Ar), 7.16–7.14 (m, 2H, Ar), 2.82 (q, *J* = 7.2 Hz, 2H, CH_2), 2.41 (s, 3H, $ArCH_3$), 1.11 (t, *J* = 7.2 Hz, 3H, CH_3). ^{13}C-NMR (100 MHz, $CDCl_3$) δ(ppm): 205.04 (C=O), 138.01 ($C^{1/2}_{Ar}$), 137.76 ($C^{1/2}_{Ar}$), 131.81 (C_{Ar}), 130.97 (C_{Ar}), 128.19 (C_{Ar}), 125.56 (C_{Ar}), 34.64 (CH_2), 21.19 ($ArCH_3$), 8.31 (CH_3).

1-Phenylpentan-3-one (**3g**) [44]: Colorless oil (77.1 mg, 95%). ^{1}H-NMR (400 MHz, $CDCl_3$) δ(ppm): 7.33–7.29 (m, 2H, Ph), 7.23–7.20 (m, 3H, Ph), 2.93 (t, *J* = 7.6 Hz, 2H, $PhCH_2CH_2CO$), 2.76 (t, *J* = 7.6 Hz, 2H, $PhCH_2$), 2.43 (q, *J* = 7.2 Hz, 2H,CO CH_2CH_3), 1.07 (t, *J* = 7.2 Hz, 3H, CH_3). ^{13}C-NMR (100 MHz, $CDCl_3$) δ(ppm): 210.61 (C=O), 141.10 (C^1_{Ph}), 128.41 (Ph), 128.24 (Ph), 126.00 (C^4_{Ph}), 43.81($PhCH_2CH_2CO$), 36.05 ($COCH_2CH_3$), 29.77 ($PhCH_2$), 7.69 (CH_3).

1-(4-Methoxyphenyl)pentan-3-one (**3h**) [45]: Colorless oil (94.3 mg, 98%). ^{1}H-NMR (400 MHz, $CDCl_3$) δ(ppm): 7.09 (d, *J* = 8.8 Hz, 2H, Ar), 6.81 (d, *J* = 8.8 Hz, 2H, Ar), 3.77 (s, 3H, OMe), 2.84 (t, *J* = 7.6 Hz, 2H, $PhCH_2CH_2CO$), 2.64 (t, *J* = 7.6 Hz, 2H, $PhCH_2$), 2.39 (q, *J* = 7.6 Hz, 2H, $COCH_2CH_3$), 1.03 (t, *J* = 7.6 Hz, 3H, CH_3). ^{13}C-NMR (100 MHz, $CDCl_3$) δ(ppm): 210.80 (C=O), 157.85 (C^4_{Ar}), 133.12 (C^1_{Ar}), 129.16 (C_{Ar}), 113.79 (C_{Ar}), 55.17 (OMe), 44.09 ($PhCH_2CH_2CO$), 36.08 ($COCH_2CH_3$), 28.93 ($PhCH_2$), 7.68 (CH_3).

1-(4-Methoxyphenyl)pentanone (**3i**) [46]: Colorless oil (93.5 mg, 97%). ^{1}H-NMR (400 MHz, $CDCl_3$) δ(ppm): 7.93 (d, *J* = 8.8 Hz, 2H, Ar), 6.91 (d, *J* = 9.2 Hz, 2H, Ar), 3.85 (s, 3H, OMe), 2.90 (t, *J* = 7.6 Hz, 2H, $COCH_2$), 1.73–1.66 (m, 2H, $COCH_2CH_2$), 1.44–1.34 (m, 2H, $COCH_2CH_2CH_2$), 0.94 (t, *J* = 7.6 Hz, 3H). ^{13}C-NMR (100 MHz, $CDCl_3$) δ(ppm): 199.20 (C=O), 163.23 (C^4_{Ar}), 130.25 ($C^{2,6}_{Ar}$), 130.11(C^1_{Ar}), 113.59 ($C^{3,5}_{Ar}$), 55.37 (OMe), 37.95 ($COCH_2$), 26.68 (CH_2), 22.48 (CH_2), 13.90 (CH_3).

1-(4-Methoxyphenyl)nonanone (**3j**) [47]: Colorless oil (123.0 mg, 99%, with tri(*n*-octyl)borane; 115 mg, 93% with hydroboration of octene with 9-BBN; 113 mg, 91% with hydroboration of octene with BH_3. ^{1}H-NMR (400 MHz, $CDCl_3$) δ(ppm): 7.86 (d, *J* = 9.2 Hz, 2H, Ar), 6.84 (d, *J* = 9.2 Hz, 2H, Ar), 3.78 (s, 3H, OMe), 2.82 (t, *J* = 7.6 Hz, 2H, $COCH_2$), 1.63 (t, *J* = 7.6 Hz, 2H, $COCH_2CH_2$), 1.29–1.19 (m, 10H, $(CH_2)_5$), 0.80 (t, *J* = 7.2Hz, 3H, CH_3). ^{13}C-NMR (100 MHz, $CDCl_3$) δ(ppm): 199.20 (C=O), 163.23 (C^4_{Ar}), 130.25

($C^{2,6}{}_{Ar}$), 130.13 ($C^{1}{}_{Ar}$), 113.60 ($C^{3,5}{}_{Ar}$), 55.37 (OMe), 38.26 ($COCH_2$), 31.79 (CH_2), 29.41 (CH_2), 29.39 (CH_2), 29.14 (CH_2), 24.59 (CH_2), 22.61 (CH_2), 14.05 (CH_3).

1-(4-Methoxyphenyl)-3-phenylbutanone (**3k**) [46]: White powder (125.0 mg, 98%), mp 84–86 °C. ^{1}H-NMR (400 MHz, $CDCl_3$) δ(ppm): 7.93 (d, J = 8.8 Hz, 2H, Ar), 7.34–7.30 (m, 4H, Ph), 7.23–7.19 (m, 1H, Ph), 6.92 (d, J = 8.8 Hz, 2H, Ar), 3.86 (s, 3H, OMe), 3.55–3.47 (m, 1H, CH), 3.26 (dd, J_1 = 16.4 Hz, J_2 = 5.6 Hz, 1H, CH_2), 3.14 (dd, J_1 = 16.0 Hz, J_2 = 8.4 Hz, 1H, CH_2), 1.34 (d, J = 7.8, 3H, CH_3). ^{13}C-NMR (100 MHz, $CDCl_3$) δ(ppm): 197.59 (C=O), 163.31 ($C^{4}{}_{Ar}$), 146.65 ($C^{1}{}_{Ph}$), 130.28 (C_{Ar}/C_{Ph}), 130.21 ($C^{1}{}_{Ar}$), 128.43 (C_{Ph}), 126.79 (C_{Ph}), 126.15 ($C^{4}{}_{Ph}$), 113.61 ($C^{3,5}{}_{Ar}$), 55.37 (OMe), 46.59 ($COCH_2$), 35.67 (CH), 21.79 (CH_3).

1-Cyclohexyl-3-phenylbutanone (**3l**): Colorless oil (103.7 mg, 90%). ^{1}H-NMR (400 MHz, $CDCl_3$) δ(ppm): 7.33–7.29 (m, 2H, Ph), 7.25–7.19 (m, 3H, Ph), 3.41–3.32 (m, 1H, PhCH), 2.77 (dd, J_1 = 16.8 Hz, J_2 = 6.4 Hz, 1H,CO CH), 2.69 (dd, J_1 = 16.4 Hz, J_2 = 8.0 Hz, 1H, $COCH_2$), 2.29–2.22 (m, 1H, COCH), 1.83–1.65 (m, 5H, Cy), 1.35–1.17 (m, 8H, Cy overlapped with CH_3). ^{13}C-NMR (100 MHz, $CDCl_3$) δ(ppm): 212.83 (C=O), 146.54 ($C^{1}{}_{Ph}$), 128.39 (C_{Ph}), 126.75 (C_{Ph}), 126.11 ($C^{4}{}_{Ph}$), 51.19 (CO*CH*), 49.10 (CO*CH*$_2$), 35.02 (CH), 28.22 (Cy), 28.05 (Cy), 25.78 (Cy), 25.61(Cy), 25.53 (Cy), 21.79 (CH_3). HRMS (ESI) m/z calcd for $C_{16}H_{22}ONa$ $[M + Na]^+$ 253.1568, found 253.1569.

3-Phenyl-1-(thiophen-2-yl)butanone (**3m**) [48]: Colorless oil (111.5 mg, 97%). ^{1}H-NMR (400 MHz, $CDCl_3$) δ(ppm): 7.70–7.69 (m, 1H, thiophenyl), 7.64–7.63 (m, 1H, thiophenyl), 7.36–7.28 (m, 4H, Ph), 7.25–7.23 (m, 1H, Ph), 7.14–7.12 (m, 1H, thiophenyl), 3.56–3.51 (m, 1H, PhCH), 3.24 (dd, J_1 = 16.0 Hz, J_2 = 6.0 Hz, 1H, CH_2CO), 3.14 (dd, J_1 = 15.6 Hz, J_2 = 8.4 Hz, 1H, CH_2CO), 1.38 (d, J = 6.8 Hz, 3H, CH_3). ^{13}C-NMR (100 MHz, $CDCl_3$) δ(ppm): 191.94 (C=O), 146.18 ($C^{1}{}_{Ph}$), 144.60 (C^2 thiophenyl), 133.56 (thiophenyl), 131.81 (thiophenyl), 128.49(C_{Ph}), 128.01(thiophenyl), 126.78 (C_{Ph}), 126.30 ($C^{4}{}_{Ph}$), 47.78 ($COCH_2$), 35.93 (CH), 21.64 (CH_3).

1-Phenylnonanone (**3n**) [47]: Colorless oil (106.8 mg, 98%). ^{1}H-NMR (400 MHz, $CDCl_3$) δ(ppm): 7.97–7.95 (m, 2H, Ph), 7.57–7.53 (m, 1H, Ph), 7.47–7.44 (m, 2H, Ph), 2.96 (t, J = 7.6 Hz, 2H, $COCH_2$), 1.77–1.69 (m, 2H, $COCH_2$*CH*$_2$), 1.39–1.25 (m, 10H, $(CH_2)_5$), 0.88 (t, J = 6.8 Hz, 3H, CH_3). ^{13}C-NMR (100 MHz, $CDCl_3$) δ(ppm): 200.62 (C=O), 137.05 $C^{1}{}_{Ph}$), 132.83 ($C^{4}{}_{Ph}$), 128.51 (C_{Ph}), 128.02 (C_{Ph}), 38.61 (CO*CH*$_2$), 31.81 (CH_2), 29.42 (CH_2), 29.36 (CH_2), 29.15 (CH_2), 24.36 (CH_2), 22.63 (CH_2), 14.08 (CH_3).

4. Conclusions

In summary, we have developed a highly efficient acylative cross-coupling of trialkylboranes with activated amides by using 1,3-bis(2,6-diisopropyl)phenylimidazolylidene and 3-chloropyridine co-supported palladium chloride, the PEPPSI catalyst, under mild conditions. Bases appeared to play a key role in the reaction, among which K_2CO_3 performed best in MTBE at room temperature. The reaction proceeded to give alkyl ketones in good to excellent yields, tolerating a variety of functional groups in the amide counterpart. Unlike the high-order arylboron compounds, in which all the aryl groups react effectively, only one of the three primary alkyl groups in trialkylboranes could be used as alkyl source for the acyl alkylation. The trialkylboranes generated in situ by hydroboration of olefins with BH_3 or 9-BBN performed comparably to those separately prepared. This protocol complements ketone synthesis via palladium-catalyzed acylative cross-coupling of amides, providing a feasible access to both monoalkyl and dialkyl ketones.

Author Contributions: Conceptualization, G.Z.; Investigation, W.S.; Writing-Original Draft Preparation, W.S.; Writing-Review & Editing, G.Z.Funding: This research was funded by National Natural Science Foundation of China [21472041].

References

1. Hie, L.; Nathel, N.F.F.; Shah, T.K.; Baker, E.L.; Hong, X.; Yang, Y.-F.; Liu, P.; Houk, K.N.; Garg, N.K. Conversion of amides to esters by the nickel-catalysed activation of amide C-N bonds. *Nature* **2015**, *524*, 79–83. [CrossRef] [PubMed]
2. Meng, G.; Szostak, M. Sterically controlled Pd-catalyzed chemoselective ketone synthesis via N-C cleavage in twisted amides. *Org. Lett.* **2015**, *17*, 4364–4367. [CrossRef] [PubMed]
3. Li, X.; Zou, G. Acylative Suzuki coupling of amides: Acyl-nitrogen activation via synergy of independently modifiable activating groups. *Chem. Commun.* **2015**, *51*, 5089–5092. [CrossRef] [PubMed]
4. Pace, V.; Holzer, W.; Meng, G.; Shi, S.; Lalancette, R.; Szostak, R.; Szostak, M. Structures of highly twisted amides relevant to amide N-C cross-coupling: Evidence for ground-state amide destabilization. *Chem. Eur. J.* **2016**, *22*, 14494–14498. [CrossRef] [PubMed]
5. Osumi, Y.; Liu, C.; Szostak, M. *N*-Acylsuccinimides: Twist-controlled, acyl-transfer reagents in Suzuki–Miyaura cross-coupling by N–C amide bond activation. *Org. Biomol. Chem.* **2017**, *15*, 8867–8871. [CrossRef] [PubMed]
6. Cui, M.; Chen, Z.; Liu, T.; Wang, H.; Zeng, Z. *N*-Acylsuccinimides: Efficient acylative coupling reagents in palladium-catalyzed Suzuki coupling via C-N cleavage. *Tetrahedron Lett.* **2017**, *58*, 3819–3822. [CrossRef]
7. Boit, T.B.; Weires, N.A.; Kim, J.; Garg, N.K. Nickel-catalyzed Suzuki-Miyaura coupling of aliphatic amides. *ACS Catal.* **2018**, *8*, 1003–1008. [CrossRef] [PubMed]
8. Hie, L.; Baker, E.L.; Anthony, S.M.; Desrosiers, J.-N.; Senanayake, C.; Garg, N.K. Nickel-catalyzed esterification of aliphatic amides. *Angew. Chem. Int. Ed.* **2016**, *55*, 15129–15132. [CrossRef] [PubMed]
9. Lanigan, R.M.; Sheppard, T.D. Recent developments in amide synthesis: Direct amidation of carboxylic acids and transamidation reactions. *Eur. J. Org. Chem.* **2013**, 7453–7465. [CrossRef]
10. Baker, E.L.; Yamano, M.M.; Zhou, Y.; Anthony, S.M.; Garg, N.K. A two-step approach to achieve secondary amide transamidation enabled by nickel catalysis. *Nat. Commun.* **2016**, *7*, 11554. [CrossRef] [PubMed]
11. Lei, P.; Meng, G.; Ling, Y.; An, J.; Nolan, S.P.; Szostak, M. General method for the Suzuki–Miyaura cross-coupling of primary amide-derived electrophiles enabled by [Pd(NHC)(cin)Cl] at room temperature. *Org. Lett.* **2017**, *19*, 6510–6513. [CrossRef] [PubMed]
12. Shi, S.; Szostak, M. Nickel-catalyzed diaryl ketone synthesis by N-C cleavage: Direct Negishi cross-coupling of primary amides by site-selective N, N-di-Boc activation. *Org. Lett.* **2016**, *18*, 5872–5875. [CrossRef] [PubMed]
13. Meng, G.; Szostak, M. Site-selective C-H/C-N activation by cooperative catalysis: Primary amides as arylating reagents in directed C–H arylation. *ACS Catal.* **2017**, *7*, 7251–7256. [CrossRef]
14. Liu, C.; Liu, Y.; Liu, R.; Lalancette, R.; Szostak, R.; Szostak, M. Palladium-catalyzed Suzuki–Miyaura cross-coupling of N-mesylamides by N-C cleavage: Electronic effect of the mesyl group. *Org. Lett.* **2017**, *19*, 1434–1437. [CrossRef] [PubMed]
15. Wang, C.; Huang, L.; Wang, F.; Zou, G. Highly efficient synthesis of aryl ketones by PEPPSI-palladium catalyzed acylative Suzuki coupling of amides with diarylborinic acids. *Tetrahedron Lett.* **2018**, *59*, 2299–2301. [CrossRef]
16. Liu, C.; Meng, G.; Liu, Y.; Liu, R.; Lalancette, R.; Szostak, R.; Szostak, M. N-Acylsaccharins: Stable electrophilic amide-based acyl transfer reagents in Pd-catalyzed Suzuki-Miyaura coupling via N-C cleavage. *Org. Lett.* **2016**, *18*, 4194–4197. [CrossRef] [PubMed]
17. Wu, H.; Li, Y.; Cui, M.; Jian, J.; Zeng, Z. Suzuki coupling of amides via palladium-catalyzed C-N cleavage of N-acylsaccharins. *Adv. Synth. Catal.* **2016**, *358*, 3876–3880. [CrossRef]
18. Liu, C.; Meng, G.; Szostak, M. *N*-Acylsaccharins as amide-based arylating reagents via chemoselective N-C cleavage: Pd-catalyzed decarbonylative Heck reaction. *J. Org. Chem.* **2016**, *81*, 12023–12030. [CrossRef] [PubMed]
19. Cui, M.; Wu, H.; Jian, J.; Wang, H.; Liu, C.; Daniel, S.; Zeng, Z. Palladium-catalyzed Sonogashira coupling of amides: Access to ynones via C-N bond cleavage. *Chem. Commun.* **2016**, *52*, 12076–12079. [CrossRef] [PubMed]
20. Wu, H.; Liu, T.; Cui, M.; Li, Y.; Jian, J.; Wang, H.; Zeng, Z. Rhodium-catalyzed C-H functionalization with N-acylsaccharins. *Org. Biomol. Chem.* **2017**, *15*, 536–540. [CrossRef] [PubMed]

21. Meng, G.; Szostak, R.; Szostak, M. Suzuki-Miyaura cross-coupling of *N*-acylpyrroles and pyrazoles: Planar, electronically activated amides in catalytic N-C cleavage. *Org. Lett.* **2017**, *19*, 3596–3599. [CrossRef] [PubMed]
22. Meng, G.; Lalancette, R.; Szostak, R.; Szostak, M. *N*-methylamino pyrimidyl amides (MAPA): Highly reactive, electronically-activated amides in catalytic N-C(O) cleavage. *Org. Lett.* **2017**, *19*, 4656–4659. [CrossRef] [PubMed]
23. Masson-Makdissi, J.; Vandavasi, J.K.; Newman, S.G. Switchable selectivity in the Pd-catalyzed alkylative cross-coupling of esters. *Org. Lett.* **2018**, *20*, 4094–4098. [CrossRef] [PubMed]
24. Liu, X.; Jia, J.; Rueping, M. Nickel-catalyzed C—O bond-cleaving alkylation of esters: Direct replacement of the ester moiety by functionalized alkyl chains. *ACS Catal.* **2017**, *7*, 4491–4496. [CrossRef]
25. Chatupheeraphat, A.; Liao, H.-H.; Srimontree, W.; Guo, L.; Minenkov, Y.; Poater, A.; Cavallo, L.; Rueping, M. Ligand-controlled chemoselective C(acyl)—O Bond vs C(aryl)—C bond activation of aromatic esters in nickel catalyzed $C(sp^2)-C(sp^3)$ cross-couplings. *J. Am. Chem. Soc.* **2018**, *140*, 3724–3735. [CrossRef] [PubMed]
26. Simmons, B.J.; Weires, N.A.; Dander, J.E.; Garg, N.K. Nickel-catalyzed alkylation of amide derivatives. *ACS Catal.* **2016**, *6*, 3176–3179. [CrossRef]
27. Liu, X.; Hsiao, C.-C.; Guo, L.; Rueping, M. Cross-coupling of amides with alkylboranes via nickel-catalyzed C-N bond cleavage. *Org. Lett.* **2018**, *20*, 2976–2979. [CrossRef] [PubMed]
28. Li, X.; Zou, G. Palladium-catalyzed acylative cross-coupling of amides with diarylborinic acids and sodium tetraarylborates. *J. Organomet. Chem.* **2015**, *794*, 136–145. [CrossRef]
29. Si, S.; Wang, C.; Zhang, N.; Zou, G. Palladium-catalyzed room-temperature acylative Suzuki coupling of high-order aryl borons with carboxylic acids. *J. Org. Chem.* **2016**, *81*, 4364–4370. [CrossRef] [PubMed]
30. O'Brien, C.J.; Kantchev, E.A.B.; Valente, C.; Hadei, N.; Chass, G.A.; Lough, A.; Hopkinson, A.C.; Organ, M.G. Easily prepared air- and moisture-stable Pd–NHC (NHC=N-heterocyclic carbene) complexes: A reliable, user-friendly, highly active palladium precatalyst for the Suzuki-Miyaura reaction. *Chem. Eur. J.* **2006**, *12*, 4743–4748. [CrossRef] [PubMed]
31. Fortman, G.C.; Nolan, S.P. *N*-Heterocyclic carbene (NHC) ligands and palladium in homogeneous cross-coupling catalysis: A perfect union. *Chem. Soc. Rev.* **2011**, *40*, 5151–5169. [CrossRef] [PubMed]
32. Huang, J.; Nolan, S.P. Efficient cross-coupling of aryl chlorides with aryl Grignard reagents (Kumada reaction) mediated by a palladium/imidazolium chloride system. *J. Am. Chem. Soc.* **1999**, *121*, 9889–9890. [CrossRef]
33. Kantchev, E.A.B.; Ying, J.Y. Practical one-pot, three-component synthesis of N-heterocyclic carbene (NHC) ligated palladacycles derived from *N,N*-dimethylbenzylamine. *Organometallics* **2009**, *28*, 289–299. [CrossRef]
34. Brahmachari, G. *Room Temperature Organic Synthesis*; Elsevier: Amsterdam, The Netherlands, 2015; ISBN 9780128011386.
35. Brown, H.C.; Sharp, R.L. Hydroboration. XXIV. Directive effects in the hydroboration of some substituted styrenes. *J. Am. Chem. Soc.* **1966**, *88*, 5851–5854. [CrossRef]
36. Brown, H.C.; Racherla, U.S. Organoboranes. 44. A convenient, highly efficient synthesis of triorganylboranes via a modified organometallic route. *J. Org. Chem.* **1986**, *51*, 427–432. [CrossRef]
37. Samanta, S.; Mishra, B.K.; Pace, T.C.S.; Sathyamurthy, N.; Bohne, C.; Moorthy, J.N. *β*-Phenyl quenching of triplet excited ketones: How critical is the geometry for deactivation? *J. Org. Chem.* **2006**, *71*, 4453–4459. [CrossRef] [PubMed]
38. Rao, M.L.N.; Venkatesh, V.; Banerjee, D. Atom-efficient cross-coupling reactions of triarylbismuths with acyl chlorides under Pd(0) catalysis. *Tetrahedron* **2007**, *63*, 12917–12926. [CrossRef]
39. Vu, M.D.; Das, M.; Liu, X.-W. Direct aldehyde $C_{sp}{}^2$-H functionalization through visible light mediated photoredox catalysis. *Chem. Eur. J.* **2017**, *23*, 15899–15902. [CrossRef] [PubMed]
40. Zimbron, J.M.; Seeger-Weibel, M.; Hirt, H.; Gallou, F. Development of a robust and practical process for the Darzens condensation and α, β-epoxide rearrangement: Scope and limitations of the methodology. *Synthesis* **2008**, *8*, 1221–1226. [CrossRef]
41. Liu, M.; Hyder, Z.; Sun, Y.; Tang, W.; Xu, L.; Xiao, J. Efficient synthesis of alkyl aryl ketones & ketals via palladium-catalyzed regioselective arylation of vinyl ethers. *Org. Biomol. Chem.* **2010**, *8*, 2012–2015. [CrossRef] [PubMed]
42. Suchand, B.; Satyanarayana, G. KO*t*Bu-mediated domino isomerization and functionalization of aromatic allylic alcohols. *Eur. J. Org. Chem.* **2017**, 3886–3895. [CrossRef]

43. Uyanik, M.; Suzuki, D.; Yasui, T.; Ishihara, K. In situ generated (hypo)iodite catalysts for the direct α-oxyacylation of carbonyl compounds with carboxylic acids. *Angew. Chem. Int. Ed.* **2011**, *50*, 5331–5334. [CrossRef] [PubMed]
44. Li, X.; Li, L.; Tang, Y.; Ling, Z.; Cun, L.; Zhu, J.; Liao, J.; Deng, J. Chemoselective conjugate reduction of α, β-unsaturated ketones catalyzed by rhodium amido complexes in aqueous media. *J. Org. Chem.* **2010**, *75*, 2981–2988. [CrossRef] [PubMed]
45. Molander, G.A.; Petrillo, D.E. Suzuki-Miyaura cross-coupling of potassium trifluoroboratohomoenolates. *Org. Lett.* **2008**, *10*, 1795–1798. [CrossRef] [PubMed]
46. Colbon, P.; Ruan, J.; Purdie, M.; Xiao, J. Direct acylation of aryl chlorides with aldehydes by palladium−pyrrolidine co-catalysis. *Org. Lett.* **2010**, *12*, 3670–3673. [CrossRef] [PubMed]
47. Vautravers, N.R.; Regent, D.D.; Breit, B. Inter- and intramolecular hydroacylation of alkenes employing a bifunctional catalyst system. *Chem. Commun.* **2011**, *47*, 6635–6637. [CrossRef] [PubMed]
48. Malanga, C.; Aronica, L.A.; Lardicci, L. Direct Ni^0 mediated synthesis of ketones from acyl bromides and Grignard reagents. *Tetrahedron Lett.* **1995**, *36*, 9185–9188. [CrossRef]

4

Asymmetric Primaquine and Halogenaniline Fumardiamides as Novel Biologically Active Michael Acceptors

Zrinka Rajić [1,*], Maja Beus [1], Hana Michnová [2], Josipa Vlainić [3], Leentje Persoons [4], Ivan Kosalec [1], Josef Jampílek [2], Dominique Schols [4], Toma Keser [1] and Branka Zorc [1,*]

1 Faculty of Pharmacy and Biochemistry, University of Zagreb, A. Kovačića 1, 10000 Zagreb, Croatia; mbeus@pharma.hr (M.B.); ikosalec@pharma.hr (I.K.); tkeser@pharma.hr (T.K.)
2 Department of Pharmaceutical Chemistry, Faculty of Pharmacy, Comenius University, Odbojárov 10, 83232 Bratislava, Slovakia; michnova.hana@gmail.com (H.M.), josef.jampilek@gmail.com (J.J.)
3 Laboratory for Advanced Genomics, Division of Molecular Medicine, Rudjer Bošković Institute, Bijenička cesta 54, 10000 Zagreb, Croatia; josipa.vlainic@irb.hr
4 Laboratory of Virology and Chemotherapy, Rega Institute for Medical Research, KU Leuven, Herestraat 49, 3000 Leuven, Belgium; leentje.persoons@rega.kuleuven.be (L.P.); dominique.schols@rega.kuleuven.be (D.S.)
* Correspondence: zrajic@pharma.hr (Z.R.); bzorc@pharma.hr (B.Z.)

Academic Editor: Michal Szostak

Abstract: Novel primaquine (PQ) and halogenaniline asymmetric fumardiamides **4a–f**, potential Michael acceptors, and their reduced analogues succindiamides **5a–f** were prepared by simple three-step reactions: coupling reaction between PQ and mono-ethyl fumarate (**1a**) or mono-methyl succinate (**1b**), hydrolysis of PQ-dicarboxylic acid mono-ester conjugates **2a,b** to corresponding acids **3a,b**, and a coupling reaction with halogenanilines. 1-[bis(Dimethylamino)methylene]-1*H*-1,2,3-triazolo[4,5-*b*]pyridinium 3-oxide hexafluorophosphate (HATU) was used as a coupling reagent along with Hünig's base. Compounds **4** and **5** were evaluated against a panel of bacteria, several *Mycobacterium* strains, fungi, a set of viruses, and nine different human tumor cell lines. *p*-Chlorofumardiamide **4d** showed significant activity against *Staphylococcus aureus, Streptococcus pneumoniae* and *Acinetobacter baumannii*, but also against *Candida albicans* (minimum inhibitory concentration (MIC) 6.1–12.5 µg/mL). Together with *p*-fluoro and *p*-CF_3 fumardiamides **4b,f**, compound **4d** showed activity against *Mycobacterium marinum* and **4b,f** against *M. tuberculosis*. In biofilm eradication assay, most of the bacteria, particularly *S. aureus*, showed susceptibility to fumardiamides. *m*-CF_3 and *m*-chloroaniline fumardiamides **4e** and **4c** showed significant antiviral activity against reovirus-1, sindbis virus and Punta Toro virus (EC_{50} = 3.1–5.5 µM), while **4e** was active against coxsackie virus B4 (EC_{50} = 3.1 µM). *m*-Fluoro derivative **4a** exerted significant cytostatic activity (IC_{50} = 5.7–31.2 µM). Acute lymphoblastic leukemia cells were highly susceptible towards *m*-substituted derivatives **4a,c,e** (IC_{50} = 6.7–8.9 µM). Biological evaluations revealed that fumardiamides **4** were more active than succindiamides **5** indicating importance of Michael conjugated system.

Keywords: fumardiamide; primaquine; succindiamide; Michael acceptor; biofilm eradication; antibacterial screening; antiviral activity; cytostatic activity

1. Introduction

Compounds bearing α,β-unsaturated carbonyl groups are Michael acceptors capable of conjugate addition, also known as Michael addition. The simplest and the best Michael acceptors are α,β-unsaturated carbonyl compounds with exposed unsaturated β-carbon atoms, such as exomethylene ketones and lactones or vinyl ketones [1]. These fragments are often used in the design of new anticancer drugs, together with others (6-methylhept-5-ene-1,4-dione, propiolamide, 4-(dimethylamino)but-2-enamide) (Figure 1). They assure the irreversible covalent binding to a cysteine residue of a specific protein and may modulate selectivity and potency of the drug candidate. The targeted covalent modification has emerged as a validated approach to drug discovery with the drug candidate canertinib [2], approved drugs afatinib, neratinib and osimertinib (inhibitors of human epidermal growth factor receptors) and ibrutinib (Bruton's tyrosine kinase inhibitor) [3,4]. A comprehensive review published by Jackson et al. gives an overview of biological activity and applicability of various Michael acceptors [5].

Michael acceptors have been explored in a prodrug strategy for cancer cell-specific targeting. In the review published by Zhang et al., two doxorubicin prodrugs with maleimide moieties have been described [6]. The maleimide component is responsible for the binding to human serum albumin. Once the drug carrier arrives at the targeted cancer tissue, doxorubicin is released from the carrier by the cleavage of hydrazone or glycosidic bond in the acidic environment of cancer cells. The prodrugs demonstrate superior anticancer efficacy than the parent drug.

Michael acceptors are present in other classes of drugs. Examples of such drugs are entacapone (antiparkinsonic) [7], dimethyl fumarate (antipsoriatic; since 2013 used in treatment of multiple sclerosis) [5], rupintivir (experimental antiviral drug against human rhinoviruses) [8], exemestane (cytostatic) [9], and ethacrynic acid (diuretic) [10].

α,β-Unsaturated carbonyl group is also a motif found in plant and microbial metabolites and their semisynthetic/synthetic derivatives. Many of them are used in clinical practice or are still under the evaluation in clinical trials (vernolepin, helenalin, curcumin, pyrrocidine, fumaric and angelic acid derivatives), but some are classified as toxins [11]. Vernolepin and helenalin are sesquiterpenes with exomethylene lactones responsible for irreversible DNA polymerase inhibition. Pyrrocidine A is a 13-membered macrocyclic alkaloid produced by endophytic fungi, which directly binds to *N*-acetyl-L-cysteine methyl ester through the Michael-type addition and exerts both antimicrobial and cytostatic effect on leukemia HL60 cells [12]. Angelic acid ester ingenol mebutate has been identified as the most active component of *Euphorbia peplus* L. latex sap, effective against human nonmelanoma skin cancer [13] and actinic keratosis [14]. A gel formulation of ingenol mebutate has been recently approved for the treatment of actinic keratosis [15] and fumaric acid esters are used for the management of psoriasis [16,17]. Curcumin is a symmetric α,β-unsaturated β-diketone extracted from *Curcuma longa* L., a tropical Southeast Asian plant used as a spice and in traditional Indian medicine [18]. Currently, there are 17 open clinical trials involving curcumin, of which mainly evaluate the combination of curcumin with other substances used in anticancer therapy [19].

In this paper, we report design and preparation of novel Michael acceptors, fumaric acid diamides **4a–f**. We have chosen α,β-unsaturated amides because they are less electrophilic than analogous esters and better Michael acceptors [1]. Based on our previous findings [20,21], one of the amide bonds was achieved with a terminal amino group of primaquine (PQ), while the other with halogenanilines (Figure 1). Sharing the same conjugated C=C-CO system and a benzene ring, our compounds are similar to cinnamic acid (*trans*-3-phenyl-2-propenoic acid) derivatives as well. Taking literature data into account [22–24] and our previous experience with PQ derivatives [20,21,25–28], we have assumed that the designed compounds have a high pharmacological potential. Here we report their synthesis, evaluation of antimicrobial activity on a wide spectrum of bacteria, fungi and viruses, their biofilm eradication ability, and finally, cytostatic activity against several human tumor cell lines.

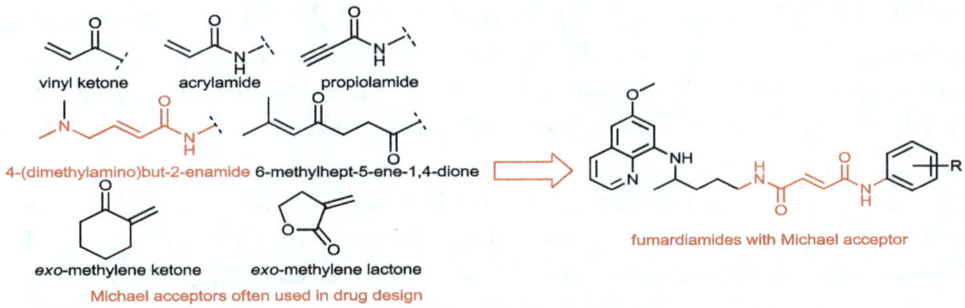

Figure 1. Design of novel fumardiamides.

2. Results and Discussion

2.1. Chemistry

In this paper, we aimed to prepare and biologically evaluate new Michael acceptors **4a–f**, asymmetric diamides of fumaric acid. To prove the significance of the conjugated Michael system for activity, we have also prepared a series of analogous compounds without a double bond, e.g., a series of succindiamides **5a–f** (compound **5a** is a reduced derivative of **4a**, **5b** of **4b** etc.). In both series of compounds, one of the amide bonds was achieved with the primary amino group of PQ and the other one with a selected halogenaniline. In the first reaction step, mono-ethyl fumarate (**1a**) and mono-methyl succinate (**1b**) were coupled with PQ to give derivatives **2a,b** using 1-[bis(dimethylamino)methylene]-1*H*-1,2,3-triazolo[4,5-*b*]pyridinium 3-oxide hexafluorophosphate (HATU) as a coupling reagent, along with *N,N*-diisopropylethylamine (DIEA) [29]. However, the transformation of **1a** to carboxylic acid chloride and amidation with PQ gave better yields of product **2a**. Hydrolysis of **2a,b** by lithium hydroxide afforded the corresponding acids **3a,b**, which were again coupled with halogenanilines in the presence of HATU/DIEA. The following anilines were used: 3-fluoroaniline, 4-fluoroaniline, 3-chloroaniline, 4-chloroaniline, 3-trifluoromethylaniline and 4-trifluoromethylaniline. Scheme 1 shows the synthetic pathway leading to compounds **4** and **5**.

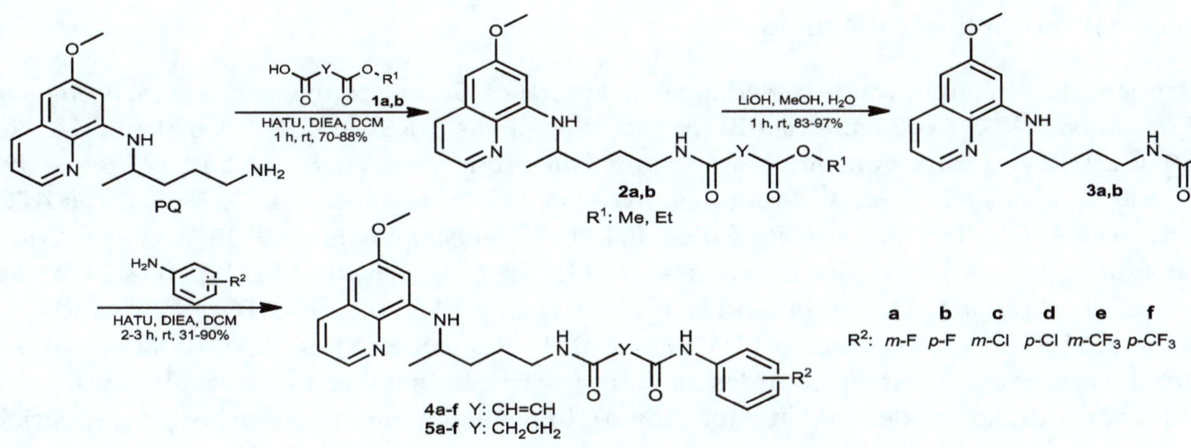

Scheme 1. Synthesis of fumardiamides **4a–f** and succindiamides **5a–f**.

New compounds are fully characterized by MS, IR, ^{1}H and ^{13}C NMR spectroscopic methods and elemental analyses. Spectral data are consistent with the proposed structures and are given in short in the Materials and Methods and in detail in the Supplementary Material. The presence of carbonyl functional groups in compounds **4** and **5** was indicated by the appearance of strong stretching vibration bands in IR spectra between ν 1677 and 1628 (amide I) and 1555 and 1515 cm^{-1} (amide II). PQ residue showed characteristic signals in ^{1}H NMR spectra: hydrogen atom CH-15 occurred between δ 8.52 and 8.55, methoxy group at δ 3.82, the hydrogen attached to chiral carbon

(CH-10) as a multiplet at δ 3.57–3.65, a methyl group at δ 1.19–1.22. NH-1 appeared as a singlet between δ 9.98 and 10.75 ppm, NH-6 as a triplet at δ 7.86–8.55, while NH-12 as a doublet at δ 6.11–6.15 ppm. ^{13}C spectra showed characteristic PQ signals at δ 54.92–54.98 (methoxy group), 46.19–47.02 (C-10), 38.46–39.23 (C-7), 33.37–33.45 (C-9), 25.70–25.99 (C-8), 20.14–20.21 (C-11) ppm and corresponding signals in aromatic region. Two signals of carbonyl groups in fumardiamides **4a–f** appeared between 161.17 and 163.34 ppm, while in succindiamides **5a–f** between 170.35 and 171.18 ppm. The other two succinic acid carbons (C-3 and C-4) appeared in the aliphatic region between 30.10 and 31.75, and fumaric acid carbons connected by double bond were located very high between 132.11 and 134.66 ppm. CF_3 groups in compounds **4e**, **4f**, **5e** and **5f** showed characteristic quartets at 121.32–127.05 with average J = 272 Hz, and the closest C-atom to trifluoromethyl quartets at 124–129 ppm with J = 31 Hz. *m*-Fluorophenyl derivatives **4a** and **5a** showed doublets at 162 ppm and *p*-fluorophenyl derivatives **4b** and **5b** doublets at 158 ppm (C-atom bearing fluorine) with a similar *J*-coupling constant of 240 Hz, while two neighbouring C-atoms appeared at 105–110 ppm (average J = 21 Hz). C-atom substituted with chloro atom appeared at 133 ppm (*m*-Cl) and 127 ppm (*p*-Cl). Chemical structures of new compounds were also supported by mass spectroscopy. Molecular ion peaks corresponding to expected relative molecular masses were obtained for all compounds.

Diamides **4** and **5** were subjected to in silico analysis to evaluate the diversity of the set of compounds against topological polar surface area (TPSA) calculations and relevant drug-like properties: number of atoms, molecular weight (MW), partition coefficient (log P), H-bond donor (HBD), H-bond acceptor (HBA) and molecular refractivity (MR). The parameters are calculated with the Chemicalize.org program [30] and presented in Table S1. All compounds are fully in agreement with the Lipinski and Gelovani rules for prospective small molecular drugs (MW $\leq$ 500, log $P \leq 5$, number of H-bond donors $\leq$ 5, number of H-bond acceptors $\leq$ 10, TPSA < 140 Å^2, MR within the range of 40 and 130 cm^3/mol, the number of atoms 20–70), although **4e,f** and **5e,f** have the highest permitted relative molecular masses.

2.2. Biological Evaluation

2.2.1. Antibacterial and Antifungal Activity

The antimicrobial screening was assessed against a panel of Gram-positive bacteria (*S. pneumoniae* MFBF 10373, *S. aureus* ATCC 6538, methicillin-resistant *S. aureus* MRSA 63718, SA 630 and SA 3202, *E. faecalis* ATCC 29212 and three vancomycin-resistant *Enterococci* VRE 342 B, 365 and 725 B, *B. cereus* ATCC 11778, and *B. subtilis* ATCC 6633), Gram-negative species (*P. aeruginosa* ATCC 27853, *E. coli* ATCC 10536, *S. marcescens* ATCC 10905, *P. mirabilis* MFBF 10430, *A. baumannii* MFBF 10913, and *S. enteritidis* MFBF 11945), four *Mycobacterium* strains (*M. tuberculosis* H37Ra, *M. smegmatis* ATCC 700084, *M. kansasii* DSM 44162, and *M. marinum* CAMP 5644) and fungi (two strains of *C. albicans* ATCC 90028 and CCM 8361, *C. krusei* CCM 8271, *C. parapsilosis* CCM 8260 and *A. brasiliensis* ATCC 16404). The results of the antimicrobial screening of fumardiamides **4a–f** are presented in Table 1 (the results for inactive fumardiamides and succindiamides **5a–f** are not shown). In general, Gram-positive bacteria, especially *S. pneumoniae* and *S. aureus* ATCC 6538 were susceptible to **4b–f**. Compounds **4a,b** were active against *B. cereus* as well. However, Gram-negative bacterium *A. baumannii* was the most susceptible among all tested microorganisms: all compounds, except *m*-fluoro derivative **4a**, showed selective antimicrobial activity against this bacterium strain. Comparison of *meta* and *para* derivatives, e.g., **a** vs. **b**, **c** vs. **d**, **e** vs. **f** revealed that in the most cases *p*-substituted derivatives were more active than the analogous *m*-derivatives. *p*-Chlorofumardiamide **4d** showed significant activity against three bacterial strains (*S. aureus, S. pneumoniae* and *A. baumannii*), but also against *C. albicans* ATCC 90028, with minimum inhibitory concentration (MIC) values ranging from 6.1 to 12.5 μg/mL. This compound, together with *p*-fluoro and *p*-CF_3 fumardiamides **4b,f** showed antitubercular activity against *M. marinum*, while **4b,f** were active against *M. tuberculosis* as well.

2.2.2. Biofilm Eradication Assay

We tested the susceptibility of different bacterial and yeast strains to fumardiamides **4a–f** and succindiamides **5a–f** by determining minimum biofilm eradication concentrations (MBECs). The results are presented in Table 2. Again, fumardiamides **4** were much more active than the analogous succindiamides **5**. The most active compounds were **4a** and **4b**. Two microorganisms, namely *E. faecalis* and *S. aureus* showed high susceptibility to all fumardiamides, while *E. coli*, *S. pneumoniae* and *P. aeruginosa* were susceptible to five out of six fumardiamides. A high biofilm eradication potential of fumardiamides might be explained by the reaction of Michael acceptors with cysteine thiol, which could prevent disulfide bond formation. It is a well-known fact that cysteine homeostasis impacts biofilm formation and production of extracellular matrix components, as well as folding and stability of extracytoplasmic proteins [31]. They are also crucial for dental plaque formation, autolysis, extracellular DNA release, genetic competence, bacteriocin production and stabilization of outer membrane porin proteins [32].

2.2.3. Antiviral Evaluation

Compounds **4a–f** and **5a–f** were evaluated against a broad variety of viral infections including herpes simplex viruses, vaccinia virus, adenovirus-2, human coronavirus (229E), vesicular stomatitis virus, coxsackie virus B4, respiratory syncytial virus, para-influenza-3 virus, reovirus-1, sindbis virus, coxsackie virus B4, Punta Toro virus and yellow fever virus. Only fumardiamides with *m*-chloro and *m*-trifluoromethyl aniline residues, namely **4c** and **4e**, showed significant antiviral activity against reovirus-1, sindbis virus and Punta Toro virus with EC_{50} = 3.1–5.5 μM (EC_{50} = concentration required to reduce virus-induced cytopathogenicity by 50%). Compound **4e** was also active against coxsackie virus B4 (EC_{50} = 3.1 μM). However, their selectivity index (SI), e.g., minimum cytotoxicity concentration (MCC) and EC_{50} ratio, was quite low (1.8–3.7). Cytotoxicity and antiviral activity of fumardiamides **4a–f** in Vero cell cultures are displayed in Table 3 (data for HEL (human erythroleukemia cell line), Hela (cervical carcinoma cell line) and Madin-Darby Canine Kidney cells (MDCK) cultures are not shown, as well as data for inactive succindiamides **5a–f**).

Table 1. Antimicrobial susceptibility assay in vitro (minimum inhibitory concentration (MIC) determination) for compounds **4a–f**.

Compd.	MIC (µg/mL)								
	Staphylococcus aureus ATCC 6538	*Streptococcus pneumoniae* MFBF 10373	*Enterococcus faecalis* ATCC 29212	*Bacillus cereus* ATCC 11778	*Bacillus subtilis* ATCC 6633	*Acinetobacter baumannii* MFBF 10913	*Mycobacterium tuberculosis* H37Ra	*Mycobacterium marinum* CAMP 5644	*Candida albicans* ATCC 90028
4a	>100	>100	128	>100	50	>100	128	>256	>100
4b	12.5	12.5	64	25	25	25	32	64	>100
4c	12.5	25	256	25	50	25	128	128	>100
4d	6.1	12.5	64	>100	>100	12.5	128	64	12.5
4e	50	50	128	>100	>100	25	128	256	>100
4f	50	50	64	>100	>100	25	64	64	>100
PQ [1]	50	50	128	70	80	50	256	256	>100
TC [2]	0.3	0.3	–	0.3	0.3	3	–	–	–
CIP [3]	–	–	–	–	–	–	16.0	0.3	–
INH [4]	–	–	–	–	–	–	5.0	64.0	–
RIF [5]	–	–	–	–	–	–	8.7	2.2	–
Amph [6]	–	–	–	–	–	–	–	–	0.5

[1] PQ—primaquine; [2] TC—tetracycline; [3] CIP—ciprofloxacin; [4] INH—isoniazid; [5] RIF—rifampicin; [6] Amph—amphotericin B.

Table 2. Sensitivity of microbial strains to fumardiamides **4a–f** expressed as minimum biofilm eradication concentrations (MBEC).

Compd.	MBEC (µg/mL)										
	Staphylococcus aureus ATCC 6538	*Streptococcus pneumonia* MFBF 10373	*Enterococcus faecalis* ATCC 29212	*Bacillus cereus* ATCC 11778	*Bacillus subtilis* ATCC 6633	*Escherichia coli* ATCC 10536	*Pseudomonas aeruginosa* ATCC 27853	*Serratia marcescens* ATCC 10905	*Proteus mirabilis* MFBF 10430	*Salmonella enteritidis* MFBF 11945	*Acinetobacter baumannii* MFBF 10913
4a	12.5	12.5	25	>100	50	100	25	25	>100	>100	6.3
4b	12.5	25	25	25	25	50	12.5	>100	25	>100	25
4c	12.5	100	25	25	50	50	50	>100	>100	>100	100
4d	25	50	50	>100	>100	50	>100	>100	>100	>100	100
4e	50	50	50	>100	>100	50	50	>100	>100	>100	100
4f	25	50	50	>100	>100	50	25	>100	>100	>100	100
PQ [1]	50	70	>100	70	80	65	65	>100	>100	70	50
Gen [2]	12.5	25	50	50	50	25	50	50	50	50	25

[1] PQ—primaquine; [2] Gen—gentamycin.

Table 3. Cytotoxicity and antiviral activity of fumardiamides **4a–f** evaluated in Vero cell cultures.

Compd.	Cytotoxicity		EC_{50} [1] (μM)											
			Para-Influenza-3 Virus		Reovirus-1		Sindbis Virus		Coxsackie Virus B4		Punta Toro Virus		Yellow Fever Virus	
	CC_{50} [2]	MCC [3]	MTS	Visual CPE Score	MTS	Visual CPE Score	MTS	Visual CPE Score	MTS	Visual CPE Score	MTS	Visual CPE Score	MTS	Visual CPE Score
4a	46.9	–	>50	–	>50	–	>50	–	>50	–	>50	–	>50	–
4b	>50	–	>50	–	>50	–	>50	–	>50	–	>50	–	>50	–
4c	>50	10	>50	>50	3.8	>50	3.0	>50	>50	>50	5.5	>50	>50	>50
4d	>50	–	>50	–	>50	–	>50	–	>50	–	>50	–	>50	–
4e	>50	10	>50	>50	3.1	>50	5.3	4.2	3.1	2.7	5.4	>50	>50	>50
4f	>50	–	>50	–	>50	–	>50	–	>50	–	>50	–	>50	–
PQ	>50	–	>50	–	>50	–	>50	–	>50	–	>50	–	>50	–
DS-10.000 [4]	>100	>100	>100	>100	>100	>100	4.0	10	>100	34	16	8.9	1.6	>100
RIB [5]	>250	>250	73	111	107	126	>250	11	>250	>250	85	111	119	>250
MPA [6]	>100	>100	1.0	0.8	0.6	0.8	12	1.7	>100	>100	11	8.9	0.5	0.8

[1] Concentration required to reduce virus-induced cytopathicity by 50%, as determined by visual scoring of the cytopathic effect (CPE), or by measuring the cell viability with the colorimetric formazan-based 3-(4,5-dimethylthiazol-2-yl)-5-(3-carboxymethoxyphenyl)-2-(4-sulfophenyl)-2*H*-tetrazolium (MTS) assay; [2] Concentration that causes 50% cytotoxic effect, as determined by measuring the cell viability with the colorimetric formazan-based MTS assay; [3] Minimum cytotoxic concentration, i.e., the minimum concentration that causes a microscopically detectable alteration of normal cell morphology; [4] Concentration in μg/mL; [5] RIB—ribavirin; [6] MPA—mycophenolic acid. All experiments were performed in duplicate.

2.2.4. Cytostatic Activity

To gain insight into cytotoxicity of newly synthesized compounds **4a–f** and **5a–f**, their cytostatic activity was evaluated in vitro against a panel of nine different human cancer cell lines, representing various solid tumor types including pancreatic adenocarcinoma (Capan-1), chronic myeloid leukemia (Hap1), colorectal carcinoma (HCT-116), lung carcinoma (NCI-H460), acute lymphoblastic leukemia (DND-41), acute myeloid leukemia (HL-60), chronic myeloid leukemia (K-562), multiple myeloma (MM.1S) and non-Hodgkin lymphoma (Z-138). Succindiamides **5a–f** were completely inactive, while free fumardiamides showed activity towards the selected tumor cell lines (Table 4). Fumardiamide **4a** with *m*-fluoroaniline moiety showed cytostatic activity against all tested cell lines, with IC_{50} values from 5.7 to 31.2 μM (IC_{50} = the lowest concentration resulting in 50% growth inhibition). DND-41 cell line was susceptible to all three *m*-substituted derivatives, e.g., **4a**, **4c** and **4f** (IC_{50} values between 6.7 and 8.9 μM). HL-60 and Z-138 cell lines were also susceptible to compound **4a** (IC_{50} values 5.7 and 8.4 μM, respectively).

Table 4. Antiproliferative screening of fumardiamides **4a–f** in diverse human tumor cell lines.

Compd.	IC_{50} (μM)								
	Capan-1	Hap1	HCT-116	NCI-H460	DND-41	HL-60	K-562	MM.1S	Z-138
4a	16.9	19.3	22.3	15.5	6.7	5.7	31.2	13.0	8.4
4b	50.5	66.4	>100	29.1	>100	>100	70.5	>100	>100
4c	69.9	>100	73.0	43.7	8.4	>100	>100	>100	46.5
4d	91.8	>100	>100	>100	>100	>100	>100	>100	>100
4e	56.5	>100	>100	34.2	8.9	>100	71.2	>100	68.4
4f	79.4	>100	>100	>100	>100	>100	>100	>100	>100
PQ [1]	18.7	42.7	30.9	52.6	11.4	2.2	35.2	28.3	7.1
DXT [2]	0.75	1.17	7.66	1.30	0.94	1.24	1.22	3.38	5.42
EPEG [3]	0.15	0.04	1.35	0.09	0.03	0.03	0.01	0.97	0.02
STS [4]	0.66	3.56	0.78	1.66	6.96	13.10	0.23	1.61	0.40

[1] PQ—primaquine; [2] DXT—docetaxel (nM); [3] EPEG—etoposide; [4] STS—staunoporin (nM). All compounds were tested with duplicate data points and averaged.

2.2.5. Interaction with Glutathione (GSH)

Glutathione (GSH) is an important compound present in most mammalian cells, a tripeptide, with the central amino acid cysteine bearing thiol group, which scavenges carcinogenic compounds by conjugate addition and protects from oxidative damage [1]. In the drug development process, it is usual to evaluate the Michael acceptor–GSH interaction [33]. That is why the final step in our research was to study the interaction of fumarmides with GSH. Fumardiamide **4b** (1.25 μM) was incubated with GSH (125 μM) in ammonium formate buffer (pH = 7.4) containing 10% acetonitrile at 37 °C for 216 h. The MS analysis confirmed the consumption of **4b** with GSH. However, the rate of GSH addition was slow and incomplete, as only 18.3% of **4b** reacted in the monitored period (See Supporting Material).

3. Materials and Methods

3.1. Chemistry

3.1.1. Materials and General Methods

Melting points were determined on an SMP3 apparatus (Barloworld Scientific, UK) in open capillaries and were uncorrected. IR spectra were recorded on Spectrum One FT-IR (Perkin-Elmer, UK) and UV-Vis spectra on Lambda 20 double-beam spectrophotometers (Perkin-Elmer, UK). NMR ^{1}H and ^{13}C spectra were recorded at 25 °C on an NMR Avance 600 spectrometer (Bruker, Germany) at 300.13 or 600.13 and 75.47 or 150.9 MHz for ^{1}H and ^{13}C nuclei, respectively. Chemical shifts (δ) were reported in parts per million (ppm) relative to tetramethylsilane in the ^{1}H and the dimethyl sulfoxide residual peak as a reference in the ^{13}C NMR spectra (39.51 ppm). Coupling constants (*J*) were reported in hertz (Hz). Mass spectra were collected on an HPLC-MS/MS instrument (HPLC, Agilent Technologies 1200 Series; MS, Agilent Technologies 6410 Triple Quad) using electrospray ionization in positive mode. Elemental analyses were performed on a CHNS LECO analyzer (LECO Corporation, USA). All compounds were routinely checked by thin-layer chromatography (TLC) with Merck silica gel 60F-254 glass plates using appropriate solvent systems. Spots were visualized by short-wave UV light and iodine vapour. Column chromatography was performed on silica gel 0.063–0.200 mm. All chemicals and solvents were of analytical grade and purchased from commercial sources. PQ diphosphate, 4-methoxy-4-oxobutanoic acid (mono-methyl succinate), (*E*)-4-ethoxy-4-oxobut-2-enoic acid (mono-ethyl fumarate), 3-fluoroaniline, 4-fluoroaniline, 3-chloroaniline, 4-chloroaniline, 3-trifluoromethylaniline, 4-trifluoromethylaniline, DIEA, and HATU were purchased from Sigma-Aldrich. PQ was prepared from PQ diphosphate prior to use. All reactions with PQ were run light protected.

3.1.2. General Procedure for the Synthesis of Esters **2a,b**

PQ-succinamide and fumaramide monoesters **2a,b** were prepared by condensation of PQ base with mono-methyl succinate (**1a**) or mono-ethyl fumarate (**1b**), following the previously described procedure [29].

3.1.3. General Procedure for the Synthesis of Carboxylic Acids **3a,b**

Carboxylic acids **3a,b** were prepared by hydrolysis of esters **2a,b** with lithium hydroxide following the previously described procedure [29].

3.1.4. General Procedure for the Synthesis of Fumardiamides **4a–f** and Succindiamides **5a–f**

A solution of 0.27 mmol of **3a** or **3b**, 0.068 g (0.54 mmol) DIEA and 0.103 g (0.27 mmol) HATU in 6 mL of dichloromethane was stirred at room temperature. After 10 min, 0.297 mmol of the corresponding halogenaniline was added. The reaction mixture was stirred for 2–3 h at room temperature, evaporated under reduced pressure, dissolved in 8 mL ethyl acetate and extracted 3 times with water. The organic layer was dried over sodium sulfate, filtered and evaporated under reduced pressure.

*(2E)-N-(3-Fluorophenyl)-N′-{4-[(6-methoxyquinolin-8-yl)amino]pentyl}but-2-enediamide (**4a**).* From the reaction of 0.096 g acid **3a** and 0.033 g (0.297 mmol) 3-fluoroaniline and after purification by column chromatography (mobile phase dichloromethane/methanol 9.5:0.5) and crystallization from ether, 0.041 g (34%) of **4a** was obtained; mp 203–204 °C; IR (ATR): ν_{max} 3388, 3319, 3269, 3080, 2961, 2935, 2866, 1630, 1554, 1520, 1452, 1387, 1334, 1201, 1158, 782, 680 cm^{-1}; ^{1}H NMR (DMSO-d_6) δ 10.63 (s, 1H), 8.55–8.53 (dd, 1H, *J* = 1.6, 4.2), 8.52 (t, 1H, *J* = 5.4), 8.09–8.06 (dd, 1H, *J* = 1.5, 8.3), 7.69 (d, 2H, *J* = 11.7), 7.45–7.41 (m, 1H), 7.39–7.36 (m, 2H), 7.06–7.96 (m, 2H), 6.96–6.92 (m, 1H), 6.47 (d, 1H, *J* = 2.4), 6.28 (d, 1H, *J* = 2.4), 6.15 (d, 1H, *J* = 8.8), 3.82 (s, 3H), 3.70–3.61 (m, 1H), 3.24–3.18 (m, 2H), 1.74–1.64, 1.63–1.52 (2m, 4H), 1.22 (d, 3H, *J* = 6.3); ^{13}C NMR (DMSO-d_6) δ 163.25, 162.62, 162.09 (d, *J* = 242.0), 159.00, 144.63, 144.23, 140.50 (d, *J* = 11.5), 134.79, 134.52, 134.40, 132.28, 130.50 (d, *J* = 9.5), 129.57, 122.09, 115.14, 110.28, 106.13 (d, *J* = 27.2), 96.14, 91.62, 54.96, 46.97, 38.95, 33.45, 25.77, 20.21; MS/MS *m/z* 451.1 (M + 1)$^+$; Anal. Calcd. for $C_{25}H_{27}FN_4O_3$: C, 66.65; H, 6.04; N, 12.44. Found: C, 66.32; H, 6.30; N, 12.49.

*(2E)-N-(4-Fluorophenyl)-N′-{4-[(6-methoxyquinolin-8-yl)amino]pentyl}but-2-enediamide (**4b**).* From the reaction of 0.096 g acid **3a** and 0.033 g (0.297 mmol) 4-fluoroaniline and after purification by column chromatography (mobile phase dichloromethane/methanol 9.5:0.5) and crystallization from ether, 0.071 g (58%) of **4b** was obtained; mp 226–227 °C; IR (ATR): ν_{max} 3386, 3294, 3072, 2963, 2928, 2863, 1635, 1548, 1513, 1452, 1391, 1330, 1212, 1160, 1051, 973, 829, 673 cm^{-1}; ^{1}H NMR (DMSO-d_6) δ 10.49 (s, 1H), 8.55–8.53 (dd, 1H, *J* = 1.6, 4.2), 8.52 (t, 1H, *J* = 5.4), 8.09–8.06 (dd, 1H, *J* = 1.5, 8.3), 7.72–7.68 (dd, 2H, *J* = 5.0, *J* = 8.9), 7.45–7.41 (m, 1H), 7.18 (t, 2H, *J* = 8.8), 7.05–6.93 (m, 2H), 6.47 (d, 1H, *J* = 2.1), 6.28 (d,

1H, J = 2.1), 6.15 (d, 1H, J = 8.8), 3.82 (s, 3H), 3.69–3.61 (m, 1H), 3.24–3.18 (m, 2H), 1.74–1.64, 1.63–1.52 (2m, 4H), 1.22 (d, 3H, J = 6.3); ^{13}C NMR (DMSO-d_6) δ 163.34, 162.21, 159.00, 158.26 (d, J = 240.0), 144.63, 144.23, 135.22, 134.79, 134.53, 133.98, 132.52, 129.58, 122.09, 121.05 (d, J = 7.6), 115.44 (d, J = 21.9), 96.14, 91.62, 54.96, 46.97, 38.95, 33.45, 25.79, 20.21; MS/MS m/z 451.1 (M + 1)$^+$; Anal. Calcd. for $C_{25}H_{27}FN_4O_3$: C, 66.65; H, 6.04; N, 12.44. Found: C, 66.47; H, 6.38; N, 12.35.

*(2E)-N-(3-Chlorophenyl)-N′-{4-[(6-methoxyquinolin-8-yl)amino]pentyl}but-2-enediamide (**4c**).* From the reaction of 0.096 g acid **3a** and 0.038 g (0.297 mmol) 3-chloroaniline and after purification by column chromatography (mobile phase dichloromethane/methanol 9.5:0.5) and crystallization from ether, 0.053 g (42%) of **4c** was obtained; mp 187–188 °C; IR (ATR): ν_{max} 3381, 3298, 3068, 2959, 2928, 2863, 1635, 1591, 1521, 1465, 1419, 1386, 1331, 1210, 1163, 976, 821, 783, 670 cm^{-1}; ^{1}H NMR (DMSO-d_6) δ 10.59 (s, 1H), 8.54–8.53 (dd, 1H, J = 1.5, 4.1), 8.50 (t, 1H, J = 5.5), 8.08–8.06 (dd, 1H, J = 1.4, 8.2), 7.90 (s, 1H), 7.51 (d, 1H, J = 8.1), 7.43–7.41 (m, 1H), 7.37 (t, H, J = 8.1), 7.16–7.14 (dd, 1H, J = 1.2, J = 8.0), 7.00 (q, 2H, J = 15.1), 6.47 (d, 1H, J = 2.4), 6.28 (d, 1H, J = 2.3), 6.14 (d, 1H, J = 8.8), 3.82 (s, 3H), 3.67–3.63 (m, 1H), 3.23–3.20 (m, 2H), 1.73–1.67, 1.63–1.53 (2m, 4H), 1.22 (d, 3H, J = 6.3); ^{13}C NMR (DMSO-d_6) δ 163.21, 162.60, 158.97, 144.60, 144.19, 140.19, 134.74, 134.50, 134.41, 133.10, 132.19, 130.48, 129.54, 123.45, 122.03, 118.74, 117.72, 96.11, 91.63, 54.93, 46.98, 39.23, 33.44, 25.72, 20.18; MS/MS m/z 467.0 (M+1)$^+$; Anal. tCalcd. for $C_{25}H_{27}ClN_4O_3$: C, 64.30; H, 5.83; N, 12.00. Found: C, 64.21; H, 6.05; N, 11.78.

*(2E)-N-(4-Chlorophenyl)-N′-{4-[(6-methoxyquinolin-8-yl)amino]pentyl}but-2-enediamide (**4d**).* From the reaction of 0.096 g acid **3a** and 0.038 g (0.297 mmol) 4-chloroaniline and after purification by column chromatography (mobile phase cyclohexane/ethyl acetate/methanol 3:1:0.5) and crystallization from ether, 0.062 g (49%) of **4d** was obtained; mp 223–226 °C; IR (ATR): ν_{max} 3381, 3289, 3071, 2959, 2931, 2864, 1640, 1526, 1452, 1388, 1331, 1210, 1163, 1094, 1049, 973, 822, 787, 686, 631, 507 cm^{-1}; ^{1}H NMR (DMSO-d_6) δ 10.56 (s, 1H), 8.55–8.53 (dd, 1H, J = 1.7, 4.2), 8.50 (t, 1H, J = 5.6), 8.09–8.06 (dd, 1H, J = 1.6, 8.3), 7.72–7.68 (m, 2H), 7.45–7.37 (m, 3H), 7.06–6.94 (m, 2H), 6.47 (d, 1H, J = 2.5), 6.27 (d, 1H, J = 2.4), 6.14 (d, 1H, J = 8.8), 3.82 (s, 3H), 3.69–3.60 (m, 1H), 3.24–3.18 (m, 2H), 1.74–1.64, 1.63–1.51 (2m, 4H), 1.22 (d, 3H, J = 6.3); ^{13}C NMR (DMSO-d_6) δ 163.29, 162.43, 159.00, 144.63, 144.23, 137.76, 134.79, 134.52, 134.21, 132.38, 129.57, 128.75, 127.36, 122.09, 120.84, 96.14, 91.61, 54.96, 46.97, 38.82, 33.44, 25.77, 20.21; MS/MS m/z 467.0 (M+1)$^+$; Anal. Calcd. for $C_{25}H_{27}ClN_4O_3$: C, 64.30; H, 5.83; N, 12.00. Found: C, 64.21; H, 5.56; N, 11.83.

*(2E)-N′-{4-[(6-Methoxyquinolin-8-yl)amino]pentyl}-N-[3-(trifluoromethyl)phenyl]but-2-enediamide (**4e**).* From the reaction of 0.096 g acid **3a** and 0.048 g (0.297 mmol) 3-trifluoroaniline and after purification by column chromatography (mobile phase dichloromethane/methanol 9.5:0.5) after crystallization from acetone/water, 0.042 g (31%) of **4e** was obtained; mp 149–150 °C; IR (ATR): ν_{max} 3399, 3357, 3282, 3094, 2960, 2935, 2867, 1651, 1621, 1563, 1526, 1452, 1388, 1331, 1168, 1122, 973, 893, 788, 694 cm^{-1}; ^{1}H NMR (DMSO-d_6) δ 10.75 (s, 1H), 8.55–8.52 (m, 2H), 8.18 (s, 1H), 8.08–8.06 (dd, 1H, J = 1.1, 8.2), 7.83 (d, 1H), 7.59 (t, 1H, J = 7.9), 7.45–7.42 (m, 2H), 7.00–6.98 (q, 2H, J = 15.1), 6.47 (d, 1H, J = 2.4), 6.28 (d, 1H, J = 2.0), 6.15 (d, 1H, J = 8.7), 3.82 (s, 3H), 3.67–3.63 (m, 1H), 3.23–3.20 (m, 2H), 1.73–1.67, 1.63–1.53 (2m, 4H), 1.22 (d, 3H, J = 6.3); ^{13}C NMR (DMSO-d_6) δ 163.19, 162.81, 158.99, 144.62, 144.21, 139.54, 134.77, 134.55, 134.52, 132.15, 130.09, 129.56, 129.82–129.19 (q, J = 31.7), 126.73–121.32 (q, J = 273.3), 122.88, 122.07, 120.09, 115.33, 96.13, 91.62, 54.95, 46.19, 39.23, 33.45, 25.76, 20.20; MS/MS m/z 501.1 (M+1)$^+$; Anal. Calcd. for $C_{26}H_{27}F_3N_4O_3$: C, 62.39; H, 5.44; N, 11.19. Found: C, 62.25; H, 5.76; N, 11.08.

*(2E)-N′-{4-[(6-Methoxyquinolin-8-yl)amino]pentyl}-N-[4-(trifluoromethyl)phenyl]but-2-enediamide (**4f**).* From the reaction of 0.096 g acid **3a** and 0.048 g (0.297 mmol) 4-trifluoroaniline and after crystallization from ether, 0.046 g (34%) of **4f** was obtained; mp 189–191 °C; IR (ATR): ν_{max} 3387, 3309, 3071, 2963, 2932, 1636, 1527, 1457, 1417, 1390, 1328, 1213, 1166, 1122, 1065, 970, 832, 681 cm^{-1}; ^{1}H NMR (DMSO-d_6) δ 10.75 (s, 1H), 8.54–8.53 (m, 1H), 8.51 (t, 1H, J = 5.3), 8.07 (d, 1H, J = 8.2), 7.88 (d, 2H, J = 8.5), 7.71 (d, 2H, J = 8.5), 7.43–7.41 (m, 1H), 7.07–6.99 (q, 2H, J = 15.1), 6.47 (d, 1H, J = 2.2), 6.28 (d, 1H,

J = 2.2), 6.14 (d, 1H, J = 8.7), 3.82 (s, 3H), 3.67–3.63 (m, 1H), 3.23–3.20 (dd, 2H, J = 6.1, 12.1), 2.59 (t, 2H, J = 7.0), 1.73–1.67, 1.61–1.53 (2m, 4H), 1.22 (d, 3H, J = 6.3); ^{13}C NMR (DMSO-d_6) δ 161.17, 162.83, 158.97, 144.60, 144.17, 142.28, 134.72, 134.66, 134.50, 132.11, 129.53, 126.95–121.55 (q, J = 271.6), 126.11–126.04 (q, J = 3.0), 124.02–123.39 (q, J = 31.7), 122.02, 119.26, 96.10, 91.63, 54.92, 46.96, 38.80, 33.43, 25.70, 20.17; MS/MS m/z 501.1 (M+1)$^+$; Anal. Calcd. for $C_{26}H_{27}F_3N_4O_3$: C, 62.39; H, 5.44; N, 11.19. Found: C, 62.17; H, 5.61; N, 11.40.

*N-(3-Fluorophenyl)-N′-{4-[(6-methoxyquinolin-8-yl)amino]pentyl}butanediamide (**5a**).* From the reaction of 0.097 g acid **3b** and 0.033 g (0.297 mmol) 3-fluoroaniline and after purification by column chromatography (mobile phase dichloromethane/methanol 9.5:0.5), 0.077 g (63%) of **5a** was obtained; mp 140–142 °C; IR (ATR): ν_{max} 3391, 3283, 3145, 3080, 2959, 2926, 1744, 1646, 1615, 1555, 1507, 1387, 1224, 1202, 1167, 1157, 838, 815, 786, 681 cm^{-1}; ^{1}H NMR (DMSO-d_6) δ 10.13 (s, 1H), 8.54–8.53 (dd, 1H, J = 1.7, 4.2), 8.08–8.06 (dd, 1H, J = 1.6, 8.3), 7.86 (t, 1H, J = 5.6), 7.60–7.58 (m, 1H), 7.43–7.41 (m, 1H), 7.31–7.27 (m, 2H), 6.85–6.81 (m, 1H), 6.47 (d, 1H, J = 2.5), 6.26 (d, 1H, J = 2.5), 6.11 (d, 1H, J = 8.7), 3.82 (s, 3H), 3.64–3.59 (m, 1H), 3.10–3.03 (m, 2H), 2.55 (t, 2H, J = 7.2), 2.37 (t, 2H, J = 7.2), 1.67–1.63, 1.55–1.45 (2m, 4H), 1.19 (d, 3H, J = 6.3); ^{13}C NMR (DMSO-d_6) δ 170.86, 162.10 (d, J = 240.1), 158.98, 144.61, 144.19, 141.03 (d, J = 13.1), 134.75, 134.51, 130.19 (d, J = 9.0), 129.55, 122.05, 114.55, 109.22 (d, J = 20.1), 105.61, 96.07, 91.59, 54.94, 46.99, 38.46, 33.38, 31.75, 30.16, 25.95, 20.15; MS/MS m/z 453.3 (M+1)$^+$; Anal. Calcd. for $C_{25}H_{29}FN_4O_3$: C, 66.35; H, 6.46; N, 12.38. Found: C, 66.25; H, 6.50; N, 12.20.

*N-(4-Fluorophenyl)-N′-{4-[(6-methoxyquinolin-8-yl)amino]pentyl}butanediamide (**5b**).* From the reaction of 0.097 g acid **3b** and 0.033 g (0.297 mmol) 4-fluoroaniline and after purification by column chromatography (mobile phase dichloromethane/methanol 9.5:0.5) and crystallization from ether, 0.098 g (80%) of **5b** was obtained; mp 118–120 °C; IR (ATR): ν_{max} 3391, 3283, 3145, 3080, 2959, 2926, 1744, 1646, 1615, 1555, 1507, 1387, 1224, 1202, 1167, 1157, 838, 815, 786, 681 cm^{-1}; ^{1}H NMR (DMSO-d_6) δ 9.98 (s, 1H), 8.55–8.53 (dd, 1H, J = 1.6, 4.2), 8.09–8.06 (dd, 1H, J = 1.5, 8.3), 7.87 (t, 1H, J = 5.5), 7.62–7.57 (m, 2H), 7.45–7.40 (m, 1H), 7.14–7.08 (m, 2H), 6.47 (d, 1H, J = 2.4), 6.26 (d, 1H, J = 2.4), 6.12 (d, 1H, J = 8.8), 3.82 (s, 3H), 3.65–3.57 (m, 1H), 3.10–3.04 (m, 2H), 2.54 (t, 2H, J = 7.1), 2.37 (t, 2H, J = 7.0), 1.69–1.59, 1.58–1.43 (2m, 4H), 1.19 (d, 3H, J = 6.3); ^{13}C NMR (DMSO-d_6) δ 170.96, 170.35, 159.00, 157.72 (d, J = 246.3), 144.63, 144.22, 135.76, 134.79, 134.52, 129.57, 122.09, 120.53 (d, J = 7.8), 115.14 (d, J = 22.1), 96.09, 91.58, 54.97, 47.00, 38.48, 33.39, 31.66, 30.34, 25.99, 20.17; MS/MS m/z 453.4 (M+1)$^+$; Anal. Calcd. for $C_{25}H_{29}FN_4O_3$: C, 66.35; H, 6.46; N, 12.38. Found: C, 66.53, H, 6.29; N, 12.55.

*N-(3-Chlorophenyl)-N′-{4-[(6-methoxyquinolin-8-yl)amino]pentyl}butanediamide (**5c**).* From the reaction of 0.097 g acid **3b** and 0.038 g (0.297 mmol) 3-chloroaniline and after purification by column chromatography (mobile phase cyclohexane/ethyl acetate/methanol 3:1:0.5) and crystallization from ether, 0.062 g (49%) of **5c** was obtained; mp 156–158 °C; IR (ATR): ν_{max} 3390, 3287, 3242, 3180, 3102, 3074, 2962, 2922, 2856, 1738, 1650, 1612, 1591, 1572, 1539, 1516, 1422, 1388, 1202, 1066, 816, 787, 698, 682 cm^{-1}; ^{1}H NMR (DMSO-d_6) δ 10.11 (s, 1H), 8.54–8.53 (dd, 1H, J = 1.6, 4.2), 8.08–8.06 (dd, 1H, J = 1.6, 8.3), 7.86 (t, 1H, J = 5.5), 7.81 (t, 1H, J = 2.0), 7.43–7.41 (m, 2H), 7.31–7.28 (m, 1H), 7.07–7.05 (m, 1H), 6.47 (d, 1H, J = 2.5), 6.25 (d, 1H, J = 2.5), 6.11 (d, 1H, J = 8.2), 3.82 (s, 3H), 3.63–3.59 (m, 1H), 3.10–3.03 (m, 2H), 2.55 (t, 2H, J = 7.1), 2.37 (t, 2H, J = 7.2), 1.67–1.63, 1.56–1.44 (2m, 4H), 1.19 (d, 3H, J = 6.3); ^{13}C NMR (DMSO-d_6) δ 170.89, 170.85, 158.98, 144.60, 144.19, 140.73, 134.77, 134.49, 132.97, 130.29, 129.55, 122.51, 122.06, 118.31, 117.19, 96.08, 91.59, 54.95, 46.99, 38.46, 33.38, 31.74, 30.16, 25.96, 20.16; MS/MS m/z 469.2 (M+1)$^+$; Anal. Calcd. for $C_{25}H_{29}ClN_4O_3$: C, 64.03; H, 6.23; N, 11.95. Found: C, 63.91; H, 6.33; N, 11.70.

*N-(4-Chlorophenyl)-N′-{4-[(6-methoxyquinolin-8-yl)amino]pentyl}butanediamide (**5d**).* From the reaction of 0.097 g acid **3b** and 0.038 g (0.297 mmol) 3-chloroaniline and after purification by column chromatography (mobile phase dichloromethane/methanol 9.5:0.5) and crystallization from ether, 0.076 g (60%) of **5d** was obtained; mp 152–153 °C; IR (ATR): ν_{max} 3390, 3287, 3242, 3180, 3102, 3074, 2962, 2922, 2856, 1738, 1650, 1612, 1591, 1572, 1539, 1516, 1422, 1388, 1202, 1066, 816, 787, 698, 682 cm^{-1}; ^{1}H NMR (DMSO-d_6) δ 10.07 (s, 1H), 8.55–8.53 (dd, 1H, J = 1.5, 4.2), 8.10–8.07 (dd, 1H, J = 1.4,

8.3), 7.88 (t, 1H, J = 5.4), 7.61 (d, 2H, J = 8.9), 7.45–7.41 (m, 1H), 7.32 (d, 2H, J = 8.9), 6.48 (d, 1H, J = 2.3), 6.26 (d, 1H, J = 2.3), 6.13 (bs, 1H), 3.81 (s, 3H), 3.63–3.57 (m, 1H), 3.09–3.03 (m, 2H), 2.55 (t, 2H, J = 7.4), 2.39 (t, 2H, J = 7.0), 1.70–1.59, 1.58–1.43 (2m, 4H), 1.19 (d, 3H, J = 6.3); ^{13}C NMR (DMSO-d_6) δ 170.92, 170.64, 159.01, 144.55, 144.18, 138.28, 134.89, 134.42, 129.59, 128.51, 126.33, 122.09, 120.37, 96.19, 91.63, 54.98, 47.02, 38.47, 33.37, 31.73, 30.23, 25.98, 20.15 (11); MS/MS m/z 469.3 (M+1)$^+$; Anal. Calcd. for $C_{25}H_{29}ClN_4O_3$: C, 64.03; H, 6.23; N, 11.95. Found: C, 64.35; H, 6.57; N, 12.13.

*N′-{4-[(6-Methoxyquinolin-8-yl)amino]pentyl}-N-[3-(trifluoromethyl)phenyl]butanediamide (**5e**).* From the reaction of 0.097 g acid **3b** and 0.048 g (0.297 mmol) 3-trifluoromethylaniline and after purification by column chromatography (mobile phase dichloromethane/methanol 9.5:0.5) and crystallization from ether, 0.122 g (90%) of **5e** was obtained; mp 146–149 °C; IR (ATR): ν_{max} 3394, 3302, 3263, 3216, 3164, 3097, 2963, 2928, 2859, 1651, 1617, 1562, 1518, 1450, 1389, 1331, 1265, 1173, 1127, 1064, 894, 816 cm^{-1}; ^{1}H NMR (DMSO-d_6) δ 10.26 (s, 1H), 8.54–8.53 (dd, 1H, J = 1.6, 4.2), 8.11 (s, 1H), 8.08–8.06 (dd, 1H, J = 1.6, 8.3), 7.88 (t, 1H, J = 5.5), 7.74 (d, 1H, J = 8.1), 7.52 (t, 1H, J = 8.0), 7.43–7.41 (m, 1H), 7.36 (d, 1H, J = 7.8), 6.47 (d, 1H, J = 2.5), 6.25 (d, 1H, J = 2.5), 6.11 (d, 1H, J = 8.7), 3.82 (s, 3H), 3.63–3.58 (m, 1H), 3.10–3.04 (m, 2H), 2.57 (t, 2H, J = 7.2), 2.41 (t, 2H, J = 7.2), 1.67–1.63, 1.56–1.44 (2m, 4H), 1.19 (d, 3H, J = 6.3); ^{13}C NMR (DMSO-d_6) δ 171.11, 170.84, 158.98, 144.61, 144.19, 140.03, 134.75, 134.50, 129.83, 129.55, 129.67–129.04 (q, J = 31.9), 126.81–121.40 (q, J = 269.8), 122.33, 122.05, 119.14, 114.88, 96.06, 91.58, 54.94, 46.98, 38.46, 33.38, 31.72, 30.11, 25.96, 20.14; MS/MS m/z 503.3 (M+1)$^+$; Anal. Calcd. for $C_{26}H_{29}F_3N_4O_3$: C, 62.14; H, 5.82; N, 11.15. Found: C, 62.25; H, 5.99; N, 11.08.

*N′-{4-[(6-Methoxyquinolin-8-yl)amino]pentyl}-N-[4-(trifluoromethyl)phenyl]butanediamide (**5f**).* From the reaction of 0.097 g acid **3b** and 0.048 g (0.297 mmol) 4-trifluoromethylaniline and after purification by column chromatography (mobile phase dichloromethane/methanol 9.5:0.5), 0.068 g (50%) of **5f** was obtained; mp 163–165 °C; IR (ATR): ν_{max} 3387, 3287, 3256, 3198, 3123, 3071, 2963, 2925, 2859, 1652, 1614, 1547, 1515, 1452, 1419, 1389, 1327, 1264, 1169, 1124, 1064, 846, 784 cm^{-1}; ^{1}H NMR (DMSO-d_6) δ 10.31 (s, 1H), 8.54–8.52 (dd, 1H, J = 1.5, 4.2), 8.09–8.06 (dd, 1H, J = 1.5, 8.3), 7.89 (t, 1H, J = 5.3), 7.79 (d, 2H, J = 8.5), 7.63 (d, 2H, J = 8.7), 7.44–7.40 (m, 1H), 6.47 (d, 1H, J = 2.4), 6.25 (d, 1H, J = 2.4), 6.11 (d, 1H, J = 8.7), 3.82 (s, 3H), 3.65–3.57 (m, 1H), 3.09–3.04 (m, 2H), 2.59 (t, 2H, J = 7.0), 2.40 (t, 2H, J = 7.0), 1.71–1.58, 1.57–1.43 (2m, 4H), 1.19 (d, 3H, J = 6.2); ^{13}C NMR (DMSO-d_6) δ 171.18, 170.86, 158.99, 144.61, 144.20, 142.83, 134.77, 134.51, 129.55, 125.95, 127.05–121.68 (q, J = 274.2), 123.15-122.52 (q, J = 28.2), 122.06, 118.69, 96.08, 91.59, 54.95, 46.99, 38.47, 33.38, 31.79, 30.10, 25.96, 20.14; MS/MS m/z 503.3 (M+1)$^+$; Anal. Calcd. for $C_{26}H_{29}F_3N_4O_3$: C, 62.14; H, 5.82; N, 11.15. Found: C, 62.39; H, 5.76; N, 11.41.

3.2. Biological Evaluation

3.2.1. In Vitro Antibacterial Susceptibility Assay (MIC Determination)

(a) *Staphylococci* and *Enterococci*: The synthesized compounds were evaluated for in vitro antibacterial activity against representatives of multidrug-resistant bacteria and clinical isolates of methicillin-resistant *Staphylococcus aureus* (MRSA) 63718, SA 630 and SA 3202, that were obtained from the National Institute of Public Health (Prague, Czech Republic). *S. aureus* ATCC 29213 was used as a reference and quality control strain. In addition, all the compounds were tested for their activity against vancomycin-susceptible *Enterococcus faecalis* ATCC 29212 as a reference strain and three isolates from American crows of vanA-carrying vancomycin-resistant *E. faecalis* (VRE) 342B, 368 and 725B [34]. Ampicillin (Amp) and ciprofloxacin (CIP) (Sigma-Aldrich, St. Louis, MO, USA) were used as standards. Prior to testing, each strain was passaged onto nutrient agar (Oxoid, Basingstoke, UK) with 5% of bovine blood, and bacterial inocula were prepared by suspending a small portion of the bacterial colony in sterile phosphate-buffered saline (pH 7.2–7.3). The cell density was adjusted to 0.5 McFarland units using a densitometer (Densi-La-Meter, LIAP, Riga, Latvia). This inoculum was diluted to reach the final concentration of bacterial cells 5 × 10^5 CFU/mL in the wells. The compounds were dissolved in DMSO (Sigma-Aldrich, St. Louis, MO, USA), and the final concentration of DMSO in the Cation Adjusted

Mueller-Hinton (CaMH) broth (Oxoid) for *Staphylococci* or brain-heart infusion for *Enterococci* did not exceed 2.5% of the total solution composition. The final concentrations of the evaluated compounds ranged from 256 to 0.008 μg/mL. The broth dilution micro-method, modified according to NCCLS (National Committee for Clinical Laboratory Standards) guidelines [35] in MH broth for *Staphylococcus* strains and CaMH for *Enterococcus* strains, was used to determine MIC. Drug-free controls, sterility controls, and controls consisting of MH and CaMH broths and DMSO alone were included. The determination of results was performed visually after 24 h of static incubation in darkness at 37 °C in an aerobic atmosphere.

(b) Other Gram-positive (*Streptococcus pneumoniae* MFBF 10373, *Bacillus cereus* ATCC 11778, and *Bacillus subtilis* ATCC 6633) and Gram-negative bacteria species (*Pseudomonas aeruginosa* ATCC 27853, *Escherichia coli* ATCC 10536, *Serratia marcescens* ATCC 10905, *Proteus mirabilis* MFBF 10430, *Acinetobacter baumannii* MFBF 10913, and *Salmonella enteritidis* MFBF 11945): Serial microdilution broth assay was used to determine MIC of compounds **4a–f** and **5a–f** [36]. Cell suspensions were prepared from stock cultures using phosphate-buffered saline (PBS) (Gibco Laboratories, USA) pH 7.4 and adjusted to 0.5 McFarland units using a nephelometer (ATB 1550, BioMérieux, France). Stock solutions of the tested compounds were prepared in DMSO (10 mg/mL). Testing was performed using serial dilution in microtiter flat bottom 96 well plates with compounds ranging from 100 to 0.78125 μg/mL. After inoculation (10^7 CFU/mL) and incubation (18 h, 35 °C, aerobically in darkness), MICs for bacterial species were determined by addition of 0.5 mg/mL TTC (2,3,5-triphenyl-2*H*-tetrazolium chloride). The absorbance was recorded at 540 nm. As a positive quality control (susceptibility of strains) tetracycline hydrochloride (TC) (Sigma-Aldrich) was added into wells. MICs were determined as non-linear regression using GraphPad Prism as the lowest concentrations resulting in 50% growth inhibition of growth in comparison to control.

(c) *Mycobacteria: Mycobacterium tuberculosis* H37Ra ATCC 25177 was grown in Middlebrook broth (MB), supplemented with Oleic-Albumin-Dextrose-Catalase (OADC) supplement (Difco, Lawrence, KS, USA) and mycobactin J (2 μg/mL). At log phase growth, a culture sample (10 mL) was centrifuged at 15,000 rpm/20 min using a bench-top centrifuge (MPW-65R, MPW Med Instruments, Poland). Following removal of the supernatant, the pellet was washed in fresh Middlebrook 7H9GC broth and re-suspended in fresh, ODAC-supplemented MB (10 mL). The turbidity was adjusted to match McFarland standard No. 1 (3×10^8 CFU) with MB broth. A further 1:20 dilution of the culture was then performed in MB broth. The antimicrobial susceptibility of *M. tuberculosis* was investigated in a 96-well plate format. In these experiments, sterile deionised water (300 μl) was added to all outer-perimeter wells of the plates to minimize evaporation of the medium in the test wells during incubation. Each evaluated compound (100 μL) was incubated with *M. tuberculosis* (100 μl). Dilutions of each compound were prepared in duplicate. For all synthesized compounds, final concentrations ranged from 1000 μg/mL to 8 μg/mL. All compounds were dissolved in DMSO, and subsequent dilutions were made in supplemented MB. The plates were sealed with parafilm and incubated at 37 °C for 7 days. Following incubation, 10% of alamarBlue (Difco, Lawrence, KS, USA) was mixed into each well, and readings at 570 nm and 600 nm were taken, initially for background subtraction and subsequently after 24 h reincubation. The background subtraction is necessary for strongly coloured compounds, where the colour may interfere with the interpretation of any colour change. For non-interfering compounds, a blue colour in the well was interpreted as the absence of growth, and a pink colour was scored as growth.

The evaluation of the in vitro antimycobacterial activity of the compounds was additionally performed against *M. smegmatis* ATCC 700084, *M. marinum* CAMP 5644 and *M. kansasii* DSM 44162. The broth dilution micro-method in Middlebrook 7H9 medium (Difco, Lawrence, KS, USA) supplemented with ADC Enrichment (Becton, Dickinson and Comp.) was used to determine MIC as previously described [37]. The compounds were dissolved in DMSO (Sigma-Aldrich), and the final concentration of DMSO did not exceed 2.5% of the total solution composition. Final concentrations of the evaluated compounds ranging from 256 μg/mL to 0.125 μg/mL were obtained by twofold serial dilution of the stock solution in a microtiter plate with sterile medium. Bacterial inocula were prepared by transferring colonies from culture to sterile water. The cell density was adjusted to 0.5 McFarland units using a densitometer (Densi-La-Meter, LIAP, Riga, Latvia). The final inoculum was made by 1:1000 dilution of the suspension with sterile water. Drug-free controls, sterility controls and controls consisted of medium and DMSO (Sigma-Aldrich, St. Louis, MO, USA) alone were included. The determination of results was performed visually after 3 days of static incubation in darkness at 37 °C in an aerobic atmosphere for *M. smegmatis*, after 7 days of static incubation in darkness at 37 °C in an aerobic atmosphere for *M. kansasii* and after 21 days of static incubation in darkness at 28 °C in an aerobic atmosphere for *M. marinum*. Ciprofloxacin (CIP), rifampicin (RIF) and isoniazid (INH) (Sigma-Aldrich, St. Louis, MO, USA) were used as the standards.

3.2.2. In Vitro Antifungal Susceptibility Testing

Microdilution method was used for testing the antifungal activities of newly synthesized compounds against *Candida albicans* CCM 8261 and ATCC 90028, *C. krusei* CCM 8271, *C. parapsilosis* CCM 8260 (Czech Collection of Microorganisms, Brno, Czech Republic) [38]. Tested compounds were diluted in RPMI-1640 (Sigma) broth to concentrations 128–0.016 μg/mL. Flucytosine (FLU) (Sigma) and amphotericin B (Amph) were used as positive controls. The plates were inoculated by an inoculum prepared in RPMI-1640 broth. The final concentration of fungal cells was 5×10^2–2.5×10^3 CFU/mL in each well. The plates were incubated at 37 °C for 24 (*C. albicans*, *C. krusei*) or 48 (*C. parapsilosis*) hours. Drug-free controls were included. MIC was defined as 80% or greater (IC_{80}) reduction of growth in comparison with the control [35].

MIC determination for *Aspergillus brasiliensis* was performed in RPMI 1640 broth with glutamine supplemented with 2% glucose, following the same scheme as for bacteria, procedure b). After incubation period (48 h, 35 °C, aerobically in dark), XTT (2*H*-tetrazolium, 2,3-bis(2-methoxy-4-nitro-5-sulfophenyl)-5-[(phenylamino)carbonyl]-hydroxide) (10 mg/mL) in combination with menadione (1 mg/mL in acetone) (7:1, *v*/*v*) was added [36]. Absorbance was read at 540 nm. Amph was used as a positive control and solvent and media (no microorganisms added) as negative controls.

3.2.3. Minimum Biofilm Eradication Assay

Biofilm eradication screening was performed on the following microorganisms: *S. aureus* ATCC 6538, *S. pneumoniae* MFBF 10373, *E. faecalis* ATCC 29212, *B. cereus* ATCC 11778, *B. subtilis* ATCC 6633, *E. coli* ATCC 10536, *P. aeruginosa* ATCC 27853, *S. marcescens* ATCC 10905, *P. mirabilis* MFBF 10430, *S. enteritidis* MFBF 11945, *A. baumannii* MFBF 10913, and *C. albicans* ATCC 90028. MBECs of fumardiamides **4a**–**f** and succindiamides **5a**–**f** were determined as follows [39]. Each well (96-well plate) was filled with 100 μl of bacterial (10^7 CFU/mL) or yeast (5×10^6 CFU/mL) suspension. When the inhibition of yeast biofilm formation was tested, the wells were pre-treated with fetal bovine serum (FBS) (250 μl per well). Negative controls contained broth only. Positive controls were performed using standard antimicrobial drugs gentamycin (Gen) and Amph, respectively. The plates were covered and incubated aerobically for 24 h (bacteria) or 48 h (yeast) at 37 °C. Following incubation period, each well was aspirated, washed three times and vigorously shaken to remove all non-adherent bacteria/yeast. The remaining attached cells were fixed with methanol (15 min) and the plates were left to dry overnight. Formed biofilm was stained with crystal violet (1%, 5 min). Excess stain was rinsed by placing the plate under running tap water and the plates were left to dry. Adherent cells were solubilized using ethanol. The absorbance was read at 570 nm. The MBEC value represents the lowest dilution of a compound at which bacteria fail to grow.

3.2.4. Antiviral Evaluation

Antiviral activity of compounds **4a**–**f** and **5a**–**f** was determined as described previously [40]. Cytotoxicity and antiviral activity assay towards herpes simplex virus (HSV) strains HSV-1 KOS, HSV-2 G, HSV-1 TK-KOS ACVr, vaccinia virus, adenovirus-2, human coronavirus (229E) in HEL cell cultures, vesicular stomatitis virus, coxsackie virus B4, respiratory syncytial virus in Hela cell cultures, para-influenza-3 virus, reovirus-1, sindbis, coxsackie virus B4, Punta Toro virus, yellow fever virus in Vero cell cultures, while influenza A/H1N1 A/Ned/378/05, influenza A/H3N2 A/HK/7/87 and influenza B B/Ned/537/05 viruses in Madin-Darby Canine Kidney (MDCK) cell cultures were performed. On the day of the infection, growth medium was aspirated and replaced by serial dilutions of the test compounds. Virus was then added to each well, diluted to obtain a viral input of 100 $CCID_{50}$ ($CCID_{50}$ being the virus dose that is able to infect 50% of the cell cultures). Mock-treated cell cultures receiving solely the test compounds were included, to determine their cellular cytotoxicity. After 3 to 10 days of incubation the virus-induced cytopathicity was determined by visual scoring of the cytopathic effect (CPE) (light microscopic evaluation of the virus-induced CPE and inhibition of evaluated compounds), as well as by measuring the cell viability with the colorimetric formazan-based MTS assay (3-(4,5-dimethylthiazol-2-yl)-5-(3-carboxymethoxyphenyl)-2-(4-sulfophenyl)-2*H*-tetrazolium). All experiments were performed in duplicate. Antiviral activity was expressed as EC_{50}. The activities were compared with the activities of the parent drug PQ, DS-10.000 (dextran sulfate, approx. MW = 10.000) and standard antiviral drugs: ribavirin (Rib), mycophenolic acid (MPA), brivudin, cidofovir, acyclovir, gancyclovir, zalcitabine, alovudine, UDA, zanamivir, amantadine and rimantadine.

3.2.5. Cytostatic Activity

Cytostatic activity was evaluated in vitro on nine different types of human tumor cell lines: Capan-1, Hap1, HCT-116, and NCI-H460, as well as hematological tumors such as DND-41, HL-60, K-562, MM.1S and Z-138 as described previously [41]. All human tumor cell lines were acquired from the American Type Culture Collection (ATCC, Manassas, VA, USA), except for the DND-41 cell line which was purchased from the Deutsche Sammlung von Mikroorganismen und Zellkulturen (DSMZ Leibniz-Institut, Braunschweig, Germany) and the Hap1 cell line which was purchased from Horizon Discovery (Waterbeach, UK). Capan-1, Hap1, HL-60, K-562, Z-138, MM.1S and DND-41 were cultured in Iscove's Modified Dulbecco's Medium (IMDM, Gibco Life Technologies, Gaithersburg, MD, USA), HCT-116 were grown in McCoy's 5A Medium (Gibco Life Technologies) and NCI-H460 were cultured in RPMI (Gibco Life Technologies). All media were supplemented with 10% FBS (HyClone, GE Healthcare Life Sciences, USA). Adherent cell lines HCT-116, Hap1, NCI-H460, and Capan-1 cells were seeded at a density between 400 and 1250 cells per well, in 384-well, black-walled, clear-bottomed tissue culture plates (Greiner Bio-One, Kremsmünster, Germany). After overnight incubation, cells were treated with test compounds at four different concentrations ranging from 100 to 0.8 μM. Suspension cell lines HL-60, K-562, Z-138, MM.1S, and DND-41 were seeded at densities ranging from 3000 to 10,000 cells per well in 384-well, black-walled, clear-bottomed tissue culture plates containing test compounds at the same four concentrations. The plates were incubated and monitored at 37 °C for 72 h in an IncuCyte (Essen BioScience Inc., Ann Arbor, MI, USA) for real-time imaging. Images were taken every 3 h, with one field image per well under 10× magnification. All compounds were tested with duplicate data points and averaged. The activities were compared with the activities of the parent drug PQ and standard anticancer drugs docetaxel (DXT), etoposide (EPEG) and staurosporine (STS).

3.2.6. Interaction with Glutathione (GSH)

Fumardiamide **4b** (1.25 μM) was incubated with GSH (125 μM) in ammonium formate buffer (pH = 7.4) containing 10% acetonitrile at 37 °C for 216 h [33]. The progress of the reactions was monitored with the percent of remaining fumardiamide determined by MS using an internal standard (*N*-(benzyloxy)-*N*′-{4-[(6-methoxyquinolin-8-yl)amino]pentyl}butanediamide). Aliquots of the reaction mixture (taken after 0, 4.5, 26, 52, 124 and 216 h) were analysed with Synapt G2-Si ESI-QTOF-MS system (Waters, Milford, USA). The aliquots were diluted 10 times with acetonitrile and sprayed at a flow rate of 50 μL/min using the fluidics system of the instrument. MS conditions were set as follows: positive ion mode, capillary voltage 3 kV, sampling cone voltage 10 V, source temperature

120 °C, desolvation temperature 350 °C, desolvation gas flow 800 L/h. Mass spectra were recorded from 100–1000 *m/z* at a frequency of 1 Hz. Data were acquired and analysed with Waters MassLynx v4.1 software.

4. Conclusions

Twelve novel PQ-derivatives of diamide type were designed and synthesized. These compounds differ in the type of spacer and/or halogen atom in aniline region. Compounds **4a–f** are fumardiamides and **5a–f** succindiamides. Compounds **4a,b**, **5a,b** are fluoro, **4c,d**, **5c,d** chloro and **4e,f**, **5e,f** are trifluoroderivatives. All new compounds were screened for antibacterial, antitubercular, antiviral and cytostatic activity as well as biofilm eradication ability. In all biological assays, fumardiamides **4** were superior to succindiamides **5**, which indicates that the double bond conjugated to the carbonyl was important for the activity. With their high bioactivity, low cytotoxicity and convenient drug-like properties, *p*-substituted derivatives **4b,d,f** provide a strong basis for further research and optimization of novel agents useful in the treatment of bacterial and biofilm-associated infections, while *m*-substituted derivatives **4a,c,e** could be potential leads for the development of antitumor agents.

Author Contributions: Conceptualization, Z.R. and B.Z.; Synthesis and characterization of compounds M.B., Z.R.; Antibacterial, antifungal and biofilm eradication assay, H.M., J.V., I.K., J.J.; Antiviral and cytostatic evaluaton, L.P., D.S.; Interaction with GSH T.K.; Writing of Manuscript, Z.R. and B.Z.

References

1. Clayden, J.; Greeves, N.; Warren, S. *Organic Chemistry*, 2nd ed.; Oxford University Press: New York, NY, USA, 2012.
2. Smaill, J.B.; Rewcastle, G.W.; Loo, J.A.; Greis, K.D.; Chan, O.H.; Reyner, E.L.; Lipka, E.; Showalter, H.D.; Vincent, P.W.; Elliott, W.L.; et al. Tyrosine kinase inhibitors. 17. Irreversible inhibitors of the epidermal growth factor receptor: 4-(phenylamino)quinazoline- and 4-(phenylamino)pyrido[3,2-*d*]pyrimidine-6-acrylamides bearing additional solubilizing functions. *J. Med. Chem.* **2000**, *43*, 1380–1397. [CrossRef] [PubMed]
3. Minami, Y.; Shimamura, T.; Shah, K.; LaFramboise, T.; Glatt, K.A.; Liniker, E.; Borgman, C.L.; Haringsma, H.J.; Feng, W.; Weir, B.A.; et al. The major lung cancer-derived mutants of ERBB2 are oncogenic and are associated with sensitivity to the irreversible EGFR/ERBB2 inhibitor HKI-272. *Oncogene* **2007**, *26*, 5023–5027. [CrossRef] [PubMed]
4. Baselga, J.; Coleman, R.E.; Cortés, J.; Janni, W. Advances in the management of HER2-positive early breast cancer. *Crit. Rev. Oncol. Hematol.* **2017**, *119*, 113–122. [CrossRef] [PubMed]
5. Jackson, P.A.; Widen, J.C.; Harki, D.A.; Brummond, K.M. Covalent modifiers: A chemical perspective on the reactivity of α,β-unsaturated carbonyls with thiols via hetero-Michael addition reactions. *J. Med. Chem.* **2017**, *60*, 839–885. [CrossRef] [PubMed]
6. Zhang, X.; Li, X.; You, Q.; Zhang, X. Prodrug strategy for cancer cell-specific targeting: A recent overview. *Eur. J. Med. Chem.* **2017**, *139*, 542–563. [CrossRef] [PubMed]
7. Compound Summary for CID 5281081. Available online: https://pubchem.ncbi.nlm.nih.gov/compound/entacapone#section=Top (accessed on 25 May 2018).
8. Matthews, D.A.; Dragovich, P.S.; Webber, S.E.; Fuhrman, S.A.; Patick, A.K.; Zalman, L.S.; Hendrickson, T.F.; Love, R.A.; Prins, T.J.; Marakovits, J.T.; et al. Structure-assisted design of mechanism-based irreversible inhibitors of human rhinovirus 3C protease with potent antiviral activity against multiple rhinovirus serotypes. *Proc. Natl. Acad. Sci. USA* **1999**, *96*, 11000–11007. [CrossRef] [PubMed]
9. Buzdar, A.U.; Robertson, J.F.; Eiermann, W.; Nabholtz, J.M. An overview of the pharmacology and pharmacokinetics of the newer generation aromatase inhibitors anastrozole, letrozole, and exemestane. *Cancer* **2002**, *95*, 2006–2016. [CrossRef] [PubMed]
10. Somberg, J.C.; Molnar, J. The pleiotropic effects of ethacrynic acid. *Am. J. Ther.* **2009**, *16*, 102–104. [CrossRef] [PubMed]
11. Dawson, R.M. The toxicology of microcystins. *Toxicon* **1998**, *36*, 953–962. [CrossRef]

12. Eusugi, S.; Fujisawa, N.; Yoshida, J.; Watanabe, M.; Dan, S.; Yamori, T.; Shiono, Y.; Kimura, K.; Pyrrocidine, A. A metabolite of endophytic fungi, has a potent apoptosis-inducing activity against HL60 cells through caspase activation via the Michael addition. *J. Antibiot.* **2016**, *69*, 133–140. [CrossRef]
13. Ramsay, J.R.; Suhrbier, A.; Aylward, J.H.; Ogbourne, S.; Cozzi, S.J.; Poulsen, M.G.; Baumann, K.C.; Welburn, P.; Redlich, G.L.; Parsons, P.G. The sap from *Euphorbia peplus* is effective against human nonmelanoma skin cancers. *Br. J. Dermatol.* **2011**, *164*, 633–636. [CrossRef] [PubMed]
14. Lebwohl, M.; Swanson, N.; Anderson, L.L.; Melgaard, A.; Xu, Z.; Berman, B. Ingenol mebutate gel for actinic keratosis. *N. Engl. J. Med.* **2012**, *366*, 1010–1019. [CrossRef] [PubMed]
15. Picato®gel–FDA. Available online: https://www.accessdata.fda.gov/drugsatfda_docs/label/2012/202833lbl.pdf. (accessed on 9 June 2018).
16. Sebök, B.; Bonnekoh, B.; Geisel, J.; Mahrle, G. Antiproliferative and cytotoxic profiles of antipsoriatic fumaric acid derivatives in keratinocyte cultures. *Eur. J. Pharmacol.* **1994**, *270*, 79–87. [CrossRef]
17. Smith, D. Fumaric acid esters for psoriasis: A systematic review. *Ir. J. Med. Sci.* **2017**, *186*, 161–177. [CrossRef] [PubMed]
18. Kocaadam, B.; Şanlier, N. Curcumin, an active component of turmeric (*Curcuma longa*), and its effects on health. *Crit. Ver. Food Sci. Nutr.* **2017**, *57*, 2889–2895. [CrossRef] [PubMed]
19. Seca, A.M.L.; Pinto, D.C.G.A. Plant secondary metabolites as anticancer agents: Successes in clinical trials and therapeutic application. *Int. J. Mol. Sci.* **2018**, *19*, 263. [CrossRef] [PubMed]
20. Pavić, K.; Perković, I.; Cindrić, M.; Pranjić, M.; Martin-Kleiner, I.; Kralj, M.; Schols, D.; Hadjipavlou-Litina, D.; Katsori, A.-M.; Zorc, B. Novel semicarbazides and ureas of primaquine with bulky aryl or hydroxyalkyl substituents: Synthesis, cytostatic and antioxidative activity. *Eur. J. Med. Chem.* **2014**, *86*, 502–514. [CrossRef] [PubMed]
21. Perković, I.; Antunović, M.; Marijanović, I.; Pavić, K.; Ester, K.; Kralj, M.; Vlainić, J.; Kosalec, I.; Schols, D.; Hadjipavlou-Litina, D.; et al. Novel urea and bis-urea primaquine derivatives with hydroxyphenyl and halogenphenyl substituents: Synthesis and biological evaluation. *Eur. J. Med. Chem.* **2016**, *124*, 622–636. [CrossRef] [PubMed]
22. Guzman, J.D. Natural cinnamic acids, synthetic derivatives and hybrids with antimicrobial activity. *Molecules* **2014**, *19*, 292–349. [CrossRef] [PubMed]
23. Kakwani, M.D.; Suryavanshi, P.; Ray, M.; Rajan, M.G.R.; Majee, S.; Samad, A.; Devarajan, P.; Degani, M.S. Design, synthesis and antimycobacterial activity of cinnamide derivatives: A molecular hybridization approach. *Bioorg. Med. Chem. Lett.* **2011**, *21*, 1997–1999. [CrossRef] [PubMed]
24. De, P.; Baltas, M.; Bedos-Belval, F. Cinnamic acid derivatives as anticancer agents-a review. *Curr. Med. Chem.* **2011**, *18*, 1672–1703. [CrossRef] [PubMed]
25. Pavić, K.; Perković, I.; Gilja, P.; Kozlina, F.; Ester, K.; Kralj, M.; Schols, D.; Hadjipavlou-Litina, D.; Pontiki, E.; Zorc, B. Design, synthesis and biological evaluation of novel primaquine-cinnamic acid conjugates of amide and acylsemicarbazide type. *Molecules* **2016**, *21*, 1629. [CrossRef] [PubMed]
26. Pavić, K.; Perković, I.; Pospíšilová, Š.; Machado, M.; Fontinha, D.; Prudêncio, M.; Jampilek, J.; Coffey, A.; Endersen, L.; Rimac, H.; et al. Primaquine hybrids as promising antimycobacterial and antimalarial agents. *Eur. J. Med. Chem.* **2018**, *143*, 769–779. [CrossRef] [PubMed]
27. Vlainić, J.; Kosalec, I.; Pavić, K.; Hadjipavlou-Litina, D.; Pontiki, E.; Zorc, B. Insights into biological activity of ureidoamides with primaquine and amino acid moieties. *J. Enzyme Inhib. Med. Chem.* **2018**, *33*, 376–382. [CrossRef] [PubMed]
28. Levatić, J.; Pavić, K.; Perković, I.; Uzelac, L.; Ester, K.; Kralj, M.; Kaiser, M.; Rottmann, M.; Supek, F.; Zorc, B. Machine learning prioritizes synthesis of primaquine ureidoamides with high antimalarial activity and attenuated cytotoxicity. *Eur. J. Med. Chem.* **2018**, *146*, 651–667. [CrossRef] [PubMed]
29. Beus, M.; Rajić, Z.; Maysinger, D.; Mlinarić, Z.; Antunović, M.; Marijanović, I.; Fontinha, D.; Prudêncio, M.; Held, J. SAHA-primaquine hybrids (sahaquines) as potential anticancer and antimalarial compounds. *Chem. Open*, submitted.
30. Chemicalize, 2017, ChemAxon Ltd. Available online: http://www.chemicalize.org (accessed on 5 April 2018).
31. Hufnagel, D.A.; Price, J.E.; Stephenson, R.E.; Kelley, J.; Benoit, M.F.; Chapman, M.R. Thiol starvation induces redox-mediated dysregulation of *Escherichia coli* biofilm components. *J. Bacteriol.* **2018**, *200*, e00389-17. [CrossRef] [PubMed]

32. Lee, S.F.; Davey, L. Disulfide bonds: A key modification in bacterial extracytoplasmic proteins. *J. Dent. Res.* **2017**, *96*, 1465–1473. [CrossRef] [PubMed]
33. Flanagan, M.E.; Abramite, J.A.; Anderson, D.P.; Aulabaugh, A.; Dahal, U.P.; Gilbert, A.M.; Li, C.; Montgomery, J.; Oppenheimer, S.R.; Ryder, T.; et al. Chemical and computational methods for the characterization of covalent reactive groups for the prospective design of irreversible inhibitors. *J. Med. Chem.* **2014**, *57*, 10072–10079. [CrossRef] [PubMed]
34. Oravcova, V.; Zurek, L.; Townsend, A.; Clark, A.B.; Ellis, J.C.; Cizek, A. American crows as carriers of vancomycin-resistant enterococci with vanA gene. *Environ. Microbiol.* **2014**, *16*, 939–949. [CrossRef] [PubMed]
35. Clinical and Laboratory Standards Institute. *Performance Standards for Antimicrobial Susceptibility Testing*; The 8th informational supplement document; CLSI: Wayne, PA, USA, 2012; M100-S22.
36. EUCAST, European Committee on Antimicrobial Susceptibility Testing. Determination of minimum inhibitory concentrations (MICs) of antibacterial agents by broth micro dilution. EUCAST Discussion Document. *Clin. Microbiol. Infect.* **2013**, *9*, 1–10.
37. Schwalbe, R.; Steele-Moore, L.; Goodwin, A.C. (Eds.) *Antimicrobial Susceptibility Testing Protocols*; CRC Press: Boca Raton, FL, USA, 2007.
38. Sheehan, D.J.; Espinel-Ingroff, A.; Steele, M.; Webb, C.D. Antifungal susceptibility testing of yeasts: A brief overview. *Clin. Infect. Dis.* **1993**, *17*, 494–500. [CrossRef]
39. Stepanović, S.; Vuković, D.; Hola, V.; Di Bonaventura, G.; Djukić, S.; Cirković, I.; Ruzicka, F. Quantification of biofilm in microtiter plates: Overview of testing conditions and practical recommendations for assessment of biofilm production by *Staphylococci*. *APMIS* **2007**, *115*, 891–899. [CrossRef] [PubMed]
40. Tzioumaki, N.; Manta, S.; Tsoukala, E.; Vande Voorde, J.; Liekens, S.; Komiotis, D.; Balzarini, J. Synthesis and biological evaluation of unsaturated keto and exomethylene D-arabinopyranonucleoside analogs: Novel 5-fluorouracil analogs that target thymidylate synthase. *Eur. J. Med. Chem.* **2011**, *46*, 993–1005. [CrossRef] [PubMed]
41. Li, Q.; Lescrinier, E.; Groaz, E.; Persoons, L.; Daelemans, D.; Herdewijn, P.; De Jonghe, S. Synthesis and biological evaluation of pyrrolo[2,1-f][1,2,4]triazine C-nucleosides with a ribose, 2′-deoxyribose, and 2′,3′-dideoxyribose sugar moiety. *ChemMedChem.* **2018**, *13*, 97–104. [CrossRef] [PubMed]

5

Fast Amide Bond Cleavage Assisted by a Secondary Amino and a Carboxyl Group—A Model for yet Unknown Peptidases?

Igor V. Komarov [1,*], Aleksandr Yu. Ishchenko [2], Aleksandr Hovtvianitsa [2], Viacheslav Stepanenko [2], Serhii Kharchenko [2], Andrew D. Bond [3] and Anthony J. Kirby [3,*]

1 Institute of High Technologies, Taras Shevchenko National University of Kyiv, Vul. Volodymyrska 64/13, 01601 Kyiv, Ukraine
2 Enamine Ltd., Vul. Chervonotkatska 78, 02094 Kyiv, Ukraine; ischa1986@ukr.net (A.Y.I.); cayenestar@gmail.com (A.H.); v.stepanenko@enamine.net (V.S.); serxioharchenko@ukr.net (S.K.)
3 University Chemical Laboratory, Cambridge CB2 1EW, UK; adb29@cam.ac.uk
* Correspondence: ik214@yahoo.com (I.V.K.); ajk1@cam.ac.uk (A.J.K.)

Academic Editor: Michal Szostak

Abstract: Unconstrained amides that undergo fast hydrolysis under mild conditions are valuable sources of information about how amide bonds may be activated in enzymatic transformations. We report a compound possessing an unconstrained amide bond surrounded by an amino and a carboxyl group, each mounted in close proximity on a bicyclic scaffold. Fast amide hydrolysis of this model compound was found to depend on the presence of both the amino and carboxyl functions, and to involve a proton transfer in the rate-limiting step. Possible mechanisms for the hydrolytic cleavage and their relevance to peptide bond cleavage catalyzed by natural enzymes are discussed. Experimental observations suggest that the most probable mechanisms of the model compound hydrolysis might include a twisted amide intermediate and a rate-determining proton transfer.

Keywords: amide hydrolysis; model compound; intramolecular catalysis; twisted amide; protease; intein

1. Introduction

Hydrolysis of the 1-azatricyclo[3.3.1.1^{3,7}]decan-2-one (**1**), which contains an ultimately twisted and *N*-pyramidalized amide bond, was reported to be very fast and accompanied by the formation of the "dimer" (compound **2**) and higher oligomers [1]. The use of a half equivalent of water under certain conditions led almost exclusively to formation of **2** (Scheme 1). Apparently, the product of the initial hydrolysis of **1**—the amino acid **3**—was immediately *N*-acylated by unreacted **1**, which possesses an extremely electrophilic carbonyl group. In contrast to the twisted and therefore highly reactive amide **1**, the dimer **2** has an essentially unconstrained amide bond: yet we found that it reacted with excess of water at a comparable rate, also giving the amino acid **3** (Scheme 1) [1]. Such a fast amide hydrolysis is unusual; therefore, we decided to study the compound **2** in more detail.

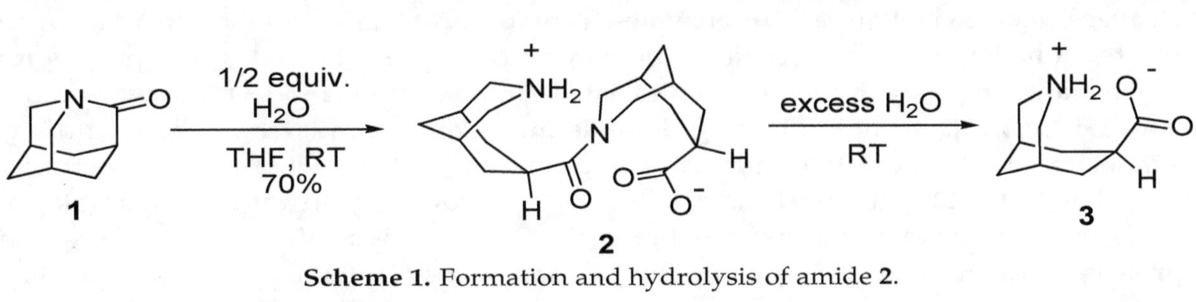

Scheme 1. Formation and hydrolysis of amide **2**.

Examples of amides hydrolysable under mild conditions are rare, but not unknown. Apart from **1** and other similar systems featuring highly deformed amide bonds (see the reviews [2–4]), there are compounds where fast amide bond hydrolysis was suggested to be facilitated by neighboring group assistance (Figure 1 lists some representative examples, [5–13]).

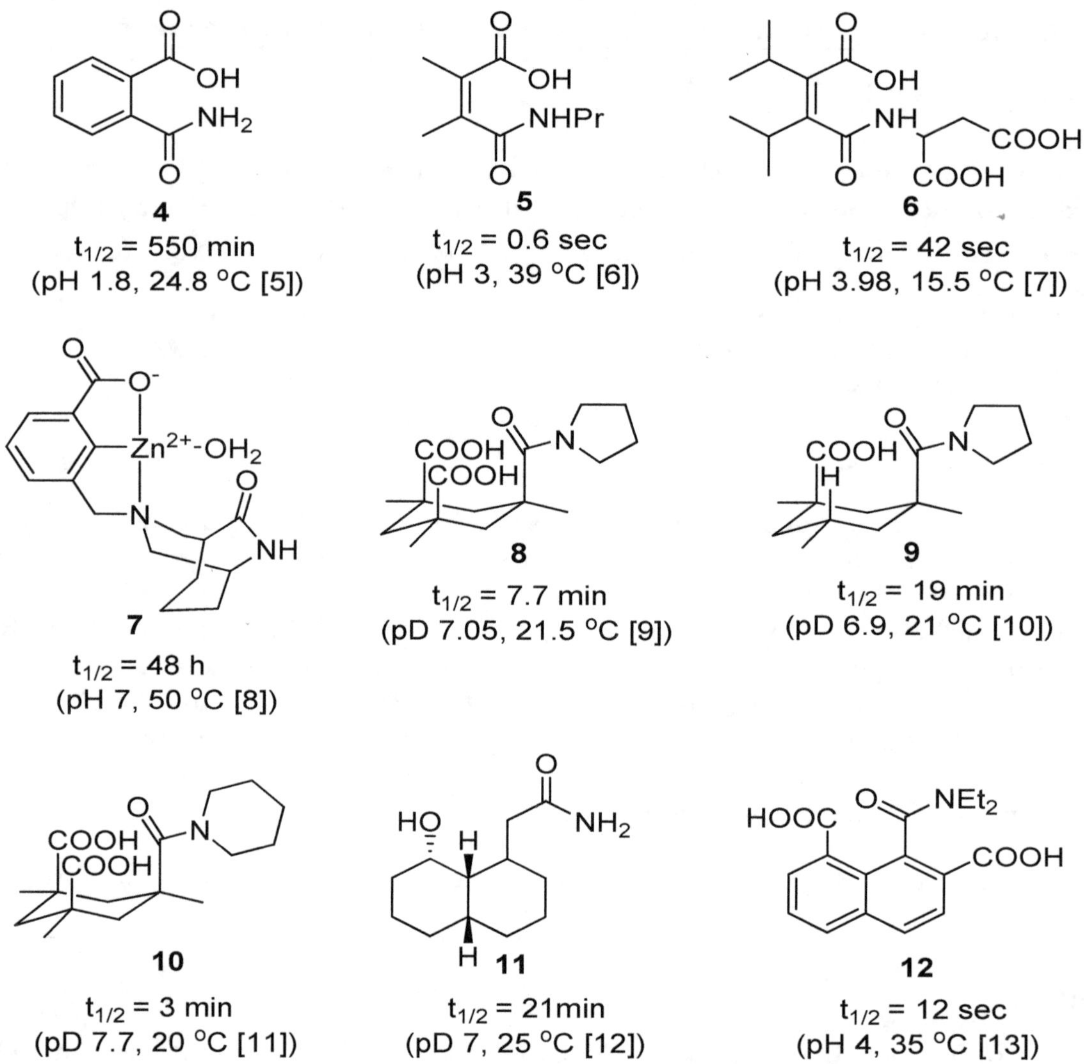

Figure 1. Selected compounds reported to undergo unusually fast amide bond hydrolysis in buffered water under mild conditions (indicated in brackets). Values of the half-life, calculated using the measured pseudo-first order rate constants k_{obs} are listed for each compound.

Compounds like **1** and **4–12** should not be regarded simply as chemical curiosities. In fact, most of them were deliberately designed and synthesized to obtain valuable information on the fundamental properties of the amide bond, and to use this information in unraveling the mechanisms of biologically relevant processes involving amide formation and cleavage: particularly mechanisms of protein cleavage catalyzed by proteolytic enzymes. Perhaps most studied in this regard are molecules mimicking the function of the side-chain carboxyl groups in the active sites of aspartic proteases. The first systematic studies with such molecules, later named enzyme models [14], were undertaken back in the 1950s [5,15]. In the following decades many more similar models were discovered. It was demonstrated that one or two carboxyl groups in close proximity to the amide function (as in compounds **4–6**, **8–10**, **12**) catalyzed the amide bond hydrolysis with rates comparable to those of the corresponding enzymatic transformations. Similar work was also published on models of metalloproteases (exemplified here by compound **7**), and serine/cysteine proteases (e.g., compound

11). Although the validity of the enzyme models and their relevance to the natural enzyme mechanisms has been extensively criticized and re-examined (see, for example, [10,11,16,17], there is no doubt that they made important contributions to our understanding of the natural catalytic systems, and will do so in the future.

Moreover, reactive model amides like compounds **1** and **4–12**, together with natural systems, inspired many chemists to design selective and efficient non-enzymatic transformations of the amide bond, not limited to hydrolysis. For example, twisting around the N−C bond, as a general concept for amide bond activation was ingeniously utilized for the diverse and unprecedented functionalization of amides. Such twisting can be easily achieved either by formation of bridgehead lactams [18], or even by modifying acyclic amides [19] without resorting to polycyclic scaffolds. Notable are the metal-catalyzed N−C cross-coupling reactions of the twisted amides, impossible for unconstrained compounds (see recent reviews on this subject, [2,20–22]. Activation by neighboring groups was also used for developing selective amide group cleavage [23] and amine group protection strategies [24,25]. There has also been significant progress towards selective biomimetic metal-ion promoted amide bond cleavage, mimicking metalloproteases and amidohydrolases [23,26–30]. It should be noted, however, that efficient artificial catalytic (in particular, organocatalytic) amide bond transformations characteristic of proteases presents a challenge so far unachieved [31].

In view of the above, compound **2** might be regarded as a promising model compound, where the observed fast amide bond hydrolysis under mild conditions could provide new information about the intrinsic reactivity of the amide linkage and its possible catalytic cleavage.

One might hypothesize that the observed fast hydrolysis of **2** is facilitated by both carboxyl and secondary amine groups flanking the amide bond. In this paper, we report experimental evidence for this hypothesis, and also discuss the hidden potential of this model system in the study of biochemical transformations and the design of artificial catalytic systems.

2. Results

We first wanted to know if the presence of only one functional group, either the amino group or the carboxylate in the structural context of the amide **2** is sufficient for the fast hydrolysis of the amide bond. The literature data (e.g., on **5** [6], **9** **[10]**, as well as on the compound described in [32]) clearly demonstrated that a single carboxyl group, if properly placed, can catalyze very fast amide bond hydrolysis intramolecularly, presumably through the proton transfer to the amide bond oxygen and subsequent nucleophilic attack at the N-CO bond carbon [6]. Involvement of the basic nitrogen of a neighboring pyrimidine substituent in the hydrolytic cleavage of the amide bond has also been well documented [33].

We synthesized two isostructural analogues of **2**, compounds **13** and **14**, one lacking the secondary amino group (bearing the –CH_2– fragment instead), and the other possessing no carboxyl group (with –CH_2– in place of –CH(COOH)–).

Compound **13** was prepared via the bicyclic carboxylic acid **20**, synthesized by the method first published in [34] and further refined by us [35]. The acid **20** was converted to the corresponding acid chloride and coupled under the Schotten–Baumann reaction conditions with amino acid **3** (Scheme 2).

The key intermediate in our synthesis of **14** was the bicyclic amine **25**, which was obtained following the steps described in [36,37] (Scheme 3). N-Cbz-protected amino acid **21** was prepared from **3** (we observed also formation of the side-product **22** during the protection) and coupled with **25** using a standard peptide coupling protocol. Cleavage of the Cbz-protecting group by hydrogenolysis afforded compound **14**.

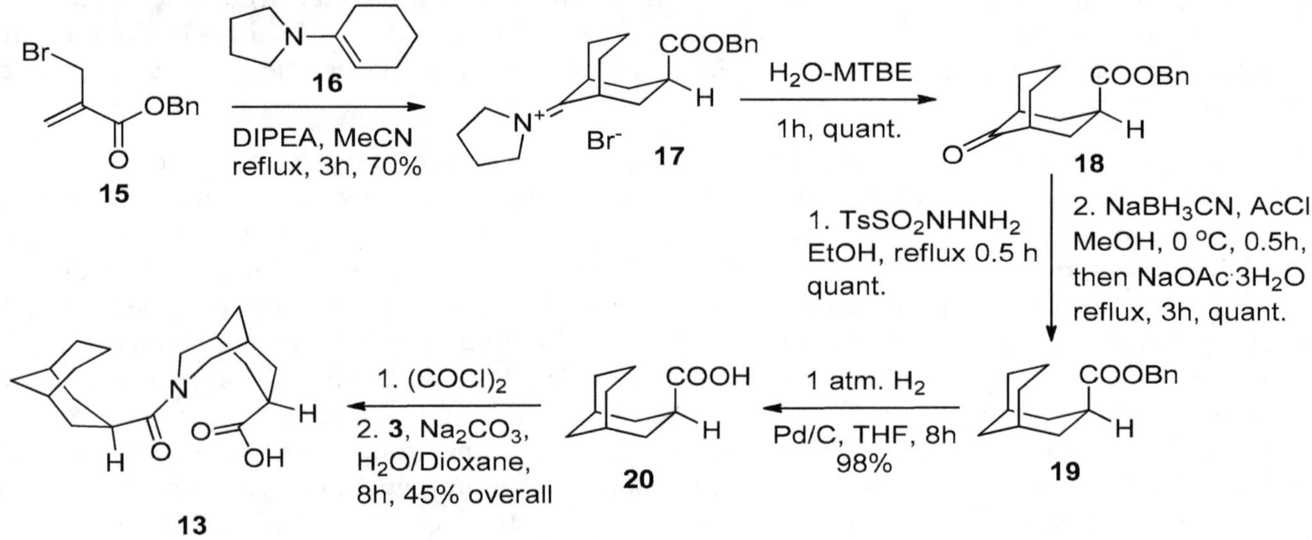

Scheme 2. Synthesis of model compound lacking the amino group.

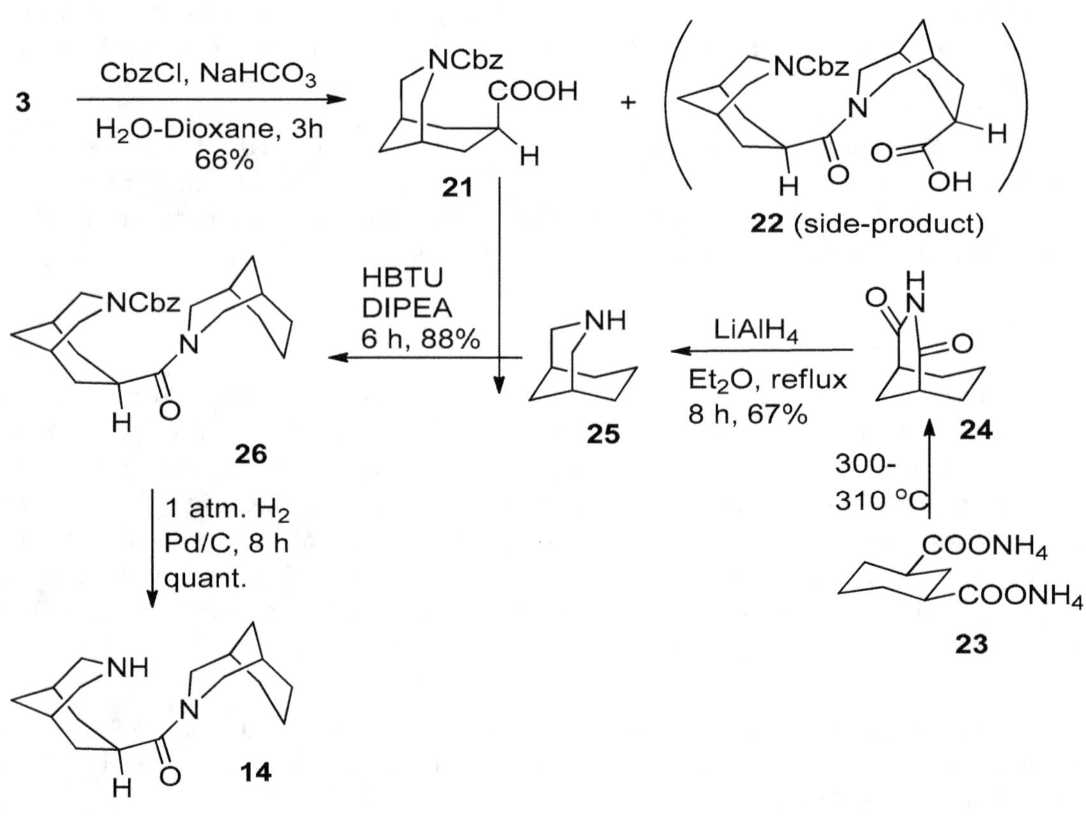

Scheme 3. Synthesis of model compound lacking the carboxyl group.

The structure of the amide **14** was confirmed by comprehensive analysis of its 2D-NMR spectra. The key NOESY correlations, which prove the structure of **14** in solution are shown in Figure 2 (see also the Supporting Information).

Compound **13** gave single crystals suitable for X-ray structure determination, which showed that the carboxyl group is indeed in close proximity to the amide bond (Figure 3) and might facilitate hydrolysis: as for example, in the known compound **9** **[10]**. However, neither **13** nor **14** was hydrolyzed at a measurable rate under the conditions used in [1] (0.084M in acetonitrile-d_3 solution in the presence of five equivalents of water), in which compound **2** was reported to be hydrolyzed within days.

We noted only very slow changes in the solutions of **14** in the presence of a large excess of water (on standing in a CD_3CN-D_2O mixture, 1:1 $v\backslash v$ at 25 °C): where formation of insoluble, presumably polymeric material was observed within several weeks. These results are also in agreement with the reported fact [1] that if the carboxyl group in compound **2** is reduced to CH_2OH, the resulting amino alcohol also withstands hydrolysis even during prolonged treatment with excess water. Therefore, one might conclude that the *amino and carboxyl groups* act together to facilitate the hydrolysis of amide **2**.

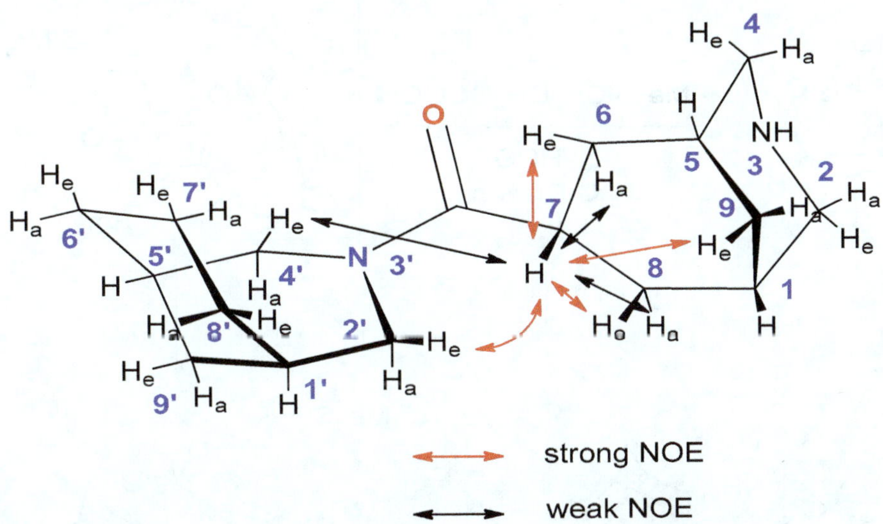

Figure 2. NOE correlations for 7-CH ^{1}H-NMR peak in the 2D-NOESY spectrum of compound **14**.

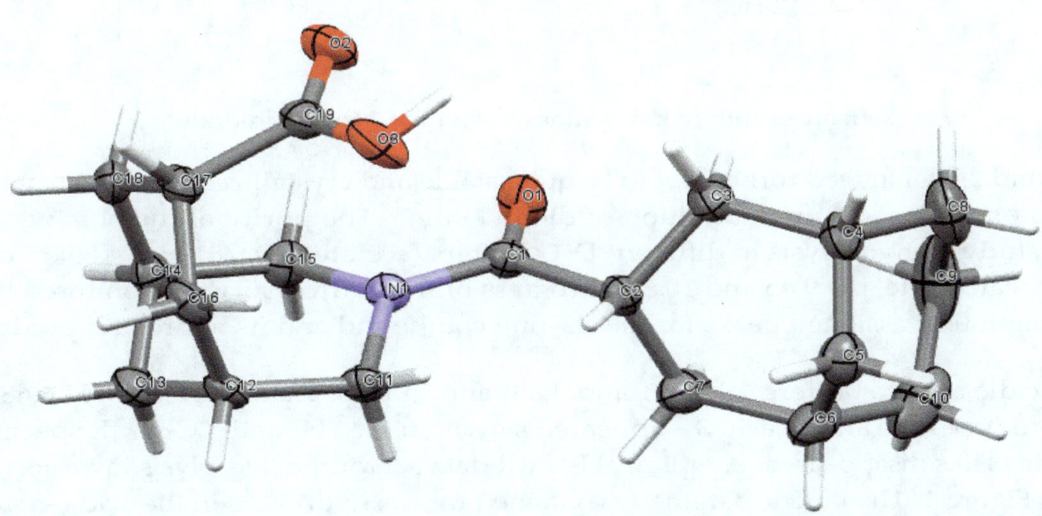

Figure 3. Molecular structure of **13** (ellipsoids are drawn at the 50% probability level).

In order to elucidate possible mechanisms for the intramolecular catalysis of the amide bond in **2**, we decided to obtain this compound as a hydrochloride by a route different from that shown in Scheme 1 in order to study the rates of its reaction with water in buffers at different pH. We reasoned that the hydrochloride salt would be more stable than the zwitterionic form of **2** obtained directly from **1**, enabling us to purify the compound. The synthetic approach was based on our observation (above) that treatment of the amino acid **3** with CbzCl gave not only the expected protected compound **21**, but also the dimer **22** as a side-product (Scheme 3, in parentheses). Using $Me_3SiCH_2CH_2OCOCl$ as protecting reagent and varying the reaction conditions, we obtained the dimer **27** in a reasonable yield. Most probably, formation of **27** proceeded through the activated intermediate **28**, which reacted with the starting amino acid (Scheme 4). Alternatively, compound **27** could be formed through the twisted amide **1**, formed upon activation of the carboxyl group. The twisted amide could acylate unreacted **3** (as shown in Scheme 1), and the dimer **2** produced could then be trapped by the N-protecting reagent. Treatment of **27** with dry HCl in dioxane afforded the hydrochloride **2**·HCl.

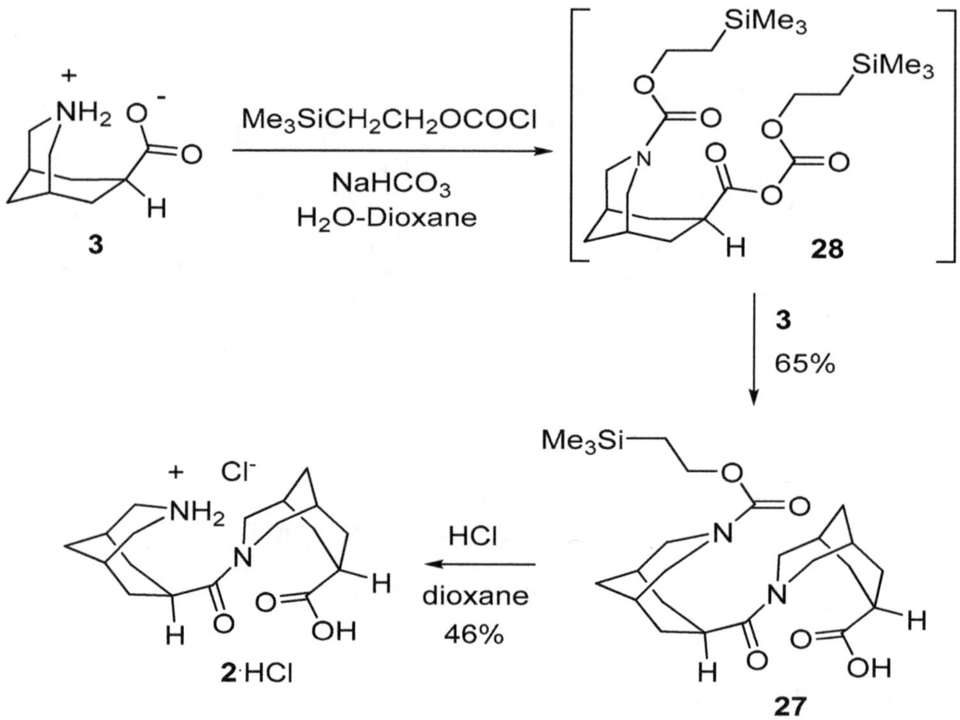

Scheme 4. Alternative synthesis of amide **2** (hydrochloride).

Compound **2**·HCl indeed turned out to be quite stable and crystallized from the reaction mixture after the deprotection of **27** upon addition of diethyl ether. The purity of the obtained **2**·HCl was sufficient to study its hydrolysis in different D_2O buffers (acetate, pD 3.81 and 4.85; phosphate, pD 6.68 and 7.95; carbonate, pD 9.45 and 10.68). Progress of the hydrolysis was monitored by ^{1}H-NMR using unambiguously assigned peaks for the starting compound and its hydrolysis product.

In the acidic acetate buffers (pD 3.81 and 4.85) and in phosphate buffer at pD 6.68, the **2**·HCl hydrolysis product—the amino acid **3**—appeared slowly in the ^{1}H- and ^{13}C-NMR spectra while the starting amide peaks disappeared. A typical ^{1}H-NMR data set for the hydrolysis in an acetate buffer is illustrated in Figure 4. The kinetic parameters obtained for the hydrolysis in the acidic-neutral buffers are listed in Table 1 (see also the Experimental section and Supporting Information for the details). As can be seen from the data in Table 1, the pseudo-first order rate constant k_{obs} of the hydrolytic amide bond cleavage in **2** gradually increases with the pD values in this pD range, with the maximum of 1.849×10^{-4} min^{-1} (half-life ~62.5 h, 23 °C) in phosphate buffer at pD 6.68.

Table 1. Kinetic parameters for the hydrolysis of **2**·HCl in acidic D_2O buffers at different pD values, 23 °C.

pD Value	Half-life of 2·HCl, min	k_{obs}, min^{-1}	Log(k_{obs})
3.81	18,138	3.821×10^{-5}	−4.418
4.85	4668	1.485×10^{-4}	−3.828
6.68	3748	1.849×10^{-4}	−3.733

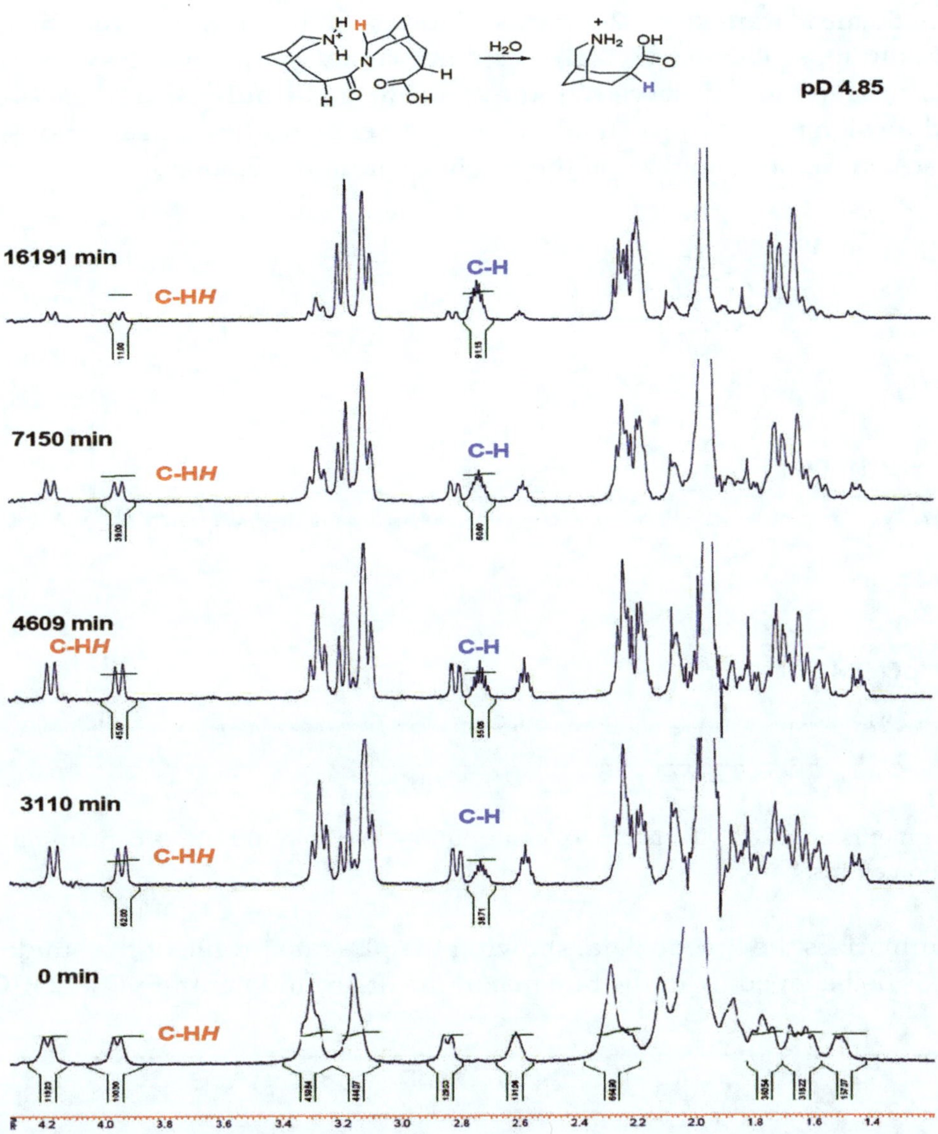

Figure 4. Representative ^{1}H-NMR spectral data set for hydrolysis of **2**·HCl in acetate D_2O buffer, pD 4.85, 23 °C.

Hydrolysis of **2**·HCl in alkaline buffers (phosphate at pD 7.95, carbonate at pD 9.45 and 10.68) was found to become slower as the pD increased. The kinetic data obtained for this pD range are shown in Table 2.

Table 2. Kinetic parameters for the hydrolysis of **2**·HCl in alkaline D_2O buffers at different pD values, 23 °C.

pD Value	Half-life of 2·HCl, min	k_{obs}, min^{-1}	$Log(k_{obs})$
7.95	18,802	3.686×10^{-5}	−4.433
9.45	91,125	7.605×10^{-6}	−5.119
10.68	126,114	5.496×10^{-6}	−5.260

Interestingly, at pD 9.45 and 10.68 we observed a duplicated set of the ^{1}H- and ^{13}C-NMR signals corresponding to the starting amide, suggesting the presence of its two forms in the solutions (Figures S33–S36 in the Supplementary Materials). These two forms were observed immediately after the compound was dissolved (the equilibrium between the forms established faster than we could run the ^{1}H-NMR spectrum, ~2 min), and their ratio remained unchanged during the whole period of the kinetic measurements (~2 months). However, the ratio between the forms depended on the pD of the solutions. The form which was dominant at the pD 9.45 had the chemical shifts of the NMR signals close to those observed for the **2**·HCl in acidic buffers; at the pD 10.68, both forms gave corresponding signals of almost equal intensity; to explain these observations, we suggest the existence of the structures **2b**$^{+-}$ and **2**$^{-}$ in the alkaline solutions (Figure 5).

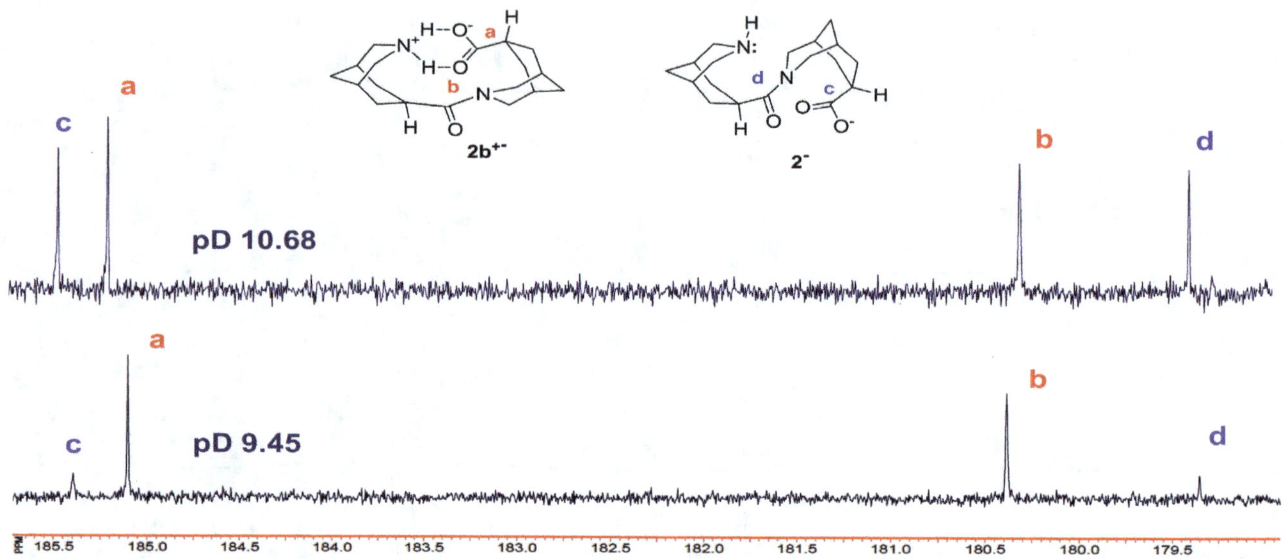

Figure 5. Fragments of ^{13}C-NMR spectra of compound **2** in different buffers containing peaks of the carbonyl carbon atoms.

Figure 6 summarizes the kinetic data, showing the pD–rate profile of the amide **2** hydrolysis in D_2O. The profile is bell-shaped, with the rate maximum at around neutral pD at 23 °C.

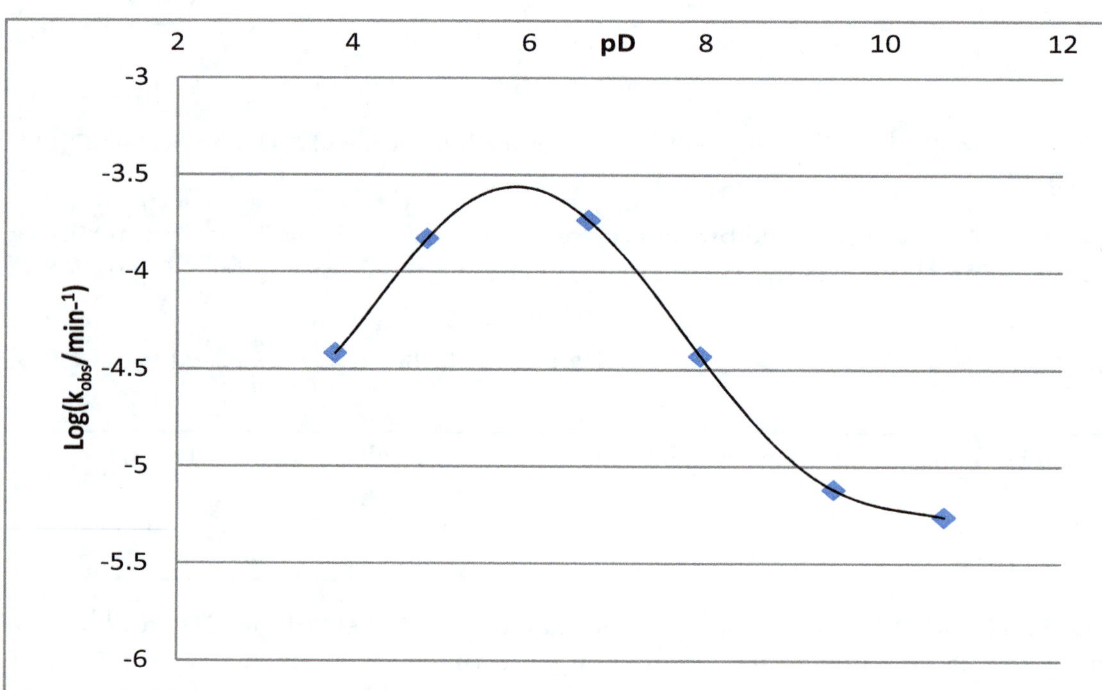

Figure 6. pD-rate profile for the hydrolysis of **2**·HCl.

3. Discussion

The bell-shaped rate profile for the hydrolytic cleavage of the amide **2** confirmed our suggestion made using the model molecules **13** and **14**: that both amino and carboxyl groups are involved in the hydrolysis. On the basis of our experimental data, one can suggest the equilibria between differently protonated **2** in different buffers, as schematically illustrated in Figure 7. Evidently, the basic nitrogen in **2** is protonated and "switched off" from intramolecular assistance of the amide bond hydrolysis at low pDs (where species $\mathbf{2}^+$ dominate). Similarly, deprotonation of the COOH group "switches off" its involvement in the reaction at high pDs (the equilibria shifting to unreactive $\mathbf{2}^-$). There is an alternative explanation of the pD rate profile—see the discussion of possible hydrolysis mechanisms below—but in all cases the amino and carboxyl groups should act synergistically.

At the high pDs, another reason for the slowdown of the reaction might be epimerization at the carbon atom next to the carboxyl group. The racemization could also explain the doubling of the signals in the solution of **2** in the alkaline buffers. However, we can exclude this possibility: the NMR signal-doubling is observed almost immediately after dissolving the amide **2** in the buffers, and the ratio of the species observed in the NMR spectra is dependent on the pD, but not on time. Epimerization in similar systems is known to proceed much more slowly and under harsher conditions, leading to almost complete conversion of the endo-COOH epimer to the exo-COOH epimer [35]. Observation of the two species in the spectra of **2** might be explained by the slow (on the NMR time scale) exchange between the species $\mathbf{2b}^{+-}$ and $\mathbf{2}^-$ at high pDs (Figure 8). This could be confirmed by the fact that one of the species (content increasing with pD, so assigned to $\mathbf{2}^-$) has the carboxyl group ^{13}C-NMR resonance at lower field then the other species (tentatively assigned to $\mathbf{2b}^{+-}$, Figure 8), in line with literature data reporting the downfield shift for the COOH ^{13}C-NMR peak upon deprotonation [38].

Two types of possible mechanisms of the hydrolysis of **2** can be suggested, if both the amino and carboxyl functions are involved. These mechanisms (which we name *N*- and *O*-mechanisms in order to distinguish them in the following discussion) differ by the roles played by the two functional groups in intramolecular catalysis. The N-mechanism, realizable either through **path a** or **b** (Figure 7) involves a nucleophilic attack of the basic nitrogen atom on the amide bond carbon while the COOH group acts as a general acid. It might be concerted (**path a**) or involve formation of a tetrahedral intermediate (**TIb**, **path b**), but in both cases should result in formation of the twisted amide **1**. Although we did not observe **1** in the reaction mixtures at any pDs by NMR, we know that its hydrolysis will be extremely fast under the reaction conditions used [1], excluding the accumulation of **1** in observable concentrations.

The O-mechanism (concerted **path c** or stepwise **path d**, Figure 7) suggests that the amino and the carboxyl group swap roles in the first step of the reaction: the deprotonated COOH is now the nucleophile and the protonated amino group acts as the general acid. This possible mechanism would involve formation of the anhydride intermediate **A** (apart from the tetrahedral intermediate **TId** in the **path d**). Degradation of **A** might again involve the nucleophilic amino group attack at one of the anhydride carbonyls and formation of the twisted amide **1**. In all cases the limiting step is the first; the following transformations should be fast—explaining why no intermediates were detected by NMR spectroscopy in the reaction mixtures. In order to prove that the rate limiting step of the hydrolysis (at least at low pH) involves a proton transfer, we set up two hydrolysis reactions of **2**·HCl running in parallel with all the conditions but one identical: the solvent for one being H_2O, for the other, D_2O. NMR monitoring of the reaction rates revealed a well-defined isotope effect: the hydrolysis in H_2O being ca 1.3–1.6 times faster than in D_2O. Figure 9 illustrates the faster appearance of the hydrolysis product in the ^{1}H-NMR spectra of the reaction mixtures run in the course of the experiment.

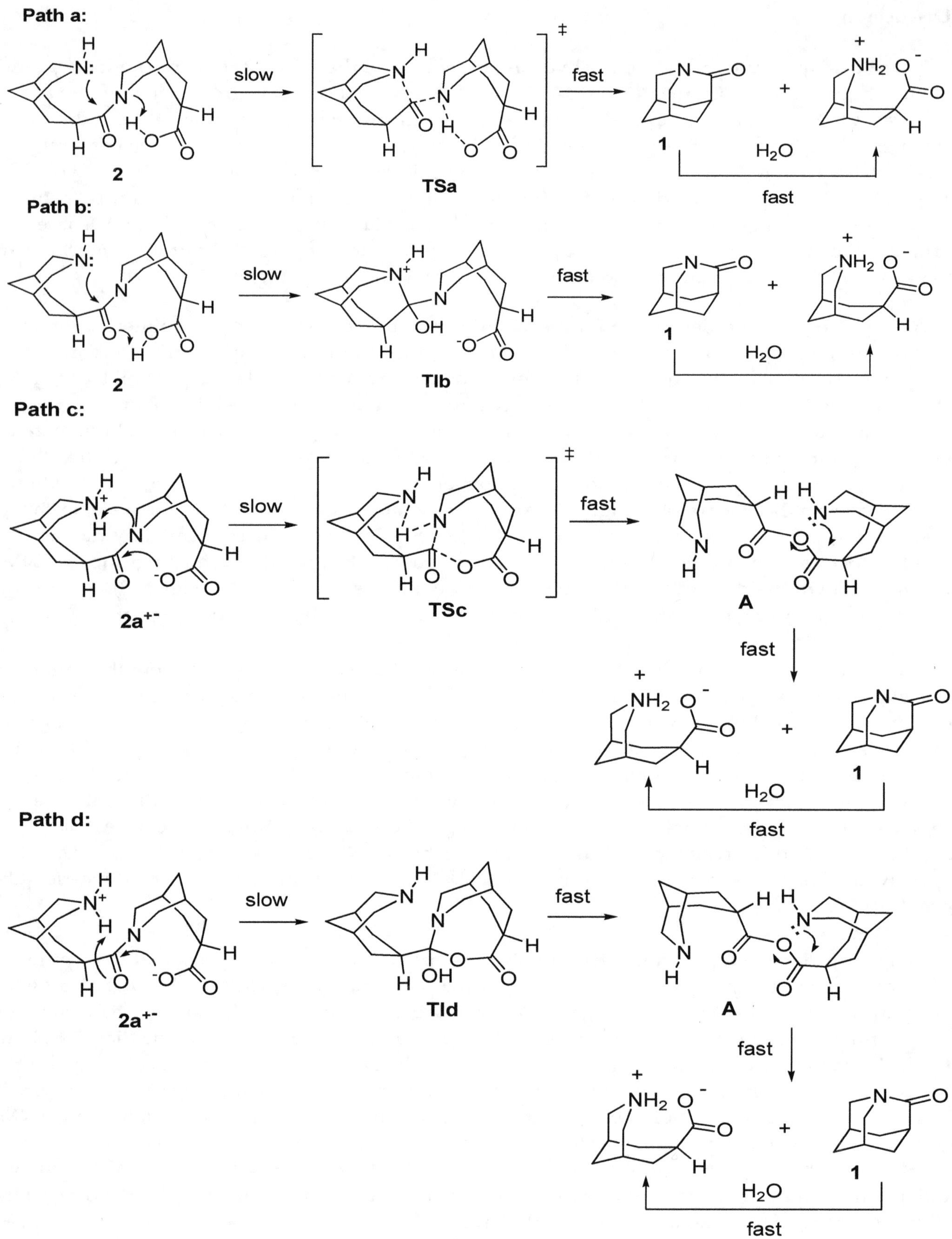

Figure 7. Possible mechanisms of the hydrolysis of the amide **2**.

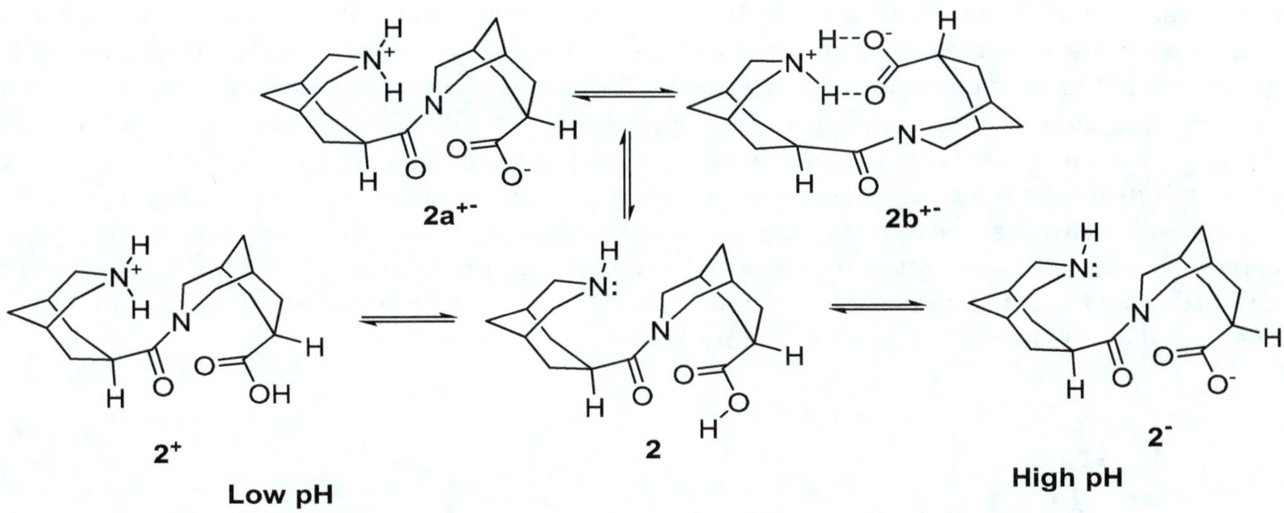

Figure 8. Equilibria established in solutions of **2** at different pH.

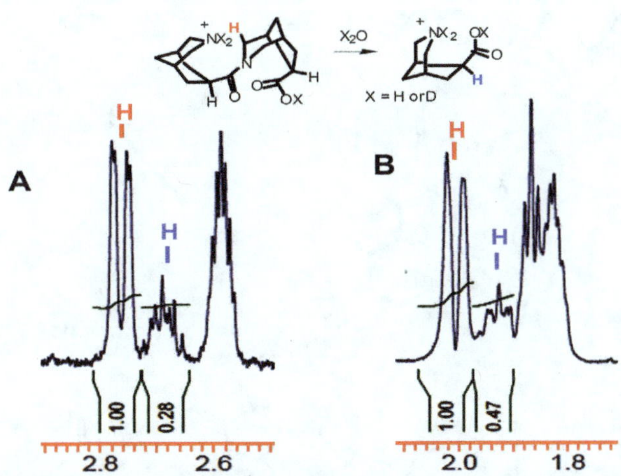

Figure 9. Characteristic regions of the ^{1}H-NMR spectra illustrating the hydrolysis rates of **2**·HCl in D_2O (**A**) and H_2O (**B**). After twelve days of reaction.

The fact that we observed none of the possible intermediates makes it difficult to choose between **paths a-d** based only on our experimental data.

If the amine nitrogen is indeed involved as the nucleophilic center in the intramolecular catalysis of the amide bond cleavage in **2**, this raises an intriguing question: why does nature not utilize nitrogen-based nucleophilic centers in enzymatic hydrolytic cleavage of the amide bond? None of the currently known peptidases (hydrolases acting on amide bonds in protein backbones [39]) utilize an N-nucleophilic attack as a part of their catalytic mechanism [40]. Ubiquitous and long-studied serine and cysteine proteases feature the side-chain –OH or –SH nucleophilic groups in their active sites. Similarly, the threonine –OH group is thought to be involved as a nucleophile in relatively recently discovered threonine proteases [41]. The catalytic mechanisms proposed for both aspartic and metalloproteases involve nucleophilic attack of a water molecule at the scissile amide bond. Finally, the rare glutamic proteases were also suggested to utilize a water molecule as the nucleophile [42]. Even if there are as yet undiscovered "XX-peptidases" (XX designating an N-nucleophile in the active site), they undoubtedly must be rare.

In fact, there are known proteolytic enzymes that cleave peptide bonds with active participation of a nucleophilic nitrogen atom. These enzymes are not peptidases but rather lyases, as their action does not include hydrolysis. The N-nucleophile thought to play the key role in the amide bond cleavage is the nominally non-nucleophilic amide nitrogen of the asparagine side chain. Named asparagine peptide lyases, these enzymes were assigned to a "seventh catalytic type of proteolytic enzymes" [43]; many families of this enzyme type are now known. Typical examples are inteins, the protein segments able to mediate their own scission from larger proteins [44,45]. This process, known as protein splicing, involves cleavage of the two amide bonds flanking the inteins, producing the intein-removed proteins built from two remaining N- and C-extein fragments. Most intein-containing proteins have a conserved asparagine residue at one of the amide-bond cleavage sites. In the key step of the splicing, the asparagine amide is activated for intramolecular nucleophilic attack at its own peptide bond, leading to its cleavage. The chain of molecular events, a part of the mechanism proposed for splicing an intein (SspDnaE) [46] is illustrated in Figure 10.

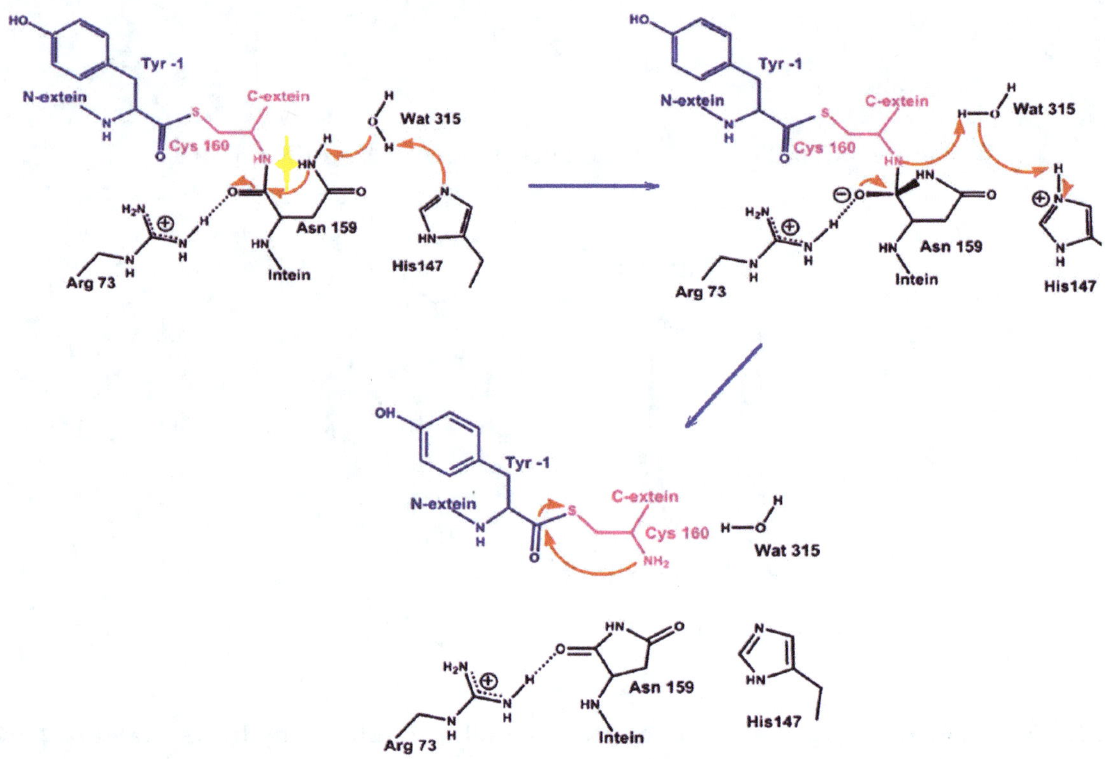

Figure 10. A part of the proposed intein SspDnaE splicing mechanism involving the Asn159 nucleophilic attack (marked with the yellow star) at the nearby amide bond carbonyl. Reprinted from [46] with permission from Elsevier.

Details of the activation of the Asn side chain amide for the nucleophilic attack are not completely understood. This could involve its tautomerization with the formation of a more reactive imidate species [47] or, alternatively, twisting the amide bond, which is known to enhance its nucleophilicity [48]. Interestingly, a mechanism involving the twisted asparagine amide group was proposed for oligosaccharyltransferases, the membrane protein complexes which catalyze asparagine-linked glycosylation [49]. Whatever the mechanism of the amide nucleophile activation, it is clear that nature avoids strong amine nucleophiles in the active sites of the proteolytic enzymes. They are present in a latent form (amide group of Asn) and only generated during the catalysis. It cannot be due to low availability of active *N*-nucleophiles in proteins. The lysine amino group, for example, acts as a nucleophile in ribulose 1,5-bisphosphate carboxylase (Rubisco), the most abundant enzyme on Earth [50]. In the catalytic protein splicing described above, where amide bonds are cleaved, however, the action of the asparagine-derived active nucleophile results in destruction of the enzyme itself. One might even question whether the splicing can be considered as an enzymatic process [43].

The prospect of self-destruction could be one of the reasons why natural proteases evolved without *N*-nucleophiles in the active site.

Our model compound **2** demonstrates that one possible way to avoid self-destruction could be structural constraint holding the intermediate *N*-acylated species in a twisted conformation, before a water molecule arrives and completes the hydrolysis cycle. There are no reasons to believe that this is impossible in natural catalytic systems, therefore, one might also believe that "XX-peptidases" are yet to be discovered. In any case, in seems worthwhile to design artificial catalytic systems exploiting the *N*-nucleophiles in the catalytic processes. Such systems could be inspired by known natural enzymes or by model molecules as described in this paper. One important step towards this goal has already been reported by Otaka et al. [51]. They developed a photo-responsive amide cleavage device, mimicking the intein-mediated protein splicing. According to their approach, photo-triggered liberation of the secondary amino group on a modified asparagine fragment activates the asparagine amide for the nucleophilic attack that subsequently cleaves the adjacent peptide bond, as illustrated in Figure 11.

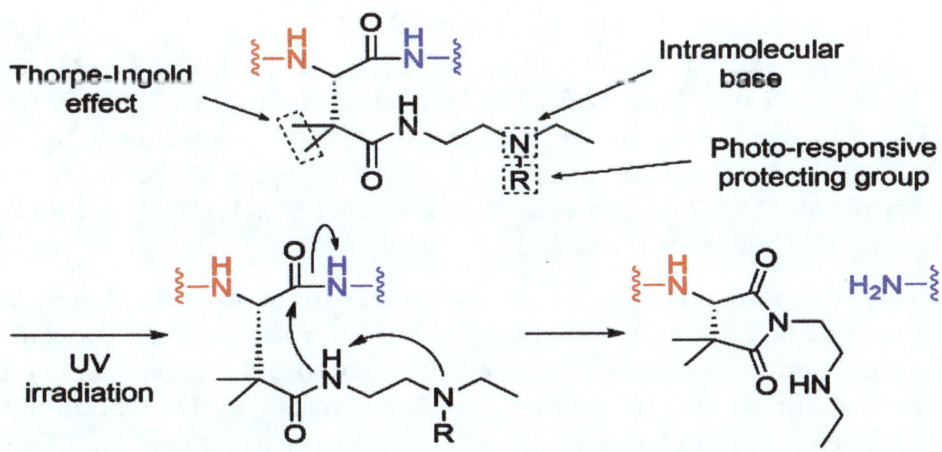

Figure 11. Photoinduced peptide bond cleavage inspired by natural protein splicing. Reprinted with permission from [51].

4. Materials and Methods

4.1. General

Solvents were purified according to standard procedures [52]. Melting points were measured on an automated melting point system and are uncorrected. Analytical TLC was performed using Polychrom SI F254 plates. Column chromatography was performed using silica gel (230–400 mesh) as the stationary phase. ^{1}H, ^{13}C NMR and all 2D NMR spectra were recorded at 499.9 or 400.4 MHz for protons and 124.9 or 100.4 MHz for carbon-13. Chemical shifts are reported in ppm downfield from TMS (^{1}H, ^{13}C) as an internal standard. Mass spectra were recorded either on an Agilent 1100 LC/MSD SL (Agilent, Santa Clara, CA, USA) instrument by chemical ionization (CI) or on a GCMS instrument with electron impact ionization (EI). CHN-analysis was done on an Elementar VarioMICRO Cube analyzer (Elementar, Langenselbold, Germany).

4.2. Synthetic Procedures

3-Endo-bicyclo[3.3.1]nonane-3-carboxylic acid (**20**). This known compound was prepared starting from 1-cyclohexenylpyrrolidine (**16**) and 2-bromomethyl-acrylic acid benzyl ester (**15**) on a 0.2 mol scale in 69% overall yield following the procedures described in [36] for the analogue, 7-endo-3-(tert-butoxycarbonyl)-3-azabicyclo [3.3.1]nonane-7-carboxylic acid. The ^{1}H-NMR spectrum of the product corresponded to that described in [53]. White crystals, m.p. 126 °C (lit. 126.5–127 °C [52]).^{1}H-NMR (δ, $CDCl_3$, 500 MHz): 11.78 (broad s, 1H, COOH), 2.48 (h, 6.0 Hz, 1H, 3-H),

1.9–2.1 (m, 4H), 1.6–1.7 (m, 1H), 1.55 (d, 13.5 Hz, 1H), 1.25–1.45 (m, 7H), 1.09 (d, 13.5 Hz, 1H). ^{13}C-NMR (δ, CD_3Cl, 124.9 MHz): 183.27 (COOH), 36.56, 32.50, 28.74, 28.48, 24.43, 15.55. LC/MS (CI, neg. scan): *m/z* = 167.2; (CI, pos. scan): *m/z* = 169.2. Anal. Calc. for $C_{10}H_{16}O_2$: C, 71.39; H, 9.59. Found: C, 71.35; H, 9.55.

3,7′-Endo-3′-(bicyclo[3.3.1]non-3-ylcarbonyl)-3′-azabicyclo[3.3.1]nonane-7′-carboxylic acid (**13**). Compound **20** (266 mg, 1.6 mmol) was dissolved in dry dichloromethane (0.5 mL) and mixed with oxallyl chloride (5 mg, 4 mmol). The mixture was stirred under protection from air moisture for 8 h. Then the volatile products were removed under vacuum (0.01 mm Hg, 40 °C, 4 h). The residue was dissolved in dioxane (1 mL) and added dropwise to a solution of the amino acid **3** (prepared as described in [1], 244 mg, 1.44 mmol) and Na_2CO_3 (0,763 g, 7.2 mmol) in water (1 mL) under stirring and cooling by an ice bath. The mixture was stirred for 8 h at ambient temperature, evaporated on a rotary evaporator, and the residue (the sodium salt of **13**) crystallized from ethanol. The crystallized and air-dried product was dissolved in a minimum amount of water (approximately 0.4 mL) and the solution was acidified by conc. HCl to afford **13** as a white crystalline material. It was filtered and dried in vacuum (0.01 mm. Hg, 40 °C, 4 h). 230 mg, 45% overall yield. Below, the data for the sodium salt of **13** (an analytical sample) are listed. m.p. 233 °C. ^{1}H-NMR (δ, D_2O, 500 MHz): 4.13 (d, 12.5 Hz, 1H), 3.82 (d, 12.5 Hz, 1H), 3.01 (d, 12.5 Hz, 1H), 2.91 (m, 1H), 2.52 (d, 12.5 Hz, 1H), 2.20 (m, 1H), 1.75–2.35 (m, 8H), 1.55–1.75 (m, 3H), 1.15–1.45 (m, 8H), 0.95—1.10 (m, 3H). ^{13}C-NMR (δ, D_2O, 124.9 MHz): 184.89, 179.73, 52.16, 48.93, 37.80, 32.41, 32.26, 32.18, 28.79, 28.75, 28.26, 28.05, 27.73, 26.35, 25.58, 25.43, 24.39, 24.11, 15.09. LC/MS (CI, neg. scan): *m/z* = 318.2; (CI, pos scan): *m/z* = 320.2. Anal. Calc. for $C_{19}H_{29}NO_3$: C, 71.44; H, 9.15; N, 4.38. Found: C, 71.40; H, 9.13; N, 4.39.

7′-Endo-3′-[(benzyloxy)carbonyl]-3-azabicyclo[3.3.1]nonane-7′-carboxylic acid (**21**). Amino acid **3** (169 mg, 1 mmol) and $NaHCO_3$ (200 mg, 2.4 mmol) were dissolved in water-dioxane mixture (10 mL, 1:1 *v/v*). To this solution was added benzyl carbonochloridate (carbobenzoxychloride, 0.2 mL) dropwise under stirring. The mixture was stirred for 3 h, and then acidified carefully by 1N HCl till pH 2. The product was extracted with MTBE (3 × 50 mL), the combined extracts were dried over Na_2SO_4 and evaporated. The product was purified by column chromatography (silica gel, MTBE as an eluent). Analytical sample was prepared by crystallization from diethyl ether. Yield: 200 mg, 66%. White crystals, m.p. 120 °C. ^{1}H-NMR (δ, $CDCl_3$, 500 MHz): 10.02 (br s, 1H, COOH), 7.40–7.70 (m, 5H, arom.), 5.12 (br s, 1H), 4.97 (br s, 1H), 4.03 (br s, 2H), 2.92 (d, 11.5 Hz, 2H), 2.60 (br s, 1H), 2.36 (d, 13.0 Hz, 2H), 1.75–2.00 (m, 4H), 1.64 (d, 11.5 Hz, 1H), 1.54 (d, 11.5 Hz, 1H). ^{13}C-NMR (δ, $CDCl_3$, 124.9 MHz): 178.67, 156.18, 136.47, 128.01, 127.72, 127.47, 66.67, 48.70, 36.96, 30.89, 29.26, 26.21. LC/MS (CI, neg. scan): *m/z* = 302.3; LC/MS (CI, pos. scan): *m/z* = 304.2. Anal. Calc. for $C_{17}H_{21}NO_4$: C, 67.31; H, 6.98; N, 4.62. Found: C, 67.36; H, 6.86; N, 4.60. Chromatographic purification of **21** also yielded a side-product assigned to the structure **22** (6 mg), on the basis of its ^{1}H-NMR and MS data.

7,7′-Endo-3′-({3-[(benzyloxy)carbonyl]-3-azabicyclo[3.3.1]non-7-yl}carbonyl)-3′-azabicyclo[3.3.1]nonane-7′-carboxylic acid (22). m.p. 170 °C: ^{1}H-NMR (δ, $(CD_3)_2SO$, 500 MHz): 11.63 (s, 1H, COOH), 7.20-7.50 (m, 5h, arom.), 5.03 (s, 2H, CH_2Ph), 4.11 (d, 13.0 Hz, 1H), 3.70-3.90 (m, 3H), 2.93 (d, 12.0 Hz, 1H), 2.55–2.75 (m, 3H), 2.32 (m, 2H), 1.35–1.95 (m, 13H), 1.33 (d, 12.0 Hz, 1H), 1.11 (d, 12.0 Hz, 1H), 0.98 (br s, 1H). LC/MS (CI, neg. scan): m/z = 453.4; LC/MS (CI, pos. scan): m/z = 455.2.

3-azabicyclo[3.3.1]nonane (**25**), *3-azabicyclo[3.3.1]nonane-2,4 dione* **24** (5 g, 32 mmol, prepared as described in [35]) was dissolved in THF (100 mL) under an argon atmosphere. Lithium aluminum hydride (3.7 g, 98 mmol) was added in portions to the solution, which was then refluxed for 5 h and stirred for an additional 8 h without heating. The excess of the $LiAlH_4$ was quenched carefully with water, the product was sublimed to afford **25** as a colorless crystalline product, 2.68 g, 67% yield.

All the spectral data were identical to those described in the literature [54]. m.p. 157 °C. For the use in the next step, the amine was converted to its hydrochloride by dissolving in an excess of 1N HCl, evaporation in vacuum, co-evaporation with water (3 × 20 mL) on a rotary evaporator, drying in a desiccator over P_2O_5 for 24 h. White crystals (crystallized from isopropanol), 3-azabicyclo[3.3.1]nonanehydrochloride (**25**·HCl): ^{1}H-NMR (δ, $(CD_3)_2SO$, 500 MHz): 9.81 (broad s, 1H, NH), 8.08 (broad s, 1H, NH), 3.14 (d, 13 Hz, 2H), 3.03 (d, 13 Hz, 2H), 1.90–2.10 (m, 3H), 1.40–1.80 (m, 7H). ^{13}C-NMR (δ, $(CD_3)_2SO$, 124.9 MHz): 46.40, 29.80, 28.96, 25.22, 19.36. Anal. Calc. for $C_8H_{16}ClN$: C, 59.43; H, 9.98; N, 8.66. Found: C, 59.40; H, 9.95; N, 8.69.

7-Endo-benzyl 7-(3′-azabicyclo[3.3.1]non-3′-ylcarbonyl)-3-azabicyclo[3.3.1]nonane-3-carboxylate (**26**). The Cbz-protected amino acid **21** (95 mg, 0.31 mmol) and DIPEA (81 mg, 0.63 mmol) were dissolved in acetonitrile (1 mL). (2-(1H-benzotriazol-1-yl)-1,1,3,3-tetramethyluronium hexafluorophosphate (HBTU, 128 mg, 0.34 mmol) was added to the solution under stirring, and the stirring continued for 5 min. Then the mixture was combined with a solution of the amine hydrochloride (**25**·HCl, prepared as described above, 61 mg, 0.37 mmol) and DIPEA (73 mg, 0.56 mmol) in acetonitrile (1 mL). The resulting mixture was stirred for 8 h, and then evaporated on a rotary evaporator. The residue was applied to a chromatographic column (silica gel) and pure **26** was eluted with hexane-ethylacetate 2:1 *v/v*). Colorless oil, crystallized on standing, 88% yield (112 mg). An analytical sample was prepared by crystallization from petroleum ether. m.p. 157 °C. ^{1}H-NMR (δ, $(CDCl_3$, 500 MHz): 7.25–7.45 (m, 5H, arom.), 5.20 and 5.09 (two d, 2H), 4.57 (m, 1H), 4.02 (d, 12.0 Hz, 1H), 3.95 (d, 12.0 Hz, 1H), 3.88 (d, 12.0 Hz, 1H), 3.29 (d, 12.0 Hz, 1H), 2.75–2.85 (m, 3H), 2.74 (broad s, 1H), 1.80–2.15 (m, 7H), 1.50–1.80 (m, 9H), 1.43 (broad s, 1H), 1.26 (m, 1H). ^{13}C-NMR (δ, $(CD_3)_2SO$, 124.9 MHz): 177.47, 177.09, 136.55, 128.15, 127.56, 127.36, 50.43, 35.75, 32.80, 30.20, 29.51, 29.00, 27.00, 26.21., 26.20, 25.97. LC/MS (CI, pos. scan): *m/z* = 411.2. Anal. Calc. for $C_{25}H_{34}N_2O_3$: C, 73.14; H, 8.35; N, 6.82. Found: C, 73.11; H, 8.39; N, 6.80.

7-Endo-3′-(3-azabicyclo[3.3.1]non-7-ylcarbonyl)-3′-azabicyclo[3.3.1]nonane (**14**). The Cbz-protected amide **26** (38 mg, 0.1 mmol) was dissolved in methanol (3 mL). Palladium on charcoal (10%, 60 mg) was added to the reaction vessel, which was then filled with hydrogen and the mixture was shaken under 1 atm H_2 for 8 h. Then the mixture was filtered, evaporated in high vacuum (0.01 mm Hg) without heating. The crude product **14** easily forms the carbonate upon handling on air, therefore it was mixed with 5% aqueous NaOH solution (5 mL) and extracted with dichloromethane (3 × 20 mL) under an argon atmosphere. The combined dichloromethane extracts were evaporated in high vacuum without heating, the residue was dried in high vacuum for 5 h. Yellowish oil, 27.6 mg (quantitative yield). ^{1}H-NMR (δ, CD_3OD, 500 MHz): 4.55 (d, 15.0 Hz, 1H), 4.08 (d, 15.0 Hz, 1H), 3.36 (d, 15.0 Hz, 1H), 3.03 (m, 1H), 2.85 (d, 15.0 Hz, 1H), 2.69 and 2.61 (two d, 14.5 Hz, 4H), 1.60–2.20 (m, 17H), 1.55 and 1.51 (two d, 14.5 Hz, 2H), 1.46 (broad s, 1H), 1.38 (d, 14.5 Hz, 1H). For the peaks assignments, see Supporting Information. ^{13}C-NMR (δ, CD_3OD, 124.9 MHz): 176.47, 51.81, 51.76, 50.71, 32.61, 31.93, 30.85, 30.36, 28.65, 28.25, 28.08, 26.88, 25.64, 25.59, 20.13. LC/MS (CI, pos. scan): *m/z* = 277.2.

7,7′-Endo-3′-[(3-{[2-(trimethylsilyl)ethoxy]carbonyl}-3-azabicyclo[3.3.1]non-7-yl)carbonyl]-3′-azabicyclo[3.3.1] nonane-7′-carboxylic acid (**27**). Amino acid **3** (100 mg, 0.59 mmol) was suspended in THF (2 mL) and triethylamine (120 mg, 1.2 mmol) was added to the suspension under stirring. The mixture was then cooled to −20 °C and 2-(trimethylsilyl)ethylcarbonochloridate (214 mg, 1.2 mmol) was slowly added under stirring, keeping the mixture temperature not higher than −10 °C. After the addition, the mixture was stirred at −10 °C for 15 min and at ambient temperature for 30 min. The mixture was then again cooled in an ice-water bath, and a suspension of **3** (100 mg, 0.59 mmol) in a triethylamine (60 mg, 0.6 mmol) solution in THF (2 mL) was added to the stirred mixture in one portion. The resulting mixture was stirred overnight at ambient temperature. The volatile products were removed in vacuum, the residue was triturated with aqueous citric acid solution (5%). The crude product was extracted with MTBE (3 × 50 mL), the combined extracts were evaporated and the residue was purified by reverse-phase HPLC to afford **27** as colorless oil which crystallized on standing. Yield: 178.2 mg, 65%. m.p. 198 °C. ^{1}H-NMR (δ, $CDCl_3$, 500 MHz): 9.83 (br. s, 1H), 4.36 (br. s, 1H), 4.13 (br. s, 2H), 3.80–4.05 (br. m, 3H), 3.12 (br. s, 1H), 2.65–2.82 (br. m, 5H), 2.53 (br. s, 1H), 2.24 (br. m, 2H), 1.75–2.15 (br. m, 9H), 1.64 (br. s, 1H), 1.57 (br. s, 2H), 1.21 (br. s, 1H), 0.98 (br. m, 2H), −0.1 (br. m, 9H). ^{13}C-NMR (δ, $CDCl_3$, 124.9 MHz): (all the signals are broadened) 117.72, 117.27, 157.53, 63.64, 50.73, 47.37, 36.83,

33.33, 33.12, 31.31, 30.02, 29.30, 29.05, 28.53, 27.26, 27.07, 26.94, 26.78, 25.81, 25.52, 25.32, -1.47. LC/MS (CI, neg. scan): m/z = 463.2, 454.2, 465.2; (CI, pos. scan): m/z = 466.2, 467.2, 468.2. Anal. Calc. for $C_{24}H_{40}N_2O_5Si$: C, 62.03; H, 8.68; N, 6.03. Found: C, 62.08; H, 8.64; N, 6.00.

7,7′-Endo-3′-(3-azabicyclo[3.3.1]non-7-ylcarbonyl)-3′-azabicyclo[3.3.1]nonane-7′-carboxylic acid hydrochloride (**2**·HCl). Compound **27** (50 mg, 0.11 mmol) was dissolved in a mixture of dry diethyl ether and dichloromethane (1:1 *v/v*, 1 mL) and a solution of HCl in dry dioxane (~10%, 0.5 mL) was added. The mixture was left standing at ambient temperature for 8 h, the formed white crystals of the product (**2**·HCl) were filtered, washed thoroughly with dry diethyl ether on the filter, dried in high vacuum (0.05 mm Hg) for 6 h at 25 °C). Yield: 18 mg (46%). m.p. 198 °C (with decomposition). ^{1}H-NMR (δ, D_2O, 500 MHz): 4.06 (d, 13.5 Hz, 1H), 3.85 (d, 13.5 Hz, 1H), 3.10–3.33 (m, 3H), 2.90–3.15 (m, 3H), 2.74 (dd, 13.5 and 2.0 Hz, 1H), 2.57 (h, 2.0 Hz, 1H), 1.85-2.20 (overlapped m, 11H), 1.73 (dd, 13.5 and 3.0 Hz, 1H), 1.50-1.70 (m, 4H), 1.34 (q, 3.0 Hz, 1H). ^{13}C-NMR (δ, D_2O, 124.9 MHz): 180.58, 179.33, 50.75, 49.01, 48.32, 47.98, 35.51, 29.83, 29.26, 29.10, 28.72, 27.11, 26.36, 26.30, 26.28, 26.09, 23.74, 23.39. Anal. Calc. for $C_{18}H_{29}ClN_2O_3$: C, 60.58; H, 8.19; N, 7.85. Found: C, 60.54; H, 8.22; N, 7.83.

4.2.1. Kinetic Measurements for Hydrolysis Reaction of Compound 2 in D_2O

Hydrolysis was carried out at 23 °C in NMR sample tubes, monitored by ^{1}H-NMR spectra (recorded by single scans). Six buffer solutions were prepared in D_2O: acetate buffers (pD 3.81 and 4.85), phosphate buffers (pD 6.68 and 7.95), and carbonate buffers (pD 9.45 and 10.68); the pD values were measured at 23 °C immediately before the measurements. The buffer concentrations were 250 mM in each case, the ionic strengths were adjusted with NaCl to 1 M. The compound **2**·HCl was weighed in an NMR sample tube (5–7 mg), and the measurements started by addition of the corresponding buffer solution to the compound (in the amount to achieve the compound concentration 25 mM). The content of the tube was vigorously shaken, starting immediately the recording of the time. The first ^{1}H-NMR spectrum was measured as quickly after the buffer addition as practical (~2 min), then the spectra were measured at intervals, depending on the reaction progress. The spectra were run as single scans, in order to avoid integral-value distortions due to relaxation effects. The unambiguously assigned non-overlapping peaks of the starting compound and its hydrolysis product were carefully integrated. The complete data sets (Tables S1–S12) were used to build the kinetic curves and calculate k_{obs} and the half-lives for each compound (see the Supporting Information).

4.2.2. Estimation of the Kinetic Isotope Effect by Comparison of the Hydrolysis Rates of 2·HCl in D_2O and H_2O

Two reactions were set up in two NMR sample tubes simultaneously by dissolving 4.2 mg of the **2**·HCl in each tube in 0.5 cm^3 D_2O or H_2O. The ^{1}H-NMR spectra of the samples were run at intervals (as single scans, in order to avoid integral-value distortions due to relaxation effects). For the spectral measurements in H_2O, excitation sculpting technique [55] was used for the water signal suppression. The unambiguously assigned non-overlapping peaks of the starting compound and its hydrolysis product were carefully integrated. Representative spectra can be seen in Figures S13 and S14.

4.2.3. X-ray Data for Compound 13

Crystal Data for $C_{19}H_{29}NO_3$ (M = 319.43 g/mol): monoclinic, space group $P2_1/c$ (no. 14), a = 6.9409(2) Å, b = 10.8409(3) Å, c = 22.2437(6) Å, β = 93.2098(12)°, V = 1671.11(8) Å^3, Z = 4, T = 180(2) K, μ(CuKα) = 0.673 mm^{-1}, D_{calc} = 1.270 g/cm^3. The diffraction pattern was indexed as two domains with 19230 measured reflections (domain 1 only), 19298 reflections (domain 2 only) and 3794 reflections (overlapped) (7.96° ≤ 2 ≤ 133.37°), R_{int} = 0.0399. 2954 unique reflections were used in all calculations. The final $R1$ was 0.0599 ($I > 2\sigma(I)$) and $wR2$ was 0.1593 (all data).

Author Contributions: Conceptualization, A.J.K. and I.V.K.; Methodology, A.J.K. and I.V.K.; Formal Analysis, I.V.K., A.Y.I., A.D.B. and A.J.K.; Investigation, I.V.K., A.Y.I., A.H., V.S., S.K. and A.D.B.; Writing: Original Draft Preparation, I.V.K.; Writing: Review & Editing, A.J.K.; Supervision, A.J.K., S.K. and I.V.K.

References

1. Komarov, I.V.; Yanik, S.; Ishchenko, A.Y.; Davies, J.E.; Goodman, J.M.; Kirby, A.J. The Most Reactive Amide As a Transition-State Mimic For cis–trans Interconversion. *J. Am. Chem. Soc.* **2015**, *137*, 926–930. [CrossRef] [PubMed]
2. Liu, C.; Szostak, M. Twisted Amides: From Obscurity to Broadly Useful Transition-Metal-Catalyzed Reactions by N–C Amide Bond Activation. *Chem. Eur. J.* **2017**, *23*, 7157–7173. [CrossRef] [PubMed]
3. Szostak, M.; Aube, J. Chemistry of bridged lactams and related heterocycles. *Chem. Rev.* **2013**, *113*, 5701–5765. [CrossRef] [PubMed]
4. Clayden, J.; Moran, W.J. The Twisted Amide 2-Quinuclidone: 60 Years in the Making. *Angew. Chem. Int. Ed.* **2006**, *45*, 7118–7120. [CrossRef] [PubMed]
5. Bender, M.L. General acid-base catalysis in the intramolecular hydrolysis of phthalamic acid. *J. Am. Chem. Soc.* **1957**, *79*, 1258–1259. [CrossRef]
6. Kirby, A.J.; Lancaster, P.W. Structure and efficiency in intramolecular and enzymic catalysis. Catalysis of amide hydrolysis by the carboxy-group of substituted maleamic acids. *J. Chem. Soc. Perkin Trans. 2* **1972**, *9*, 1206–1214. [CrossRef]
7. Kirby, A.J.; McDonald, R.S.; Smith, C.R. Intramolecular catalysis of amide hydrolysis by two carboxy-groups. *J. Chem. Soc. Perkin Trans. 2* **1974**, *12*, 1495–1504. [CrossRef]
8. Groves, J.T.; Olson, J.R. Models of zinc-containing proteases. Rapid amide hydrolysis by an unusually acidic Zn^{2+}-OH_2 complex. *Inorg.Chem.* **1985**, *24*, 2715–2717. [CrossRef]
9. Menger, F.M.; Ladika, M. Fast hydrolysis of an aliphatic amide at neutral pH and ambient temperature. A peptidase model. *J. Am. Chem. Soc.* **1988**, *110*, 6794–6796. [CrossRef]
10. Curran, T.P.; Borysenko, C.W.; Abelleira, S.M.; Messier, R.J. Intramolecular acylolysis of amide derivatives of Kemp's triacid: Strain effects and reaction rates. *J. Org. Chem.* **1994**, *59*, 3522–3529. [CrossRef]
11. Dougan, M.L.; Chin, J.L.; Solt, K.; Hansen, D.E. Rapid cleavage of cyclic tertiary amides of Kemp's triacid: Effects of ring structure. *Bioorg. Med. Chem. Lett.* **2004**, *14*, 4153–4156. [CrossRef] [PubMed]
12. Fernandes, N.M.; Fache, F.; Rosen, M.; Nguyen, P.L.; Hansen, D.E. Rapid Cleavage of Unactivated, Unstrained Amide Bonds at Neutral pH. *J. Org. Chem.* **2008**, *73*, 6413–6416. [CrossRef] [PubMed]
13. Souza, B.S.; Mora, J.R.; Wanderlind, E.H.; Clementin, R.M.; Gesser, J.C.; Fiedler, H.D.; Nome, F.; Menger, F.M. Transforming a stable amide into a highly reactive one: Capturing the essence of enzymatic catalysis. *Angew. Chem. Int. Ed.* **2017**, *56*, 5345–5348. [CrossRef] [PubMed]
14. Kirby, A.J. Enzyme Mechanisms, Models, and Mimics. *Angew. Chem. Int. Ed.* **1996**, *35*, 707–724. [CrossRef]
15. Bender, M.L.; Chow, Y.-L.; Chloupek, F. Intramolecular Catalysis of Hydrolytic Reactions. II. The Hydrolysis of Phthalamic Acid. *J. Am. Chem. Soc.* **1958**, *80*, 5380–5384. [CrossRef]
16. Jindal, G.; Warshel, A. Misunderstanding the preorganization concept can lead to confusions about the origin of enzyme catalysis. *Proteins* **2017**, *85*, 2157–2161. [CrossRef]
17. Kubyshkin, V.; Budisa, N. Amide rotation trajectories probed by symmetry. *Org. Biomol. Chem.* **2017**, *15*, 6764–6772. [CrossRef]
18. Hu, F.; Nareddy, P.; Lalancette, R.; Jordan, F.; Szostak, M. σ N–C Bond Difunctionalization in Bridged Twisted Amides: Sew-and-Cut Activation Approach to Functionalized Isoquinolines. *Org. Lett.* **2017**, *19*, 2386–2389. [CrossRef]
19. Meng, G.; Shi, S.; Lalancette, R.; Szostak, R.; Szostak, M. Reversible Twisting of Primary Amides via Ground State N–C (O) Destabilization: Highly Twisted Rotationally Inverted Acyclic Amides. *J. Am. Chem. Soc.* **2018**, *140*, 727–734. [CrossRef]
20. Meng, G.; Shi, S.; Szostak, M. Cross-coupling of amides by N–C bond activation. *Synlett* **2016**, *27*, 2530–2540. [CrossRef]
21. Dander, J.E.; Garg, N.K. Breaking amides using nickel catalysis. *ACS Catal.* **2017**, *7*, 1413–1423. [CrossRef] [PubMed]
22. Adachi, S.; Kumagai, N.; Shibasaki, M. Conquering amide planarity: Structural distortion and its hidden reactivity. *Tetrahedron Lett.* **2018**, *59*, 1147–1158. [CrossRef]

23. Wybon, C.C.; Mensch, C.; Hollanders, K.; Gadais, C.; Herrebout, W.A.; Ballet, S.; Maes, B.U. Zn-catalyzed tert-butyl nicotinate-directed amide cleavage as a biomimic of metallo-exopeptidase activity. *ACS Catal.* **2018**, *8*, 203–218. [CrossRef]
24. Balachandra, C.; Sharma, N.K. Instability of Amide Bond Comprising the 2-Aminotropone Moiety: Cleavable under Mild Acidic Conditions. *Org. Lett.* **2015**, *17*, 3948–3951. [CrossRef] [PubMed]
25. Hutchby, M.; Houlden, C.E.; Haddow, M.F.; Tyler, S.N.; Lloyd-Jones, G.C.; Booker-Milburn, K.I. Switching Pathways: Room-Temperature Neutral Solvolysis and Substitution of Amides. *Angew. Chem. Int. Ed.* **2012**, *51*, 548–551. [CrossRef] [PubMed]
26. Milović, N.M.; Kostić, N.M. Palladium (II) complexes, as synthetic peptidases, regioselectively cleave the second peptide bond "upstream" from methionine and histidine side chains. *J. Am. Chem. Soc.* **2002**, *124*, 4759–4769. [CrossRef]
27. Hohage, O.; Sheldrick, W.S. Cisplatin mediates selective downstream hydrolytic cleavage of Met-(Gly) n-His segments (n= 1, 2) in methionine-and histidine-containing peptides: The role of ammine loss trans to the initial Pt-S (Met) anchor in facilitating amide hydrolysis. *J. Inorg. Biochem.* **2006**, *100*, 1506–1513. [CrossRef] [PubMed]
28. Rajković, S.; Živković, M.D.; Kállay, C.; Sóvágó, I.; Djuran, M.I. A study of the reactions of a methionine-and histidine-containing tetrapeptide with different Pd (II) and Pt (II) complexes: Selective cleavage of the amide bond by platination of the peptide and steric modification of the catalyst. *Dalton Trans.* **2009**, *39*, 8370–8377. [CrossRef]
29. Barrios, A.M.; Lippard, S.J. Interaction of Urea with a Hydroxide-Bridged Dinuclear Nickel Center: An Alternative Model for the Mechanism of Urease. *J. Am. Chem. Soc.* **2000**, *122*, 9172–9177. [CrossRef]
30. Uprety, B.; Arderne, C.; Bernal, I. Catalytic Cleavage of the Amide Bond in Urea Using a Cobalt(III) Amino-Based Complex. *Eur. J. Inorg. Chem.* **2018**, 5058–5067. [CrossRef]
31. MacDonald, M.J.; Lavis, L.D.; Hilvert, D.; Gellman, S.H. Evaluation of the Ser-His dipeptide, a putative catalyst of amide and ester hydrolysis. *Org. Lett.* **2016**, *18*, 3518–3521. [CrossRef] [PubMed]
32. Menger, F.M.; Ladika, M. Remote enzyme-coupled amine release. *J. Org. Chem.* **1990**, *55*, 3006–3007. [CrossRef]
33. Blagoeva, I.B.; Kirby, A.J. Intramolecular nucleophilic catalysis of anilide hydrolysis by pyrimidine nitrogen. *J. Chem. Soc. Perkin Trans. 2* **1985**, *7*, 1017–1020. [CrossRef]
34. McEuten, J.M.; Nelson, R.P.; Lawton, R.G. The α,α' Annelation of Cyclic Ketones. Synthesis and Conformational Properties of Bicyclo[3.3.l]nonanone Derivatives. *J. Org. Chem.* **1970**, *35*, 690–696. [CrossRef]
35. Ishchenko, A.Y.; Yanik, S.; Rusanov, E.B.; Komarov, I.V.; Kirby, A.J. An Expedient and Practical Approach to Functionalized 3-Aza-, 3-Oxa-, and 3-Thiabicyclo [3.3.1] nonane Systems. *Synthesis* **2015**, *47*, 367. [CrossRef]
36. Komppa, G. Über ein neues bicyclisches Imin, das Iso-granatanin. *Ber. Dtsch. Chem. Ges.* **1932**, *65*, 792–793. [CrossRef]
37. Rossi, S.; Valvo, C. Several derivatives of 3-azabicyclo-(3, 3, 1)-nonane; pharmacological activity. I. *Il Farmaco* **1957**, *12*, 1008–1015.
38. Wimley, W.C.; Gawrisch, K.; Creamer, T.P.; White, S.H. Direct measurement of salt-bridge solvation energies using a peptide model system: Implications for protein stability. *Proc. Natl. Acad. Sci. USA* **1996**, *93*, 2985–2990. [CrossRef]
39. Webb, E.C. *Enzyme Nomenclature 1992. Recommendations of the Nomenclature Committee of the International Union of Biochemistry and Molecular Biology on the Nomenclature and Classification of Enzymes*, 6th ed.; Academic Press: San Diego, CA, USA, 1992; p. 863, ISBN 0122271653.
40. Price, N.; Stevens, L. *Fundamentals of Enzymology: The Cell and Molecular Biology of Catalytic Proteins*, 3rd ed.; OUP Oxford: Oxford, UK, 2000; p. 496, ISBN 9780198502296.
41. Seemuller, E.; Lupas, A.; Stock, D.; Lowe, J.; Huber, R.; Baumeister, W. Proteasome from *Thermoplasma acidophilum*: A threonine protease. *Science* **1995**, *268*, 579–582. [CrossRef]
42. Fujinaga, M.; Cherney, M.M.; Oyama, H.; Oda, K.; James, M.N. The molecular structure and catalytic mechanism of a novel carboxyl peptidase from *Scytalidium lignicolum*. *Proc. Natl. Acad. Sci. USA* **2004**, *101*, 3364–3369. [CrossRef]
43. Rawlings, N.D.; Barrett, A.J.; Bateman, A. Asparagine peptide lyases: A seventh catalytic type of proteolytic enzymes. *J. Biol. Chem.* **2011**, *286*, 38321–38328. [CrossRef] [PubMed]

44. Paulus, H. Protein splicing and related forms of protein autoprocessing. *Annu. Rev. Biochem.* **2000**, *69*, 447–496. [CrossRef] [PubMed]
45. Noren, C.J.; Wang, J.; Perler, F.B. Dissecting the chemistry of protein splicing and its applications. *Angew. Chem. Int. Ed.* **2000**, *39*, 450–466. [CrossRef]
46. Sun, P.; Ye, S.; Ferrandon, S.; Evans, T.C.; Xu, M.Q.; Rao, Z. Crystal structures of an intein from the split dnaE gene of Synechocystis sp. PCC6803 reveal the catalytic model without the penultimate histidine and the mechanism of zinc ion inhibition of protein splicing. *J. Mol. Biol.* **2005**, *353*, 1093–1105. [CrossRef] [PubMed]
47. Liu, Z.; Frutos, S.; Bick, M.J.; Vila-Perelló, M.; Debelouchina, G.T.; Darst, S.A.; Muir, T.W. Structure of the branched intermediate in protein splicing. *Proc. Natl. Acad. Sci. USA* **2014**, *111*, 8422–8427. [CrossRef] [PubMed]
48. Kirby, A.J.; Komarov, I.V.; Feeder, N. Synthesis, structure and reactions of the most twisted amide. *J. Chem. Soc. Perkin Trans. 2* **2001**, *4*, 522–529. [CrossRef]
49. Lizak, C.; Gerber, S.; Numao, S.; Aebi, M.; Locher, K.P. X-ray structure of a bacterial oligosaccharyltransferase. *Nature* **2011**, *474*, 350. [CrossRef] [PubMed]
50. Cleland, W.W.; Andrews, T.J.; Gutteridge, S.; Hartman, F.C.; Lorimer, G.H. Mechanism of Rubisco: The carbamate as general base. *Chem. Rev.* **1998**, *98*, 549–562. [CrossRef]
51. Komiya, C.; Aihara, K.; Morishita, K.; Ding, H.; Inokuma, T.; Shigenaga, A.; Otaka, A. Development of an intein-inspired amide cleavage chemical device. *J. Org. Chem.* **2015**, *81*, 699–707. [CrossRef]
52. Armarego, W.L.F.; Chai, C. *Purification of Laboratory Chemicals*, 7th ed.; Butterworth-Heinemann: Oxford, UK, 2013; p. 1024, ISBN 9780123821614.
53. Peters, J.A.; Van Der Toorn, J.M.; Van Bekkum, H. 3, 7-disubstituted bicyclo [3.3.1] nonanes—III: Synthesis and conformation of bicyclo[3.3.1]nonane-3α, 7α-dicarboxylic acid, its dimethyl ester and some other 3,7-disubstituted bicyclo[3.3.1]nonanes; adamantane as an integrated holding system. *Tetrahedron* **1975**, *31*, 2273–2281. [CrossRef]
54. Połoński, T.; Pham, M.; Milewska, M.J. Structure, Conformation, and Stereodynamics of N-Nitroso-2,4-diaryl-3-azabicyclo[3.3.1]nonanes and N-Nitroso-2,4-diaryl-3-azabicyclo[3.3.1]nonan-9- ones. *J. Org. Chem.* **1996**, *61*, 3766–3772. [CrossRef]
55. Hwang, T.-L.; Shaka, A.J. Water Suppression that Works. Excitation Sculpting Using Arbitrary Waveform and Pulsed Field Gradient. *J. Magn. Res. Ser. A* **1995**, *112*, 275–279. [CrossRef]

6

Amide Bond Activation of Biological Molecules

Sriram Mahesh, Kuei-Chien Tang and Monika Raj *

Department of Chemistry and Biochemistry, Auburn University, Auburn, AL 36849, USA;
mzs0154@auburn.edu (S.M.); kzt0026@tigermail.auburn.edu (K.-C.T.)
* Correspondence: mzr0068@auburn.edu

Academic Editor: Michal Szostak

Abstract: Amide bonds are the most prevalent structures found in organic molecules and various biomolecules such as peptides, proteins, DNA, and RNA. The unique feature of amide bonds is their ability to form resonating structures, thus, they are highly stable and adopt particular three-dimensional structures, which, in turn, are responsible for their functions. The main focus of this review article is to report the methodologies for the activation of the unactivated amide bonds present in biomolecules, which includes the enzymatic approach, metal complexes, and non-metal based methods. This article also discusses some of the applications of amide bond activation approaches in the sequencing of proteins and the synthesis of peptide acids, esters, amides, and thioesters.

Keywords: peptide bond cleavage; amide bond resonance; twisted amides; enzymes; metal complexes; catalysts

1. Introduction

The amide bond is one of the most abundant chemical bonds and widely exists in many organic molecules and biomolecules [1–6]. Nature has used amide bonds to make these important biomolecules because of the high stability of amide bonds towards various reaction conditions (acidic and basic conditions), high temperature, and the presence of other chemicals [7]. The high stability of amide bonds is attributed to its tendency to form a resonating structure, which provides a double bond character to the amide CO-N bond (Figure 1) [8–10]. The resonance of these amide bonds forms a planar structure and hinders the free rotation around the CO-N bond, thus, it is responsible for 3D structures adopted by proteins and other biomolecules. These 3D structures of biomolecules are responsible for various important biological functions.

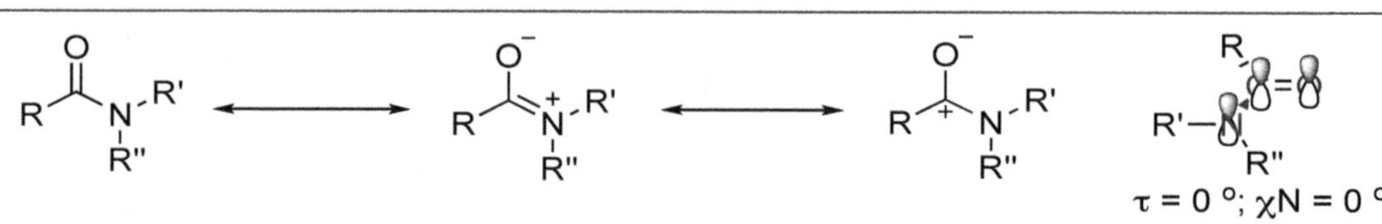

Figure 1. Classical amide bond resonance.

Hansen et al. carried out the rate studies on the hydrolysis of amide bonds at various pH conditions [11]. The study concluded that at pH 7, the rate of hydrolysis is due to the direct attack of water on peptide and measured as kH_2O. The rate constant showed that the half-life of the amide bonds is 267 years, similar to the value determined by Radzicka and Wolfenden [12]. This study also showed that the rates of acid (kH_3O^+) and base hydrolysis (kOH^-) are identical, therefore, the rate of the hydrolysis of the peptide bond is dominated by kH_2O throughout the pH range from pH 5–9.

Recently, various methods have been reported in the literature to activate the amide bonds towards a variety of nucleophiles or electrophiles for the synthesis of other organic compounds. This includes the use of enzymes, metal complexes, and non-metal based methods [13–15]. One widely reported approach for the activation of amide bonds involves the distortion of amide bonds, thus, the amide bond is no longer able to form a resonating structure, loses its double bond character, and becomes more susceptible to nucleophilic or electrophilic attack. A higher distortion of the amide bond from the planar structure makes it more reactive, as evidenced by various twisted amide bonds present in cyclic nonplanar bridged lactams, as demonstrated by Stoltz [16,17], Kirby [18–20], and others [21–23] (Figure 2). One of the special cases to achieve maximum rotational inversion of the amide bond so that it remains in the twisted conformation is the use of *N*-acyl-glutarimides [24–29] and *N*,*N*-substituted amide bonds [30,31] (Figure 2). It is this strong distortion of amide bonds that provides amide bonds with a high reactivity toward a variety of nucleophiles and electrophiles.

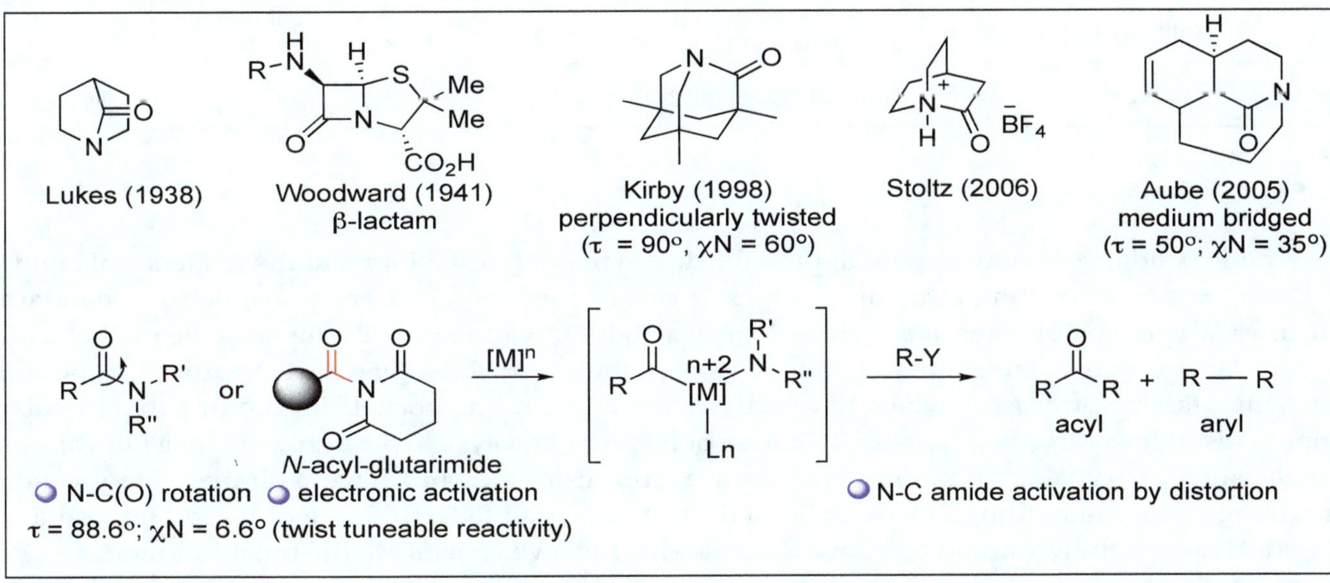

Figure 2. Twisted amides for activation of amide bonds.

There are already some excellent review articles in the literature covering the reactivity of twisted/activated amide bonds for the synthesis of the variety of different organic molecules such as ketones, esters, acids, and alcohols, by cross-coupling reactions [24–31]. The main focus of this review is to summarize the methods for the activation of less reactive amide bonds present in biomolecules such as peptides, proteins, glycopeptides, nucleotides in DNA and RNA and various other peptide bioconjugates, toward attack by various nucleophiles. This task was accomplished by various methods such as by using biological molecules, metal complexes, and non-metal based methods and is discussed below.

2. Biomolecules for the Activation of Amide Bonds—The Enzyme-Directed Hydrolysis of Amides

We have summarized different kinds of enzymes, their mechanisms of hydrolysis of unactivated peptide bonds, and the point of cleavages in Table 1.

Table 1. Enzymatic directed hydrolysis of peptide bonds.

Entry	Enzyme	Method of Hydrolysis	Point of Cleavage	Ref.
1	Serine and Cysteine Proteases	Oxyanion binding hole with catalytic triad	-	[32–37]
2	Metallo-endopeptidase	Thermolysine with Zn^{2+} binds to His 142, His 146, Glu 166	Internal peptide bonds on the N-terminal side of large hydrophobic amino acids	[38–51]
3	Metalloexopeptidase	Carboxypeptidase A with Zn^{2+} through Lewis acid activation	C-terminus comprising large hydrophobic amino acids	[52–54]
4	*O*-GlcNAc transferase	Glycosylation followed by enzyme catalyzed pyroglutamate formation	N-terminal glutamic acid	[55,56]
5	Nicotinamidase	Enzyme Chelation to Zn^{2+} and catalytic triad	Nicotinamide	[57–60]
6	Flavoenzyme	Flavin hydroperoxide intiated oxidative mechanism	Unactivated amide bond in uracil	[61–63]
7	Antibody Fab-BL 125	Catalyzes unactivated primary amide bond hydrolysis	Primary amide bond of L-isomer of peptides	[64,65]
8	RNA	Mg^{2+} catalyzed mechanism	Unactivated alkyl amide of DNA analog	[66]

2.1. Serine Proteases

Amide bonds are widely present in proteins due to their high stability and the tendency of amide bonds to exist in resonating structures, which is one of the key factors responsible for secondary structures adopted by proteins and their biological activities. Nature has developed some methods for the cleavage of highly stable amide bonds to control their functions. One such approach is the use of enzymes (serine proteases), which have active sites, and binding pockets for binding to particular amino acids followed by the activation of amide bonds for hydrolysis. These enzymes exist in various families such as trypsin, chymotrypsin, elastase, subtilisin, etc., but have a similar catalytic site containing oxyanion binding hole with Ser, His and Asp triad [32,33]. Some of the proteases have catalytic dyads with two amino acids at the active site, however, triads are the most common.

All these enzymes based on the binding pocket prefer to bind to particular amino acids but the mechanism by which they hydrolyze the amide bond is similar. During the catalysis, these enzymes form an oxyanion hole made up of three amino acids—His, Asp, and Ser—which work in a synergistic manner to break the amide bond (Figure 3). First, the side chain of Asp makes a hydrogen bond with histidine, thus making it more nucleophilic. Second, histidine forms a strong H-bond with the hydroxyl group of serine and abstracts the proton from the hydroxyl group (OH) of serine which in turn attacks amide bond to form a tetrahedral transition state (TS). This TS eventually collapses resulting in the hydrolysis of the amide bond by acid–base catalysis. Wells et al. demonstrated the importance of these residues at the active site by mutating it to alanine [32,33]. They showed that any mutation in the catalytic triad greatly reduces the turnover number which is a consequence of the changes in the enzyme mechanism. Residues in the catalytic triad function in a strongly synergistic manner and contribute a factor of 2×10^6 to the rate enhancement. The study concluded that enzymes increase the rate of amide bond hydrolysis at by least 10^9 to 10^{10} times that of the non-enzymatic hydrolysis of amide bonds.

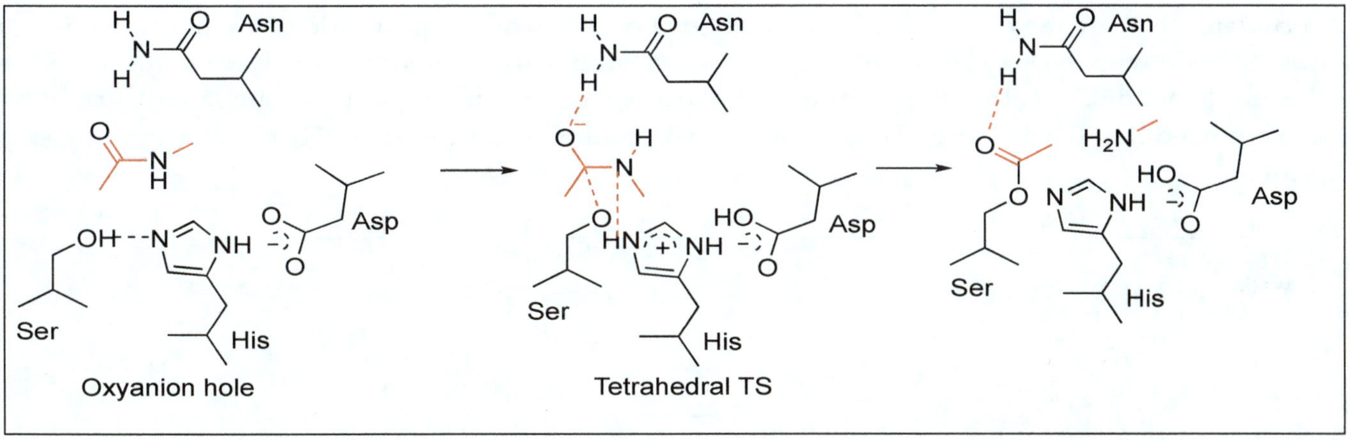

Figure 3. General pathway of serine proteases directed amide bond hydrolysis.

2.2. *Cysteine Proteases*

Cysteine proteases (CPs) hydrolyze the peptide bonds with maximum efficiency at pH 4–6.5 [14]. The thiol group of cysteine protease is susceptible to oxidation so the environment of the enzyme is reducing in nature. Till now 21 families of CPs have been discovered [34–37]. CPs form a triad at the active site during the hydrolysis of the peptide bond made up of Cys-His-Asn residues. First, Asn forms the hydrogen bond with His, then His abstracts the proton from Cys to generate a nucleophilic thiolate ion (S^-) similar to enolate ion generated by serine proteases (Figure 3). Next, the thiolate ion (S^-) attacks the carbonyl group of the peptide resulting in the formation of a tetrahedral intermediate TS followed by the hydrolysis of the amide bond [34–37].

2.3. *Metalloproteases*

Metalloproteases are members of a class of proteases that require a metal ion cofactor at the active site for the hydrolysis of peptide bonds [38]. The most common metal ion cofactor present in metalloproteases is the zinc ion (Zn^{2+}) [39]. Other transition metals such as Co^{2+} and Mn^{2+} are capable of restoring the functions in zinc-metalloproteases where the Zn^{2+} core has been removed [39]. Metalloproteases are divided into two major families: metalloendopeptidases and metalloexopeptidases. The names of these families are based on the site of the hydrolysis of the peptide bonds [40,41]. Metalloendopeptidases cleave the internal amide bonds whereas metalloexopeptidases cleave the amide bonds present at the C- or N-terminus of peptides.

2.3.1. Metalloendopeptidase: Thermolysin

Thermolysin (TLN) catalyzes the cleavage of the internal peptide bond at the amino-side of large hydrophobic amino acids, such as leucine, isoleucine, or phenylalanine. TLN and TLN-like proteins require Zn^{2+} as a metal ion cofactor for the cleavage of amide bonds [42–47].

TLN-mediated hydrolysis of the peptide bond is a two-step process (Figure 4) [48–51]. The active site of TLN contains three residues—His142, His146, Glu166—and a water molecule, which are bound to the Zn^{2+} ion. First, the carbonyl group of the peptide coordinates with Zn^{2+} and displaces the hydrogen of a water molecule to form an H-bond with Glu143 and the oxygen of the water molecule remains associated to the Zn^{2+} ion, resulting in the formation of the enzyme-substrate complex (ES). Second, the oxygen of the water attached to Zn^{2+} attacks the carbonyl carbon of the peptide, resulting in the formation of transition state 1 (TS1). TS1 is stabilized by the formation of the H-bond with Asp226 and His231 at the carbonyl oxygen of the peptide followed by the formation of intermediate gem-diolate (INT) by the breakage of hydroxyl OH bond of water. The amide of the peptide forms a hydrogen bond with the H of H_2O. Third, the carbonyl bond rearrangement in TS2 leads to the breakage of the amide bond (CONH) of the peptide and releases the N-terminal peptide. The rate-determining studies showed that the collapse of a zwitterionic tetrahedral intermediate (INT) is a rate-limiting step (Figure 4).

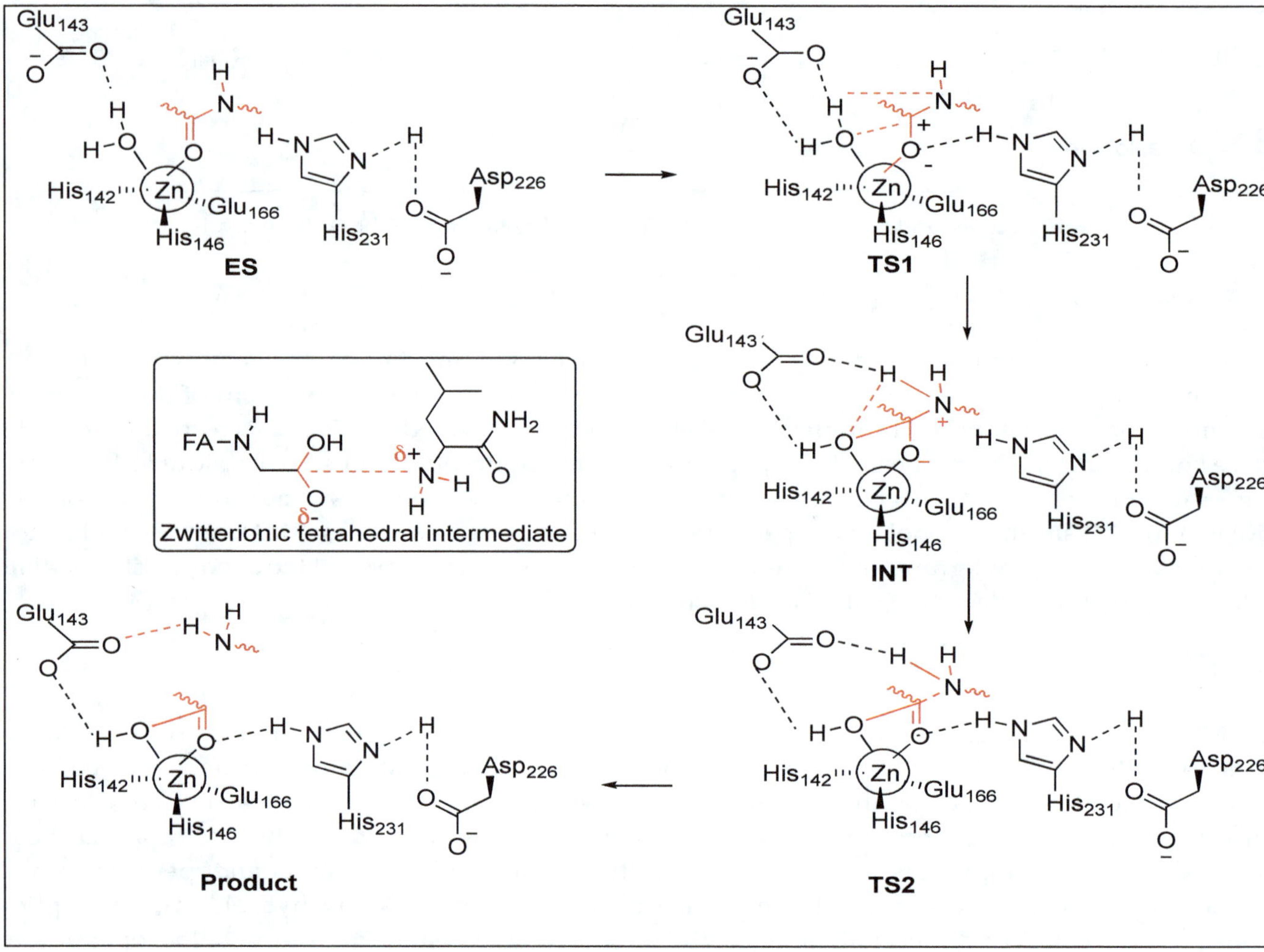

Figure 4. Thermolysin Mechanistic pathway.

2.3.2. Metalloexopeptidase: Carboxypeptidase A

Carboxypeptidase A (CPA) is a 35 kDa metalloenzyme and contains a Zn^{2+} ion cofactor in its active site [52–54]. CPA is an exopeptidase, which catalyzes the hydrolysis of amide bonds present at the C-terminus comprising large hydrophobic side chains. Two different mechanisms have been proposed for the cleavage by these metalloproteases (Figure 5) which showed the importance of Lewis acid catalysis for the activation of amide bonds [53,54]. One involves the Lewis-acid activation of the carbonyl group of the amide bond by Zn^{2+}, followed by the attack of water (Figure 5). The second involves the Lewis-acid activation of H_2O by Zn^{2+} ions followed by the attack of the hydroxide ion of the water on the carbonyl group of the amide bond (Figure 5).

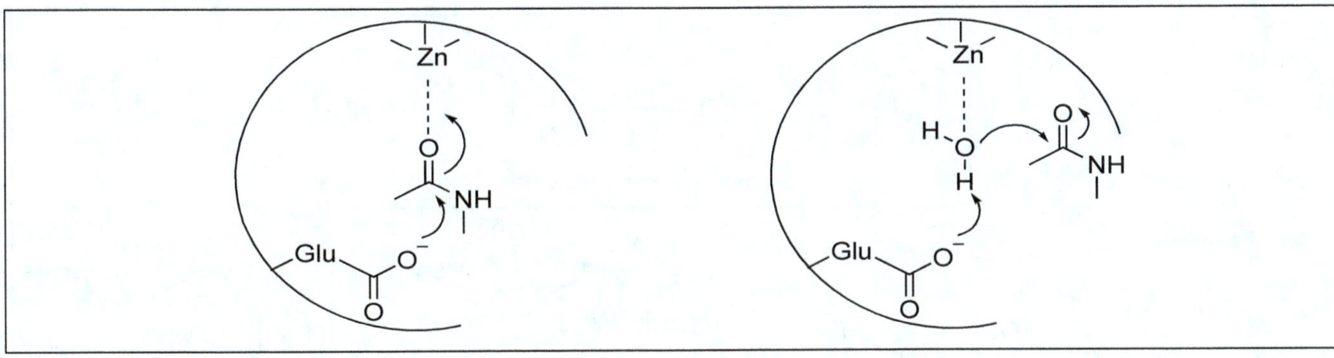

Figure 5. Mechanisms of carboxypeptidase A.

2.3.3. Glutamate Glycosylation for Amide Bond Cleavage

Human enzyme *O*-GlcNAc transferase (OGT) is essential for the cleavage of amide bonds in host cell factor-1 (HCF-1). HCF-1 cleavage takes place at the N-terminal glutamic acid by the glycosylation which is catalyzed by enzyme OGT. Mechanistic studies showed that the glycosylation of the glutamate side chain (intermediate 1, Figure 6) leads to the formation of an enzyme-catalyzed internal pyroglutamate formation (intermediate 2, Figure 6) with the amidic nitrogen of the peptide backbone chain, which then undergoes spontaneous hydrolysis (Figure 6) [55]. Detailed mechanistic studies showed that the rate of conversion of glycopeptide to internal pyroglutamate was an order of magnitude slower than observed in the presence of OGT, thus, it was concluded that both the first and second steps occurred while the peptide is bound to OGT (Figure 6). Hydrolysis likely occurs after dissociation from the enzyme. It has also been reported that glycosylation on Thr next to glutamate also prevents the cleavage at the glutamate (Glu) because of the steric hindrance and thus the enzyme is unable to carry out the glycosylation of glutamate [56].

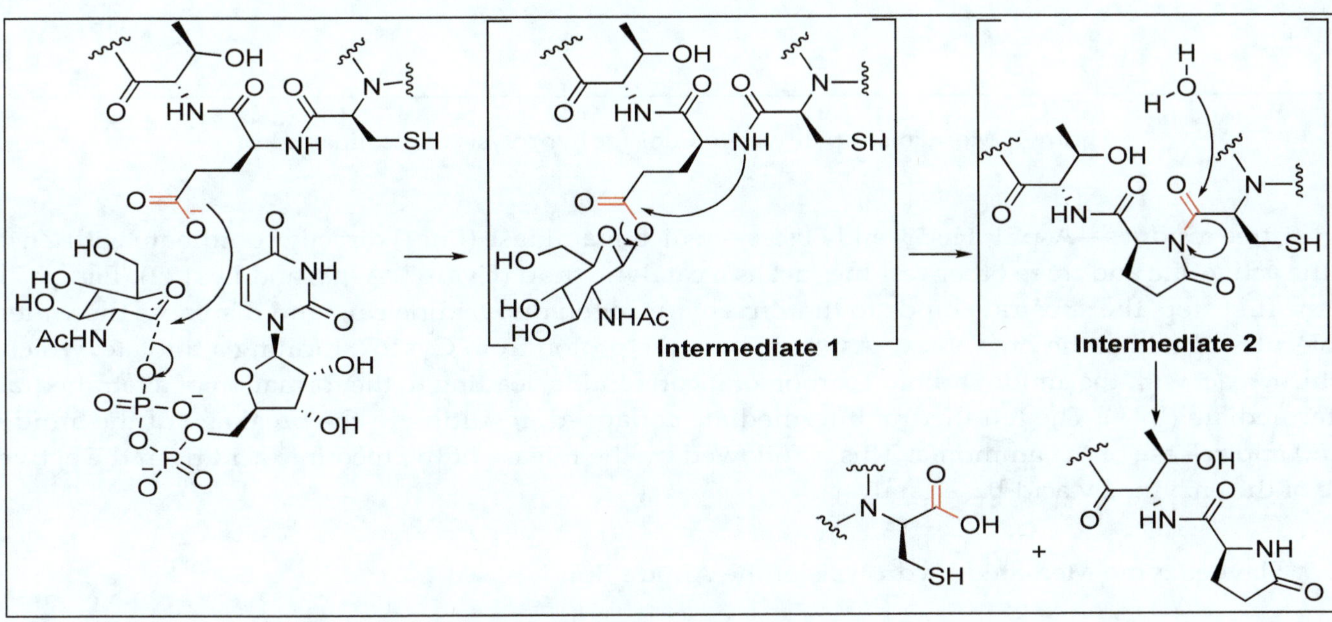

Figure 6. Glycosylation pathway.

2.3.4. Nicotinamidase (Pnc1) for the Hydrolysis of the Amide Bond of Nicotinamide

Nicotinamidases catalyze the cleavage of nicotinamide, which is a critically important part of NAD$^+$ and NADH, to nicotinic acid and ammonia (Figure 7). A detailed study showed that both the carbonyl oxygen and the ring nitrogen of nicotinamide are critical for binding to the nicotinamidases and reactivity [57–60].

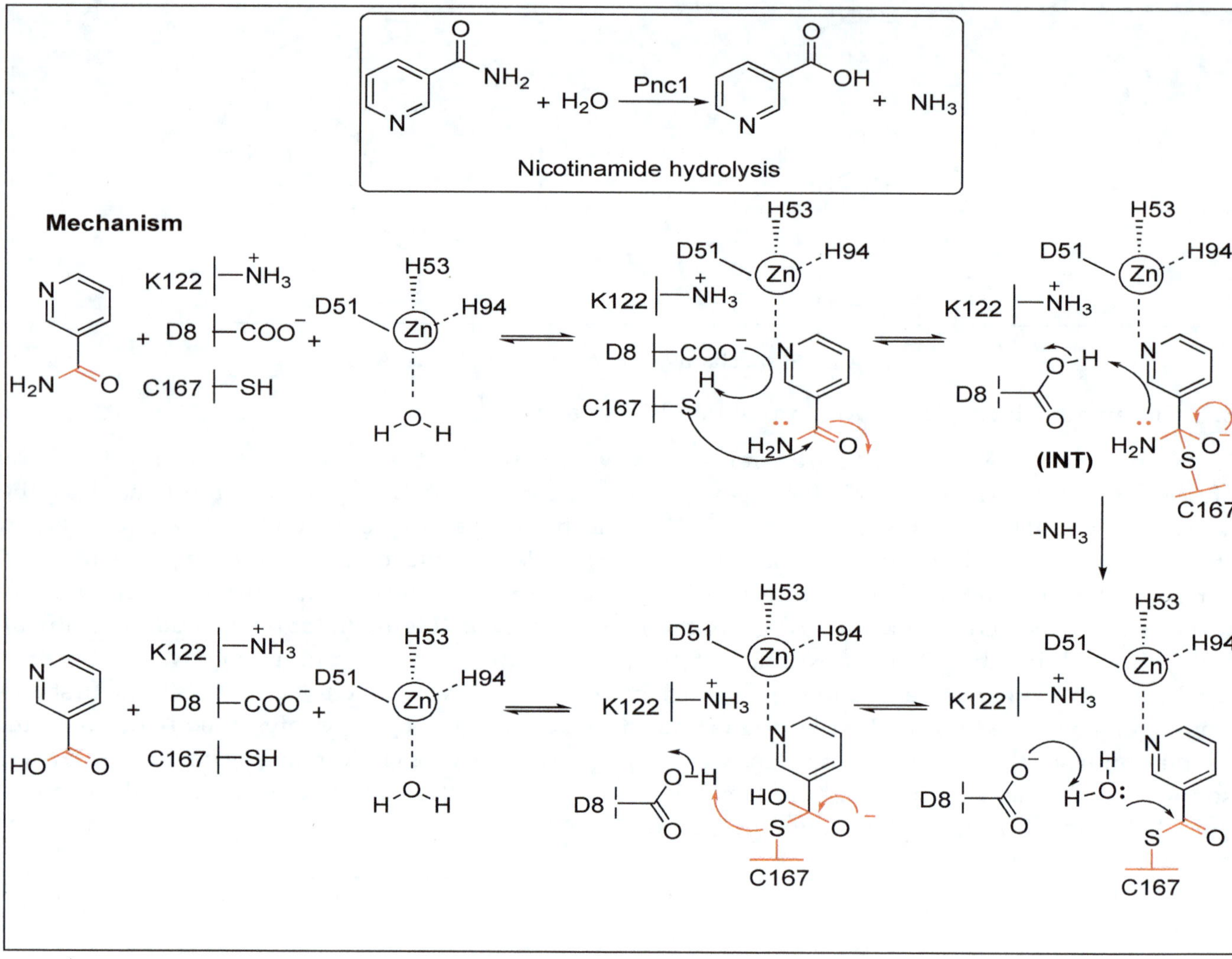

Figure 7. Mechanistic pathway of Pnc1 for hydrolysis of nicotinamide.

Three residues—Asp51, His53, and His94—in nicotinamidase (Pnc1) directly coordinate with Zn^{2+} at the active site and three other residues act as a catalytic triad (Cys167, Asp8, and Lys122) (Figure 7). In the first step, the substrate binds to the Zn^{2+} by nitrogen of pyridine ring and displaces the water molecules ligated to the Zn^{2+}. Next, Asp8 removes the proton from Cys167, forming a thiolate, which, in turn, react with the amide carbonyl carbon of nicotinamide, leading to the formation of a tetrahedral intermediate (INT). The tetrahedral intermediate collapsed, resulting in the breakage of the amide bond and release of the ammonia. This is followed by the release of the nicotinic acid from the active site of the enzyme by acid-base catalysis.

2.3.5. Flavoenzyme-Mediated Hydrolysis of the Amide Bond

Begley et al. demonstrated the role of flavoenzyme in the cleavage of the unactivated amide bond in uracil, a building block for RNA (Figure 8) [61–63]. The detailed mechanistic analysis showed that the reaction takes place through the oxidative mechanism that is initiated by the addition of a flavin hydroperoxide to the C* carbonyl of uracil, forming a tetrahedral intermediate (INT) (Figure 8). This is followed by the collapsing of the tetrahedral intermediate (INT), leading to the cleavage of an amide bond in uracil. This was the first example where such chemistry was shown by flavin hydroperoxides.

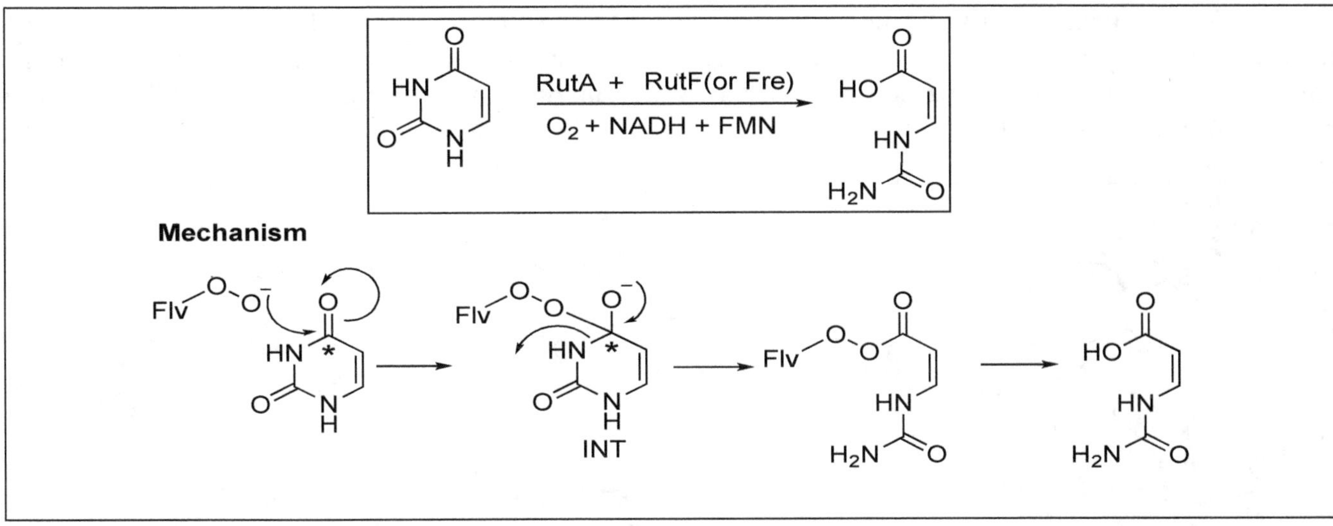

Figure 8. Flavoenzyme-mediated hydrolysis of amide bond.

2.3.6. Primary Amide Bond Hydrolysis by Antibodies

The antibody Fab-BL125 catalyzes the hydrolysis of the unactivated primary amide bond of the L-isomer of peptides to generate free peptide acid (Figure 9). The antibody showed high regio- and diastereoselectivity since the D-proline primary amide diastereoisomer did not undergo any hydrolysis. The antibody Fab-BL125 decreases the half-life of the peptide from 17.5 years to only 3.9 h. Such an antibody was obtained by using the α-amino boronic acid Hapten by a direct selection strategy from the antibody combinatorial libraries [64,65].

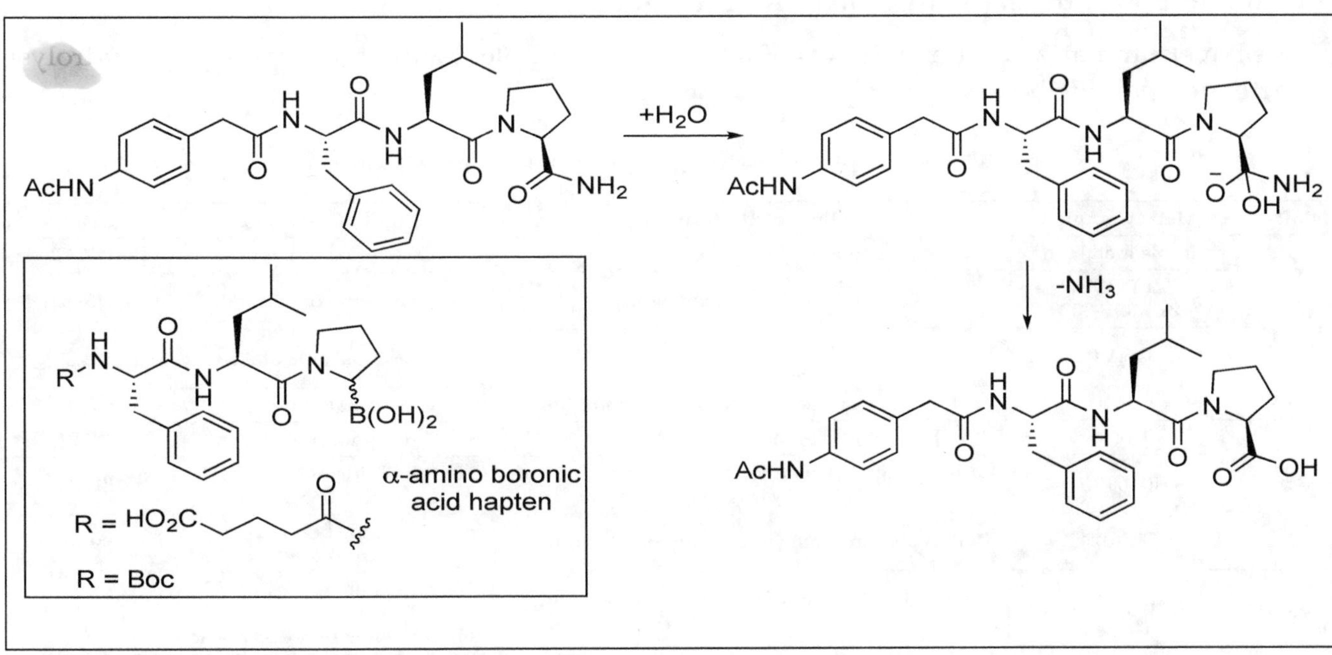

Figure 9. Antibody Fab catalyzed primary amide bond hydrolysis.

2.3.7. RNA-Assisted Cleavage of Amide Bonds

A group I RNA obtained by in vitro evolution catalyzes the cleavage of unactivated alkyl amides of DNA analog. This includes substrates with an amide bond that joins either two DNAs, or a DNA with a short peptide. The RNA increases the rate of hydrolysis by more than 10^3 in comparison to the uncatalyzed reaction. The RNA-catalyzed amide bond cleavage was entirely dependent on Mg^{2+} where Mg^{2+} acts as a Lewis acid thus activating the carbonyl group of the amide bond for the nucleophilic attack by the hydroxy group of RNA (Figure 10). No amide bond cleavage was detected in the presence of other metal ions such as Zn^{2+}, Ca^{2+}, or Sr^{2+}. A trace amount of amide bond cleavage was observed in the presence of $MnCl_2$ which is in contrast to the RNA and DNA cleavage reactions [66].

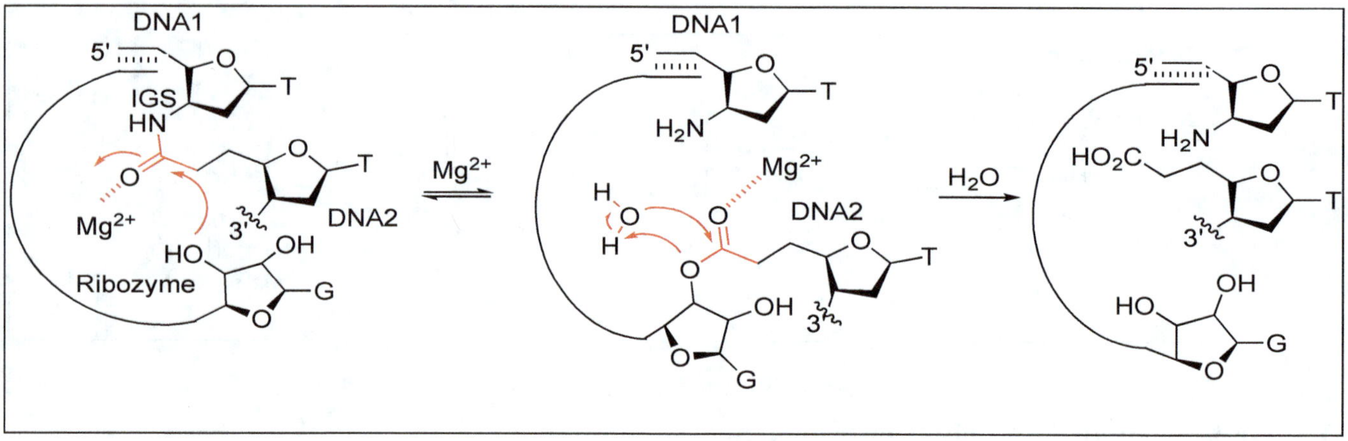

Figure 10. RNA catalyzed amide bond cleavage.

To determine the generality of the amide bond cleavage by RNA, a series of substrates were explored. If a short DNA is attached to a peptide or an amino acid by using an amide bond, the immediate cleavage at the amide bond was observed in the presence of the RNA. The cleavage of the amide bond between the amino acid residues was not observed.

3. Metal Complexes for the Activation of Amide Bonds

We have summarized different kinds of metal-based complexes, their mechanisms of hydrolysis of unactivated peptide bonds and point of cleavages in Table 2.

Table 2. Metal catalyzed hydrolysis of peptide bonds.

Entry	Metal Complex	Method of Hydrolysis	Point of Cleavage	Ref.
1	Simple metal ions	Lewis acidity of metal ion	C-terminal of peptide	[67–72]
2	Zr POMs, Zr MOF-808	Lewis acidity of metal ion	C-terminal of peptide	[73–82]
3	Mo(VI)	Lewis acidity of metal ion and formation of 5 membered ring	C-terminal side of Asp	[81,82]
4	Co(III)	N-terminal amine intiated tertiary complex	C-terminal of peptide	[83–88]
5	Mo(II)	Favorable six-membered chelate ring	C-terminal side of Cys	[89]
6	Pd(II), Pt(II)	Carboxylic group of amino acid and side chain of amino acid anchoring metal complex	C-terminal side of Met, His, Cys and S-MeCys	[90–98]
7	Pd(0)	Methionine side chain anchoring metal complex	Second amide bond upstream from Met	[99–102]
8	Co(III) and Cu(II)	Lewis acidity activation and PNA for selectivity towards a particular protein	Mb = Leu 89 – Ala 90 PDF = Gln 152 – Arg 153 BSA = Solvent exposed portion	[103–111]
9	Ni(II)	Non Lewis acid based N,O acyl rearrangement	N-terminal side of Ser/Thr	[112,113]
10	Sc(III)	Lewis acid based N,O acyl rearrangement	N-terminal side of Ser/Thr	[114]

Another approach for the cleavage of peptide bonds involves the use of metal complexes. This metal complex has potential applications in the field of chemical biology, biochemistry, and bioengineering. A variety of metal ion complexes has been utilized for testing the reactivity with substrates such as peptides, and proteins [67]. Most of the metal catalyzed reactions reported so far are based on the activation of amide carbonyl or water by the Lewis acid mechanism of the metal ion. Another metal ion hydrolysis mechanism involves the formation of a square planar complex of the metal ions Cu(II), Ni(II), or Pd(II) with the Ser/Thr-His or Ser/Thr-Xaa-His sequence leading to the N→O rearrangement of the acyl moiety resulting in the cleavage of the peptide bond (Figure 11). In this section, we will provide few examples of both Lewis acid and the N→O acyl rearrangement for the cleavage of peptide bonds.

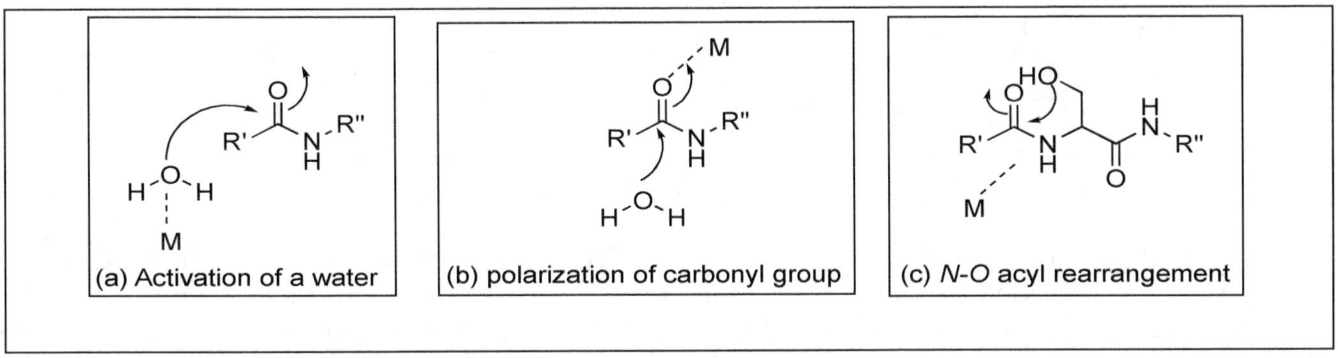

Figure 11. Metal-assisted peptide bond hydrolysis.

3.1. Lewis Acid Mediated Hydrolysis

3.1.1. Simple Metal Ions

Yashiro et al. reported that the rate of hydrolysis of peptide bonds increases by almost all the metal ions and the highest conversion was observed for Zn(II) [68–72]. The active intermediate was metal complexes where metal binds to both the carbonyl group and the N-terminal amino group of the peptide bond. Cleavage of the peptide bond in the presence of the metal takes place through the mechanism in Figure 11b.

3.1.2. Oxo Metal Ions

Parac-Vogt et al. proved that polyoxometalates oxo-metal compounds such as MoO_2^{-4}, WO_2^{-4}, CrO_2^{-4}, and VO_2^{-4} cleave the peptide bonds in various dipeptides [73–77]. They have shown that these oxo-metal compounds also hydrolyze the amide bonds in their regular oligomeric forms, polyoxometalates (POMs). A POM is a unit where one or more atoms can be replaced by a metal center leading to the change in the coordination properties of POM. POMs were utilized for the Zr(IV)- and Ce(IV)-assisted peptide bond hydrolysis because they are homogenous in nature [78,79].

Zirconium Complex Mediated Hydrolysis of Peptide Bonds

Parac-Vogt et al. reported the hydrolysis of peptide bonds catalyzed by a polyoxometalate complex for the first time. They demonstrated the role of metal-substituted Wells-Dawson type polyoxometalates $K_{15}H[Zr(\alpha_2\text{-}P_2W_{17}O_{61})_2]\cdot 25H_2O$ for the hydrolysis of peptide bonds in diglycine, triglycine, tetraglycine, and pentaglycine (GG), yielding glycine as a final product (Figure 12) [80]. A detailed mechanistic investigation by NMR showed that the free amino terminus and both carbonyl functionalities of GG interact with polyoxometalates $K_{15}H[Zr(\alpha_2\text{-}P_2W_{17}O_{61})_2]\cdot 25H_2O$ either by the formation of metal ion coordination complex or by the non-covalent interactions of the protonated amino group with the negatively charged surface of POMs, thus responsible for the activation of amide bonds towards hydrolysis. These POMs selectively cleave the C-terminal amide bond of glycylglycyl amide ($GGNH_2$), resulting in the formation of GG. No free glycine amide (GNH_2) was detected during the course of the reaction.

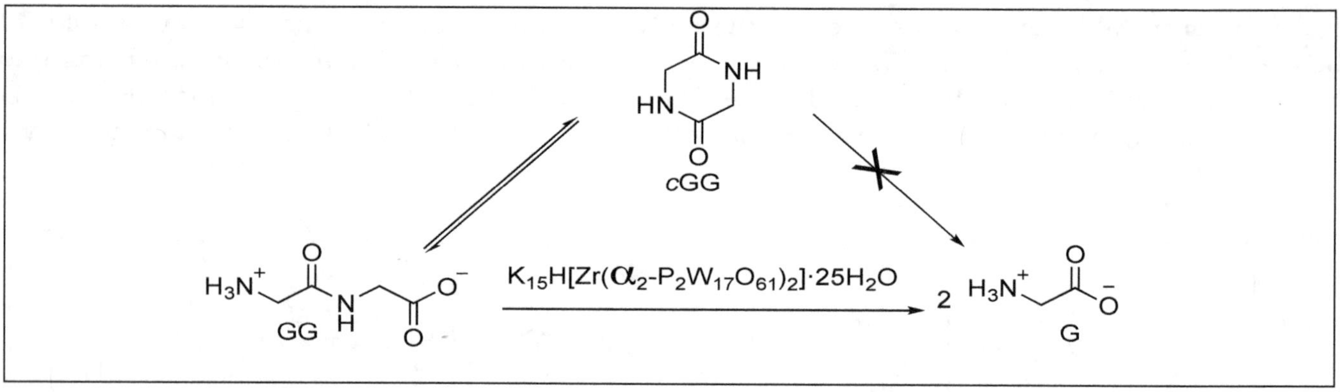

Figure 12. Peptide hydrolysis catalyzed by a polyoxometalate complex.

Recently, Vogt et al. utilized the MOF-808, a Zr(IV)-based metal–organic complex for the hydrolysis of the peptide bond in a wide range of peptides and proteins such as hen egg white lysozyme (HEWL) under physiological conditions [81,82]. The MOF-88 is heterogeneous in nature and is thus a reusable catalyst. The experimental studies and calculations showed that MOF-808 hydrolyzed the Gly-Gly bond by the formation of the active complexes with two adjacent Zr(IV) centers of the $\{Zr_6O_8\}$ core by coordination with amide oxygen and the amine nitrogen atoms. The catalytic efficiency of MOF-808 towards the hydrolysis of peptides is dependent on the bulkiness and nature of the side chain amino acid residues. Dipeptides with small or hydrophilic residues undergo cleavage faster as compared to those with bulky and hydrophobic residues.

Asp-Xaa Selective Hydrolysis of the Peptide Bond by Oxo-Metal Ions

It has been reported that oxomolybdate(VI) catalyzes the cleavage of various peptides containing aspartic acid (Asp) with cleavage at the C-terminal side of the Asp residue. This is due to the attack of the side chain of the Asp on the amidic carbonyl which is activated by the coordination with oxomolybdate(VI), resulting in the formation of the five-membered ring and simultaneous cleavage of the amide bonds (Figure 13) [81,82].

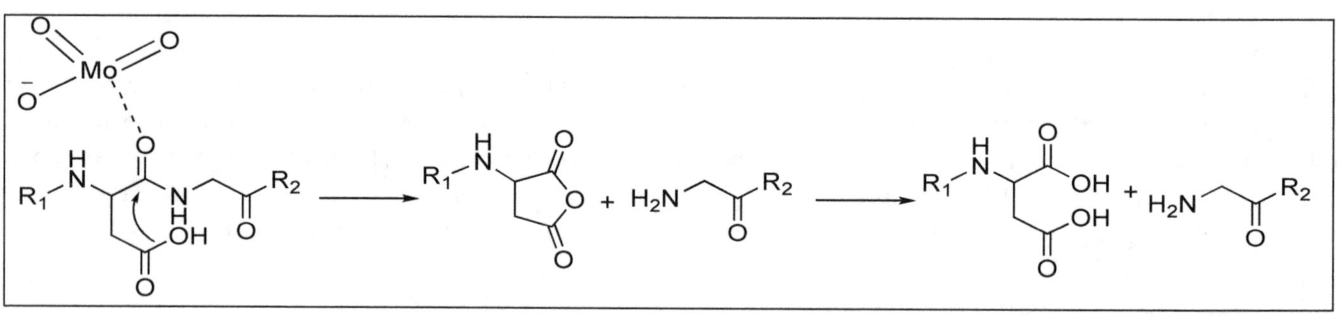

Figure 13. Oxomolybdate(VI) catalyzed cleavage of peptide bonds.

Various Other Metal Complexes

The Westheimer and Trapmann groups showed that the metal complexes containing Co(II), Cu(II), and Ni(II) ions cleave various dipeptides [83,84]. Out of various metal complexes, the Co(III) complex $[Co(trien)OH(H_2O)]^{2+}$ is one of the widely studied metal ion complexes and carried out the rapid hydrolysis of the peptide bond [85–88]. The detailed mechanistic investigation showed that first, the metal complex $[Co(trien)OH(H_2O)]^{2+}$ forms a tertiary complex with a dipeptide by the replacement of an equatorially coordinated water molecule in the octahedral Co(III) complex by the N-terminal amine of the peptide. This mode of coordination brings the axially bound hydroxyl group in close proximity to the peptide carbonyl resulting in the hydrolysis of the peptide bond (Figure 14).

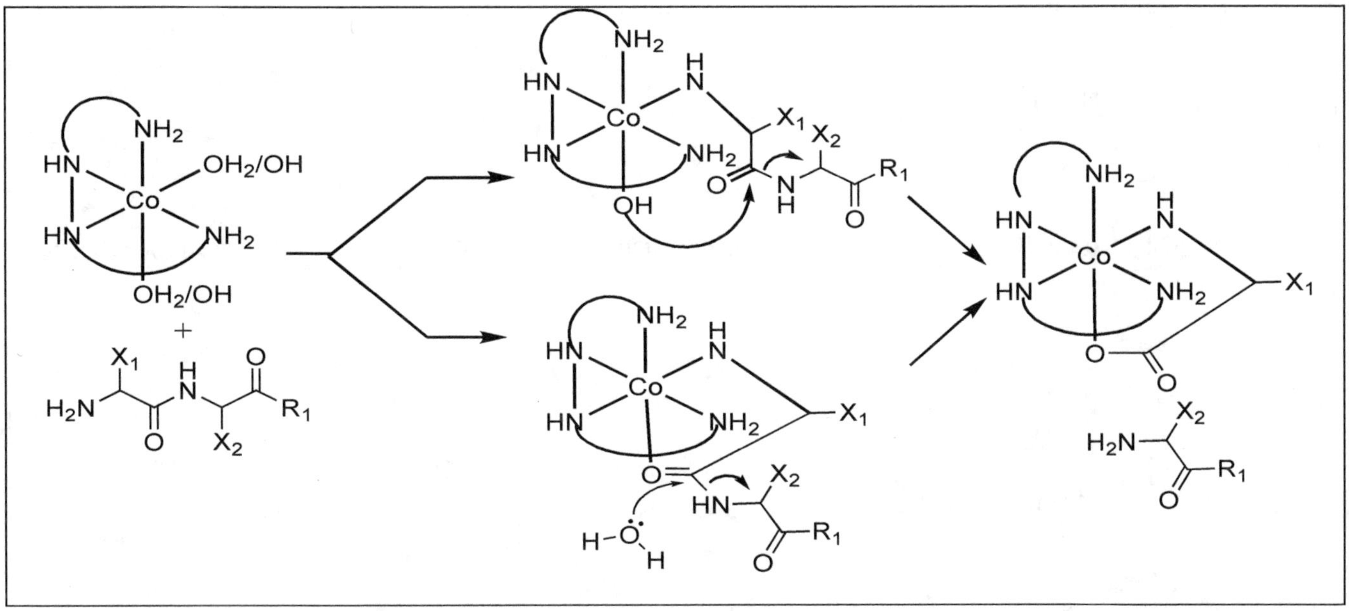

Figure 14. Co(III) Complex catalyzed peptide hydrolysis.

Anchoring at Cys Side Chains: Molybdocene

Erxleben et al. showed that molybdocene dichloride, Cp_2MoCl_2, cleaves the amide bond at the C-terminal side of cysteine to generate Cys-Gly from Gly-Cys-Gly [89]. The cleavage is highly selective for the C-side of Cys because it leads to the formation of the favorable six-membered ring. The mechanism of this reaction is illustrated in Figure 15 [89].

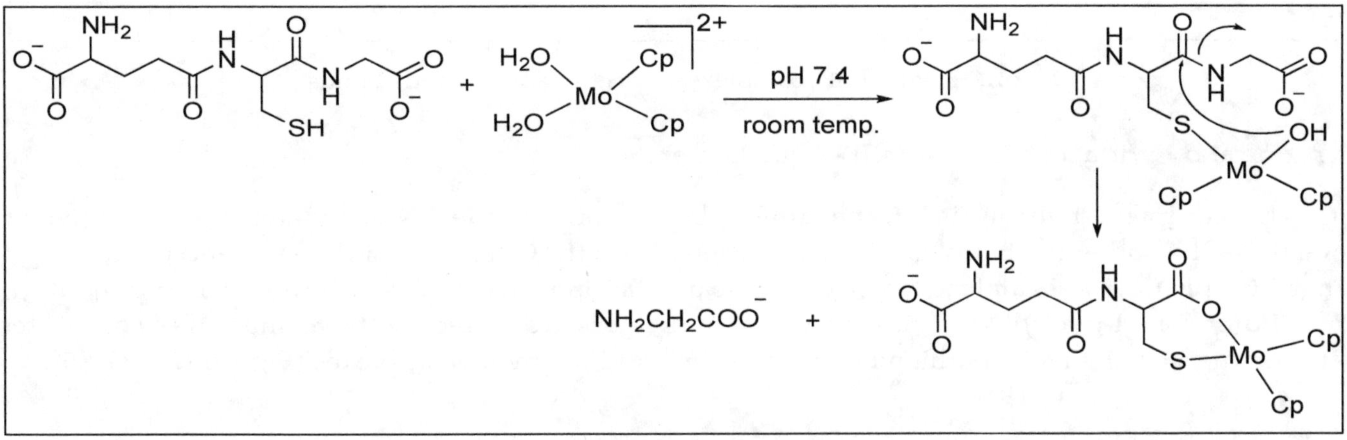

Figure 15. Mo catalyzed peptide hydrolysis.

3.1.3. Anchoring at Met, His, and Cys Side Chains: Palladium(II) Complexes for the Cleavage of Amide Bonds

Figure 16 shows the examples of platinum(II) and palladium(II) complexes which are known for the cleavage of amide bonds under mild conditions. These platinum(II) and palladium(II) complexes attach to the sulfur atom of cysteine, *S*-methylcysteine, and methionine in peptides, thus, promoting the selective cleavage of the unactivated amide bonds at the C-side of the amino acid (Figure 16) [90–92].

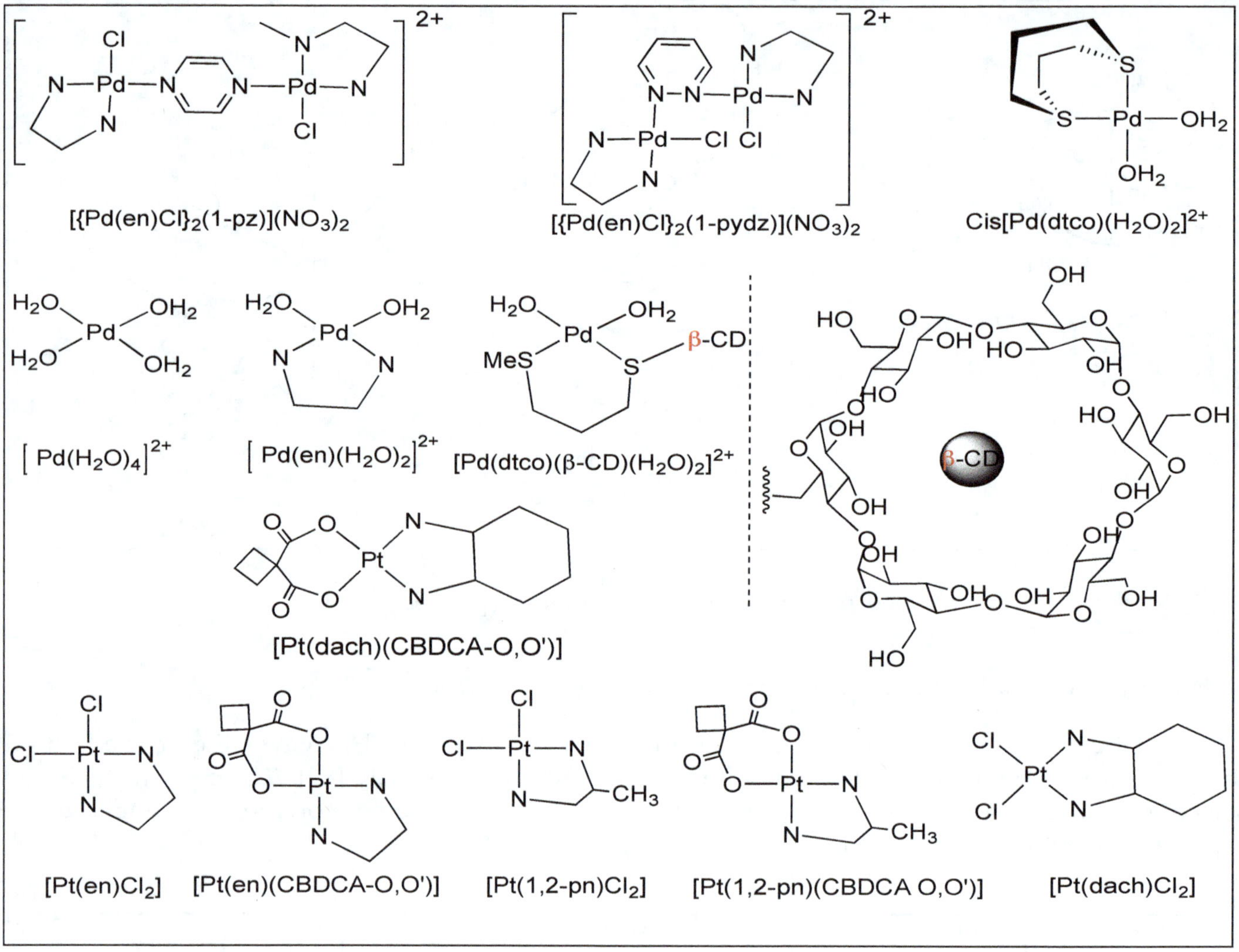

Figure 16. Pd and Pt complexes for activation of amide bonds.

Pyrazine and Pyridazine Palladium(II)-Aqua Dimers

The complete hydrolysis of the amide bonds of peptides *N*-acetylated-L-histidylglycine (Ac-L-His-Gly) and -L-methionylglycine (Ac-L-Met-Gly) at the C-terminal side of Met and His in the pH range $2.0 < pH < 2.5$ was catalyzed by two dinuclear palladium(II) complexes, [{Pd(en)Cl}$_2$(l-pz)](NO$_3$)$_2$ and [{Pd(en)Cl}$_2$(l-pydz)](NO$_3$)$_2$ at 37 °C. The hydrolysis is assisted by the formation of complexes between the side chains of methionine and histidine and the metal complexes (Figure 17) [93–95].

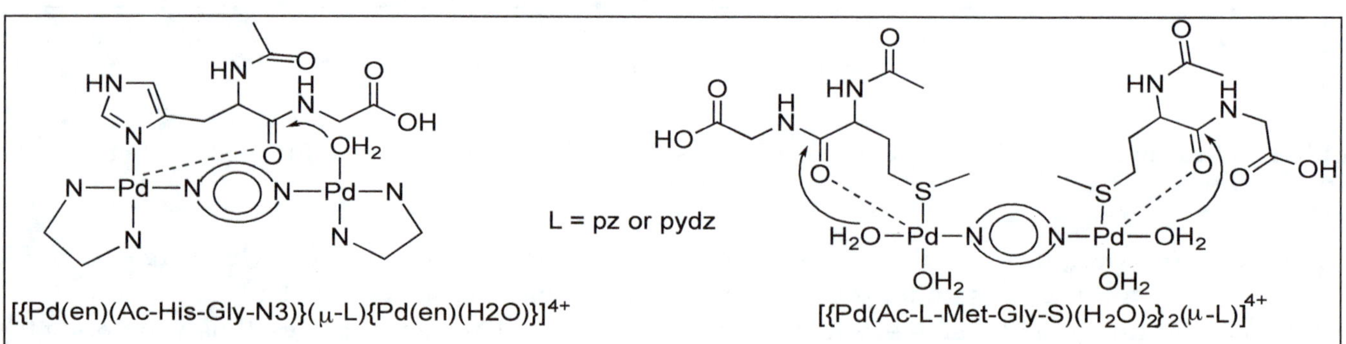

Figure 17. Pd(II) aqua dimers.

Platinum Complexes for the Cleavage of the Amide Bond

^{1}H-NMR investigation of the cleavage reactions between various Pt(II) complexes of the type [Pt(L)Cl_2] and [Pt(L)(CBDCA-*O*,*O′*] (L = ethylenediamine-en; (±)-*trans*-1,2-diaminocyclohexane-dach; (±)-1,2-propylenediamine-1,2-pn and CBDCA is the 1,1-cyclobutanedicarboxylic anion) and the *N*-acetylated-L-methionylglycine dipeptide (MeCOMet-Gly) were reported by Djuran and co-workers [93–96]. The comparison of the rate studies of these Pt complexes for the cleavage of *N*-acetylated-L-methionylglycine dipeptide (MeCOMet-Gly) showed that the rate of hydrolysis decreases with the increase in the steric bulk of the CBDCA and chlorido Pt(II) complexes (en > 1,2-pn > dach) (Figure 18).

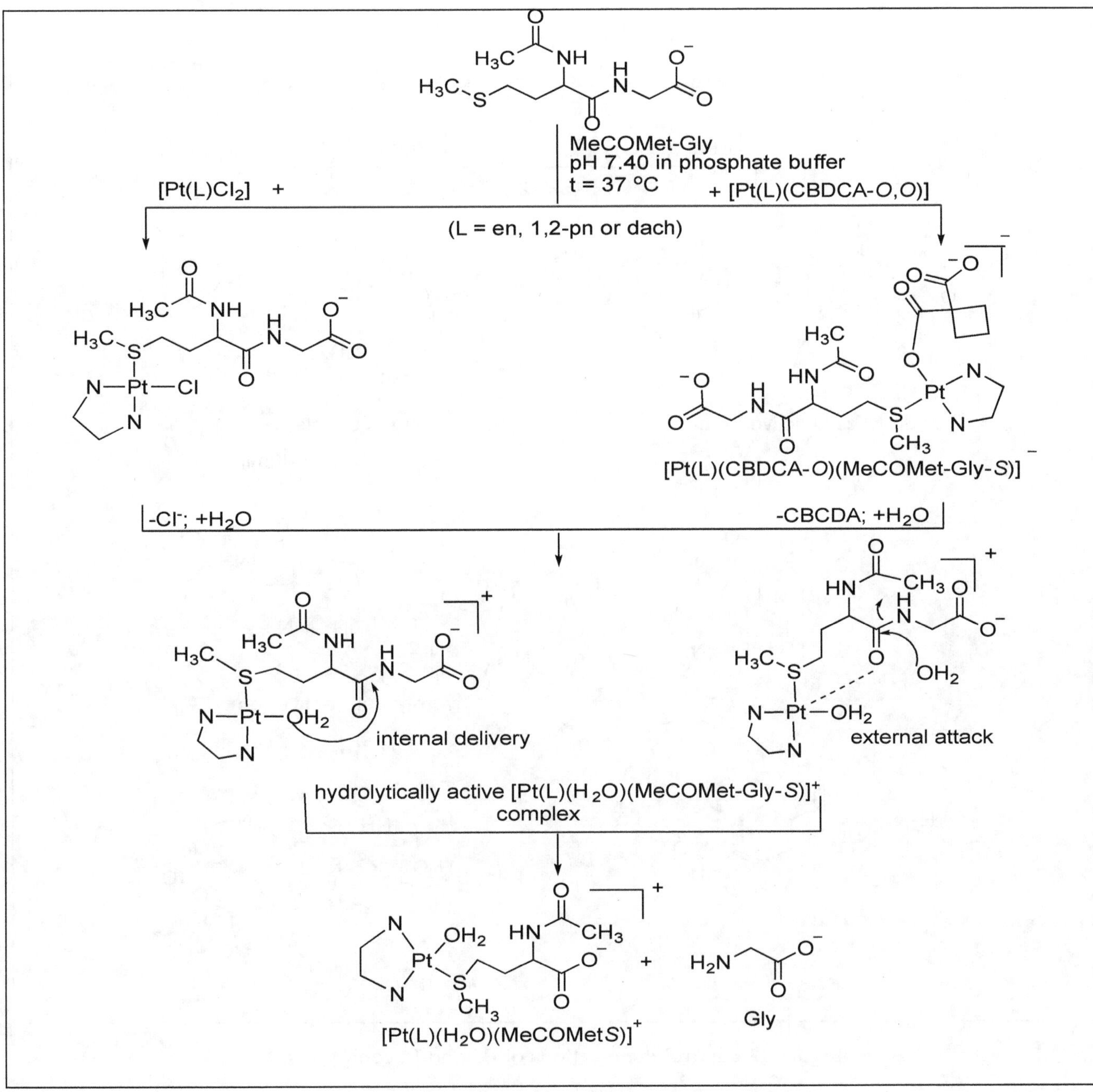

Figure 18. Hydrolytic reactions of Pt complexes.

Later, these $[Pd(en)(H_2O)_2]^{2+}$ and $[Pt(en)(H_2O)_2]^{2+}$ complexes were used for the cleavage of tetrapeptide (MeCOMet-Gly-His-GlyNH$_2$) in the pH range of 1.5–2.0 and at 60 °C. The study showed that these complexes are highly selective for the cleavage of the amide bond at the C-terminal side of methionine (Figure 19) [97]. The high selectivity for the Met-Gly amide bond compared to the other amides is due to the high affinity of the Pt(II) and Pd(II) ions for the sulfur atom on Met. Two different mechanisms for the cleavage of tetrapeptide at the C-terminus of Met residue by Pd complexes have been proposed. One involves the direct coordination of the Pd complex with Met followed by the cleavage. The second involves the formation of macrochelate with both His and Met followed by the hydrolysis of the amide bond.

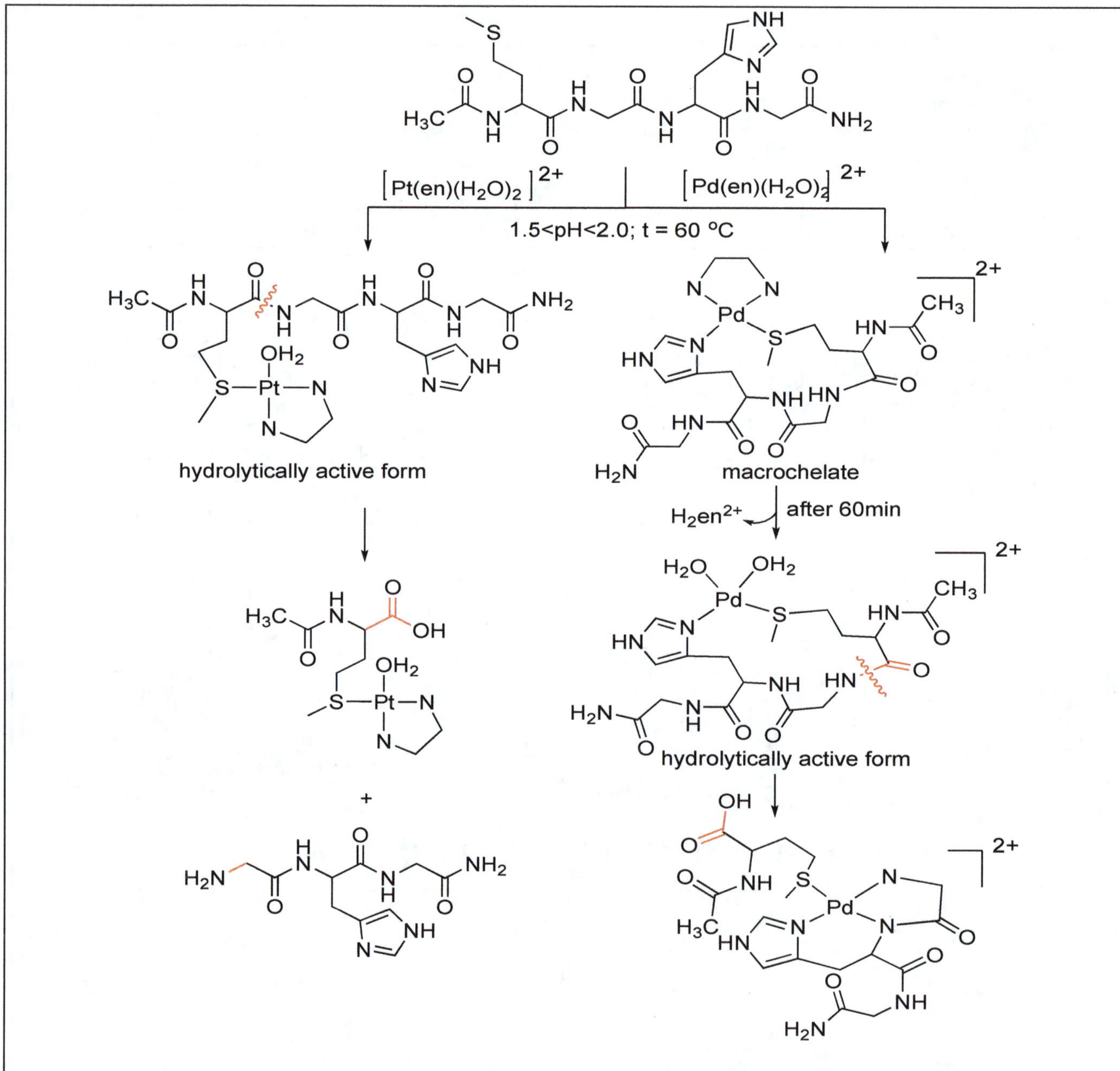

Figure 19. Hydrolytic reactions of Pd and Pt complexes.

Cis[Pd(dtco)(H$_2$O)$_2$]$^{2+}$ leads to the selective cleavage of the amide bond at the C-terminal of Met (Figure 20) [98]. Coordination of the metal complex promotes hydrolysis by two different mechanisms. The first involves the formation of a six-membered complex by the chelation of a metal atom with both the sulfur of Met and the carbonyl of the amide bond, thus, activating the amide bond toward cleavage by an external attack from water. This mechanism is favorable in the case of platinum(II) promoters and substrates with smaller anchoring side chains. Second, the mechanism involves the chelation of metal with sulfur only, followed by the internal attack of water molecules to the amide bond. This mechanism is favorable with palladium(II) promoters and substrates with longer anchoring side chains (Figure 20) [98].

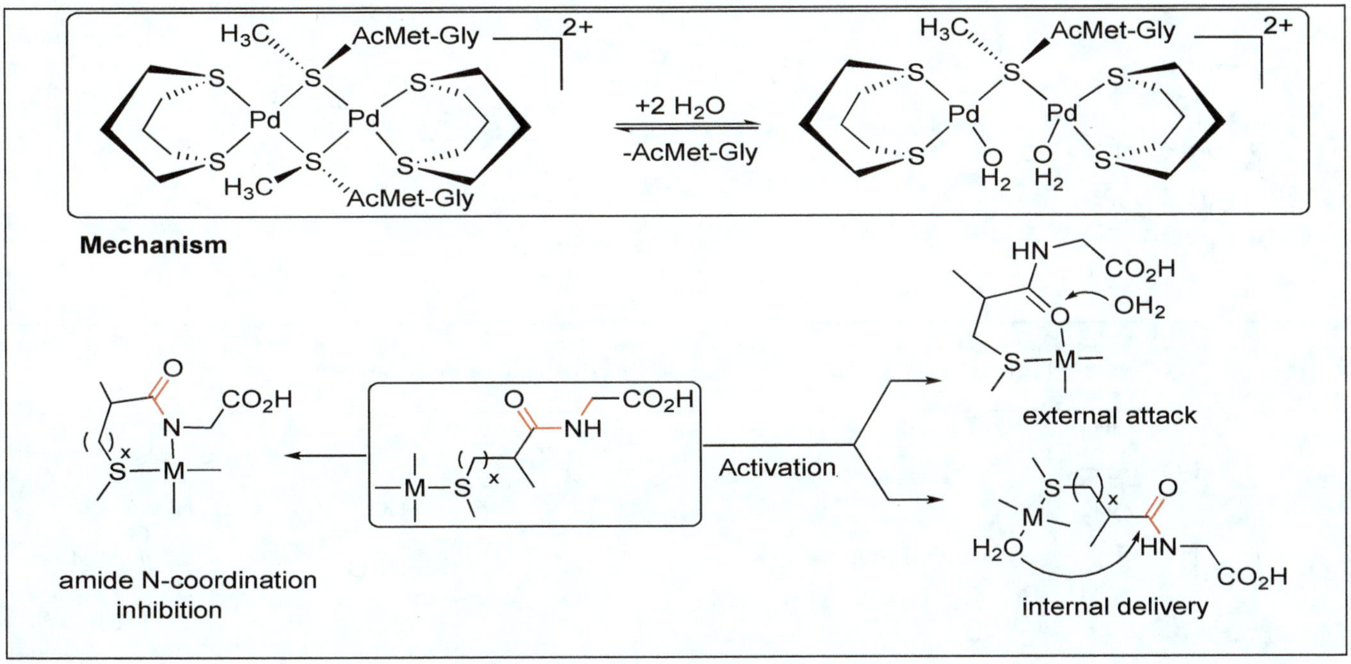

Figure 20. $[Pd(dtco)(H_2O)_2]^{2+}$ mediated hydrolysis of amide bond.

Kostic et al. utilized the $Pd(H_2O)_4$ complex for the cleavage of decapeptide (Ac-AKYGGMAARA) under acidic conditions (pH 2.3) at 60 °C [99]. The cleavage of the peptide bond took place at the Gly residue next to Met on the N-terminal side and generated two fragments after 24 h. The $Pd(H_2O)_4$ complex binds to the S of the Met residue leading to the formation of two different complexes under the reaction conditions (active and inactive), which are in equilibrium with each other (Figure 21). The active complex led to the formation of two fragments through hydrolysis but the inactive complex did not undergo any hydrolysis. The coordination of two amide nitrogen atoms in the inactive complex quench the Lewis acidity of Pd(II), thus no hydrolysis was observed.

Next, Kostic et al. used this metal complex $Pd(H_2O)_4$ for the cleavage of the amide bond under neutral conditions in different peptide substrates such as Gly-Met, Sar-Met, and Pro-Met. This study showed that the rate of hydrolysis at a neutral pH is slow compared to that at a low pH [44]. Interestingly, the complete cleavage of Sar-Met and Pro-Met was observed but no cleavage was observed for Gly-Met at a neutral pH. This is due to the difference in the equilibrium position of the Gly-Met compared to that of Pro-Met. In Gly-Met, equilibrium is more shifted towards the inactive form at neutral pH due to the ability of Gly to form a strong coordinate bond with Pd. In contrast, for the peptide containing Sar-Met or Pro-Met, the equilibrium is shifted towards the hydrolytically active form because they are unable to form a strong coordination with Pd due to the tertiary amide of the Pro or Sar residue (Figure 22).

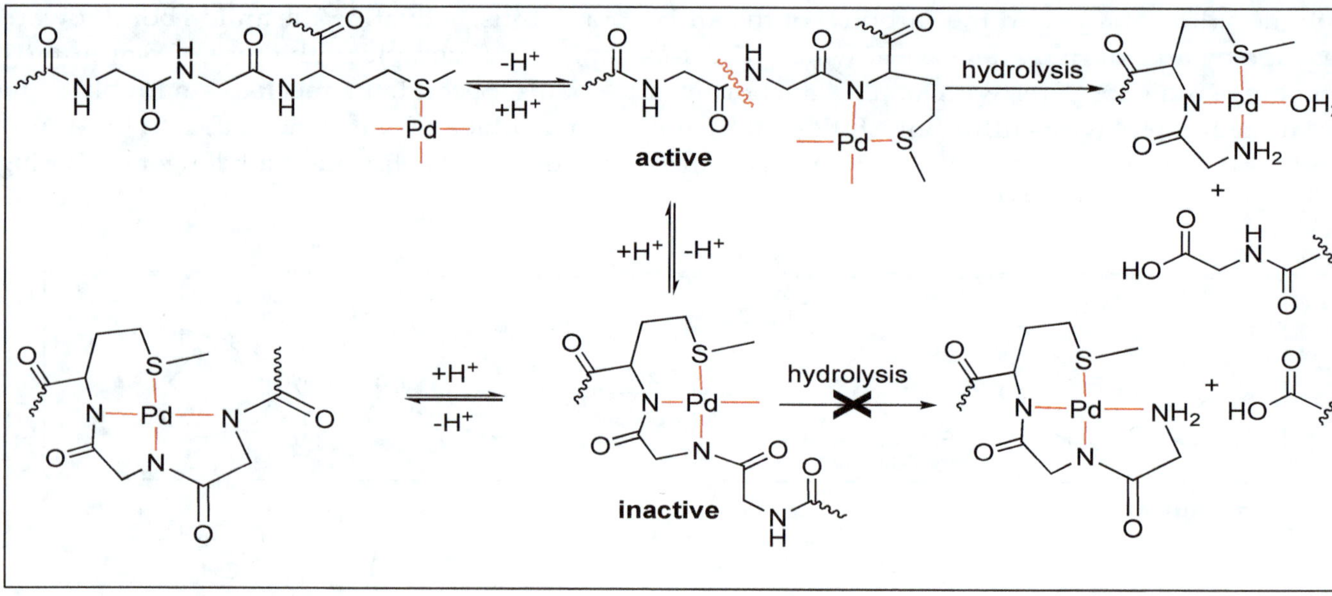

Figure 21. $[Pd(H_2O)_4]^{2+}$ mediated hydrolysis of amide bond.

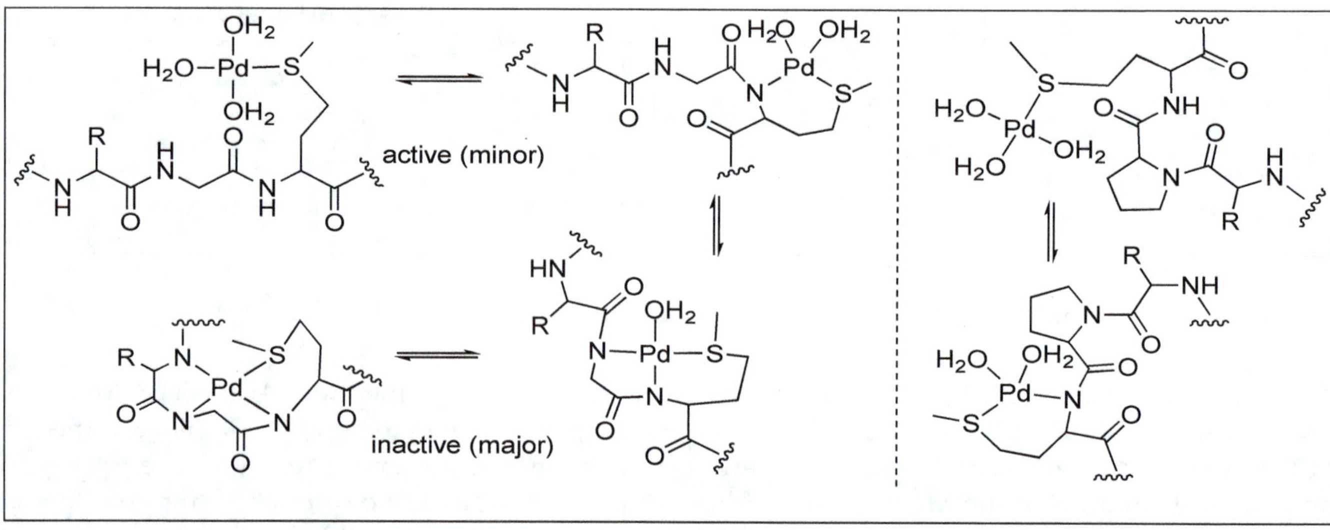

Figure 22. Pd triggered pH dependent hydrolysis of amide bond.

In the case of Gly-Pro-Met, the hydrolysis of the amide bond can take place either through the external or internal attack of water molecules depending upon the cis/trans conformation adopted by proline (Figure 23). The ROESY NMR studies showed that the hydrolysis of the Gly-Pro-Met peptide takes place by the external attack of water on trans-Gly-Pro [100].

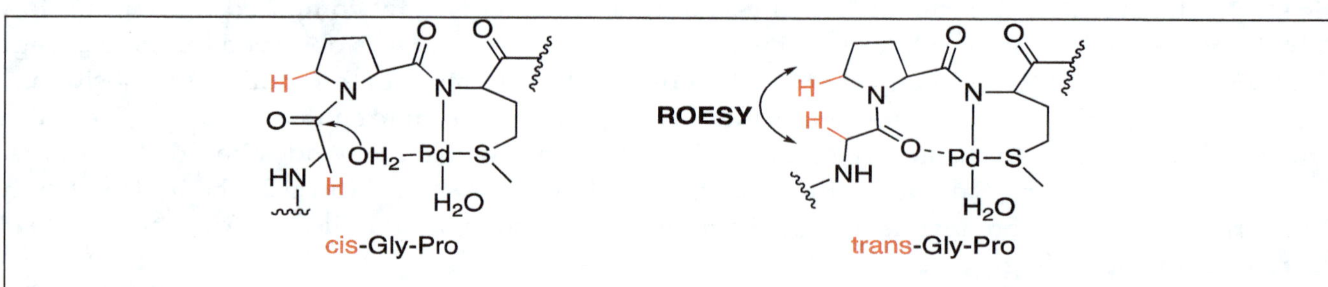

Figure 23. Cis/Trans conformations of Pd complexes.

Overall, the $Pd(H_2O)_4$ complex is residue selective under acidic conditions and cleaves the second amide bond upstream from the Met residue. However, the same complex is sequence-specific under neutral conditions and cleaves the amide bond only at the Pro-Met or Sar-Met residue with no cleavage observed at Gly-Met peptide (Figure 24).

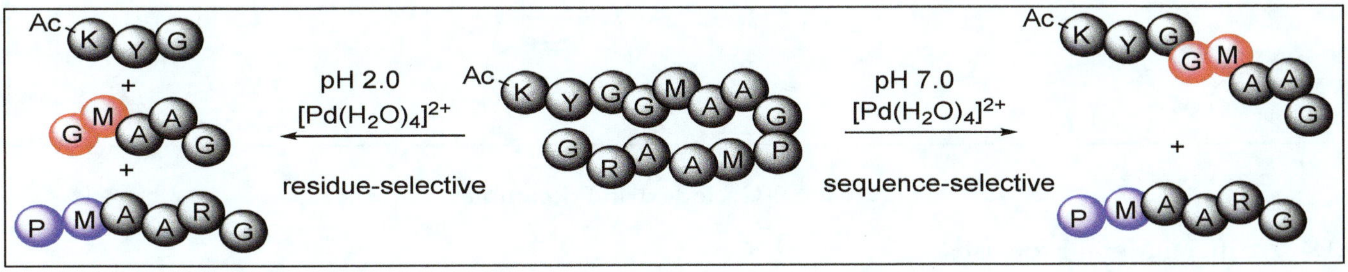

Figure 24. pH selective Pd catalyst for the hydrolysis.

Next, the same group utilized the $Pd(en)(H_2O)_2$ complex for the acidic hydrolysis of the B-chain of bovine insulin containing two histidine residues. These complexes promoted the cleavage of the second amide bond upstream from histidine. The detailed mechanistic analysis showed the selective cleavage by the coordination of the Pd complexes with the histidine side chain and amidic nitrogen followed by the internal attack of water on the amide bond (Figure 25) [101].

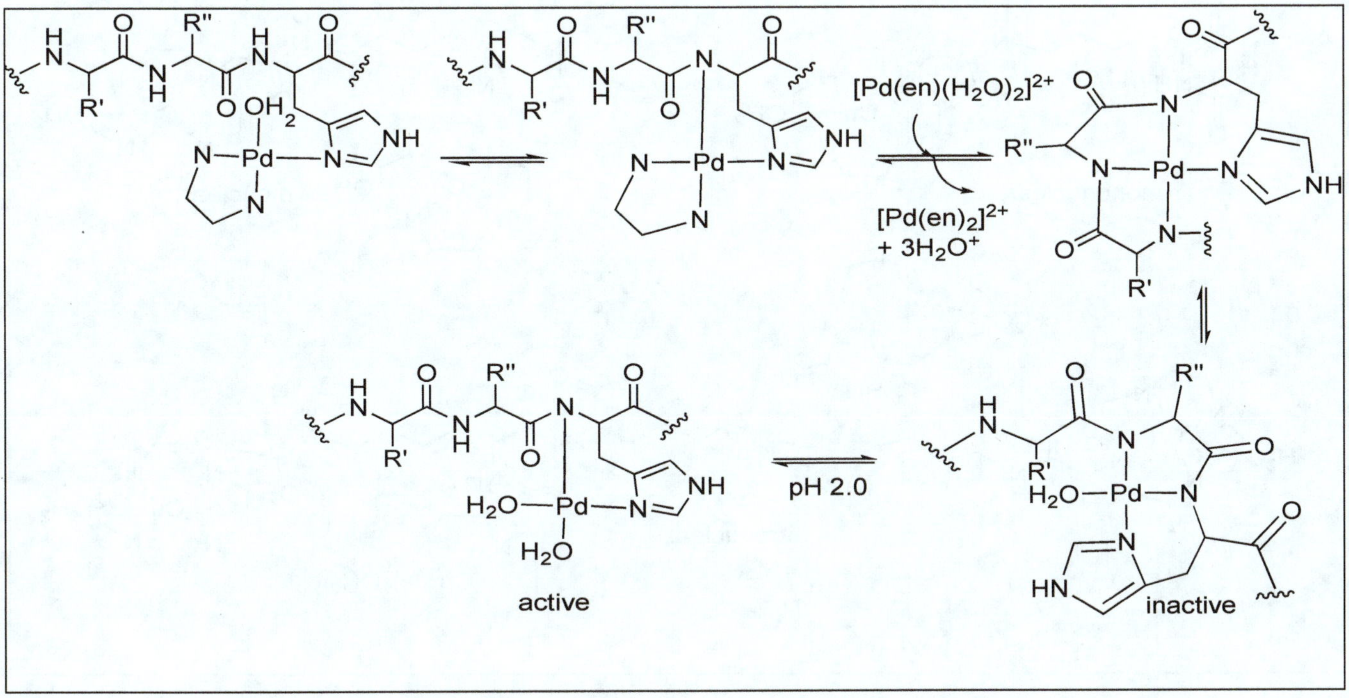

Figure 25. Acidic hydrolysis of amide bonds.

Next, they used the β-cyclodextrin Pd complex for the cleavage of the amide bond at the first amide bond upstream from the Pro-Phe sequence. The role of the β-CD complex was to bind to the hydrophobic residue (Phe) in an aqueous medium followed by the activation of the amide carbonyl group by the metal coordination, and the attack by the external water molecule leading to the cleavage of the amide bond (Figure 26) [102].

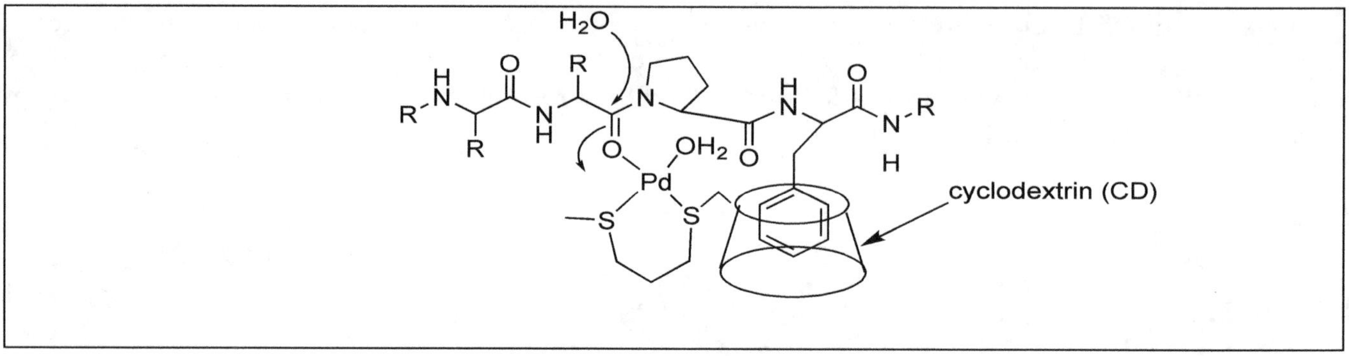

Figure 26. β-Cyclodextrin Pd-complex.

3.1.4. Artificial Metal Proteases

Suh et al. designed metal complexes for the cleavage of amide bonds in proteins at specific locations [103–107]. These metal complexes are highly selective with their protein partners similar to some natural proteases. They also demonstrated that the catalytic rate of hydrolysis increased by the formation of the complex between the substrate and catalyst. This is achieved by increasing the multinuclear metal centers, which provide extra metal centers as substrate binding sites. They showed that the rate of hydrolysis of the myoglobin protein increased with the increase in the number of metal centers in the mono (half-life for hydrolysis 24 h, Figure 27), dinuclear (half-life for hydrolysis 3.5 h), and tetranuclear metal centers (half-life for hydrolysis 1.3 h) [103].

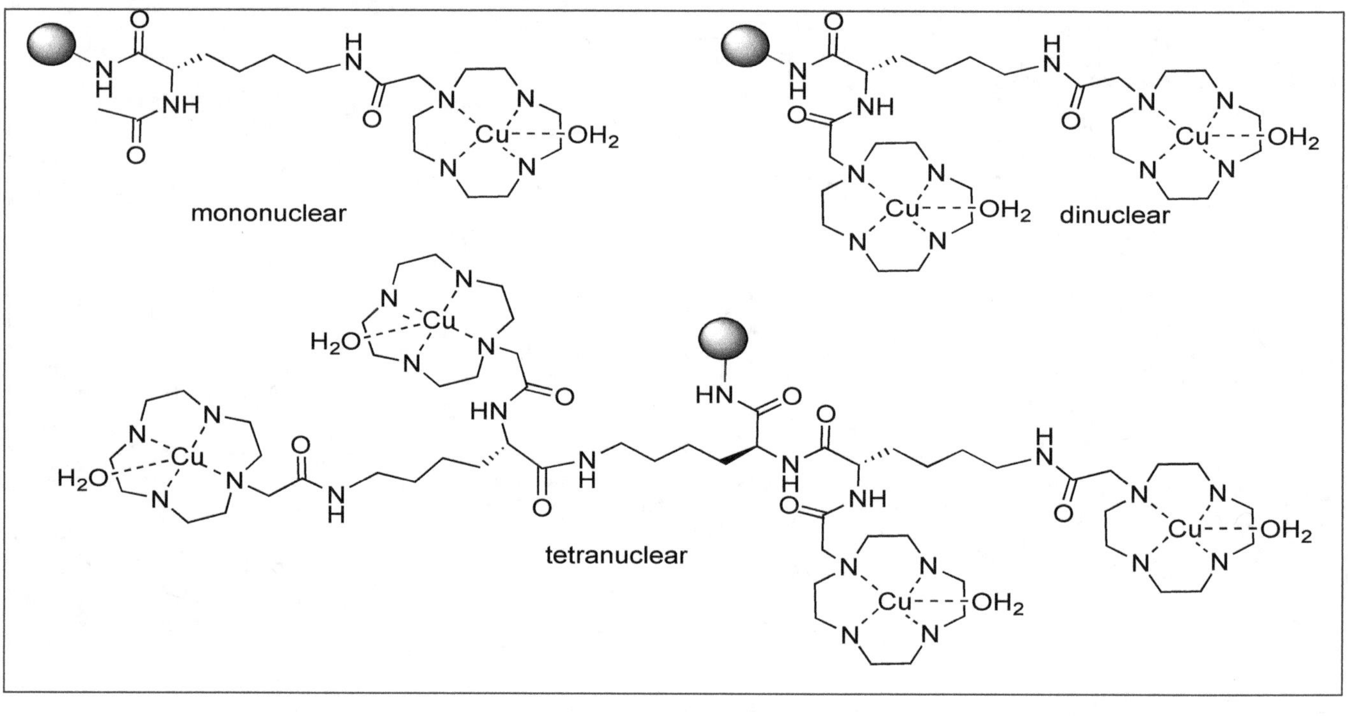

Figure 27. Artificial metal complexes with different numbers of metal centers.

They showed that the selectivity towards a specific protein was achieved by the addition of organic pendants such as peptide nucleic acid (PNA), which are responsible for selective binding to a particular protein. Overall, the goal of this research was to synthesize peptide-cleavage agents selective for the hydrolysis of pathogenic proteins responsible for Alzheimer's disease, type 2 diabetes mellitus, and Parkinson's disease.

Aldehyde Pendant for the Cleavage of Proteins

Ammonium groups are abundant on the surface of proteins and form imine with aldehyde faster than the peptide bond cleavage and, through this process, brings the reactive metal catalytic center in close proximity to the protein/peptide chain (Figure 28). The rapid formation of imine makes the attack of the metal center on the peptide bond an intramolecular process, thus, increases the rate of hydrolysis. Based on this concept, several artificial metalloproteases have been developed by incorporating an aldehyde handle close to the metal center and has been successfully been applied for the faster cleavage of peptide bonds [104].

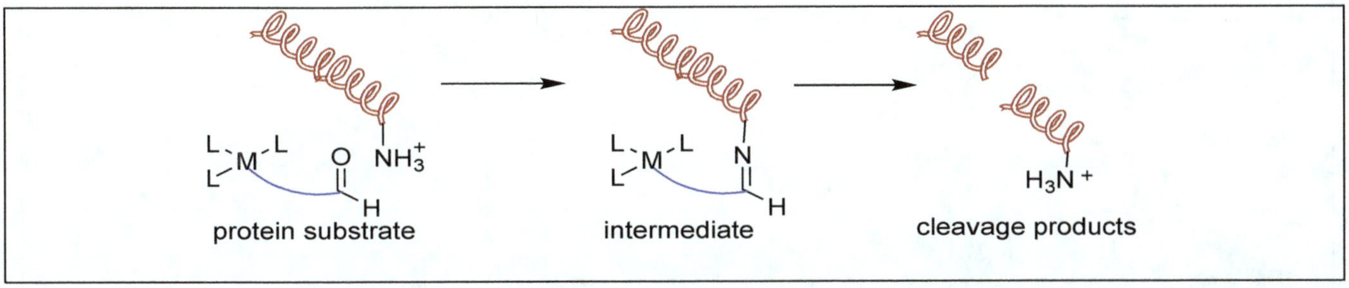

Figure 28. Aldehyde pendant mediated cleavage of peptide bonds.

Mb-Selective Artificial Protease

The catalyst for the cleavage of myoglobin (Mb) was designed by attaching the cyclen metal complex containing Cu(II) or Co(III) to the peptide nucleic acid (PNA) monomers, which act as binding sites for the Mb (Figure 29) [105]. Varieties of linkers were inserted between the PNA binding site and the catalytic cyclen site for the formation of Mb-cleaving catalysts. MALDI-TOF MS showed that the cleavage of the peptide backbone chain of Mb takes place at Leu89-Ala90. Various chelating ligands were tested for determining the activity of the Mb-cleaving catalyst but only cyclen and its triaza-monooxo analog showed efficient catalytic activity.

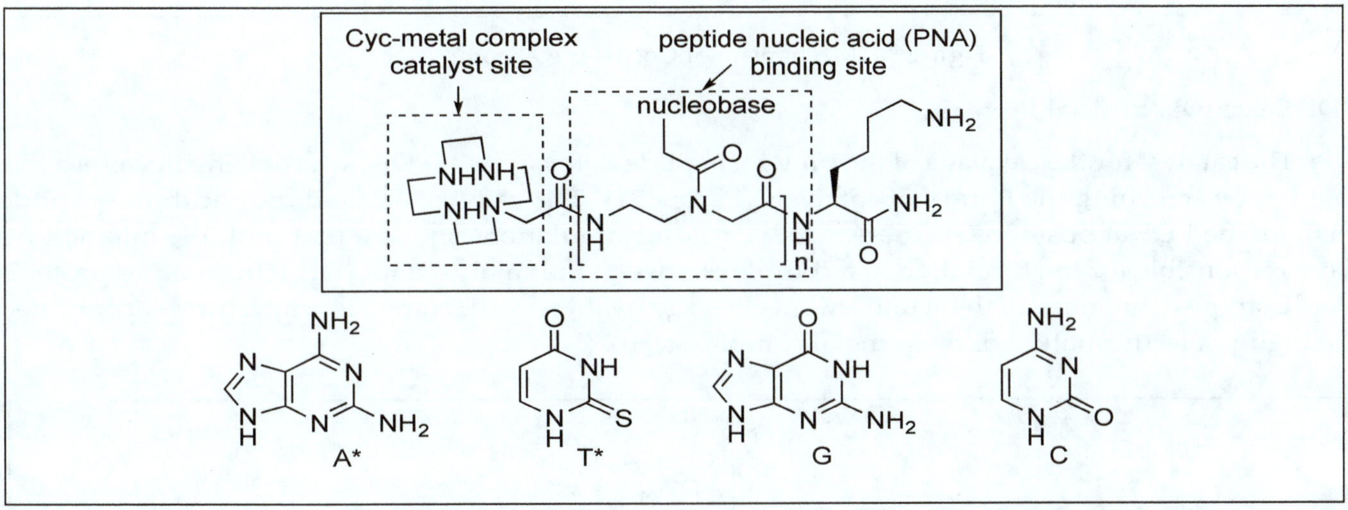

Figure 29. Catalyst design for protein cleavage.

Mechanism: The first step involves the binding of the carbonyl group of the amide to the Co(III) metal center to form a CS complex (Figure 30). The selectivity of one amide carbonyl over the rest of the amidic carbonyls is based on the other half of the catalyst, which is responsible for the recognition of a particular protein and a particular site. This Co(III) carbonyl coordination activates the amidic carbonyl for a nucleophilic attack by the hydroxide ion on the metal center, resulting in the formation of tetrahedral intermediate T. This is followed by the collapse of the tetrahedral intermediate (T), resulting in the breakage of the amide bond to generate a peptide amine and corresponding peptide acid [106].

Suh and co-workers showed that the cleaving capability of the Co(III)/Cu(II) complexes of cyclen increased by the replacement of one nitrogen atom of cyclen with an oxygen atom. The Co(III)-oxacyclen complexes (1-oxa-4,7,10-triazacyclododecane Co(III)) cleaved the proteins such as BSA, HEWL, Mb, and bovine serum-globulin with a 4–14 times higher catalytic efficiency compared to the Co(III)-cyclen complexes [106].

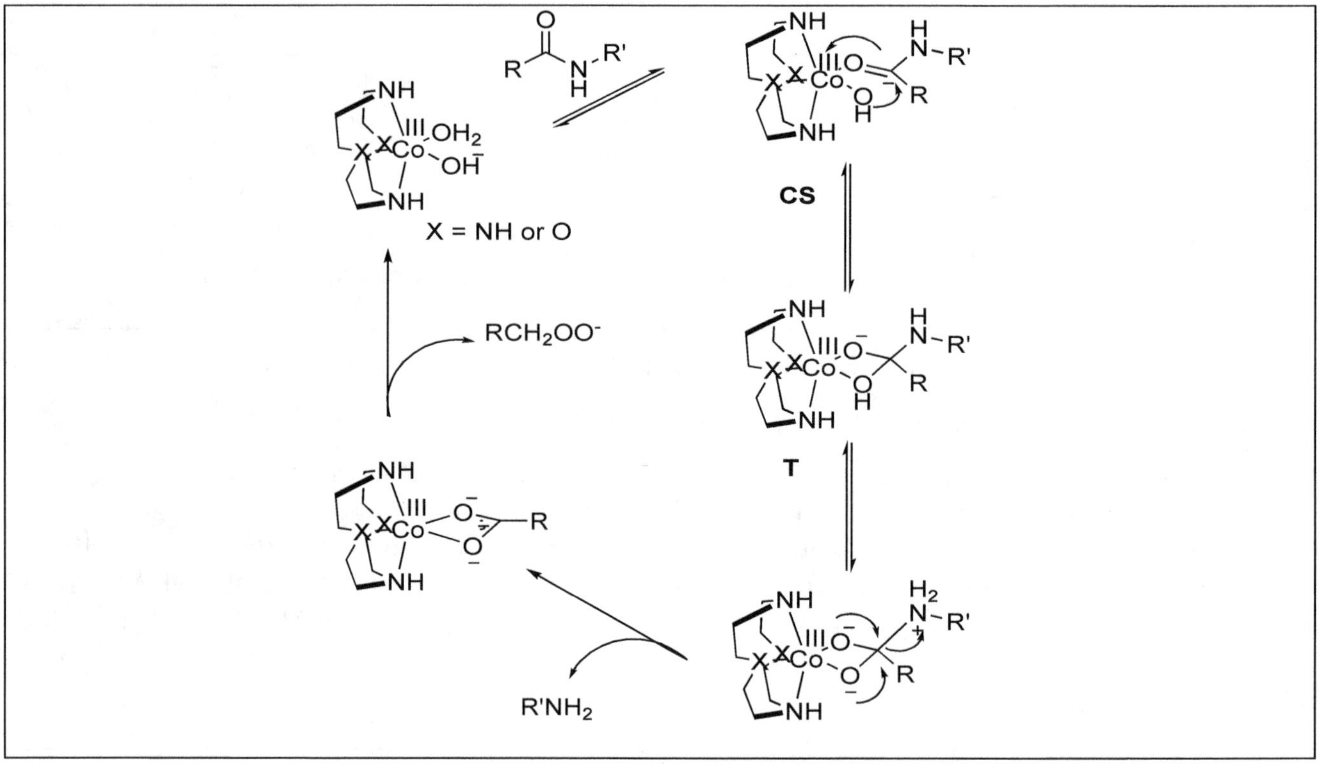

Figure 30. Mechanism of Co(III)-metal complexes.

PDF-Selective Artificial Protease

The catalyst for the cleavage of the protein, peptide deformylase (PDF), was obtained in a selective manner by screening the library of catalysts (Figure 31). The catalyst cleaved the backbone peptide chain of the PDF at position Gln152-Arg153. Docking simulations showed that multiple interactions were responsible for the formation of a complex between the catalyst and PDF. Fifteen other proteins were examined, but none of them underwent cleavage by this Co(III) complex, which further confirmed the highly selective nature of these metal complexes [107].

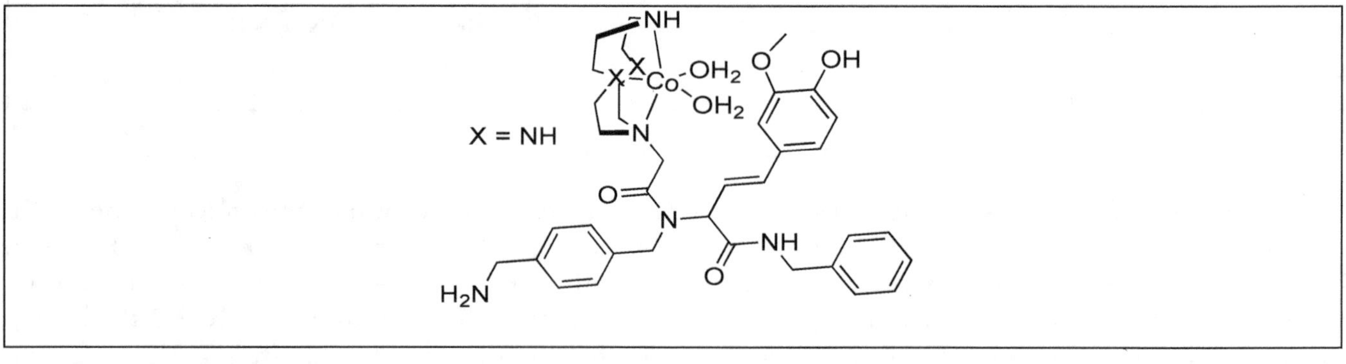

Figure 31. Metal-assisted catalyst for PDF.

AmPs-Selective Artificial Protease

AmPs are amyloidogenic peptides or proteins which lack active sites and are related to diseases such as Alzheimer's and type 2 diabetes. Based on the above concept, Suh et al. synthesized various other metal complexes for the selective cleavage of amide bonds in these proteins. The main advantage of this work is that such amyloidogenic proteins cannot be targeted by conventional approaches due to the lack of an active site (Figure 32) [108–110].

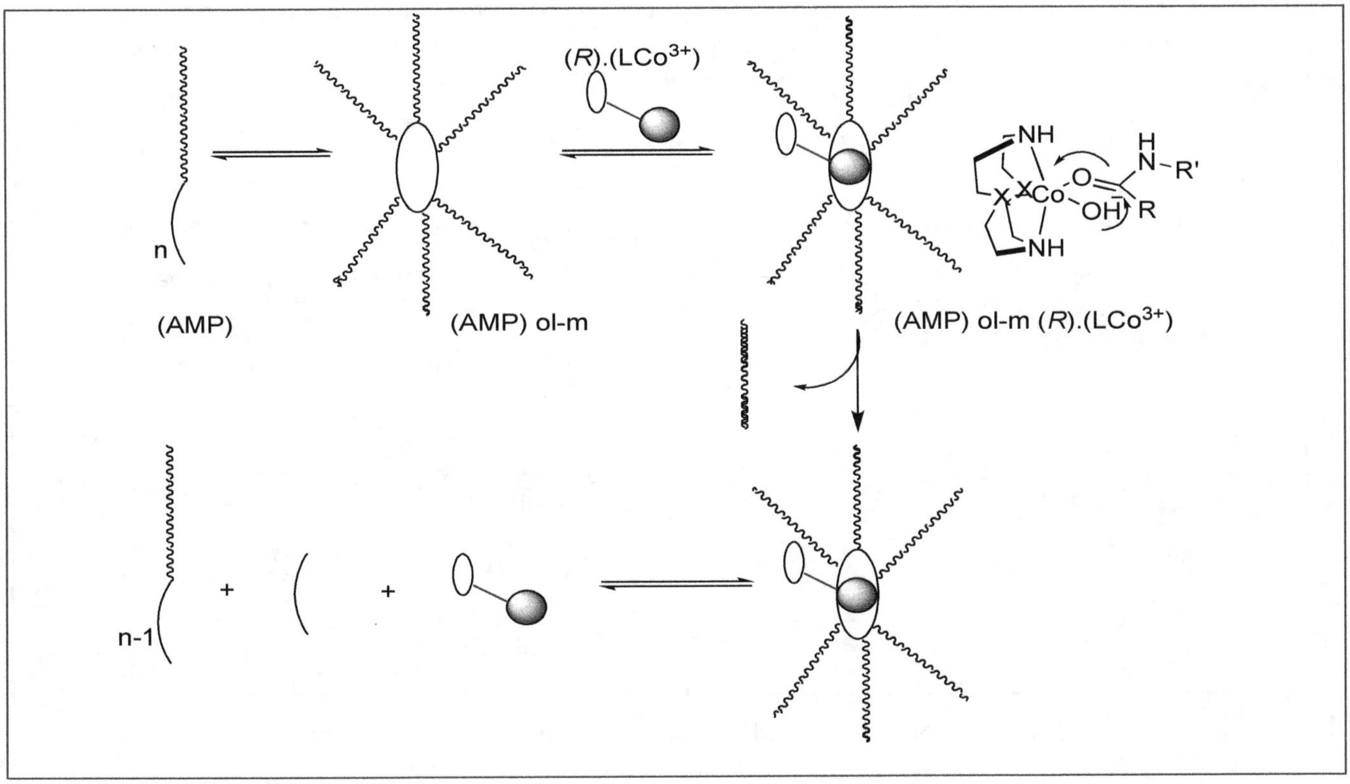

Figure 32. Metal complexes for the cleavage of amyloidgenic peptides.

Soares et al. utilized mononuclear copper(II) complexes $[Cu(HL^1)Cl_2]$ and $[Cu(L^1)Cl]$ for the cleavage of unactivated amide bonds of the proteins bovine serum albumin (BSA) and Taq DNA polymerase, under mild pH and temperature conditions (Figure 33). The cleavage occurred at the specific site on the solvent- exposed portions of the protein to generate particular proteolytic fragments [111].

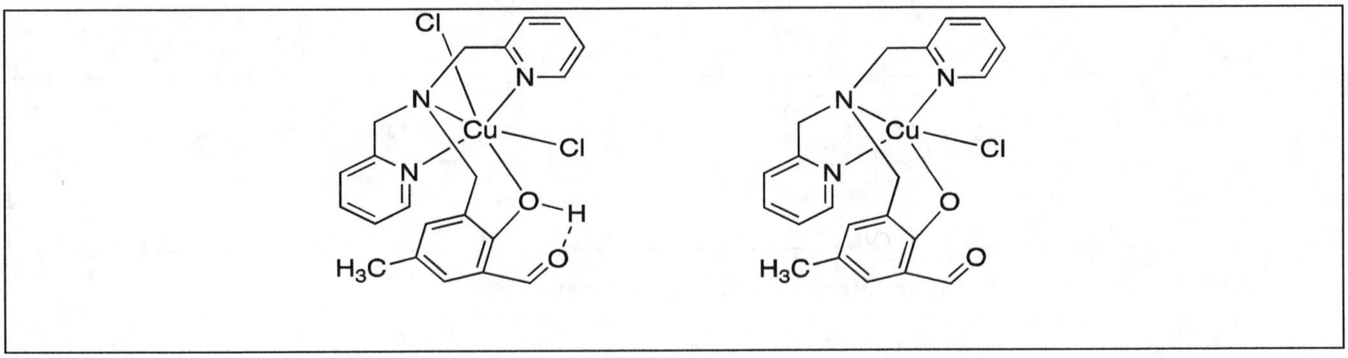

Figure 33. Cu complexes for activation of amide bonds.

3.2. The Non-Lewis Acid Reaction Mechanisms Based on the N→O Rearrangement

In some cases, metal catalysts showed a high rate of hydrolysis of the peptide backbone chain at the N-terminus of the serine and threonine residues. Such a cleavage was catalyzed by the N→O rearrangement and does not employ the Lewis acid properties of the metal atom. Based on the proposed mechanism, the first step involves the formation of the Ni(II) complex with 4 N of the backbone amide chain and the side chain of the His residue with the (Ser/Thr)-Xaa-His sequence (Figure 34). The second step involves the N→O rearrangement from the side chain of the Ser or Thr that transfers the N-terminal R1 moiety from the peptide bond to form an ester bond. This is followed by the hydrolysis of the resulting ester, leading to the formation of two reaction products, the R1 peptide acid and the Ni(II) complex with the peptide [112,113].

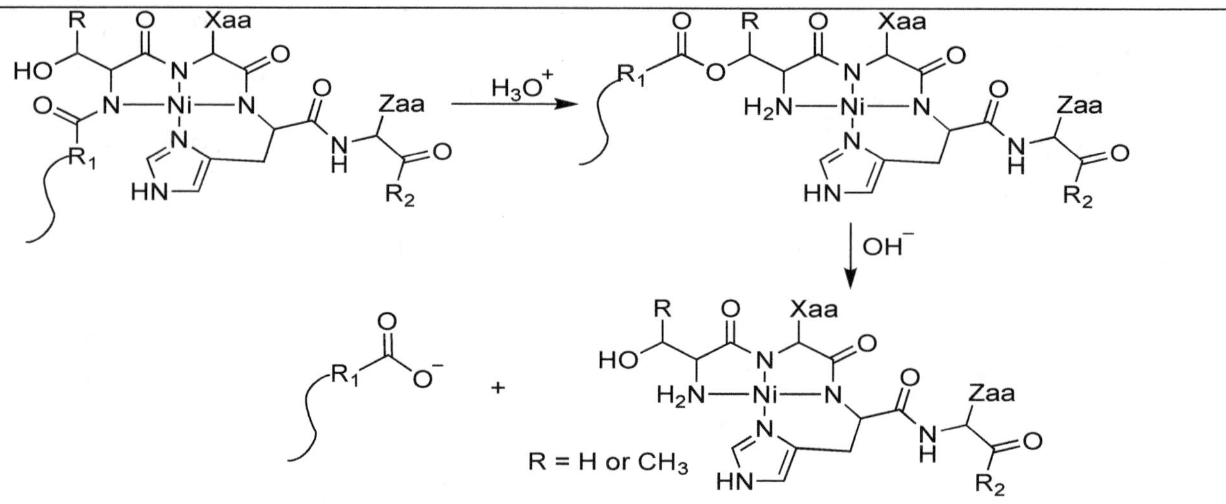

Figure 34. Non lewis acid mediated N, O Acyl rearrangement.

Scandium(III) Triflate-Promoted Serine/Threonine Selective Peptide Bond Cleavage

Kanai et al. reported the hydrolysis of the peptide bond at the N-terminus of Ser/Thr residue by using scandium triflate. This chemical cleavage relies on Sthe c triflate mediated N to O acyl rearrangement followed by the subsequent hydrolysis of the ester by heating it at 80–100 °C (Figure 35). Complete hydrolysis took place in 18–20 h. The authors have used this approach for the cleavage of various peptides including posttranslationally modified (PTM) peptides and the cleavage of native protein Aβ1-42, which is closely related to Alzheimer's disease [114].

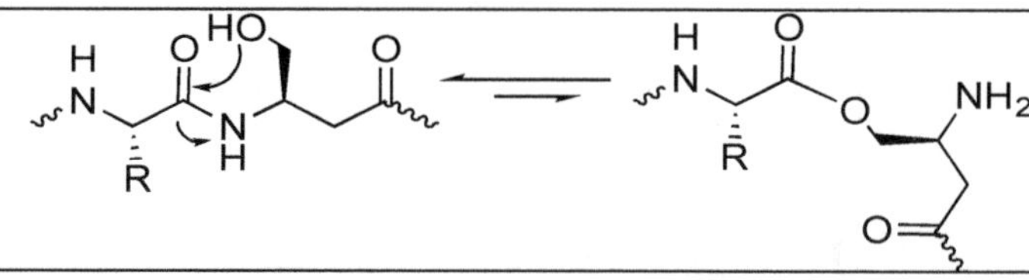

Figure 35. N, O Acyl rearrangement.

4. Organic Molecules for Activation of Amide Bonds

We have summarized different kinds of nonmetal-based methods, their mechanisms of hydrolysis of unactivated peptide bonds and the point of cleavages in Table 3.

Table 3. Organic molecules based hydrolysis of peptide bonds.

Entry	Organic Molecules	Method of Hydrolysis	Point of Cleavage	Ref.
1	Phenyl isothiocyanate	Through 5 membered cyclic phenylisothiocyanate intermediate	N-terminal side of peptide	[115]
2	Cyanogen bromide	Through 5 membered iminolactone	C-terminal side of Met	[116]
3	2-Nitro-5-thiocyano benzoic acid	Through 5 membered thialactone	N-terminal side of Cys	[117]
4	2-iodosobenzoic acid	Through iminospirolactone	C-terminal side of Trp	[118,119]
5	TBC	Through Oxindole	C-terminal side of Trp	[118,119]
6	*N*-amidination	Through 5 membered cyclic amidine ring	N-terminal side of peptide	[120]
7	Protecting goups (Table 4)	Lactonization strategy	C-terminal side of peptide	[121–127]
8	H_2O_2 responsive protecting groups	Lactonization strategy	C-terminal side of peptide	[128]
9	PyBroP	Glutamic acid selective activation through pyroglutamyl imide	N-terminal side of Glu	[129,130]
10	DIB	Hofmann rearrangement mediated *N*-acylurea intermediate	N-terminal side of Asn	[131]
11	DSC	Cyclic urethane amide activation	N-terminal side of Ser, Thr, Cys and Glu	[132–136]
12	*o*-NBnoc	A photo responsive amide cleavage through succinimide ring	C-terminal side of Asn	[137–142]
13	Cu-organo radical conjugate	Aerobic chemoselective oxidation of Ser followed by oxalimide formation	N-terminal side of Ser	[143]
14	TFA/H_2O	N-acyl group mediated oxazolinium specie	C-terminal side of peptide	[144]
15	Hydrazine hydrate	Hydrazinolysis accelarated by addition of ammonium salts	N-terminal of peptide	[145]
16	*N*-Mecysteine	Through oxazolinium ion	C-terminal side of *N*-MeCys	[146]
17	Acylated *N*-MeAib	Through oxazolinium ion	C-terminal side of *N*-MeAib	[147]

4.1. *N-Terminal Cleavage of Amide Bonds*

4.1.1. Edman's Degradation

This approach utilizes phenyl isothiocyanate for the cleavage of the peptide bond at the N-terminus. Phenyl isothiocyanate reacted with an uncharged N-terminal amino group, under mildly alkaline conditions, to form a cyclic *phenylisothiocyanate* derivative, which undergoes cleavage as a thiazolinone derivative under acidic conditions (Figure 36). We proposed that the activation of an amide bond is due to the formation of the five-membered cyclic *phenylisothiocyanate* intermediate which creates a twist in the amide bond, thus preventing the amidic nitrogen from forming a resonating structure and making it susceptible towards hydrolysis [115].

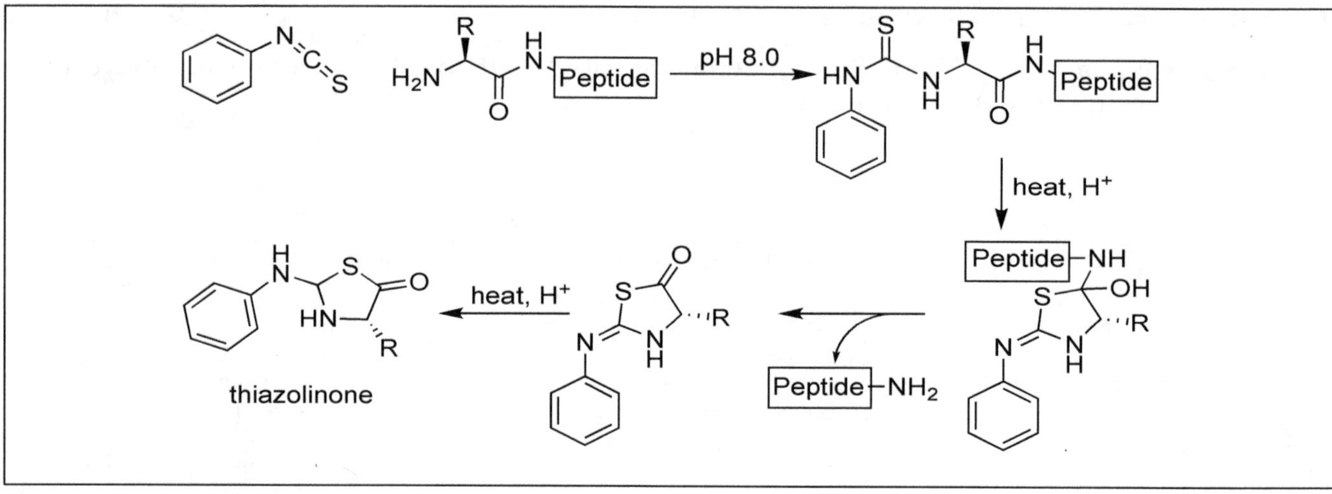

Figure 36. Edman's degradation approach for cleavage of peptide bonds.

4.1.2. Cyanogen Bromide for Cleavage at Met Residue

Cyanogen bromide led to the cleavage of the peptide bond at the C-terminus of the methionine residue in a selective manner. The first step involves the nucleophilic attack of the sulfur of methionine on cyanogen bromide (Figure 37) [116]. This displaces the bromide from cyanogen bromide, followed by the attack of the amide carbonyl on the cyano group, resulting in the formation of the five-membered ring, iminolactone, comprising a double bond in the ring between nitrogen and carbon. This double bond results in a rigid ring conformation, thus activating the amide bond towards cleavage at the C-terminus of Met, resulting in the generation of homoserine lactone. This approach has widely been utilized for the sequencing of proteins [116].

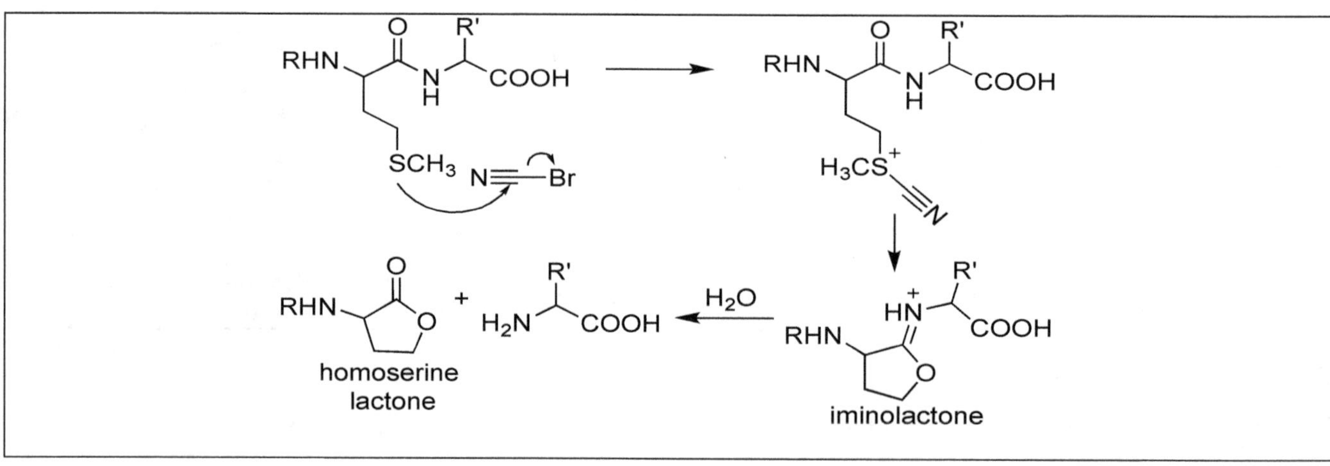

Figure 37. Cyanogen bromide for selective cleavage at Met.

4.1.3. 2-Nitro-5-Thiocyano Benzoic Acid for Cleavage at Cys

2-Nitro-5-thiocyano benzoic acid led to the hydrolysis of the amide bond at the N-terminal side of the cysteine residue. The first step is the cyanylation of the side chain of cysteine on a peptide by 2-nitro-5-thiocyano benzoic acid, which is followed by the attack of the cysteine amidic nitrogen to the cyano group on the side-chain of cysteine, resulting in the formation of the 5-membered thiolactone ring. This, in turn, activates the amide bond towards hydrolysis under basic conditions (Figure 38). This is again due to the inability of cysteine amidic nitrogen in a thiolactone to form a resonating structure with the carbonyl of peptide bond [117].

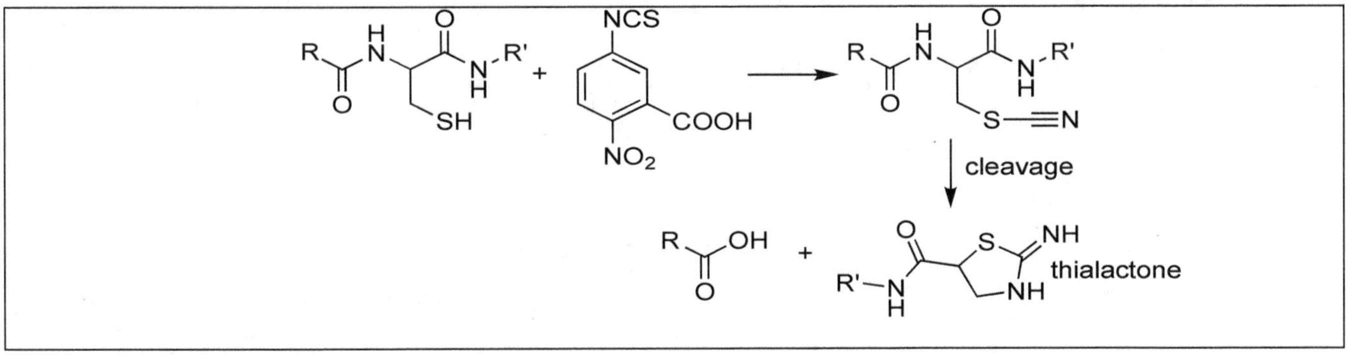

Figure 38. 2-nitro-5-thiocyano benzoic acid selective cleavage at Cys.

4.1.4. 2-Iodosobenzoic Acid for Cleavage at Trp

2-Iodosobenzoic acid has been used for the hydrolysis of the amide bond at the C-terminal side of the Trp residue. The mechanism of the cleavage is a two-step process. The first step involves the oxidation of the side-chain of tryptophan by 2-iodosobenzoic acid followed by the nucleophilic attack from the neighboring carbonyl group of the amide bond, leading to the formation of an iminospirolactone which hydrolyzes the peptide chain in the presence of water (Figure 39) [118,119].

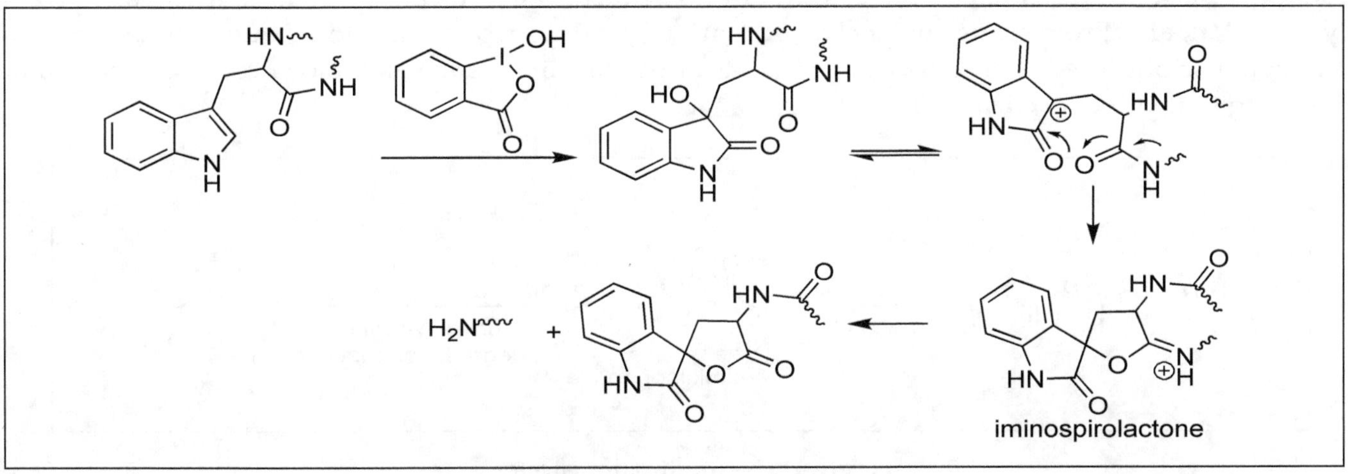

Figure 39. Iodosobenzoic acid for hydrolysis.

4.1.5. TBC for Cleavage at Trp

Tryptophanyl peptide bonds underwent selective cleavage by 2,4,6-tribromo-4-methylcyclohexadienone (TBC) at the C-terminus (Figure 40). Tyrosyl and histidyl peptide bonds which are usually cleaved by other brominating agents (such as α-bromosuccinimide, α-bromoacetamide, etc.) are stable to this reagent. Additionally, other amino acids, which are sensitive to oxidation, react with TBC but do not cleave the peptide bonds. This method was successfully applied to a variety of peptides and proteins [118,119].

According to the reaction mechanism suggested by Patchornik et al. (1960), oxidative bromine participates in the modification-cleavage reaction [118,119]. Two equivalents of bromine first brominate the indole nucleus followed by a spontaneous debromination through a series of oxidation and hydrolysis reactions (Figure 40). These reactions led to the formation of an oxindole derivative, which cleaves the peptide bond.

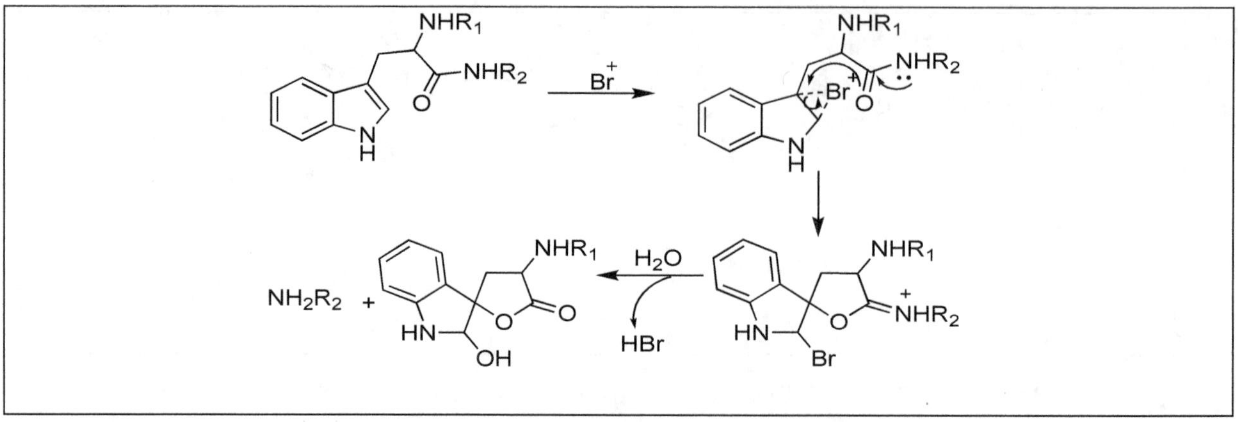

Figure 40. TBC for selective cleavage at Trp residue.

4.2. *N-Amidination for Cleavage of the N-Terminal Amide Bond*

Hamada et al. reported the cleavage of the amide bonds by the *N*-amidination of peptides. The *N*-amidination of peptides leads to the formation of a cyclic moiety which resulted in the cleavage of the amide bond at room temperature (Figure 41) [120]. The rate of cleavage was slow under ambient conditions (PBS buffer, pH 7.4) at 37 °C with $t_{1/2}$ = 35.7 h, but a rapid cleavage was observed under basic conditions (2% NaOH aq) with $t_{1/2}$ = 1.5 min. To evaluate the broad applicability of this cleavage reaction, a series of peptides with different amino acids at the N-terminus such as the Lys, Glu, Ser, Cys, Tyr, Val, and Pro residues were cleaved with $t_{1/2}$ values from 1 min to 10 min. A slightly slow cleavage was observed with bulky amino acids at the terminus such as Val or Pro, which might be hindering the path of cyclization.

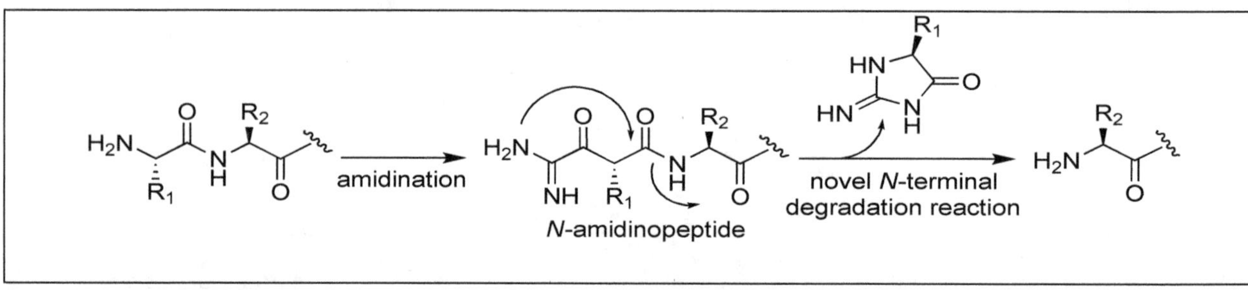

Figure 41. N-amidination strategy.

N-amidinated peptide with a Cys residue at the N-terminus also generated a five-membered ring, thiazolidine by path b (intermediate B) which did not lead to any cleavage, therefore, the $t_{1/2}$ of the peptide with Cys at the N-terminus in 2% NaOH aq at 37 °C was 3.4 min (slower than with other amino acids at the N-terminus) (Figure 42) [120].

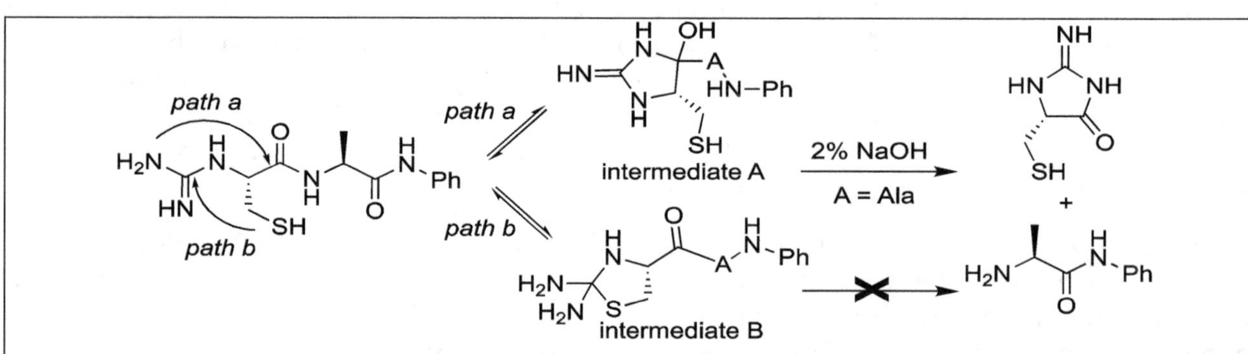

Figure 42. N-amidination strategy with N-terminal cysteine.

4.3. Lactonization Mediated Cleavage of Amide Bonds

Otaka et al. developed an auxiliary with special protecting groups (PGs), which is capable of forming a lactone with the carbonyl of an amide bond, resulting in the cleavage of the amide bond (Table 4) [121–128]. Depending on the nature of the protecting groups, an amide bond cleavage can be initiated in peptides by using different responsive reagents for the deprotection of PGs (Table 4). Table 4 showed various PGs and the corresponding responsive reagents for their deprotection. The thiol responsive reagent was applied for the cleavage of the PNA/DNA complex using thiol-responsive protecting groups [121–128].

Table 4. Lactonization mediated cleavage of amide bonds.

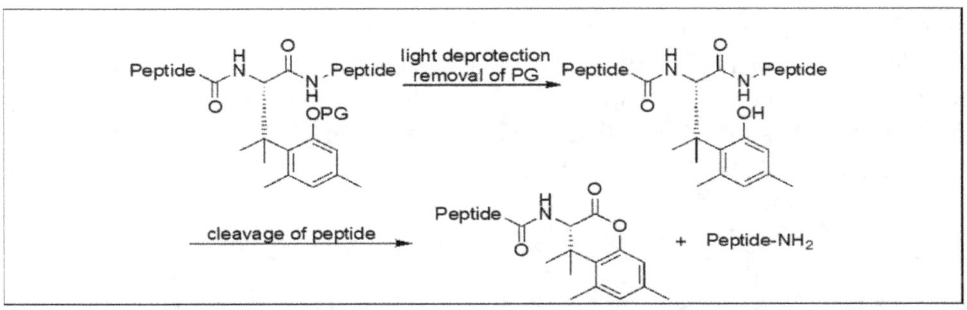

Reagent/Condition	PG	R1	R2	R3
Ultraviolet	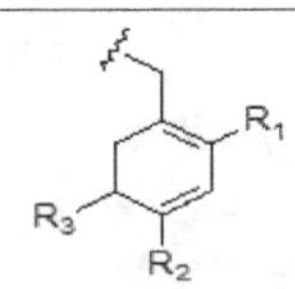	NO_2	H	H
Near-infrared		NO_2	OMe	OMe
Fluoride		H	OTBDPS	H
hypoxia		H	NO_2	H
Thiol	4-nitrobenzenesulfonyl	-	-	-
Phosohatase	phosphate	-	-	-

4.4. Hydrogen Peroxide-Induced Amide Bond Cleavage

Later, the hydrogen peroxide (H_2O_2)-responsive protecting group was introduced to the amino acid. This protecting group contains a boronate or boronic acid moiety which underwent deprotection in oxidative stress because of the release of hydrogen peroxide followed by the formation of lactone and the cleavage of the peptide bond (Figure 43) [128].

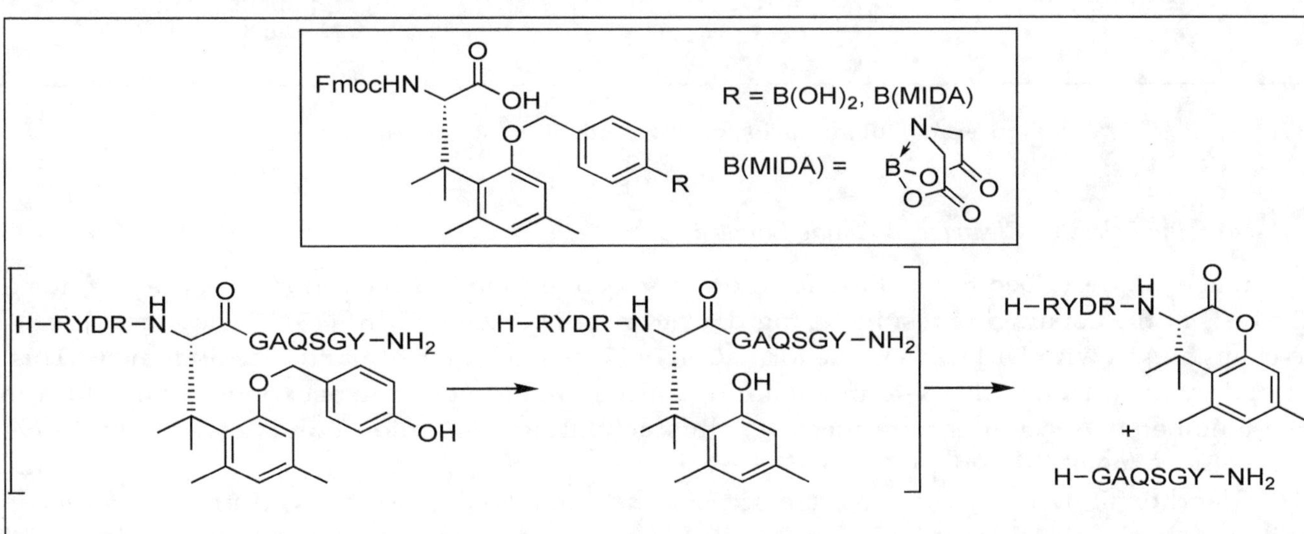

Figure 43. Hydrogen peroxide responsive probes.

4.5. Glutamic Acid Specific Activation of Amide Bonds

We have also reported a new method for the site-specific cleavage of peptide bonds at glutamic acid under physiological conditions [129,130]. The method involves the activation of the backbone peptide chain at the N-terminal side of glutamic acid by the formation of a pyroglutamyl imide (pGlu) moiety using bromotripyrrolidinophosphonium hexafluorophosphate (PyBroP) (Figure 44). This activation increases the susceptibility of the peptide bond toward various nucleophiles including thiol and water (Figure 44). We showed that this pyroglutamyl imide activated peptide chain underwent the complete cleavage of the peptide bond under neutral buffer conditions (pH 7.5). It was observed that the rate of hydrolysis increase under basic pH conditions (pH = 10.5). Although the Asp has a carboxylic group on the side chain, no cleavage was observed under the reaction conditions. Jensen et al. exposed the pGlu activated peptide bond towards thiol, resulting in the formation of peptide thioesters [129,130]. A noted feature about this approach is that it leads to the formation of epimerization free peptide acids and peptide thioesters. This method is highly specific and exhibits a broad substrate scope including the cleavage of bioactive peptides with unnatural amino acids, which are unsuitable substrates for enzymes.

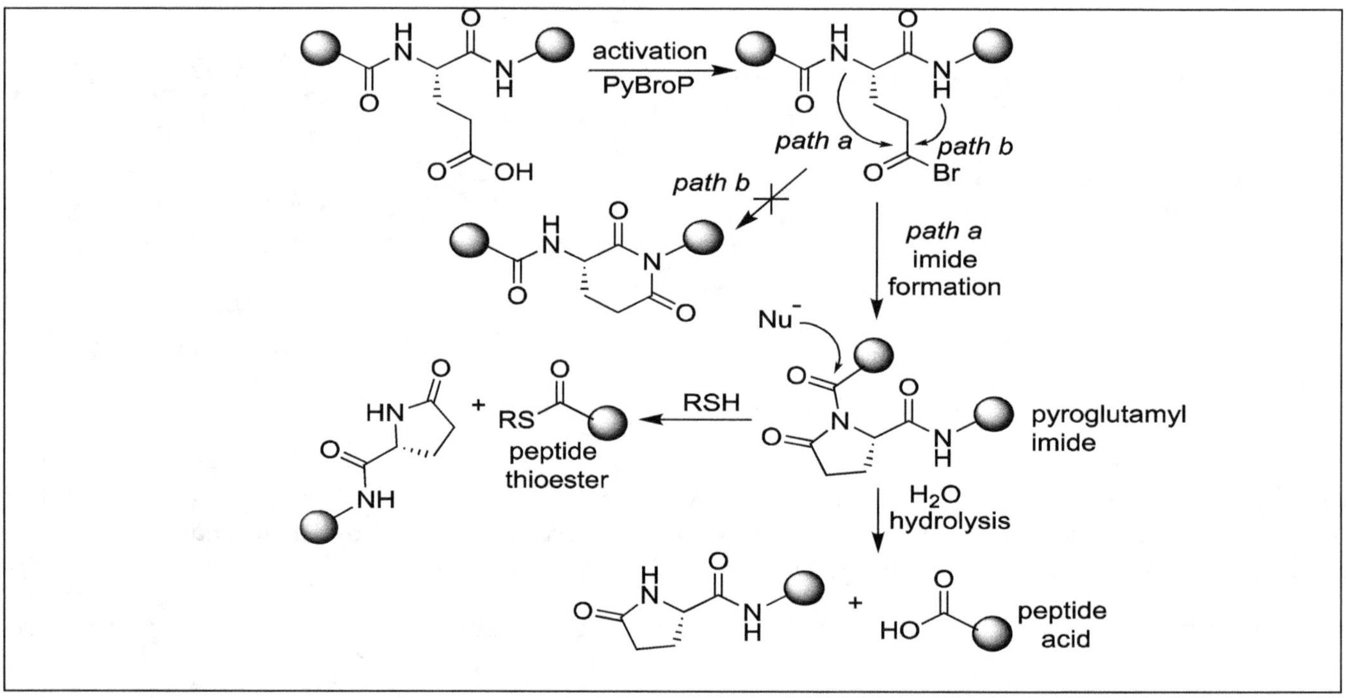

Figure 44. Glutamic acid selective activation of peptide bonds.

4.6. Asparagine Selective Cleavage of Amide Bonds

Kanai et al. described the method for the site-selective chemical activation of peptide bonds for hydrolysis at the asparagine residue using diacetoxyiodobenzene (DIB) [131]. The reaction of the side-chain of Asn with DIB leads to the formation of isocyanate by Hofmann rearrangement. This is followed by the attack of the N-terminal amidic nitrogen of the peptide backbone chain, affording a five-membered *N*-acylurea intermediate, thus activating the amide bond towards hydrolysis (Figure 45). Asn-selective peptide bond cleavage was proceeded in an aqueous neutral solution at 37 °C and exhibited a broad substrate scope. The Gln-site was not cleaved under the reaction conditions. Specifically, this method is applicable to peptides containing unnatural amino acids and/or posttranslational modifications where enzymatic cleavage is not very efficient.

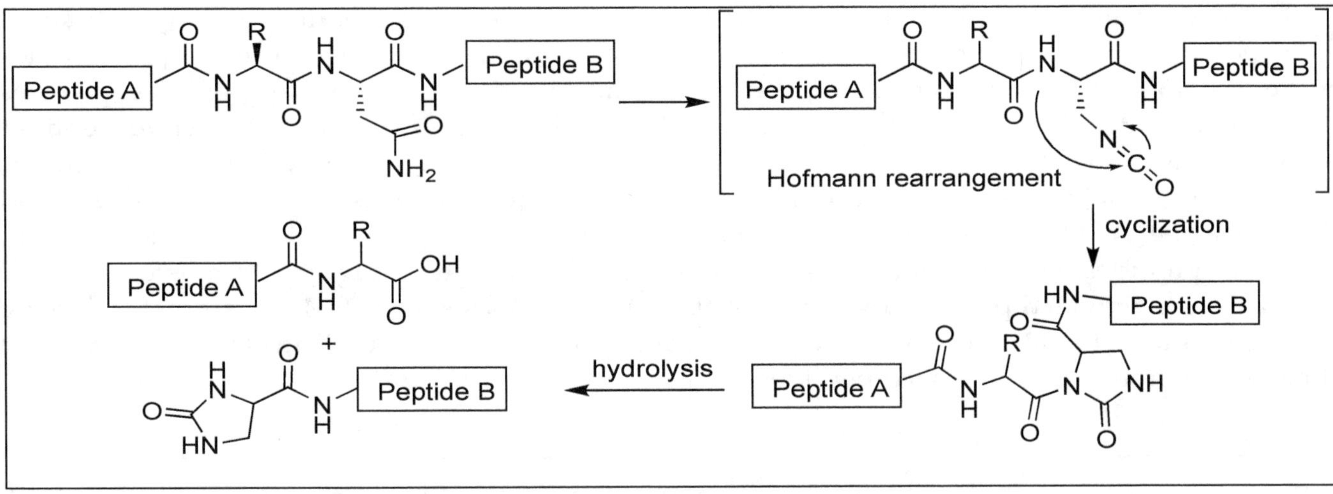

Figure 45. Asparagine selective cleavage of peptide bonds.

4.7. Cyclic Urethane Mediated Activation of Amide Bonds

We have developed a method for the cleavage of the amide backbone chain at the N-terminal side of Ser, Thr, and Cys by the formation of a five-membered cyclic urethane moiety [132]. The formation of the cyclic urethane moiety with an amide backbone makes the amidic carbonyl group susceptible to nucleophilic attack. This is presumably due to the twist in the backbone amide chain caused by the cyclic urethane moiety. Thus, it was no longer able to form a resonating structure. To achieve this goal, we screened various carbonylating reagents and a maximum conversion to cyclic urethane moiety was achieved with *N,N*-disuccinimidyl carbonate (DSC). We proposed that the hydroxymethyl group of the side chain of Ser reacted with DSC to generate an activated intermediate, A, which then undergoes nucleophilic displacement by the amidic nitrogen on the N-side of serine through the path to generate a five-membered cyclic urethane intermediate B (Figure 46). The formation of the cyclic urethane intermediate B makes the amide bond susceptible to nucleophilic attack and led to the cleavage of the amide bond in neutral aqueous conditions (room temp, pH 7.5, 12 h). There is a possibility of nucleophilic displacement of the intermediate A by the amidic nitrogen through path b. This could lead to the formation of six-membered ring B′, but we did not observe the formation of any six-membered ring as analyzed by NMR studies.

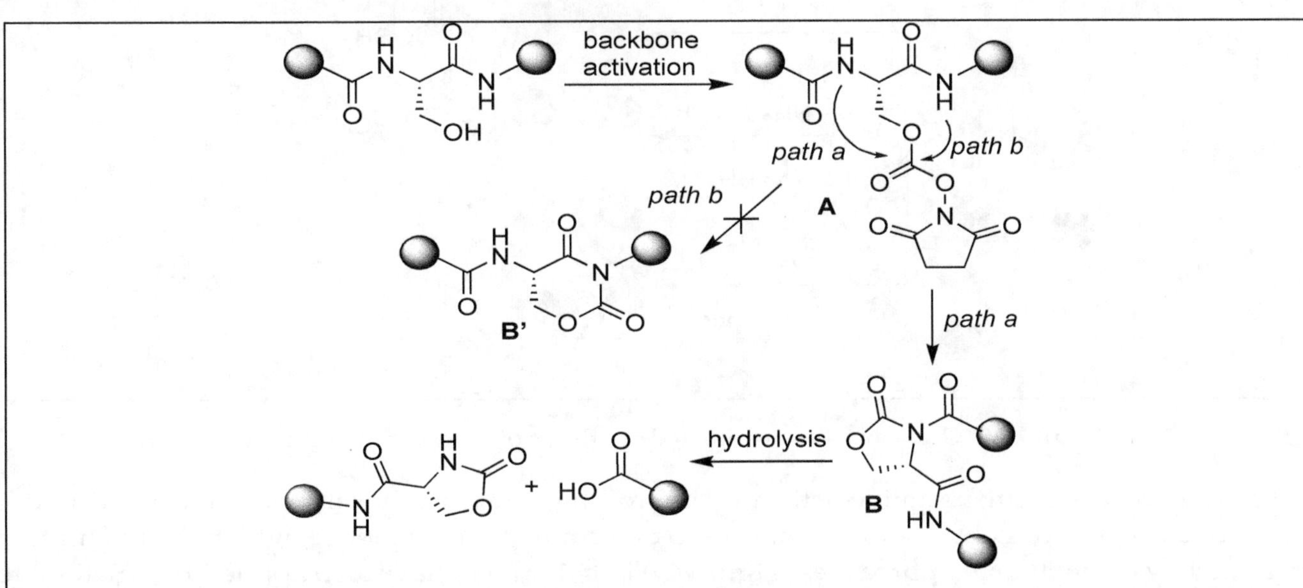

Figure 46. Cyclic urethane mediated activation of peptide bond.

The side chain of Glu on reaction with DSC also led to the formation of a pyroglutamyl imide moiety with an amide backbone chain, thus making it susceptible to nucleophilic attack (Figure 47) [132]. We have used this approach for the selective hydrolysis of peptides/proteins at the N-terminus of Ser, Thr, Cys, and Glu. This method cleaved various bioactive peptides containing posttranslational modifications (e.g., *N*-acetylation and -methylation) and mutations (D-and β-amino acids), which are not suitable substrates for enzymes, thus exhibited a broad substrate scope. We have also used this approach for the synthesis of a variety of functionalized C-terminal peptides such as esters, amides, alcohols, and thioesters (Figure 48) by exposing the cyclic urethane activated peptide towards various nucleophiles such as alcohols, amines, reducing agents, and thiols [133,134]. The attractive feature of this approach is that it leads to the formation of epimerization free C-functionalized peptides.

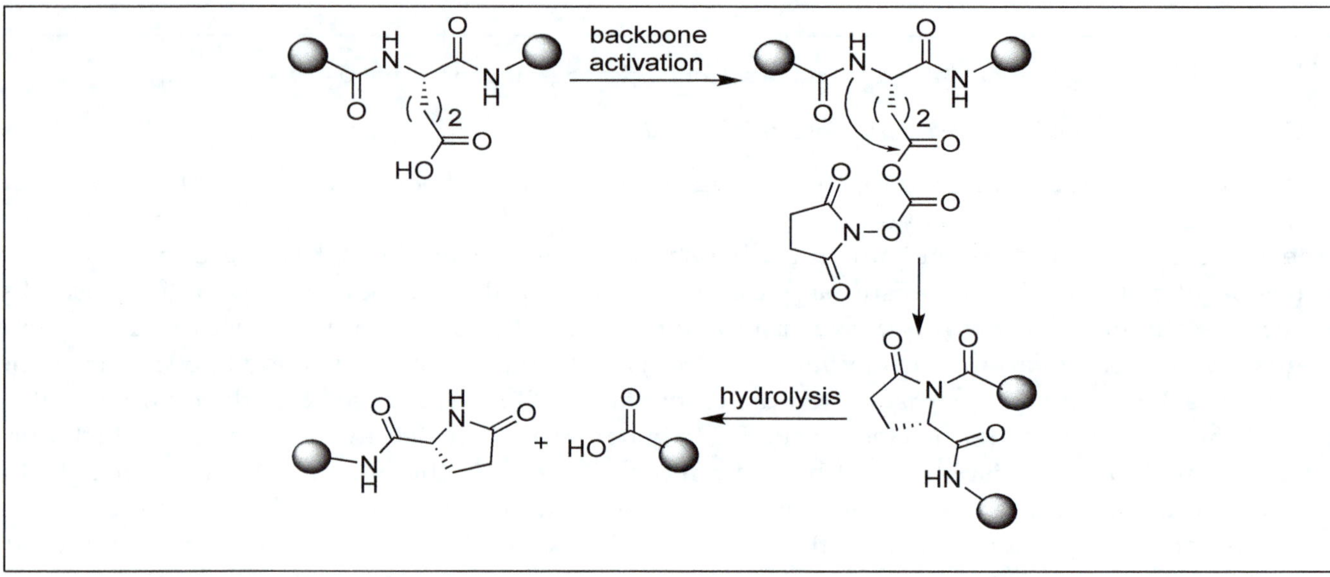

Figure 47. Pyroglutamyl imide mediated activation of peptide bond.

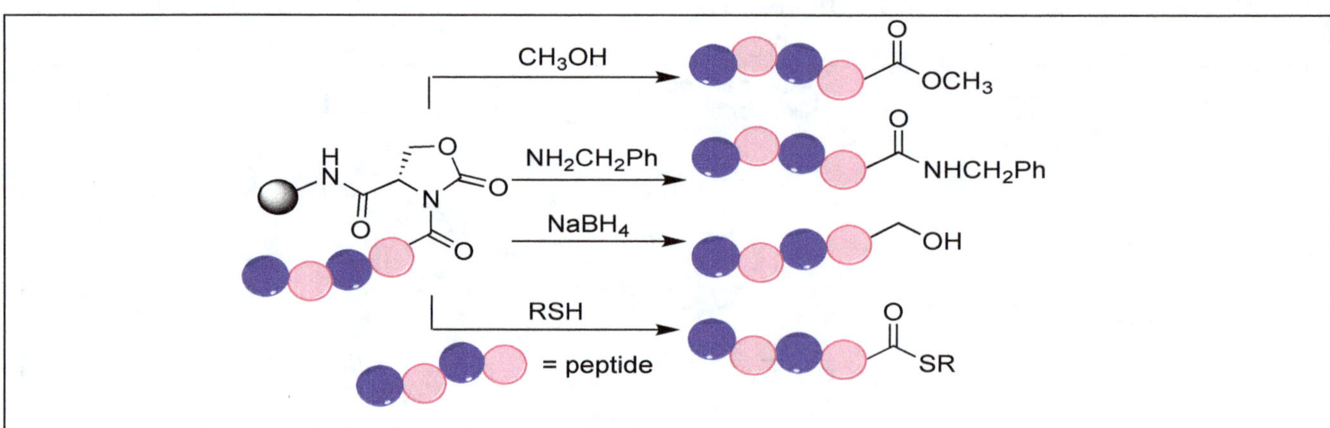

Figure 48. Cyclic urethane mediated synthesis of C-terminal peptides.

Later, this cyclic urethane amide-activation approach was applied for the cleavage of a variety of cyclic and lasso peptides obtained from nature to determine their sequence, which is difficult to be determined by conventional approaches (Figure 49) [135,136]. We have also applied this method for the synthesis of a peptide-based molecular machine (rotaxanes) for the first time (Figure 50) [135,136].

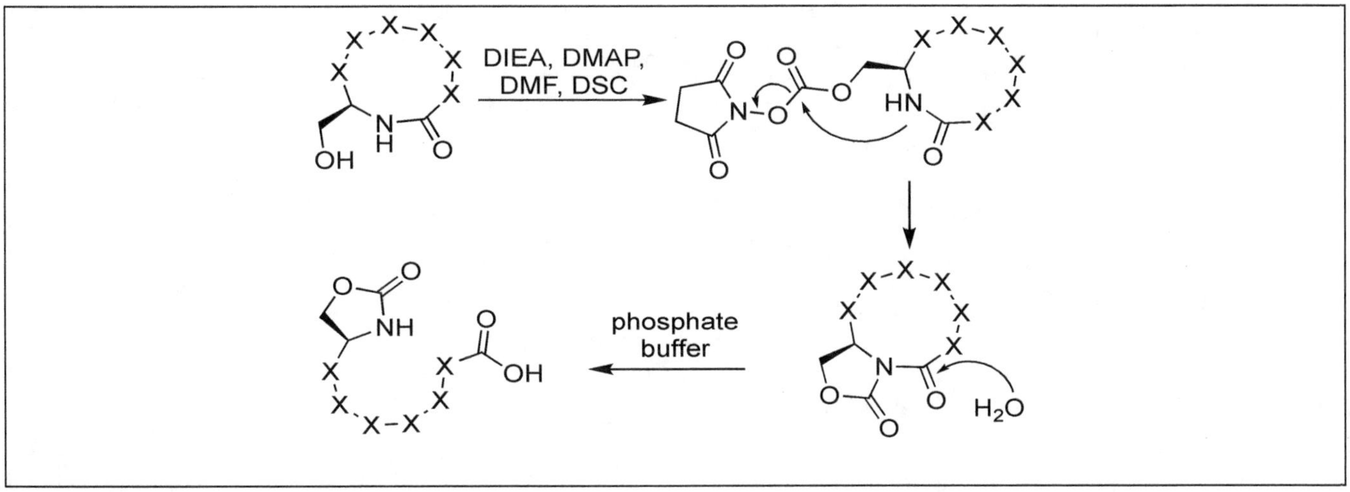

Figure 49. Cyclic urethane for cleavage of cyclic peptides.

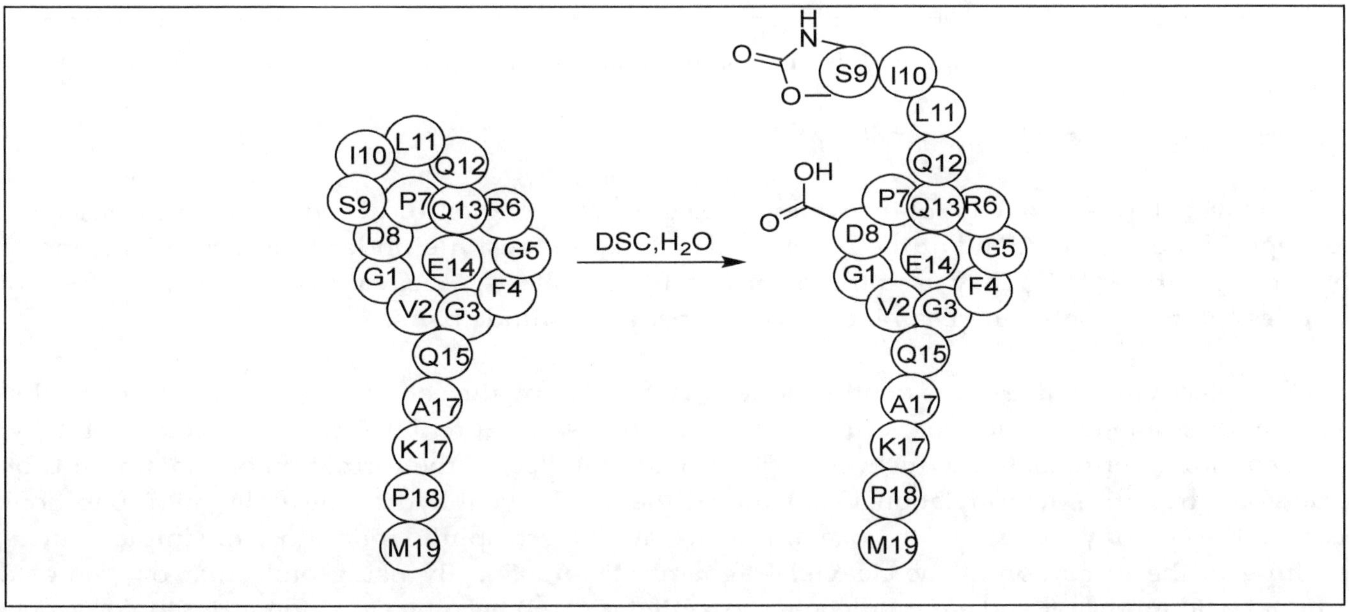

Figure 50. Synthesis of rotaxane from lasso peptide.

4.8. Intein-Inspired Amide Cleavage Chemical Device

A photoresponsive device was developed for the cleavage of the amide bond at the C-terminus of the Asn residue [137]. This approach was inspired by intein-mediated protein splicing and its chemical environment was mimicked by the incorporation of geminal dimethyl groups and a secondary amine on the asparagine scaffold.

The secondary amine acts as an intramolecular base, which enhances the nucleophilicity of the amide nitrogen (Figure 51). The geminal dimethyl groups led to a Thorpe-Ingold effect, which enhances the intramolecular attack, thus assisting in the formation of the succinimide ring [138–140]. The *o*-nitrobenzyloxycarbonyl (*o*-NBnoc) masks the basic character of the secondary amine [141,142], thus leading to the photo triggered cleavage of an amide bond by the deprotection of the secondary amine unit containing the *o*-nitrobenzyloxycarbonyl group.

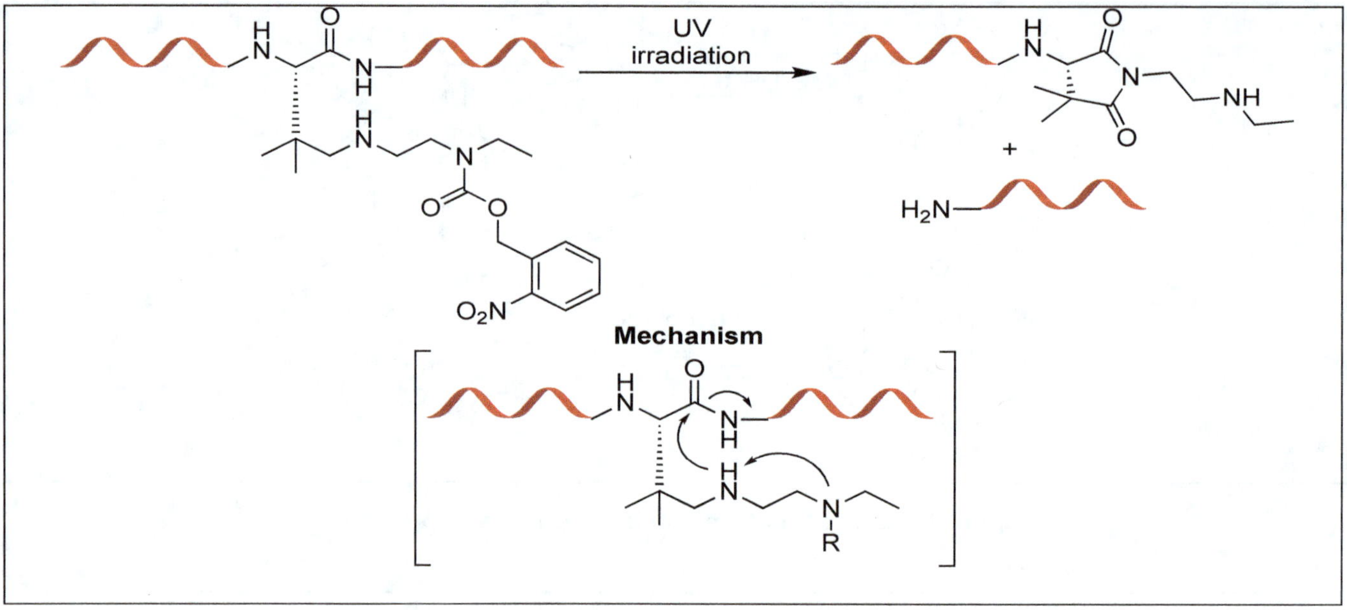

Figure 51. Intein-inspired amide bond cleavage.

4.9. Serine-Selective Aerobic Cleavage of Peptides

Kanai et al. reported a use of the water-soluble copper-organoradical conjugate for the selective cleavage of the peptide bond at the N-terminus of a serine residue under mild conditions and at room temperature [143]. They used this approach for the selective cleavage of a variety of different peptides/proteins containing D-amino acids or sensitive disulfide pairs.

Ser-selective cleavage of the peptide bond was initiated by the aerobic chemoselective oxidation of the hydroxymethyl moiety of Ser to a formyl group (**A**) (Figure 52). This produced a β-formyl glycineamide intermediate **B** which, on further oxidation, led to the formation of oxalamide **C** by undergoing oxidative deformylation. Oxalamide **C** then underwent hydrolysis under mild conditions because the carbonyl groups of the oxalamide are more electrophilic than those of simple amides, resulting in the formation of the cleaved fragments **D** and **D′**. By using molecular oxygen as a terminal oxidant, water and a C1 molecule (possibly HCO_2H) become stoichiometric side products. This strategy is widely distinct from Lewis acid, promoted by the Ser-selective peptide hydrolysis through N-to-O rearrangement.

4.10. Hydrolysis of Amide Bonds by the Formation of Oxazolinium Specie: Function of Acyl Protecting Group

Peptides containing a simple N-acyl group activates the amide bond four bonds away from an acyl group for cleavage under acidic conditions [144]. First, TFA leads to the protonation of amide carbonyl followed by the nucleophilic attack from the oxygen of the acyl carbonyl to generate a five-membered oxazolinium specie **A** in the peptide chain. Second, the collapse of the oxazolinium intermediate **A** leads to the cleavage of an unactivated amide bond (Figure 53).

The nature of the aromatic group, G, was responsible for the rate of hydrolysis of the peptide bond. Electron-donating groups increase the rate of hydrolysis whereas electron withdrawing substituents decrease it (Figure 53).

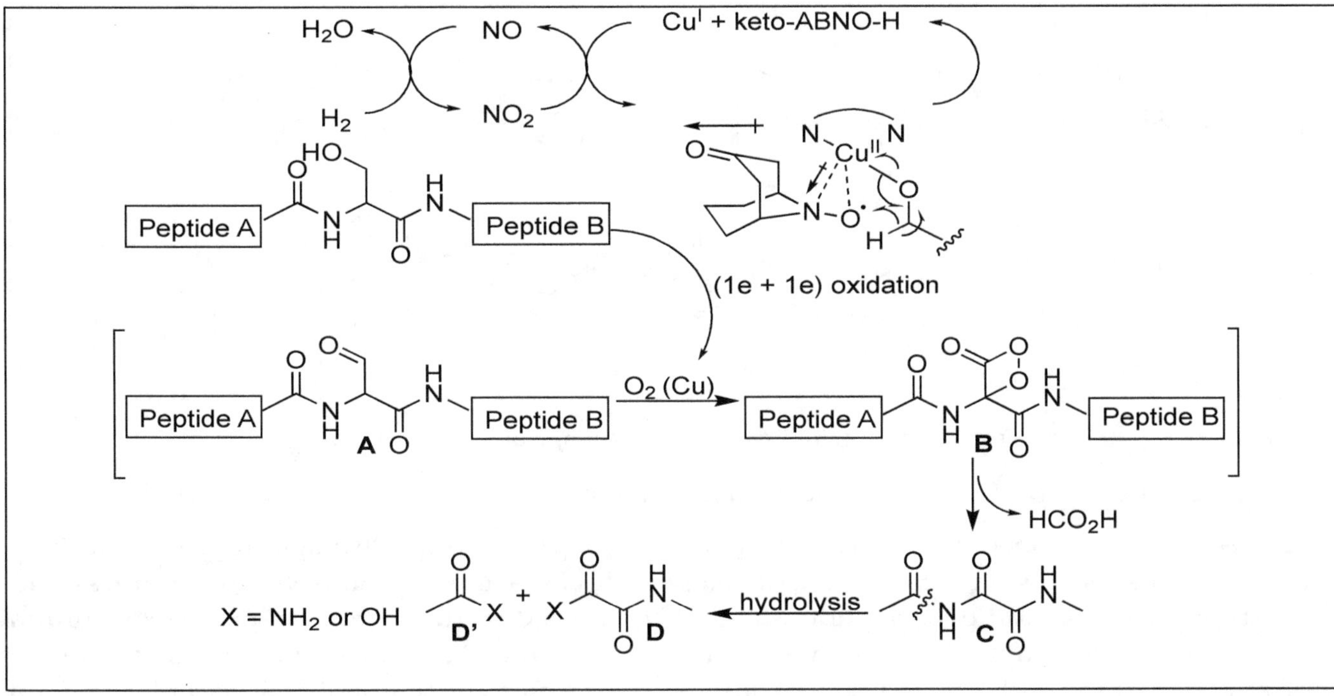

Figure 52. Serine selective aerobic cleavage of peptide bonds.

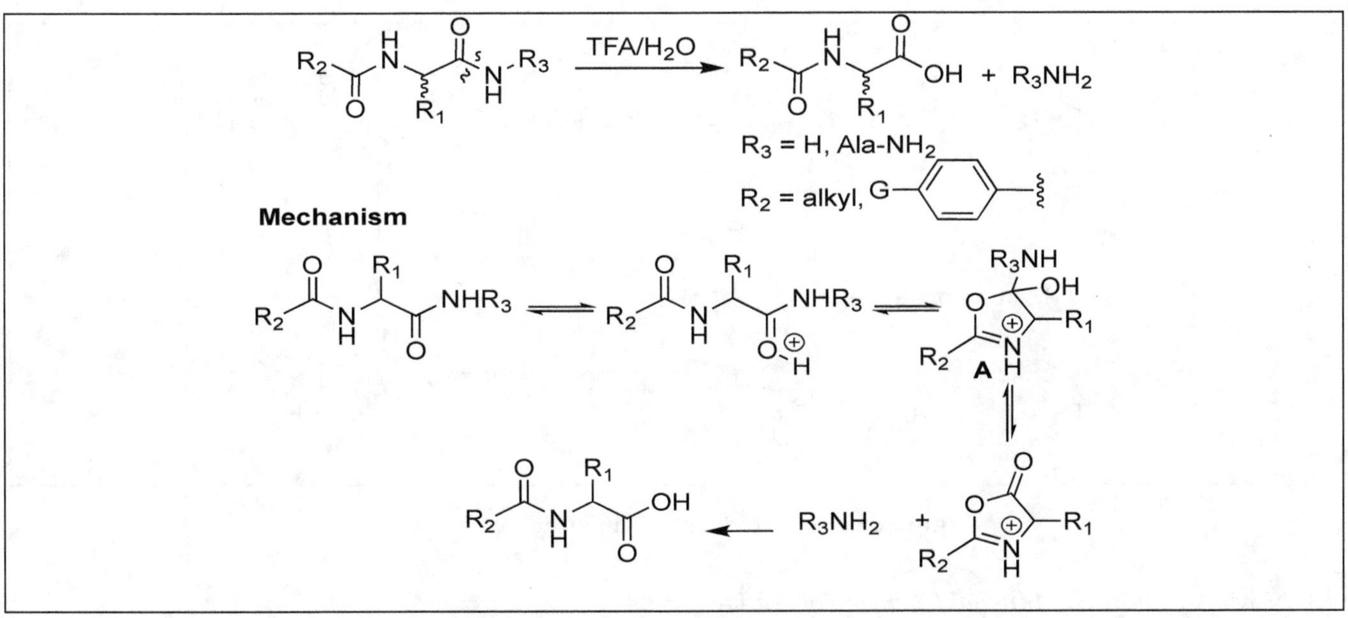

Figure 53. Oxazolinium species formation.

4.11. Hydrazinolysis for the Cleavage of Amide Bonds

The hydrazinolysis of unactivated amide bonds was accelerated by the addition of ammonium salts. The reaction proceeds at 50–70 °C to give peptide cleavage products and exhibits a broad substrate scope that out-performs existing amide bond cleavage reactions [145]. This approach was applied for the cleavage of the peptide bonds without racemization at the α-position of the amino acids (Figure 54). It was also applied to the cleavage of the *N*-acetyl group from the amino sugar.

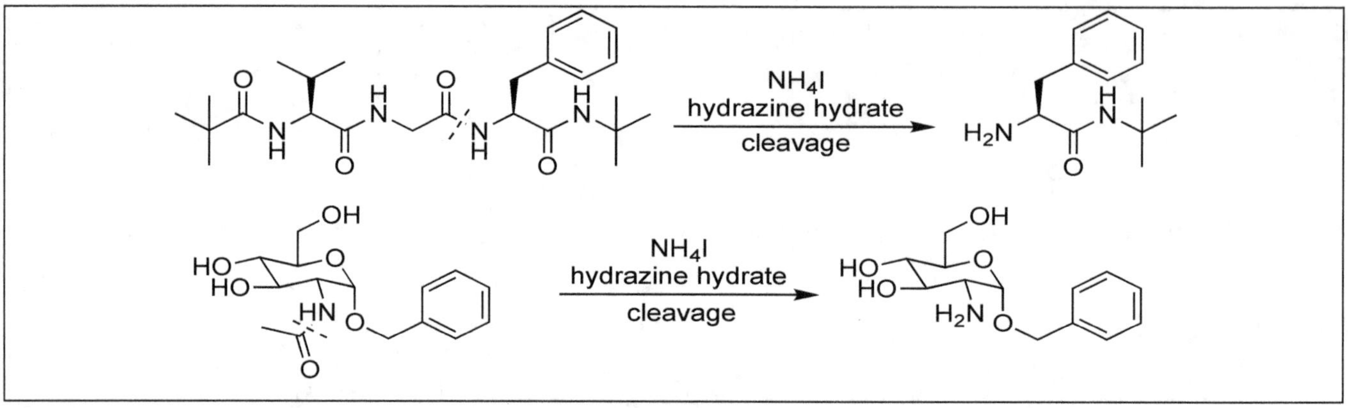

Figure 54. Hydrazinolysis for the cleavage of peptide bonds.

4.12. Amide Bond Cleavage of the N-Methylcysteinyl Peptide

Tam et al. developed a selective bi-directional peptide bond cleavage approach utilizing *N*-methylcysteine (MeCys) in the Xaa-MeCys-Yaa peptides (Xaa and Yaa, non-cysteine residues) [146]. Under strong acidic conditions, peptide Xaa-MeCys-Yaa led to the formation of an oxazolinium intermediate, resulting in the cleavage of the Xaa-MeCys bond. The oxazolinium intermediate was later trapped by thiocresol (TC) to form a Xaa-MeCys-TC thioester (Figure 55). The replacement of MeCys by Cys residue did not result in the peptide bond cleavage, suggesting the important role of *N*-methylation in MeCys residue for the formation of oxazolone.

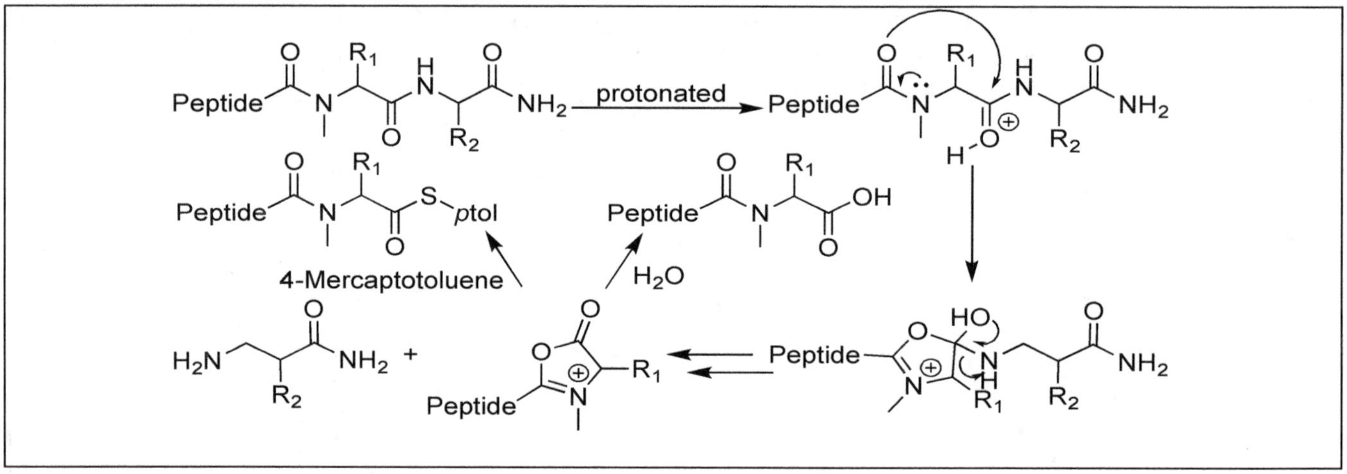

Figure 55. *N*-methylcysteinyl peptide cleavage.

4.13. N-MeAib Induced Unusual Cleavage of Amide Bonds

Peptides containing acylated *N*-methyl-aminoisobutyryl (NMeAib) residues showed unusual cleavage of amide bonds under acidic conditions. The cleavage takes place at the C-terminal side of the NMeAib residue (Figure 56) [147]. X-ray diffraction studies of the NMeAib containing molecules showed that the oxygen atom of the carbonyl group of the preceding residue is close to the carbonyl carbon of the NMeAib residue, thus acting as an internal nucleophile forming a tetrahedral intermediate. Once this tetrahedral intermediate was formed, lone pair electrons on the nitrogen of the phenylalanine were no longer be able to form resonating structures with the carbonyl group of NMeAib. In fact, the phenylalanine nitrogen becomes a proton acceptor like the amines, which resulted in the cleavage of the amide bond followed by the removal of phenylalanine and the formation of an oxazolinium ion intermediate, which further reacts with water to form a carboxylic acid product.

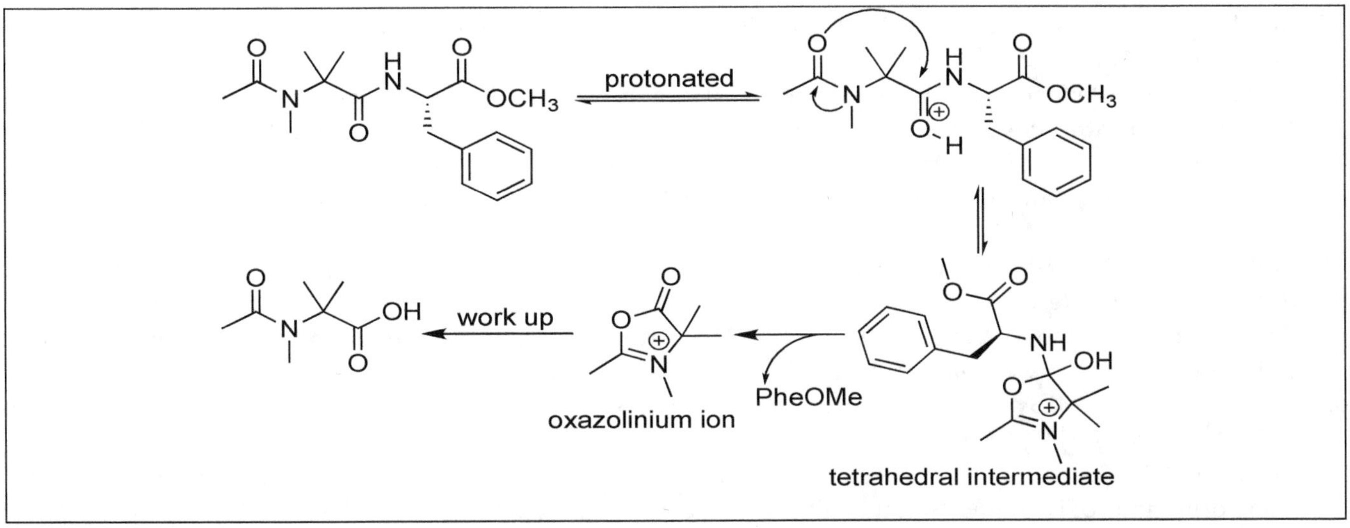

Figure 56. N-Me Aib mediated amide bond cleavage.

5. Conclusions

This mini-review highlights the methods for the activation of unactivated amide bonds in biomolecules. This review further highlights the application of these methods in the sequencing of proteins and the synthesis of peptide acids, thioesters, alcohols, and amides. These studies showed that there is still a lot to learn from enzymes catalyzed pathways and how we can develop enzyme mimetics for catalyzing the cleavage of unactivated and highly stable amide bonds at particular residues in a selective manner. We assume that these enzyme mimetics can have potential applications in various fields.

Author Contributions: M.R., S.M. and K.-C.T. collected the literature references. M.R., S.M. and K.-C.T. contributed to the manuscript writing.

References

1. Greenberg, A.; Breneman, C.M.; Liebman, J.F. *The Amide Linkage: Structural Significance, Shemistry, Biochemistry and Material Science*; Wiley: New York, NY, USA, 2000.
2. Brunton, L.; Chabner, B.; Knollman, B. *Goodman and Gilman's the Pharmacological Basis of Therapeutics*; MacGraw-Hill: New York, NY, USA, 2010.
3. Brown, D.G.; Bostrom, J. Analysis of Past and Present Synthetic Methodologies on Medicinal Chemistry: Where Have All the New Reactions Gone? *J. Med. Chem.* **2016**, *59*, 4443–4458. [CrossRef] [PubMed]
4. Pattabiraman, V.R.; Bode, J.W. Rethinking amide bond synthesis. *Nature* **2011**, *480*, 471–479. [CrossRef] [PubMed]
5. Hughes, A.B. *Amino Acids, Peptides and Proteins in Organic Chemistry*; Wiley-VCH: Weinheim, Germany, 2009.
6. Kaspar, A.A.; Reichert, J.M. Drug future directions for peptide therapeutics development. *Discov. Today* **2013**, *18*, 807–817. [CrossRef] [PubMed]
7. Thorner, J.; Emr, S.D.; Abelson, J.N. Applications of chimeric genes and hybrid proteins part A: Gene expression and protein purification. *Methods Enzymol.* **2000**, *326*, 601–617.
8. Kemnitz, C.R.; Loewen, M.J. Amide Resonance correlates with a breadth of C-N rotation barriers. *J. Am. Chem. Soc.* **2007**, *129*, 2521–2528. [CrossRef] [PubMed]
9. Mujika, J.I.; Mercero, J.M.; Lopez, X. Water-promoted hydrolysis of a highly twisted amide: Rate acceleration caused by the twist of the amide bond. *J. Am. Chem. Soc.* **2005**, *127*, 4445–4453. [CrossRef] [PubMed]
10. Wang, B.; Cao, Z. Acid-catalyzed reactions of twisted amides in water solution: Competition between hydration and hydrolysis. *Chem.-Eur. J.* **2011**, *17*, 11919–11929. [CrossRef] [PubMed]

11. Smith, R.M.; Hansen, D.E. The pH-rate profile for the hydrolysis of a peptide bond. *J. Am. Chem. Soc.* **1998**, *120*, 8910–8913. [CrossRef]
12. Radzicka, A.; Wolfenden, R. Rates of uncatalyzed peptide bond hydrolysis in neutral solution and the transition state affinities of proteases. *J. Am. Chem. Soc.* **1996**, *118*, 6105–6109. [CrossRef]
13. Arnau, J.; Lauritzen, C.; Petersen, G.E.; Pedersen, J. Current strategies for the use of affinity tags and tag removal for the purification of recombinant proteins. *Protein Express. Purif.* **2006**, *48*, 1–13. [CrossRef] [PubMed]
14. Hoyer, D.; Cho, H.; Schultz, P.G. New strategy for selective protein cleavage. *J. Am. Chem. Soc.* **1990**, *112*, 3249–3250. [CrossRef]
15. Schepartz, A.; Cuenoud, B. Site-specific cleavage of the protein calmodulin using a trifluoperazine-based affinity reagent. *J. Am. Chem. Soc.* **1990**, *112*, 3247–3249. [CrossRef]
16. Tani, K.; Stoltz, B.M. Synthesis and structural analysis of 2-quinuclidonium tetrafluoroborate. *Nature* **2006**, *441*, 731. [CrossRef] [PubMed]
17. Hayashi, M.; Bachman, S.; Hashimoto, S.; Eichman, C.C.; Stoltz, B.M. Catalytic anti-Markovnikov transformations of hindered terminal alkenes enabled by aldehyde-selective Wacker-type oxidation. *J. Am. Chem. Soc.* **2016**, *138*, 8997–9000. [CrossRef] [PubMed]
18. Kirby, A.J.; Komarov, I.V.; Wothers, P.D.; Feeder, N. The most twisted amide: Structure and reactions. *Angew. Chem. Int. Ed.* **1998**, *37*, 785–786. [CrossRef]
19. Kirby, A.J.; Komarov, I.V.; Feeder, N. Spontaneous, millisecond formation of a twisted amide from the amino acid, and the crystal structure of a tetrahedral intermediate. *J. Am. Chem. Soc.* **1998**, *120*, 7101–7102. [CrossRef]
20. Kirby, A.J.; Komarov, I.V.; Feeder, N. Synthesis, structure and reactions of the most twisted amide. *J. Chem. Soc. Perkin Trans.* **2001**, *2*, 522–529. [CrossRef]
21. Szostak, M.; Aube, J. Chemistry of bridged lactams and related heterocycles. *Chem. Rev.* **2013**, *113*, 5701. [CrossRef] [PubMed]
22. Artacho, J.; Ascic, E.; Rantanen, T.; Karlsson, J.; Wallentin, C.J.; Wang, R.; Wendt, O.F.; Harmata, M.; Snieckus, V.; Warnmark, K. Twisted amide analogues of Troger's base. *Chem. Eur. J.* **2012**, *18*, 1038–1042. [CrossRef] [PubMed]
23. Bashore, C.G.; Samardjiev, I.J.; Bordner, J.; Coe, J.W. Twisted amide reduction under Wolff–Kishner conditions: Synthesis of a benzo-1-aza-adamantane derivative. *J. Am. Chem. Soc.* **2003**, *125*, 3268. [CrossRef] [PubMed]
24. Meng, G.; Szostak, M. Sterically controlled Pd-catalyzed chemoselective ketone synthesis via N–C cleavage in twisted amides. *Org. Lett.* **2015**, *17*, 4364–4367. [CrossRef] [PubMed]
25. Meng, G.; Szostak, M. General olefin synthesis by the palladium catalyzed Heck reaction of amides: Sterically controlled chemoselective N–C activation. *Angew. Chem. Int. Ed.* **2015**, *54*, 14518–14522. [CrossRef] [PubMed]
26. Shi, S.; Meng, G.; Szostak, M. Synthesis of biaryls through nickel catalyzed Suzuki–Miyaura coupling of amides by carbon–nitrogen bond cleavage. *Angew. Chem. Int. Ed.* **2016**, *55*, 6959–6963. [CrossRef] [PubMed]
27. Shi, S.; Szostak, M. Efficient synthesis of diaryl ketones by nickel-catalyzed Negishi cross-coupling of amides by carbon–nitrogen bond cleavage at room temperature accelerated by a solvent effect. *Chem.-Eur. J.* **2016**, *22*, 10420–10424. [CrossRef] [PubMed]
28. Meng, G.; Szostak, M. Rhodium-catalyzed C–H bond functionalization with amides by double C–H/C–N bond activation. *Org. Lett.* **2016**, *18*, 796–799. [CrossRef] [PubMed]
29. Meng, G.; Szostak, M. Palladium-catalyzed Suzuki–Miyaura coupling of amides by carbon–nitrogen cleavage: General strategy for amide N–C bond activation. *Org. Biomol. Chem.* **2016**, *14*, 5690–5707. [CrossRef] [PubMed]
30. Hie, L.; Fine Nathel, N.F.; Shah, T.K.; Baker, E.L.; Hong, X.; Yang, Y.F.; Liu, P.; Houk, K.N.; Garg, N.K. Conversion of amides to esters by the nickel-catalysed activation of amide C–N bonds. *Nature* **2015**, *524*, 79–83. [CrossRef] [PubMed]
31. Li, X.; Zou, G. Acylative Suzuki coupling of amides: Acyl nitrogen activation via synergy of independently modifiable activating groups. *Chem. Commun.* **2015**, *51*, 5089–5092. [CrossRef] [PubMed]
32. Kraut, J. Serine proteases: Structure and mechanism of catalysis. *Annu. Rev. Biochem.* **1977**, *46*, 331–358. [CrossRef] [PubMed]
33. Carter, P.; Wells, J.A. Dissecting the catalytic triad of a serine protease. *Nature* **1988**, *332*, 564–568. [CrossRef] [PubMed]

34. Verma, S.; Dixit, R.; Pandey, K.C. Cysteine proteases: Modes of activation and future prospects as pharmacological targets. *Front. Pharmacol.* **2016**, *7*, 107. [CrossRef] [PubMed]
35. Otto, H.H.; Schirmeister, T. Cysteine proteases and their inhibitors. *Chem. Rev.* **1997**, *97*, 133–171. [CrossRef] [PubMed]
36. Rawlings, N.D.; Barrett, A.J. MEROPS: The peptidase database. *Nucleic Acids Res.* **1999**, *27*, 325–331. [CrossRef] [PubMed]
37. Coulombe, R.; Grochulski, P.; Sivaraman, J.; Menard, R.; Mort, J.S.; Cygler, M. Structure of human procathepsin L reveals the molecular basis of inhibition by the prosegment. *EMBO J.* **1996**, *15*, 5492–5503. [CrossRef] [PubMed]
38. Hooper, N.M. Families of zinc metalloproteases. *FEBS Lett.* **1994**, *354*, 1–6. [CrossRef]
39. Hase, C.C.; Finkelstein, R.A. Bacterial extracellular zinc containing metalloproteases. *Microbiol. Rev.* **1993**, *57*, 823–837. [PubMed]
40. Barrett, A. J. *Proteinases in Mammalian Cells and Tissues*; Elsevier/North-Holland Biomedical Press: Amsterdam, The Netherlands, 1977; pp. 181–208.
41. Garrett, R.H.; Grisham, C.M. *Biochemistry*; University of Virginia: Charlottesville, VA, USA, 2010; p. 103.
42. Hausrath, A.C.; Matthews, B.W. Thermolysin in the absence of substrate has an open conformation. *Acta Crystallogr.* **2002**, *58*, 1002–1007. [CrossRef]
43. Endo, S. Studies on protease produced by thermophilic bacteria. *J. Ferment. Technol.* **1962**, *40*, 346.
44. Kooi, C.; Sokol, P.A. Differentiation of thermolysins and serralysins by monoclonal antibodies. *J. Med. Microbiol.* **1996**, *45*, 219–225. [CrossRef] [PubMed]
45. De Kreij, A.; Venema, G.; van den Burg, B. Substrate specificity in the highly heterogeneous M4 peptidase family is determined by a small subset of amino acids. *J. Biol. Chem.* **2000**, *275*, 31115–31120. [CrossRef] [PubMed]
46. Titani, K.; Hermodson, M.A.; Ericsson, L.H.; Walsh, K.A.; Neurath, H. Amino-acid sequence of thermolysin. *Nat. New Biol.* **1972**, *238*, 35–37. [CrossRef] [PubMed]
47. Ooshima, H.; Mori, H.; Harano, Y. Synthesis of aspartame precursor by thermolysin solid in organic solvent. *Biotechnol. Lett.* **1985**, *7*, 789–792. [CrossRef]
48. Holmquist, B.; Vallee, B.L. Metal substitutions and inhibition of thermolysin: Spectra of the cobalt enzyme. *J. Biol. Chem.* **1974**, *249*, 4601–4607. [PubMed]
49. Vallee, B.L.; Auld, D.S. Zinc coordination, function, and structure of zinc enzymes and other proteins. *Biochemistry* **1990**, *29*, 5647–5659. [CrossRef] [PubMed]
50. Hangauer, D.G.; Monzingo, A.F.; Matthews, B.W. An interactive computer graphics study of thermolysin-catalyzed peptide cleavage and inhibition by *N*-carboxymethyl peptides. *Biochemistry* **1984**, *23*, 5730–5741. [CrossRef] [PubMed]
51. Pelmenschikov, V.; Blomberg, M.R.A.; Siegbahn, P.E.M. A theoretical study of the mechanism of peptide hydrolysis by thermolysine. *J. Biol. Inorg. Chem.* **2001**, *7*, 284–298. [CrossRef] [PubMed]
52. Kilshtain, A.V.; Warshel, A. On the origin of the catalytic power of carboxypeptidase A and other metalloenzymes. *Proteins* **2009**, *77*, 536–550. [CrossRef] [PubMed]
53. Quicho, F.A.; McMurray, C.H.; Lipscomb, W.N. Similarities between the Conformation of Arsanilazotyrosine 248 of Carboxypeptidase A*a* in the Crystalline State and in Solution. *Proc. Nat. Acad. Sci. USA* **1972**, *69*, 2850–2854. [CrossRef]
54. Wu, S.; Zhang, C.; Xu, D.; Guo, H. Catalysis of Carboxypeptidase A: Promoted-Water versus Nucleophilic Pathways. *J. Phys. Chem. B* **2010**, *114*, 9259–9267. [CrossRef] [PubMed]
55. Lazarus, M.B.; Jiang, J.; Kapuria, V.; Bhuiyan, T.; Janetzko, J.; Zandberg, W.F.; Vocadlo, D.J.; Herr, W.; Walker, S. HCF-1 is cleaved in the active site of O-GlcNAc transferase. *Science* **2013**, *342*, 1235–1239. [CrossRef] [PubMed]
56. Janetzko, J.; Trauger, S.A.; Lazarus, M.B.; Walker, S. How the glycosyltransferase OGT catalyzes amide bond cleavage. *Nat. Chem Biol.* **2016**, *12*, 899–901. [CrossRef] [PubMed]
57. Smith, C.S.; Anderson, M.A.; Hoadley, K.A.; Keck, J.L.; Cleland, W.W.; Denu, J. M. Structural and kinetic isotope effect studies of nicotinamidase (Pnc1) from saccharomyces cerevisiae. *Biochemistry* **2012**, *51*, 243–256. [CrossRef] [PubMed]

58. Fyfe, P.K.; Rao, V.A.; Zemla, A.; Cameron, S.; Hunter, W.N. Specificity and mechanism of acinetobacter baumanii nicotinamidase: Implications for activation of the front-line tuberculosis drug pyrazinamide. *Angew. Chem. Int. Ed.* **2009**, *48*, 9176–9179. [CrossRef] [PubMed]
59. Hu, G.; Taylor, A.B.; McAlister-Henn, L.; Hart, P.J. Crystal structure of the yeast nicotinamidase Pnc1p. *Arch. Biochem. Biophys.* **2007**, *461*, 66–75. [CrossRef] [PubMed]
60. Du, X.; Wang, W.; Kim, R.; Yakota, H.; Nguyen, H.; Kim, S.H. Crystal structure and mechanism of catalysis of a pyrazinamidase from Pyrococcus horikoshii. *Biochemistry* **2001**, *40*, 14166–14172. [CrossRef] [PubMed]
61. Mukherjee, T.; Zhang, Y.; Abdelwahed, S.; Ealick, S.E.; Begley, T.P. Catalysis of a flavoenzyme-mediated amide hydrolysis. *J. Am. Chem. Soc.* **2010**, *132*, 5550–5551. [CrossRef] [PubMed]
62. Massey, V. Activation of molecular oxygen by flavins and flavoproteins. *J. Biol. Chem.* **1994**, *269*, 22459. [PubMed]
63. Vaughn, H.L.; Robbins, M.D. Rapid procedure for the hydrolysis of amides to acids. *J. Org. Chem.* **1975**, *40*, 1187. [CrossRef]
64. Gao, C.; Lavey, B.J.; Lo, C.L.; Datta, A.; Wentworth, P., Jr.; Janda, K.D. Direct selection for catalysis from combinatorial antibody libraries using a boronic acid probe: Primary amide bond hydrolysis. *J. Am. Chem. Soc.* **1998**, *120*, 2211–2217. [CrossRef]
65. Martin, M.T.; Angeles, T.S.; Sugasawara, R.; Aman, N.I.; Napper, A.D.; Darsley, M.J.; Sanchez, R.I.; Booth, P.; Titmas, R.C. Antibody-catalyzed hydrolysis of an unsubstituted amide. *J. Am. Chem. Soc.* **1994**, *116*, 6508–6512. [CrossRef]
66. Dai, X.; de Mesmaeker, A.; Joyce, G.F. Cleavage of an amide bond by a ribozyme. *Science* **1995**, *267*, 237–241. [CrossRef] [PubMed]
67. Wezynfeld, N.E.; Frączyk, T.; Bal, W. Metal assisted peptide bond hydrolysis: Chemistry, biotechnology and toxicological implications. *Coord. Chem. Rev.* **2016**, *327–328*, 166–187. [CrossRef]
68. Yashiro, M.; Sonobe, Y.; Yamamura, A.; Takarada, T.; Komiyama, M.; Fujii, Y. Metal-ion-assisted hydrolysis of dipeptides involving a serine residue in a neutral aqueous solution. *Org. Biomol. Chem.* **2003**, *1*, 629–632. [CrossRef] [PubMed]
69. Suh, J.H. Model studies of metalloenzymes involving metal-ions as Lewis acid catalysts. *Acc. Chem. Res.* **1992**, *25*, 273–279. [CrossRef]
70. Wybon, C.C.D.; Mensch, C.; Hollanders, K.; Gadais, C.; Herrebout, W.A.; Ballet, S.; Maes, B.U.W. Zn-catalyzed *tert*-Butyl nicotinate-directed amide cleavage as a biomimic of metallo-exopeptidase activity. *ACS Catal.* **2018**, *8*, 203–218. [CrossRef]
71. Yashiro, M.; Takarada, T.; Miyama, S.; Komiyama, M. Cerium(IV)–cyclodextrin complex for peptide hydrolysis in neutral homogeneous solutions. *J. Chem. Soc. Chem. Commun.* **1994**, *15*, 1757–1758. [CrossRef]
72. Kita, Y.; Nishii, Y.; Higuchi, T.; Mashima, K. Zinc-catalyzed amide cleavage and esterification of beta-hydroxyethylamides. *Angew. Chem. Int. Ed.* **2012**, *51*, 5723–5726. [CrossRef] [PubMed]
73. Ho, P.H.; Stroobants, K.; Parac-Vogt, T.N. Hydrolysis of serine-containing peptides at neutral pH promoted by [MoO4]2–oxyanion. *Inorg. Chem.* **2011**, *50*, 12025–12033. [CrossRef] [PubMed]
74. Absillis, G.; Parac-Vogt, T.N. hydrolytic activity of vanadate toward serine-containing peptides studied by kinetic experiments and DFT theory. *Inorg. Chem.* **2012**, *51*, 9902–9910. [CrossRef] [PubMed]
75. Cartuyvels, E.; Absillis, G.; Parac-Vogt, T.N. Questioning the paradigm of metal complex promoted phosphodiester hydrolysis: [Mo7O24]6−polyoxometalate cluster as an unlikely catalyst for the hydrolysis of a DNA model substrate. *Chem. Commun.* **2008**, 85–87. [CrossRef]
76. Absillis, G.; Cartuyvels, E.; Van Deun, R.; Parac-Vogt, T.N. Hydrolytic cleavage of an RNA-Model phosphodiester catalyzed by a highly negatively charged polyoxomolybdate [Mo7O24]6−Cluster. *J. Am. Chem. Soc.* **2008**, *130*, 17400–17408. [CrossRef] [PubMed]
77. Absillis, G.; Van Deun, R.; Parac-Vogt, T.N. Polyoxomolybdate promoted hydrolysis of a DNA-model phosphoester studied by NMR and EXAFS spectroscopy. *Inorg. Chem.* **2011**, *50*, 11552–115560. [CrossRef] [PubMed]
78. Takarada, T.; Yashiro, M.; Komiyama, M. Catalytic hydrolysis of peptides by cerium(IV). *Chem. Eur. J.* **2000**, *6*, 3906–3913. [CrossRef]
79. Kassai, M.; Ravi, R.G.; Shealy, S.J.; Grant, K.B. Unprecedented acceleration of zirconium(IV)-assisted peptide hydrolysis at neutral pH. *Inorg. Chem.* **2004**, *43*, 6130–6132. [CrossRef] [PubMed]

80. Absillis, G.; Parac-Vogt, T.N. Peptide bond hydrolysis catalyzed by the Wells–Dawson Zr(α2-P2W17O61)2 polyoxometalate. *Inorg. Chem.* **2012**, *51*, 9902–9910. [CrossRef] [PubMed]
81. Ho, P.H.; Stroobants, K.; Moelants, E.; Proost, P.; Parac-Vogt, T.N. Selective hydrolysis of hen egg white lysozyme at Asp-X peptide bonds promoted by oxomolybdate. *J. Inorg. Biochem.* **2014**, *136*, 73–80.
82. Ly, H.G.T.; Fu, G.; Kondinski, A.; Bueken, B.; De Vos, D.; Parac-Vogt, T.N. Superactivity of MOF-808 toward peptide bond hydrolysis. *J. Am. Chem. Soc.* **2018**, *140*, 6325–6335. [CrossRef] [PubMed]
83. Meriwether, L.; Westheimer, F.H. Metal ion promoted hydrolysis of glycine amide and of phenylalanylglycine amide. *J. Am. Chem. Soc.* **1956**, *78*, 5119. [CrossRef]
84. Bamann, E.; Hass, J.G.; Trapmann, H. Metallionenkatalytische zerlegung der (-CO-NH-)-bindung in peptiden und *N*-Acetyl-aminosäuren. *Arch. Pharm.* **1961**, *294*, 569–580. [CrossRef]
85. Collman, J.P.; Buckingham, D.A. Hydrolytic cleavage of N-terminal peptide bonds by a cobalt chelate. *J. Am. Chem. Soc.* **1963**, *85*, 3039–3040. [CrossRef]
86. Buckingham, D.A.; Coliman, J.P.; Happer, D.A.R.; Marzilli, L.G. Hydrolysis of N-terminal peptide bonds and amino acid derivatives by the β-hydroxoaquotriethylenetetraminecobalt(III) ion. *J. Am. Chem. Soc.* **1967**, *89*, 1082–1087. [CrossRef] [PubMed]
87. Bentley, K.W.; Creaser, E.H. Qualitative determination of N-terminal amino acids of peptides and proteins with cobalt (3) chelates. *Bfochem. J.* **1973**, *135*, 507–511. [CrossRef]
88. Kimura, E. Sequential hydrolysis of peptides with β-hydroxoaquo triethylenetetraminecobalt(III) ion. *Inorg. Chem.* **1974**, *13*, 951–954. [CrossRef]
89. Erxleben, A. Interaction of molybdocene dichloride with cysteine-containing peptides: Coordination, regioselective hydrolysis, and intramolecular aminolysis. *Inorg. Chem.* **2005**, *44*, 1082–1094. [CrossRef] [PubMed]
90. Burgeson, I.E.; Kostic', N.M. Selective hydrolysis of unactivated peptide bonds, promoted by platinum(II) complexes anchored to amino acid side chains. *Inorg. Chem.* **1991**, *30*, 4299–4305. [CrossRef]
91. Rajkovic, S.; Zivkovic, M.D.; Djuran, M.I. Reactions of dinuclear platinum(II) complexes with peptides. *Curr. Protein Pept. Sci.* **2016**, *17*, 95–105. [CrossRef] [PubMed]
92. Parac, T.N.; Kostic, N.M. New selectivity and turnover in peptide hydrolysis by metal complexes. A palladium(II) aqua complex catalyzes cleavage of peptides next to the histidine residue. *J. Am. Chem. Soc.* **1996**, *118*, 51–58. [CrossRef]
93. Živković, M.D.; Rajković, S.; Glišić, B.D.; Drašković, N.S.; Djuran, M.I. Hydrolysis of the amide bond in histidine- and methionine-containing dipeptides promoted by pyrazine and pyridazine palladium(II)-aqua dimers: Comparative study with platinum(II) analogues. *Bioorg. Chem.* **2017**, *72*, 190–198. [CrossRef] [PubMed]
94. Rajkovic, S.; Asanin, D.P.; Zivkovic, M.D.; Djuran, M.I. Synthesis of different pyrazine-bridged platinum(II) complexes and ^{1}H NMR study of their catalytic abilities in the hydrolysis of the Nacetylated L-methionylglycine. *Polyhedron* **2013**, *65*, 42–47. [CrossRef]
95. Asanin, D.P.; Zivkovic, M.D.; Rajkovic, S.; Warzajtis, B.; Rychlewska, U.; Djuran, M.I. Crystallographic evidence of anion···π interactions in the pyrazine bridged {[Pt(en)Cl]$_2$(μ-pz)}Cl$_2$ complex and a comparative study of the catalytic ability of mononuclear and binuclear platinum(II) complexes in the hydrolysis of N-acetylated L-methionylglycine. *Polyhedron* **2013**, *51*, 255–262.
96. Živković, M.D.; Asanin, D.P.; Rajkovic, S.; Djuran, M.I. Hydrolysis of the amide bond in *N*-acetylated L-methionylglycine catalyzed by various platinum(II) complexes under physiologically relevant conditions. *Polyhedron* **2011**, *30*, 947–952. [CrossRef]
97. Rajkovic, S.; Zivkovic, M.D.; Kallay, C.; Sóvágó, I.; Djuran, M.I. A study of the reactions of a methionine- and histidine-containing tetrapeptide with different Pd(II) and Pt(II) complexes: Selective cleavage of the amide bond by platination of the peptide and steric modification of the catalyst. *Dalton Trans.* **2009**, *39*, 8370–8377. [CrossRef] [PubMed]
98. Zhu, L.; Kostie, N.M. Toward artificial metalloproteases: Mechanisms by which platinum(II) and palladium(II) complexes promote selective, fast hydrolysis of unactivated amide bonds in peptides. *Inorg. Chem.* **1992**, *31*, 3994–4001. [CrossRef]
99. Milovic, N.M.; Kostic, N.M. Palladium(II) complexes, as synthetic peptidases, regioselectively cleave the second peptide bond "upstream" from methionine and histidine side chains. *J. Am. Chem. Soc.* **2002**, *124*, 4759–4769. [CrossRef] [PubMed]

100. Milovic, N.; Kostic, N. Palladium(II) complex as a sequence-specific peptidase: Hydrolytic cleavage under mild conditions of X-Pro peptide bonds in X-Pro-Met and X-Pro-His segments. *J. Am. Chem. Soc.* **2003**, *125*, 781–789. [CrossRef] [PubMed]
101. Milovic, N.M.; Kostic, N.M. Interplay of terminal amino group and coordinating side chains in directing regioselective cleavage of natural peptides and proteins with palladium(II) complexes. *Inorg. Chem.* **2002**, *41*, 7053–7063. [CrossRef] [PubMed]
102. Milovic, N.M.; Badjic, J.D.; Kostic, N.M. Conjugate of palladium(II) complex and β-cyclodextrin acts as a biomimetic peptidase. *J. Am. Chem. Soc.* **2004**, *126*, 696–697. [CrossRef] [PubMed]
103. Yoo, C.E.; Chae, P.S.; Kim, J.E.; Jeong, E.J.; Suh, J. Degradation of myoglobin by polymeric artificial metalloproteases containing catalytic modules with various catalytic group densities: Site selectivity in peptide bond cleavage. *J. Am. Chem. Soc.* **2003**, *125*, 14580–14589. [CrossRef] [PubMed]
104. Yoo, S.H.; Lee, B.J.; Kim, H.; Suh, J. Artificial metalloprotease with active site comprising aldehyde group and Cu(II)Cyclen complex. *J. Am. Chem. Soc.* **2005**, *127*, 9593–9602. [CrossRef] [PubMed]
105. Jeon, J.W.; Son, S.J.; Yoo, C.E.; Hong, I.S.; Suh, J. Toward protein-cleaving catalytic drugs: Artificial protease selective for myoglobin. *Bioorg. Med. Chem.* **2003**, *11*, 2901–2910. [CrossRef]
106. Kim, H.; Jang, B.; Cheon, Y.; Suh, M.; Suh, J. Proteolytic activity of Co(III) complex of 1-oxa-4,7,10-triazacyclododecane: A new catalytic center for peptide-cleavage agents. *J. Biol. Inorg. Chem.* **2009**, *14*, 151–157. [CrossRef] [PubMed]
107. Chae, P.S.; Kim, M.-S.; Jeung, C.-S.; Lee, S.D.; Park, H.; Lee, S.; Suh, J. Peptide-cleaving catalyst selective for peptide deformylase. *J. Am. Chem. Soc.* **2005**, *127*, 2396–2397. [CrossRef] [PubMed]
108. Suh, J.; Chei, W.S.; Lee, T.Y.; Kim, M.G.; Yoo, S.H.; Jeong, K.; Ahn, J.Y. Cleavage agents for soluble oligomers of human islet amyloid polypeptide. *J. Biol. Inorg. Chem.* **2008**, *13*, 693–701. [CrossRef] [PubMed]
109. Lee, T.Y.; Suh, J. Target-selective peptide-cleaving catalysts as a new paradigm in drug design. *Chem. Soc. Rev.* **2008**, *38*, 1949–1957. [CrossRef] [PubMed]
110. Suh, J. Progress in designing artificial proteases: A new therapeutic option for amyloid diseases. *Asian J. Org. Chem.* **2014**, *3*, 18–32. [CrossRef]
111. De Oliveira, M.C.B.; Scarpellini, M.; Neves, A.; Terenzi, H.; Bortoluzzi, A.J.; Szpoganics, B.; Greatti, A.; Mangrich, A.S.; de Souza, E.M.; Fernandez, P.M.; et al. Hydrolytic protein cleavage mediated by unusual mononuclear copper(II) complexes: X-ray structures and solution studies. *Inorg. Chem.* **2005**, *44*, 921–929. [CrossRef] [PubMed]
112. Kopera, E.; Krężel, A., A.; Protas, A.M.; Belczyk, A.; Bonna, A.; Wysłouch-Cieszynska, A.; Poznanski, J.; Bal, W. Sequence-specific Ni(II)-dependent peptide bond hydrolysis for protein engineering: Reaction conditions and molecular mechanism. *Inorg. Chem.* **2010**, *49*, 6636–6645. [CrossRef] [PubMed]
113. Ariani, H.H.; Polkowska-Nowakowska, A.; Bal, W. Effect of D-amino acid substitutions on Ni(II)-assisted peptide bond hydrolysis. *Inorg. Chem.* **2013**, *52*, 2422–2431. [CrossRef] [PubMed]
114. Ni, J.; Sohma, Y.; Kanai, M. Scandium(III) triflate-promoted serine/threonine-selective peptide bond cleavage. *Chem. Commun.* **2017**, *53*, 3311–3314. [CrossRef] [PubMed]
115. Edman, P.; Begg, G.A. Protein sequenator. *Eur. J. Biochem.* **1967**, *1*, 80–91. [CrossRef] [PubMed]
116. Gross, E. The cyanogen bromide reaction. *Methods Enzymol.* **1967**, *11*, 238.
117. Degani, Y.; Patchornik, A. Cyanylation of sulfhydryl groups by 2-nitro-5-thiocyanobenzoic acid. High-yield modification and cleavage of peptides at cysteine residues. *Biochemistry* **1974**, *13*, 1. [CrossRef] [PubMed]
118. Mahoney, W.C.; Smith, P.K.; Hermodson, M.A. Fragmentation of proteins with *o*-iodosobenzoic acid: Chemical mechanism and identification of *o*-iodoxybenzoic acid as a reactive contaminant that modifies tyrosyl residues. *Biochemistry* **1981**, *20*, 443–448. [CrossRef] [PubMed]
119. Burstein, Y.; Patchornik, A. Selective chemical cleavage of tryptophanyl peptide bonds in peptides and proteins. *Biochemistry* **1972**, *11*, 4641–4650. [CrossRef] [PubMed]
120. Hamada, Y.A. Novel N-terminal degradation reaction of peptides via *N*-amidination. *Bioorg. Med. Chem. Lett.* **2016**, *26*, 1690. [CrossRef] [PubMed]
121. Yamamoto, J.; Denda, M.; Maeda, N.; Kita, M.; Komiya, C.; Tanaka, T.; Nomura, W.; Tamamura, H.; Sato, Y.; Yamauchi, A.; et al. Development of a traceable linker containing a thiol-responsive amino acid for the enrichment and selective labelling of target proteins. *Org. Biomol. Chem.* **2014**, *12*, 3821–3826. [CrossRef] [PubMed]

122. Yamamoto, J.; Maeda, N.; Komiya, C.; Tanaka, T.; Denda, M.; Ebisuno, K.; Nomura, W.; Tamamura, H.; Sato, Y.; Yamauchi, A.; et al. Development of a fluoride-responsive amide bond cleavage device that is potentially applicable to a traceable linker. *Tetrahedron* **2014**, *70*, 5122–5127. [CrossRef]
123. Shigenaga, A.; Ogura, K.; Hirakawa, H.; Yamamoto, J.; Ebisuno, K.; Miyamoto, L.; Ishizawa, K.; Tsuchiya, K.; Otaka, A. Development of a reduction-responsive amino acid that induces peptide bond cleavage in hypoxic cells. *ChemBioChem* **2012**, *13*, 968–971. [CrossRef] [PubMed]
124. Shigenaga, A.; Hirakawa, H.; Yamamoto, J.; Ogura, K.; Denda, M.; Yamaguchi, K.; Tsuji, D.; Itoh, K.; Otaka, A. Design and synthesis of caged ceramide: UV-responsive ceramide releasing system based on UV-induced amide bond cleavage followed by O–N acyl transfer. *Tetrahedron* **2011**, *67*, 3984–3990. [CrossRef]
125. Shigenaga, A.; Yamamoto, J.; Sumikawa, Y.; Furuta, T.; Otaka, A. Development and photo-responsive peptide bond cleavage reaction of two-photon near-infrared excitation-responsive peptide. *Tetrahedron Lett.* **2010**, *51*, 2868–2871. [CrossRef]
126. Shigenaga, A.; Yamamoto, J.; Hirakawa, H.; Ogura, K.; Maeda, N.; Morishita, K.; Otaka, A. Development of thiol-responsive amide bond cleavage device and its application for peptide nucleic acid-based DNA releasing system. *Tetrahedron Lett.* **2010**, *51*, 2525. [CrossRef]
127. Shigenaga, A.; Tsuji, D.; Nishioka, N.; Tsuda, S.; Itoh, K.; Otaka, A. Synthesis of a stimulus-responsive processing device and its application to a nucleocytoplasmic shuttle peptide. *ChemBioChem* **2007**, *8*, 1929–1931. [CrossRef] [PubMed]
128. Kita, M.; Yamamoto, J.; Morisaki, T.; Komiya, C.; Inokuma, T.; Miyamoto, L.; Tsuchiya, K.; Shigenaga, A.; Otaka, A. Design and synthesis of a hydrogenperoxide-responsive amino acid that induces peptide bond cleavage after exposure to hydrogen peroxide. *Tetrahedron Lett.* **2015**, *56*, 4228–4231. [CrossRef]
129. Nalbone, J.M.; Lahankar, N.; Buissereth, L.; Raj, M. Glutamic acid selective chemical cleavage of peptide bonds. *Org. Lett.* **2016**, *18*, 1186–1189. [CrossRef] [PubMed]
130. Tofteng, A.P.; Sørensen, K.K.; Conde-Frieboes, K.W.; HoegJensen, T.; Jensen, K.J. Fmoc solid-phase synthesis of C-terminal peptide thioesters by formation of a backbone pyroglutamyl imide moiety. *Angew. Chem. Int. Ed.* **2009**, *48*, 7411–7414. [CrossRef] [PubMed]
131. Tanabe, K.; Taniguchi, A.; Matsumoto, T.; Oisaki, K.; Sohma, Y.; Kanai, M. Asparagine-selective cleavage of peptide bonds through hypervalent iodine-mediated Hofmann rearrangement in neutral aqueous solution. *Chem. Sci.* **2014**, *5*, 2747–2753. [CrossRef]
132. Elashal, H.E.; Raj, M. Site-selective chemical cleavage of peptide bonds. *Chem. Commun.* **2016**, *52*, 6304–6307. [CrossRef] [PubMed]
133. Elashal, H.E.; Cohen, R.D.; Raj, M. Fmoc solid-phase synthesis of C-terminal modified peptides by formation of a backbone cyclic urethane moiety. *Chem. Commun.* **2016**, *52*, 9699–9702. [CrossRef] [PubMed]
134. Elashal, H.E.; Sim, Y.E.; Raj, M. Serine promoted synthesis of peptide thioester precursor on solid support for native chemical ligation. *Chem. Sci.* **2017**, *8*, 117–123. [CrossRef] [PubMed]
135. Elashal, H.E.; Cohen, R.D.; Elashal, H.E.; Raj, M. Oxazolidinone-mediated sequence determination of one-bead one-compound cyclic peptide libraries. *Org. Lett.* **2018**, *20*, 2374–2377. [CrossRef] [PubMed]
136. Elashal, H.E.; Cohen, R. D.; Elashal, H.E.; Zong, C.H.; Link, A.J.; Raj, M. Cyclic and lasso peptides: Sequence determination, topology analysis, and rotaxane formation. *Angew. Chem. Int. Ed.* **2018**, *57*, 6150–6154. [CrossRef] [PubMed]
137. Komiya, C.; Aihara, K.; Morishita, K.; Ding, H.; Inokuma, T.; Shigenaga, A.; Otaka, A. Development of an intein-inspired amide cleavage chemical device. *J. Org. Chem.* **2016**, *81*, 699–707. [CrossRef] [PubMed]
138. Jung, M.E.; Piizzi, G. Gem-Disubstituent effect: Theoretical basis and synthetic applications. *Chem. Rev.* **2005**, *105*, 1735–1766. [CrossRef] [PubMed]
139. Bachrach, S.M. The gem-dimethyl effect revisited. *J. Org. Chem.* **2008**, *73*, 2466–2468. [CrossRef] [PubMed]
140. Beesley, R.M.; Ingold, C.K.; Thorpe, J.F. CXIX—The formation and stability of spiro-compounds. Part I. spiro-compounds from cyclohexane. *J. Chem. Soc. Trans.* **1915**, *107*, 1080–1106. [CrossRef]
141. Bochet, C.G. Photolabile protecting groups and linkers. *J. Chem. Soc. Perkin Trans.* **2001**, *2*, 125–142.
142. Brieke, C.; Rohrbach, F.; Gottschalk, A.; Mayer, G.; Heckel, A. Light-controlled tools. *Angew. Chem. Int. Ed.* **2012**, *51*, 8446–8476. [CrossRef] [PubMed]
143. Seki, Y.; Tanabe, K.; Sasaki, D.; Sohma, Y.; Oisaki, K.; Kanai, M. Serine-selective aerobic cleavage of peptides and a protein using a water-soluble copper–organoradical conjugate. *Angew. Chem. Int. Ed.* **2014**, *53*, 6501–6505. [CrossRef] [PubMed]

144. Samaritoni, J.G.; Copes, A.T.; Crews, D.K.; Glos, C.; Thompson, A.L.; Wilson, C.; O'Donnell, M.J.; Scott, W.L. Unexpected hydrolytic instability of N-acylated amino acid amides and peptides. *J. Org. Chem.* **2014**, *79*, 3140–3151. [CrossRef] [PubMed]
145. Shimizu, Y.; Noshita, M.; Mukai, Y.; Morimoto, H.; Ohshima, T. Cleavage of unactivated amide bonds by ammonium salt-accelerated hydrazinolysis. *Chem. Commun.* **2014**, *50*, 12623–12625. [CrossRef] [PubMed]
146. Qiu, Y.; Hemu, X.; Liu, D.X.; Tam, J.P. Selective bi-directional amide bond cleavage of *N*-methylcysteinyl peptide. *Eur. J. Org. Chem.* **2014**, *20*, 4370–4380. [CrossRef]
147. Creighton, C.J.; Romoff, T.T.; Bu, J.H.; Goodman, M. Mechanistic studies of an unusual amide bond scission. *J. Am. Chem. Soc.* **1999**, *121*, 6786–6791. [CrossRef]

7

[2+2+2] Annulation of *N*-(1-Naphthyl)acetamide with Two Alkynoates via Cleavage of Adjacent C–H and C–N Bonds Catalyzed by an Electron-Deficient Rhodium(III) Complex

Jyunichi Terasawa, Yu Shibata *, Miho Fukui and Ken Tanaka *

Department of Chemical Science and Engineering, Tokyo Institute of Technology, O-okayama, Meguro-ku, Tokyo 152-8550, Japan; terasawa.j.aa@m.titech.ac.jp (J.T.); mfv1.1.174@gmail.com (M.F.)
* Correspondence: yshibata@apc.titech.ac.jp (Y.S.); ktanaka@apc.titech.ac.jp (K.T.)

Academic Editor: Michal Szostak

Abstract: It has been established that an electron-deficient cationic Cp^E-rhodium(III) complex catalyzes the non-oxidative [2+2+2] annulation of *N*-(1-naphthyl)acetamide with two alkynoates via cleavage of the adjacent C–H and C–N bonds to give densely substituted phenanthrenes under mild conditions (at 40 °C under air). In this reaction, a dearomatized spiro compound was isolated, which may support the formation of a cationic spiro rhodacycle intermediate in the catalytic cycle. The use of *N*-(1-naphthyl)acetamide in place of acetanilide switched the reaction pathway from the oxidative [2+2+2] annulation-lactamization via C–H/C–H cleavage to the non-oxidative [2+2+2] annulation via C–H/C–N cleavage. This chemoselectivity switch may arise from stabilization of the carbocation in the above cationic spiro rhodacycle by the neighboring phenyl and acetylamino groups, resulting in the nucleophilic C–C bond formation followed by β-nitrogen elimination.

Keywords: alkynes; C–H bond cleavage; C–N bond cleavage; cyclopentadienyl complexes; *N*-(1-naphthyl)acetamide; rhodium; [2+2+2] annulation

1. Introduction

The transition-metal-catalyzed [2+2+2] annulation of three unsaturated compounds is a useful method for the synthesis of six-membered carbocycles and heterocycles [1–9]. For example, the transition-metal-catalyzed [2+2+2] annulation of three alkynes is able to afford densely substituted benzenes with an atom- and step-economical manner [10–14]. For the synthesis of naphthalene derivatives, benzynes have been employed as one of the three alkynes but the use of the benzynes suffers from the redundant precursor synthesis and harsh reaction conditions [15–17]. Alternatively, several examples of the transition-metal-catalyzed oxidative [2+2+2] annulation via cleavage of adjacent two C–H bonds of benzenes, in which formally dehydrogenated benzenes are employed as benzyne equivalents, have been reported [18–22]. As such, our research group reported that an electron-deficient cationic Cp^E-rhodium(III) complex, derived from **1**, catalyzes the oxidative tandem [2+2+2] annulation-lactamization of acetanilide (**2a**) with two alkynoates **3** via cleavage of adjacent two C–H bonds at room temperature under air to give densely substituted banzo[*cd*]indolones **4a** (Scheme 1, top) [23]. In this catalysis, not only acetanilide (**2a**) but also *ortho*-substituted acetanilide **2b** was able to react with two alkynoates **3** to give the corresponding banzo[*cd*]indolones **4b** after treatment with KOH in ethanol (Scheme 1, top) [23]. In this paper, we have established that the electron-deficient cationic Cp^E-rhodium(III) complex is able to catalyze the non-oxidative [2+2+2] annulation of *N*-(1-naphthyl)acetamide (**2c**) with two alkynoates **3** via cleavage of the adjacent C–H and C–N bonds, leading to densely substituted phenanthrenes **5**, under mild conditions (at 40 °C under air) (Scheme 1, middle). Although several examples of the transition-metal-catalyzed decarboxylative and oxidative [2+2+2] annulation of benzoic acids with two alkynoates via cleavage of the adjacent C–H and C–C bonds, leading to densely substituted naphthalenes, have been reported [24–27], only a single example of the non-oxidative [2+2+2] annulation via cleavage of the adjacent C–H and C–N bonds, in which the acylamino moiety is employed as a traceless directing group, has been reported in the neutral Cp*-rhodium(III) complex-catalyzed synthesis of tetraarylnaphthalenes from *N*-acylanilines and two diarylacetylenes at elevated temperature (110 °C) (Scheme 1, bottom) [28].

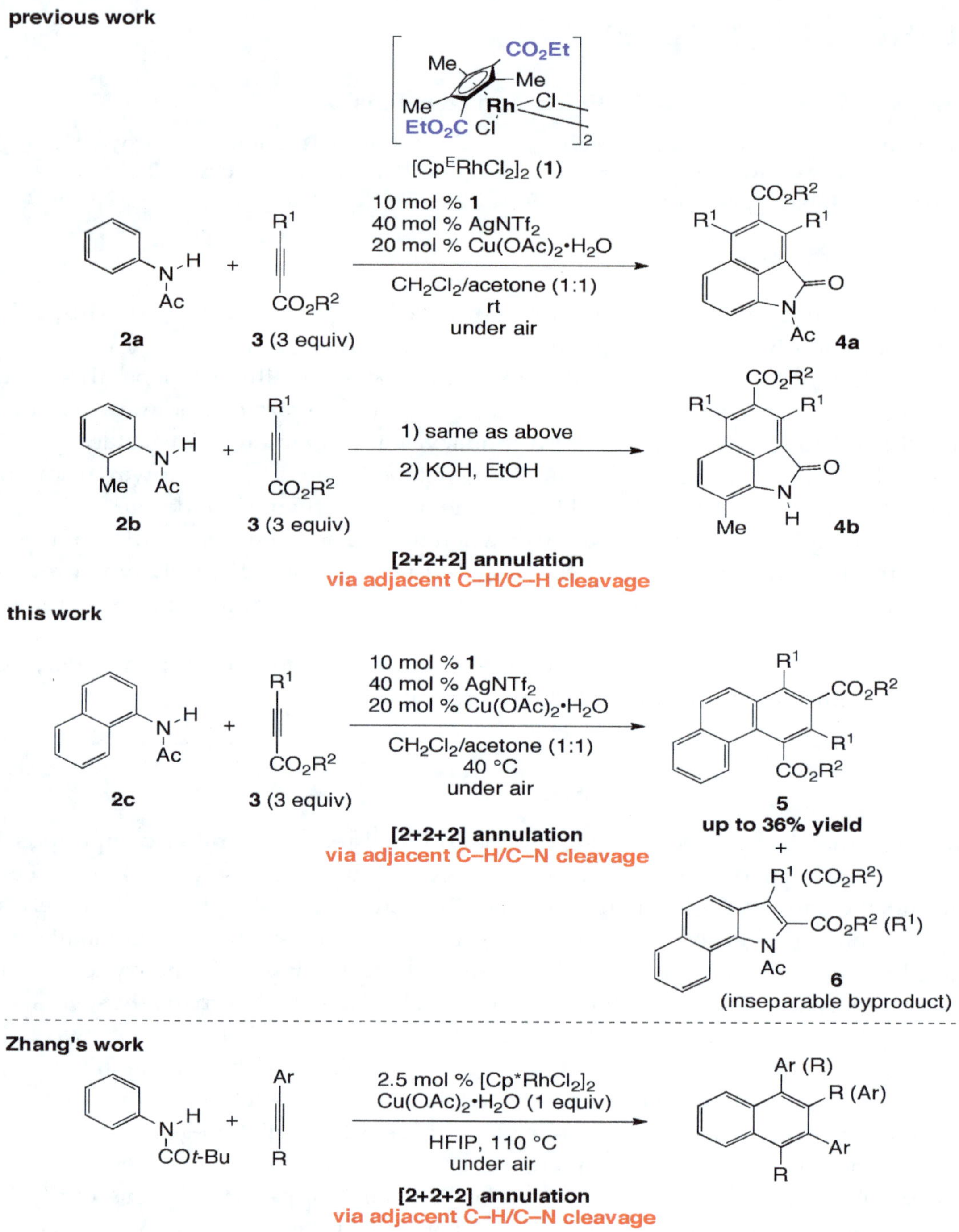

Scheme 1. Cationic Cp^E-rhodium(III) complex-catalyzed [2+2+2] annulations of acetanilides **2a**/**2b** and *N*-(1-naphthyl)acetamide (**2c**) with two alkynoates **3** via cleavage of adjacent C–H/C–H and C–H/C–N bonds, respectively.

2. Results

In the course of our study of the cationic Cp^E-rhodium(III) complex-catalyzed oxidative tandem [2+2+2] annulation-lactamization of acetanilides with two alkynoates, leading to banzo[*cd*]indolones, the reaction of 2-methyl acetanilide **2b** and ethyl 2-butynoates (**3a**) was examined. As already shown in Scheme 1, the expected banzo[*cd*]indolones **4b** were generated after treatment with KOH in ethanol [23]. Surprisingly, the use of *N*-(1-naphthyl)acetamide (**2c**) in place of **2b** failed to afford the expected naphtho[*cd*]indolone **4ca**. Instead, densely substituted phenanthrene **5ca** was generated as a major product along with a mixture of the corresponding regioisomeric [3+2] annulation products **6ca**/**6ca′** (Scheme 2). In addition to the above products, unidentified oligomerization products derived from **2c** and **3a** were generated as by-products.

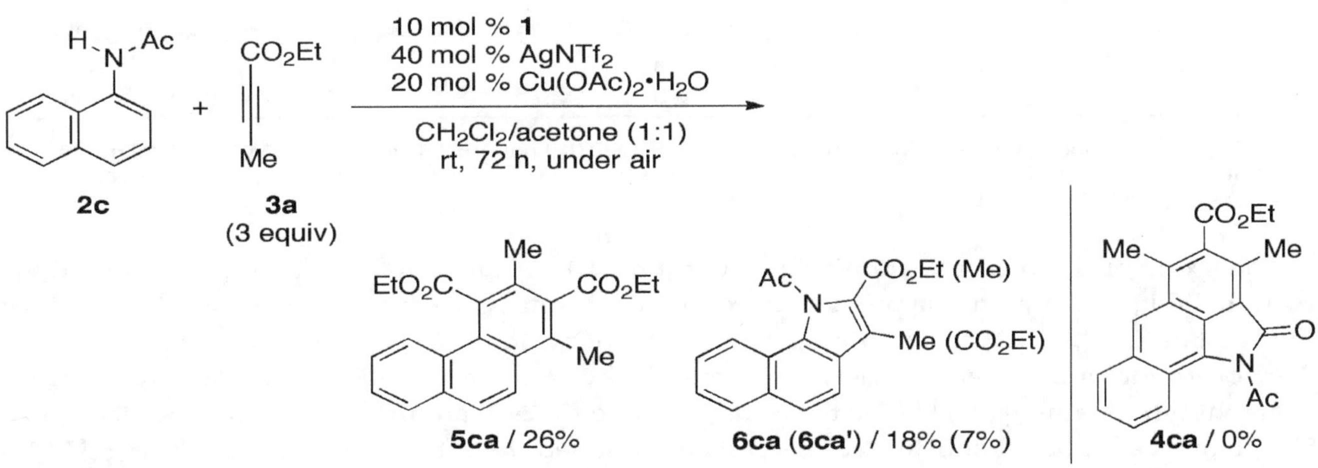

Scheme 2. Cationic Cp^E-rhodium(III) complex-catalyzed non-oxidative [2+2+2] annulation of *N*-(1-naphthyl)acetamide (**2c**) with two ethyl 2-butynoates (**3a**).

Then the screening of the reaction conditions and the acyl groups on the nitrogen of 1-aminonaphthalene was conducted as shown in Table 1. Elevating the reaction temperature (40 °C) slightly increased the yields of **5ca** and **6ca**/**6ca′** (entry 2). Increasing the amount of **3a** increased the yield of **5ca** and decreased that of **6ca**/**6ca′** (entry 3). However, this increase was very small, therefore, the conditions of entry 2 were selected as the best conditions. With respect to the acyl groups on the nitrogen of the 1-aminonaphthalene moiety, electron-poor *N*-(1-naphthyl)amide **2d**, possessing the highly acidic amide proton, was tested in place of **2c**, while no reaction was observed even at 80 °C (entry 4). Sterically demanding *N*-(1-naphthyl)amide **2e** was also tested with expectation for acceleration of reductive elimination. Unfortunately, the use of **2e** significantly increased the yield of not the [2+2+2] annulation product **5ea** but the [3+2] annulation products **6ea**/**6ea′** (entry 5). Finally, the reaction was conducted using the Cp*-rhodium(III) complex instead of the Cp^E-rhodium(III) complex **1**, which resulted in significant decrease of the yields of **5ca** and **6ca**/**6ca′** (entry 6).

Table 1. Optimization of reaction conditions for rhodium (III)-catalyzed [2+2+2] annulation of *N*-(1-naphthyl)amides **2** with two ethyl 2-butynoates (**3a**) [a].

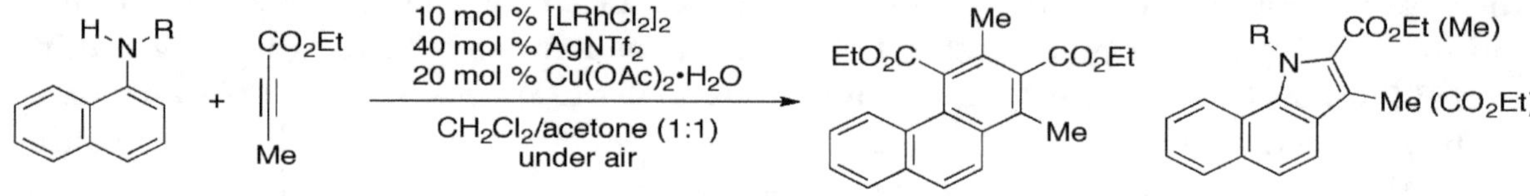

2c (R = Ac) **2d** (R = $COCF_3$) **2e** (R = CO*t*-Bu) — **3a** (3 equiv) — **5** — **6 (6')**

Entry	L	2 (R)	Conditions	3a (Equiv.)	Yield (%) [b]	
					5	6 (6′)
1	Cp^E	**2c** (Ac)	rt, 72 h	3	**5ca**/26	**6ca** (**6ca′**)/18 (7)
2	Cp^E	**2c** (Ac)	40 °C, 16 h	3	**5ca**/31	**6ca** (**6ca′**)/24 (13)
3	Cp^E	**2c** (Ac)	40 °C, 16 h	10	**5ca**/33	**6ca** (**6ca′**)/22 (10)
4 [c]	Cp^E	**2d** ($COCF_3$)	80 °C, 16 h	3	**5da**/0	**6da** (**6da′**)/0 (0)
5	Cp^E	**2e** (CO*t*-Bu)	40 °C, 16 h	3	**5ea**/0	**6ea** (**6ea′**)/55 (3)
6	Cp*	**2c** (Ac)	40 °C, 16 h	3	**5ca**/13	**6ca** (**6ca′**)/3 (<1)

[a] $[Rh_2]$ (0.020 mmol), $AgNTf_2$ (0.080 mmol), $Cu(OAc)_2{\cdot}H_2O$ (0.040 mmol), **2** (0.200 mmol), **3a** (0.60–2.00 mmol), CH_2Cl_2 (1.0 mL) and acetone (1.0 mL) were used. [b] Determined by ^1H NMR yield using C_6Me_6 as an internal standard. [c] $(CH_2Cl)_2$ (2.0 mL) was used.

The scope of the present cationic Cp^E-rhodium(III) complex-catalyzed [2+2+2] annulation is shown in Table 2. As with our previously reported cationic Cp^E-rhodium(III) complex-catalyzed oxidative tandem [2+2+2] annulation-lactamization of acetanilides with two alkynoates, leading to banzo[*cd*]indolones, a wide variety of primary alkyl-substituted alkynoates **3a–e** reacted with *N*-(1-naphthyl)acetamide (**2c**) to give the corresponding [2+2+2] annulation products **5ca–ce** (entries 1–5). However, phenyl-substituted alkynoate **3f** failed to afford the corresponding [2+2+2] annulation product **5cf** and gave the corresponding [3+2] annulation product **6cf′** in good yield with perfect regioselectivity (entry 6). This product switch is also the same as our previously reported tandem [2+2+2] annulation-lactamization of acetanilide (**2a**) with **3f** [23]. Unfortunately, the use electron-deficient alkynes did not afford the corresponding annulation products at all. For example, the use of ethyl propiolate afforded a complex mixture of products and no reaction was observed when using ethyl 2-butynoate. The structure and regiochemistry of the [2+2+2] annulation product **5ce** were unambiguously determined by the X-ray crystallographic analysis as shown in Figure 1.

Table 2. Scope of alkynoates **3** on rhodium(III)-catalyzed [2+2+2] annulation with *N*-(1-naphthyl)acetamide (**2c**) [a].

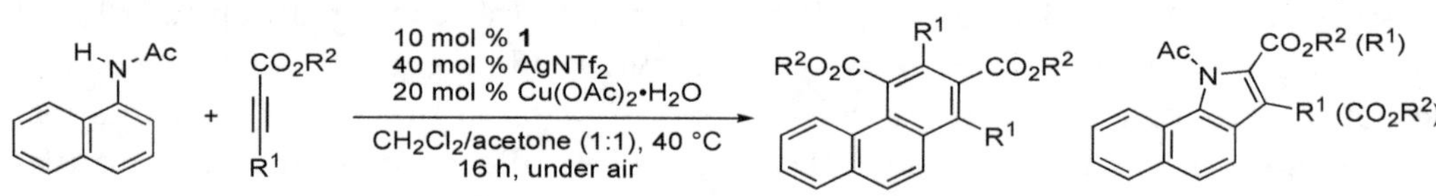

2c — **3** (3 equiv) — **5** — **6 (6')**

Entry	3	R¹	R²	Yield (%) [b]	
				5	6 (6′)
1	**3a**	Me	Et	**5ca**/33	**6ca** (**6ca′**)/22 (10)
2	**3b**	Me	Me	**5cb**/30	**6cb** (**6bb′**)/22 (8)
3	**3c**	*n*-Bu	Et	**5cc**/32	**6cc** (**6cc′**)/37 (7)
4	**3d**	$Ph(CH_2)_3$	Me	**5cd**/35	**6cd** (**6cd′**)/38 (5)
5	**3e**	$Cl(CH_2)_3$	Me	**5ce**/36	**6ce** (**6ce′**)/31 (5)
6	**3f**	Ph	Et	**5cf**/0	**6cf** (**6cf′**)/0 (69)

[a] **1** (0.020 mmol), $AgNTf_2$ (0.080 mmol), $Cu(OAc)_2{\cdot}H_2O$ (0.040 mmol), **2c** (0.200 mmol), **3** (0.600 mmol), CH_2Cl_2 (1.0 mL) and acetone (1.0 mL) were used. [b] Determined by ^1H NMR yield using C_6Me_6 as an internal standard.

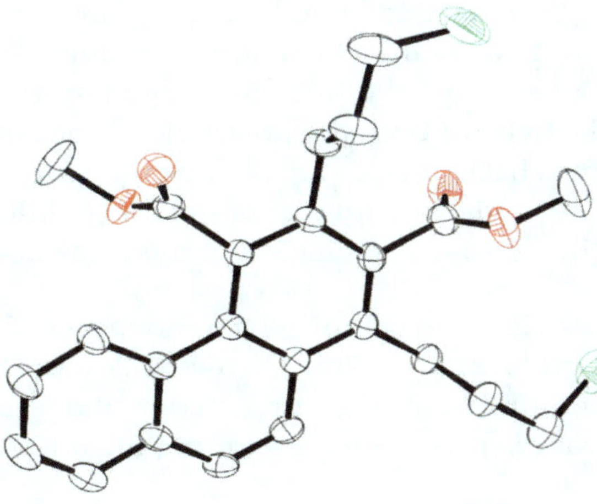

Figure 1. ORTEP diagram of **5ce** with ellipsoids at 50% probability. All hydrogen and disordered atoms are omitted for clarity.

Importantly, an interesting by-product was generated in the reaction of **2c** and **3a** as shown in Scheme 3. When using 3 equiv of **3a**, a very small amount of dearomatized spiro compound **7ca** was detected in the crude reaction mixture (3% yield). An isolable amount (13% yield) of **7ca** was generated when using an excess amount of **3a**.

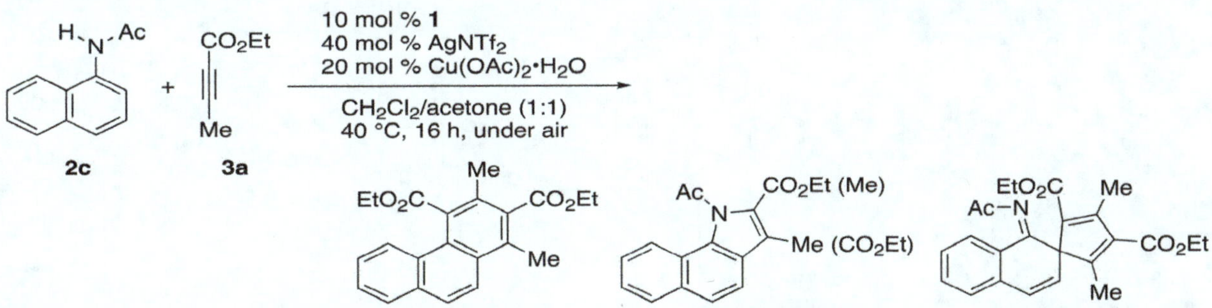

3a	5ca	6ca (6ca')	7ca
3 equiv	31%	24% (13%)	3%
10 equiv	33%	22% (10%)	13%

Scheme 3. Formation of dearomatized spiro compound **7ca**. The yields were determined by ^{1}H NMR yield using C_6Me_6 as an internal standard.

Plausible mechanisms for the formation of **5ca**, **6ca** and **7ca** from **2c** and **3a** are shown in Scheme 4. First, C–H bond cleavage of **2c** by Cp^{E}-rhodium(III) species **A** affords naphthylrhodium **B**. Next, insertion of **3a** into **B** gives alkenylrhodium **C**. The subsequent insertion of **3a** into **C** gives dienylrhodium **D**. The second alkyne insertion may not proceed in the case of sterically demanding phenyl-substituted alkynoate **3f**, thus generating [3+2] annulation product **6cf′** (Table 2, entry 6). Electrophilic metalation of the electron-rich 1-aminonaphthalene ring produces cationic spiro rhodacycle **E**, in which the carbocation is stabilized by the neighboring phenyl and acetylamino groups. Nucleophilic attack of the dienylrhodium moiety to the electrophilic 1-position of the 1-aminonaphthalene ring gives π-pentadienyl complex **F** [29]. Importantly, nucleophilic attack of the dienylrhodium moiety to the 3-position of the 1-aminonaphthalene ring in spiro rhodacycle **E′**, leading to π-pentadienyl complex **F′**, would be unfavorable due to the unstable dearomatized quinodimethane structures in **E′** and **F′**. β-Nitrogen elimination [30,31] from intermediate **F** affords **5ca** and the catalytically active rhodium(III) species **A**. However, the copper(II) co-catalyst was necessary in order to reoxidize rhodium(I) species generated through the competing oxidative [3+2] annulation giving **6ca**. In this reaction, dearomatized spiro compound **7ca** was isolated as a by-product. This compound may be generated by reductive elimination and deprotonation from spiro rhodacycle **E**. Increasing the amount of **3a** may facilitate the reductive elimination, which increased the yield of **7ca**. This result may support the intermediacy of spiro rhodacycle **E** in the catalytic cycle. Interestingly, the regioselectivity of both the present Cp^E-rhodium(III)-catalyzed [2+2+2] and [3+2] annulations of *N*-(1-naphthyl)acetamide (**2c**) is opposite to that of our previously reported Cp^E-rhodium(III)-catalyzed [2+2+2] and [3+2] annulations of acetanilides that proceed presumably via alkenylrhodium **G** [23], although the reason is not clear at the present stage.

We considered that it is possible to regenerate spiro rhodacycle **E** by oxidative addition with a neutral rhodium(I) complex and protonation of **7ca**. Therefore, **7ca** was treated with acetic acid and an in situ generated neutral Cp^E-rhodium(I) complex, prepared by the reaction of **1**, $AgNTf_2$, NaOAc, **2a** and **3a**. However, no reaction was observed, thus excluding the intermediacy of **7ca** in the catalytic cycle (Scheme 5).

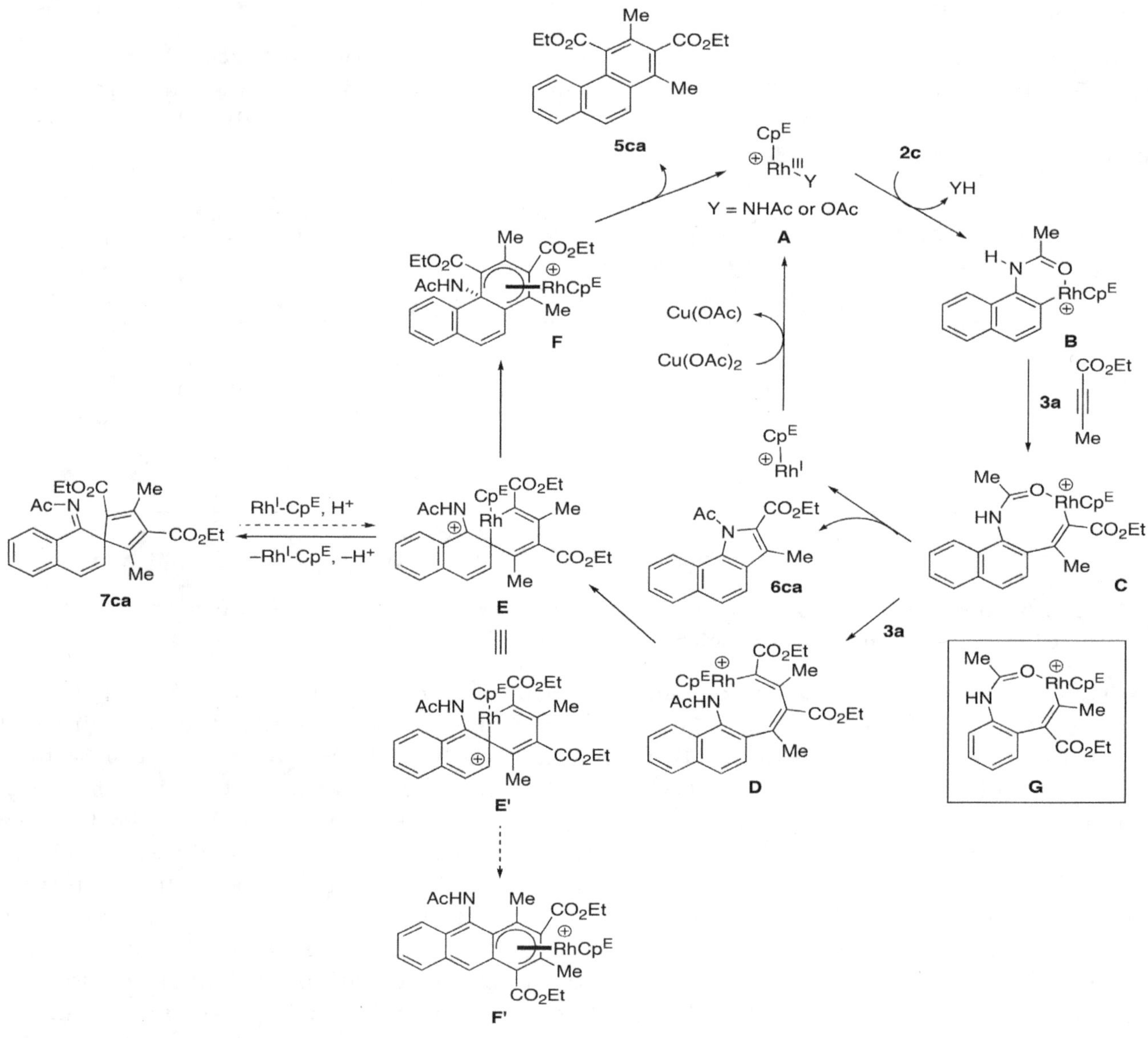

Scheme 4. Plausible mechanisms for formation of **5ca**, **6ca** and **7ca** from **2c** and **3a**.

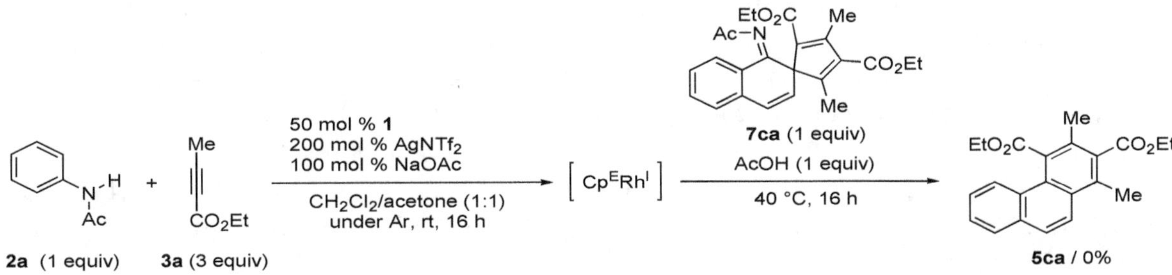

Scheme 5. Reaction of **7ca** with an in situ generated Cp^E-rhodium(III) complex and acetic acid.

3. Materials and Methods

3.1. General Information

Anhydrous acetone (No. 27,072-5) and CH_2Cl_2 (No. 130-02457) were obtained from Aldrich (St. Louis, MO, USA) and Wako (Osaka, Japan) and used as received. Solvents for the synthesis of substrates were dried over Molecular Sieves 4Å (Wako) prior to use. Anilides **2d** [32] and **2e** [33] were prepared according to the literature. Internal alkynes **3d** [34] and **3e** [35] were prepared according to the literature. All other reagents were obtained from commercial sources and used as received. ^{1}H and ^{13}C data were collected on a Bruker AVANCE III (Billerica, MA, USA) HD 400 (400 MHz) at ambient temperature. HRMS data were obtained on a Bruker micro TOF Focus II (Billerica, MA, USA). A single crystal X-ray diffraction measurement was made on Rigaku XtaLAB mini II diffractometer (Akishima, Japan) using graphite monochromated Mo-Kα radiation. All reactions were carried out in oven-dried glassware with magnetic stirring.

3.2. General Procedure for the Rhodium-Catalyzed Annulation of N-(1-Naphthyl)amides with Two Alkynoates (5ca, 6ca and 6ca′; Table 1, entry 2)

To a Schlenk tube was added $AgNTf_2$ (15.5 mg, 0.040 mmol), $[Cp^ERhCl_2]_2$ **1** (8.5 mg, 0.010 mmol), $Cu(OAc)_2 \cdot H_2O$ (4.0 mg, 0.020 mmol), *N*-(naphthalen-1-yl)acetamide (**2c**, 18.5 mg, 0.100 mmol), ethyl 2-butynoate (**3a**, 33.6 mg, 0.300 mmol), acetone (0.5 mL) and CH_2Cl_2 (0.5 mL) under air in this order. The mixture was sealed and stirred at 40 °C under air for 16 hours. The resulting mixture was diluted with ether, filtered through a silica gel pad and washed with ether. The solvent was removed under reduced pressure and the residue was purified by a preparative thin layer chromatography (TLC, hexane/AcOEt = 2:1), to give a mixture of **5ca**, **6ca** and **6ca′**. The yields of **5ca** (31%), **6ca** (22%) and **6ca′** (10%) were determined by ^{1}H NMR with hexamethylbenzene as an internal standard.

3.3. Product Characterization

Diethyl 1,3-dimethylphenanthrene-2,4-dicarboxylate (**5ca**). Analytically pure **5ca** was isolated from a mixture of **5ca**, **6ca** and **6ca′** (21.8 mg, **5ca**/**6ca**/**6ca′** = 45:41:14) by a gel permeation chromatography (GPC). The regiochemistry of the title compound was determined by the NOESY experiment as well as analogy to the ^{1}H NMR chemical shifts of **5ce**. Colorless solid, 7.2 mg, 0.0205 mmol, 21% isolated yield, mp 83.8–85.0 °C; ^{1}H NMR ($CDCl_3$, 400 MHz) δ 8.44 (d, J = 8.4 Hz, 1H), 7.95 (d, J = 9.2 Hz, 1H), 7.92–7.87 (m, 1H), 7.79 (d, J = 9.2 Hz, 1H), 7.63–7.51 (m, 2H), 4.57–7.44 (m, 4H), 2.70 (s, 3H), 2.47 (s, 3H), 1.95 (s, 3H), 1.44 (t, J = 7.1 Hz, 3H), 1.35 (t, J = 7.2 Hz, 3H); ^{13}C NMR ($CDCl_3$, 100 MHz) δ 172.0, 170.0, 134.1, 133.0, 132.3, 129.8, 129.6, 129.5, 129.1, 128.7, 127.8, 127.5, 127.0, 126.3, 125.6, 122.4, 61.8, 61.4, 17.4, 17.2, 14.3, 13.9; HRMS (ESI) calcd for $C_{22}H_{22}O_4Na$ $[M + Na]^+$ 373.1414, found 373.1410.

Ethyl 1-acetyl-3-methyl-1H-benzo[g]indole-2-carboxylate (**6ca**) *and ethyl 1-acetyl-2-methyl-1H-benzo[g]indole-3-carboxylate* (**6ca′**) [36]. An analytically pure mixture of **6ca** and **6ca′** was isolated from a mixture of **5ca**, **6ca** and **6ca′** (21.8 mg, **5ca**/**6ca**/**6ca′** = 45:41:14) by GPC. The regiochemistry of the title compounds was determined by comparison with the ^{1}H NMR chemical shifts of the literature.

Colorless solid, 7.6 mg, 0.0257 mmol, 26% isolated yield, **6ca**/**6ca′** = 75:25, mp 86.1–87.2 °C; ^{1}H NMR ($CDCl_3$, 400 MHz) δ **6ca**: 8.17–8.11 (m, 1H), 7.96–7.90 (m, 1H), 7.67 (d, J = 8.7 Hz, 1H), 7.59 (d, J = 8.7 Hz, 1H), 7.55–7.48 (m, 2H), 4.43 (q, J = 7.1 Hz, 2H), 2.76 (s, 3H), 2.65 (s, 3H), 2.04 (s, 3H), 1.45 (t, J = 7.1 Hz, 3H); **6ca′**: 8.28 (d, J = 8.7 Hz, 1H), 7.97 (d, J = 7.9 Hz, 1H), 7.74 (t, J = 7.9 Hz, 1H), 7.57–7.44 (m, 3H), 4.50–4.40 (m, 2H), 2.85 (s, 3H), 2.62 (s, 3H), 1.49 (t, J = 7.1 Hz, 3H); ^{13}C NMR ($CDCl_3$, 100 MHz) δ 177.6, 162.3, 133.5, 129.41, 129.36, 126.3, 126.2, 125.8, 125.2, 124.7, 124.4, 124.0, 123.7, 123.2, 121.8, 121.7, 121.3, 120.8, 119.0, 61.2, 30.0, 14.3, 10.4; HRMS (ESI) calcd for $C_{17}H_{15}NONa$ [M + Na]$^+$ 318.1101, found 318.1115.

Dimethyl 1,3-dimethylphenanthrene-2,4-dicarboxylate (**5cb**). The yield of **5cb** (30%) was determined by ^{1}H NMR with hexamethylbenzene as an internal standard. Analytically pure **5cb** was isolated from a mixture of **5cb**, **6cb** and **6cb′** (19.2 mg, **5cb**/**6cb**/**6cb′** = 51:37:12) by GPC. The regiochemistry of the title compound was determined by analogy to the ^{1}H NMR chemical shifts of **5ce**. Colorless solid, 8.8 mg, 0.0273 mmol, 27% isolated yield, mp 62.8–64.5 °C; ^{1}H NMR ($CDCl_3$, 400 MHz); ^{1}H NMR ($CDCl_3$, 400 MHz) δ 8.35 (d, J = 8.4 Hz, 1H), 7.95 (d, J = 9.1 Hz, 1H), 7.93–7.87 (m, 1H), 7.79 (d, J = 9.2 Hz, 1H), 7.64–7.53 (m, 2H), 4.00 (s, 3H), 3.99 (s, 3H), 2.68 (s, 3H), 2.44 (s, 3H); ^{13}C NMR ($CDCl_3$, 100 MHz) δ 172.4, 170.5, 133.9, 133.1, 132.6, 129.83, 129.79, 129.1, 129.0, 128.8, 127.9, 127.7, 127.1, 126.4, 125.3, 122.3, 52.7, 52.3 17.5, 17.4; HRMS (ESI) calcd for $C_{20}H_{18}O_4Na$ [M + Na]$^+$ 345.1097, found 345.1101.

Methyl 1-acetyl-3-methyl-1H-benzo[g]indole-2-carboxylate (**6cb**) *and methyl 1-acetyl-2-methyl-1H-benzo[g]indole-3-carboxylate* (**6cb′**). The yields of **6cb** (22%) and **6cb′** (8%) was determined by ^{1}H NMR with hexamethylbenzene as an internal standard. An analytically pure mixture of **6cb** and **6cb′** was isolated from a mixture of **5cb**, **6cb** and **6cb′** (19.2 mg, **5cb**/**6cb**/**6cb′** = 51:37:12) by GPC. The regiochemistry of the title compounds was determined by analogy to the ^{1}H NMR chemical shifts of **6ca** and **6ca′**. Colorless solid, 6.0 mg, 0.0213 mmol, 21% isolated yield, **6cb**/**6cb′** = 75:25, mp 93.7–95.0 °C; ^{1}H NMR ($CDCl_3$, 400 MHz) **6cb**: δ 8.17–8.10 (m, 1H), 7.95–7.89 (m, 1H), 7.66 (d, J = 8.7 Hz, 1H), 7.59 (d, J = 8.7 Hz, 1H), 7.57–7.44 (m, 2H), 3.97 (s, 3H), 2.76 (s, 3H), 2.64 (s, 3H); **6cb′**: δ 8.26 (d, J = 8.8 Hz, 1H), 7.97 (d, J = 8.1 Hz, 1H), 7.93–7.87 (m, 1H), 7.79 (d, J = 9.2 Hz, 1H), 7.64–7.53 (m, 2H), 4.00 (s, 3H), 3.99 (s, 3H), 2.68 (s, 3H), 2.44 (s, 3H); ^{13}C NMR ($CDCl_3$, 100 MHz) δ 171.0, 136.8, 134.9, 133.6, 130.3, 130.2, 128.7, 128.5, 125.3, 123.4, 118.6, 118.1, 116.3, 27.6, 9.2; HRMS (ESI) calcd for $C_{17}H_{15}NONa$ [M + Na]$^+$ 304.0944, found 304.0952.

Diethyl 1,3-dibutylphenanthrene-2,4-dicarboxylate (**5cc**). The yield of **5cc** (32%) was determined by ^{1}H NMR with hexamethylbenzene as an internal standard. Analytically pure **5cc** was isolated from a mixture of **5cc**, **6cc** and **6cc′** (40.0 mg, **5cc**/**6cc**/**6cc′** = 42:9:49) by GPC. The regiochemistry of the title compound was determined by analogy to the ^{1}H NMR chemical shifts of **5ce**. Pale yellow oil, 11.7 mg, 0.0269 mmol, 27% isolated yield; ^{1}H NMR ($CDCl_3$, 400 MHz) δ 8.42 (d, J = 8.6 Hz, 1H), 7.93 (d, J = 9.2 Hz, 1H), 7.90–7.86 (m, 1H), 7.77 (d, J = 9.1 Hz, 1H), 7.61–7.56 (m, 1H), 7.55–7.50 (m, 1H), 4.51–4.43 (m, 4H), 3.04–2.97 (m, 2H), 2.80–2.68 (m, 2H), 1.78–1.64 (m, 4H), 1.55–1.29 (m, 4H), 1.44 (t, J = 7.1 Hz, 3H), 1.34 (t, J = 7.2 Hz, 3H), 1.02–0.92 (m, 6H); ^{13}C NMR ($CDCl_3$, 100 MHz) δ 171.9, 170.0, 137.5, 135.0, 133.7, 132.9, 129.4, 129.2, 129.1, 128.5, 128.1, 127.7, 127.0, 126.1, 125.7, 122.3, 61.7, 61.3, 33.9, 33.5, 32.1, 31.3, 23.4, 23.3, 14.2, 13.9, 13.83, 13.79; HRMS (ESI) calcd for $C_{28}H_{34}O_4Na$ [M + Na]$^+$ 457.2349, found 457.2359.

Ethyl 1-acetyl-3-butyl-1H-benzo[g]indole-2-carboxylate (**6cc**) *and ethyl 1-acetyl-2-butyl-1H-benzo[g]indole-3-carboxylate* (**6cc′**). The yields of **6cc** (37%) and **6cc′** (7%) were determined by ^{1}H NMR with hexamethylbenzene as an internal standard. An analytically pure mixture of **6cc** and **6cc′** was isolated from a mixture of **5cc**, **6cc** and **6cc′** (40.0 mg, **5cc**/**6cc**/**6cc′** = 42:9:49) by GPC. The regiochemistry of the title compounds was determined by analogy to the ^{1}H NMR chemical shifts of **6ca** and **6ca′**. Colorless solid, 13.2 mg, 0.0391 mmol, 39% isolated yield, **6cc**/**6cc′** = 86:14, mp 61.0–62.4 °C; ^{1}H NMR ($CDCl_3$, 400 MHz) δ **6cc**: 8.16–8.09 (m, 1H), 7.95–7.88 (m, 1H), 7.67 (d, J = 8.7 Hz, 1H), 7.58 (d, J = 8.6 Hz, 1H), 7.56–7.43 (m, 2H), 4.43 (q, J = 7.1 Hz, 2H), 3.11 (t, J = 7.7 Hz,

2H), 2.77 (s, 3H), 1.75–1.63 (m, 2H), 1.52–1.40 (m, 2H), 1.44 (t, J = 7.1 Hz, 3H), 0.97 (t, J = 7.2 Hz, 3H); **6cc′**: 8.30 (d, J = 8.8 Hz, 1H), 8.00–7.95 (m, 1H), 7.77–7.71 (m, 2H), 7.56–7.43 (m, 2H), 4.50–4.39 (m, 2H), 3.25–3.18 (m, 2H), 1.75–1.63 (m, 2H), 1.52–1.40 (m, 5H), 1.00–0.93 (m, 3H); ^{13}C NMR ($CDCl_3$, 100 MHz) δ 177.7, 162.2, 133.4, 131.0, 129.4, 129.0, 126.3, 125.7, 124.8, 123.4, 123.1, 121.9, 121.6, 119.1, 61.2, 33.4, 30.0, 24.7, 22.9, 14.2, 14.0; HRMS (ESI) calcd for $C_{21}H_{23}NO_3Na$ [M + Na]$^+$ 360.1570, found 360.1575.

Dimethyl 1,3-bis(3-phenylpropyl)phenanthrene-2,4-dicarboxylate (**5cd**). The yield of **5cd** (35%) was determined by ^{1}H NMR with hexamethylbenzene as an internal standard. Analytically pure **5cd** was isolated from a mixture of **5cd**, **6cd** and **6cd′** (42.3 mg, **5cd**/**6cd**/**6cd′** = 46:7:47) by GPC. The regiochemistry of the title compound was determined by analogy to the ^{1}H NMR chemical shifts of **5ce**. Pale yellow solid, 16.4 mg, 0.0294 mmol, 29% isolated yield, mp 91.0–92.6 °C; ^{1}H NMR ($CDCl_3$, 400 MHz) δ 8.30 (d, J = 8.4 Hz, 1H), 7.89–7.83 (m, 1H), 7.78–7.68 (m, 2H), 7.61–7.44 (m, 2H), 7.36–7.14 (m, 10H), 3.84 (s, 3H), 3.65 (s, 3H), 3.00–2.90 (m, 2H), 2.78 (t, J = 7.2 Hz, 2H), 2.70 (t, J = 7.2 Hz, 2H), 2.70–2.62 (m, 2H), 2.10–1.94 (m, 4H); ^{13}C NMR ($CDCl_3$, 100 MHz) δ 172.2, 170.1, 141.8, 141.6, 137.3, 134.6, 133.5, 132.9, 129.3, 129.2, 128.9, 128.7, 128.6, 128.4, 128.3, 128.2, 128.0, 127.1 126.3, 126.0, 125.8, 125.3, 122.1, 52.5, 52.0, 36.4, 36.2, 33.2, 32.6, 32.0, 30.9; HRMS (ESI) calcd for $C_{36}H_{34}O_4Na$ [M + Na]$^+$ 553.2349, found 553.2354.

Ethyl 1-acetyl-3-(3-phenylpropyl)-1H-benzo[g]indole-2-carboxylate (**6cd**) *and ethyl 1-acetyl-2-(3-phenylpropyl)-1H-benzo[g]indole-3-carboxylate* (**6cd′**). The yields of **6cd** (38%) and **6cd′** (5%) were determined by ^{1}HNMR with hexamethylbenzene as an internal standard. An analytically pure mixture of **6cd** and **6cd′** was isolated from a mixture of **5cd**, **6cd** and **6cd′** (42.3 mg, **5cd**/**6cd**/**6cd′** = 46:7:47) by GPC. The regiochemistry of the title compounds was determined by analogy to the ^{1}H NMR chemical shifts of **6ca** and **6ca′**. Pale yellow oil, 14.1 mg, 0.0353 mmol, 35% isolated yield, **6cd**/**6cd′** = 89:11; ^{1}H NMR ($CDCl_3$, 400 MHz) δ 8.15–8.08 (m, 1H), 7.95–7.87 (m, 1H), 7.61–7.54 (m, 2H), 7.53–7.48 (m, 2H), 7.33–7.26 (m, 2H), 7.25–7.15 (m, 3H), 3.85 (s, 3H), 3.18–3.10 (m, 2H), 2.78–2.72 (m, 2H), 2.76 (s, 3H), 2.07–1.97 (m, 2H); ^{13}C NMR ($CDCl_3$, 100 MHz) δ 177.6, 162.6, 142.1, 133.4, 131.1, 129.4, 128.5, 128.3, 126.3, 125.84, 125.80, 124.7, 123.3, 123.2, 121.8, 121.6, 119.0, 51.9, 35.9, 32.5, 29.9, 24.5; HRMS (ESI) calcd for $C_{36}H_{34}O_4Na$ [M + Na]$^+$ 408.1570, found 408.1574.

Diethyl 1,3-dibutylphenanthrene-2,4-dicarboxylate (**5ce**). The yield of **5ce** (36%) was determined by ^{1}H NMR with hexamethylbenzene as an internal standard. Analytically pure **5ce** was isolated from a mixture of **5ce**, **6ce** and **6ce′** (42.6 mg, **5ce**/**6ce**/**6ce′** = 50:7:43) by GPC. The regiochemistry of the title compound was determined by the X-ray crystallographic analysis as shown in Figure 1. Colorless solid, 13.5 mg, 0.0284 mmol, 28% isolated yield, mp 108.1–109.2 °C; ^{1}H NMR ($CDCl_3$, 400 MHz) δ 8.34 (d, J = 8.4 Hz, 1H), 7.97 (d, J = 9.2 Hz, 1H), 7.93–7.88 (m, 1H), 7.82 (d, J = 9.2 Hz, 1H), 7.65–7.53 (m, 2H), 4.02 (s, 3H), 3.98 (s, 3H), 3.68 (t, J = 6.1 Hz, 2H), 3.63 (t, J = 6.1 Hz, 2H), 3.24–3.16 (m, 2H), 2.93–2.84 (m, 2H), 2.26–2.13 (m, 4H); ^{13}C NMR ($CDCl_3$, 100 MHz) δ 172.0, 170.0, 136.1, 133.9, 133.3, 133.0, 129.6, 129.5, 129.2, 128.8, 128.6, 128.4, 127.4, 126.6, 125.3, 121.8, 52.8, 52.5, 45.2, 45.0, 34.4, 33.7, 30.1, 28.9; HRMS (ESI) calcd for $C_{24}H_{24}Cl_2O_4Na$ [M + Na]$^+$ 469.0944, found 469.0948.

Methyl 1-acetyl-3-(3-chloropropyl)-1H-benzo[g]indole-2-carboxylate (**6ce**) *and methyl 1-acetyl-2-(3-chloropropyl)-1H-benzo[g]indole-3-carboxylate* (**6ce′**). The yields of **6ce** (31%) and **6ce′** (5%) were determined by ^{1}H NMR with hexamethylbenzene as an internal standard. An analytically pure mixture of **6ce** and **6ce′** was isolated from a mixture of **5ce**, **6ce** and **6ce′** (42.6 mg, **5ce**/**6ce**/**6ce′** = 50:7:43) by GPC. The regiochemistry of the title compounds was determined by analogy to the ^{1}H NMR chemical shifts of **6ce** and **6ce′**. Colorless oil, 8.6 mg, 0.0240 mmol, 24% isolated yield, **6ce**/**6ce′** = 83:17. ^{1}H NMR ($CDCl_3$, 400 MHz) **6ce**: δ 8.14–8.07 (m, 1H), 7.96–7.89 (m, 1H), 7.71 (d, J = 8.8 Hz, 1H), 7.60 (d, J = 8.8 Hz, 1H), 7.58–7.46 (m, 2H), 3.98 (s, 3H), 3.61 (t, J = 6.3 Hz, 2H), 3.32–3.26 (m, 2H), 2.79 (s, 3H), 2.30–2.13 (m, 2H); **6ce′**: δ 8.76 (d, J = 8.8 Hz 1H), 8.00–7.97 (m, 1H), 7.77–7.75 (m, 1H), 7.75–7.73 (m, 1H), 7.58–7.46 (m, 2H), 4.00 (s, 3H), 3.68 (t, J = 6.6 Hz, 2H), 3.37–3.32 (m, 2H), 2.62 (s, 3H), 2.30–2.13 (m, 2H); ^{13}C NMR ($CDCl_3$, 100 MHz) δ 177.5, 162.3, 133.5, 131.1, 129.5, 126.8, 126.5, 126.0, 125.0, 124.6, 123.5,

123.3, 121.7, 121.60, 120.8, 118.8, 52.1, 44.7, 33.7, 29.9, 22.1; HRMS (ESI) calcd for $C_{19}H_{18}O_3NClNa$ [M + Na]$^+$ 366.0867, found 366.0871.

Ethyl 2-phenyl-1-pivaloyl-1H-benzo[g]indole-3-carboxylate (**6cf′**). The yield of **6cf′** (69%) was determined by ^{1}H NMR with hexamethylbenzene as an internal standard. Analytically pure **6cf′** was isolated by repeated preparative TLCs. Pale yellow oil, 21.9 mg, 0.0612 mmol, 61% isolated yield; ^{1}H NMR ($CDCl_3$, 400 MHz) δ 8.37 (d, J = 8.8 Hz, 1H), 8.00–7.95 (m, 1H), 7.91–7.86 (m, 1H), 7.78 (d, J = 8.7 Hz, 1H), 7.55–7.44 (m, 7H), 4.22 (q, J = 7.1 Hz, 2H), 2.27 (s, 3H), 1.16 (t, J = 7.1 Hz, 3H), 1.17 (s, 9H); ^{13}C NMR ($CDCl_3$, 100 MHz) δ 176.0, 164.5, 142.4, 132.0, 131.2, 130.8, 129.43, 129.41, 129.3, 128.0, 126.1, 125.2, 125.0, 124.7, 121.7, 121.3, 120.6, 110.1, 59.9, 29.3, 14.0; HRMS (ESI) calcd for $C_{23}H_{19}NO_3Na$ [M + Na]$^+$ 380.1257, found 380.1266.

Ethyl 3-methyl-1-pivaloyl-1H-benzo[g]indole-2-carboxylate (**6ea**) *and ethyl 2-methyl-1-pivaloyl-1H-benzo[g]indole-3-carboxylate* (**6ea′**). The yields of **6ea** (55%) and **6ea′** (3%) were determined by ^{1}H NMR with hexamethylbenzene as an internal standard. An analytically pure mixture of **6ea** and **6ea′** was isolated by repeated preparative TLCs. Colorless solid; 15.5 g, 0.0459 mmol, 46% isolated yield, **6ea**. mp 128.4–129.5 °C; ^{1}H NMR ($CDCl_3$, 400 MHz) δ 8.24–8.17 (m, 1H), 7.91–7.86 (m, 1H), 7.66 (d, J = 8.7 Hz, 1H), 7.56 (d, J = 8.7 Hz, 1H), 7.52–7.42 (m, 2H), 4.42 (q, J = 7.1 Hz, 2H), 2.66 (s, 3H), 1.43 (t, J = 7.1 Hz, 3H), 1.17 (s, 9H); ^{13}C NMR ($CDCl_3$, 100 MHz) δ 187.5, 162.9, 133.4, 131.3, 129.1, 125.8, 125.7, 125.1, 124.7, 122.9, 122.8, 122.4, 122.3, 119.0, 61.2, 46.8, 28.2, 14.3, 10.4; HRMS (ESI) calcd for $C_{21}H_{23}NO_3Na$ [M + Na]$^+$ 360.1570, found 360.1570.

Diethyl (Z)-1′-(acetylimino)-3,5-dimethyl-1′H-spiro[cyclopentane-1,2′-naphthalene]-2,4-diene-2,4-dicarboxylate (**7ca**). The yield of **7ca** (13%) was determined by ^{1}H NMR with hexamethylbenzene as an internal standard. Analytically pure **7ca** was isolated by repeated preparative TLCs. The structure was determined by the HSQC and HMBC analyses. Pale yellow oil, 9.7 mg, 0.214 mmol, 11% isolated yield; ^{1}H NMR ($CDCl_3$, 400 MHz) δ 8.08 (d, J = 8.0 Hz, 1H, H^h), 7.48 (td, J = 7.5, 1.3 Hz, 1H, H^g), 7.31 (td, J = 7.7, 1.3 Hz, 1H, H^f), 7.28–7.22 (m, 1H, H^e), 6.79 (d, J = 9.4 Hz, 1H, H^d), 5.05 (d, J = 9.4 Hz, 1H, H^c), 4.31 (q, J = 7.1 Hz, 2H, $CO_2\underline{CH_2}CH_3$ on C^2), 4.04 (q, J = 7.1 Hz, 2H, $CO_2\underline{CH_2}CH_3$ on C^4), 2.50 (s, 3H, CH_3 on C^3), 2.10 (s, 3H, CH_3 on C^1), 2.01 (s, 3H, $NCO\underline{CH_3}$), 1.36 (t, J = 7.1 Hz, 3H, $CO_2CH_2\underline{CH_3}$ on C^2), 0.89 (t, J = 7.1 Hz, 3H, $CO_2CH_2\underline{CH_3}$ on C^4); ^{13}C NMR ($CDCl_3$, 100 MHz) δ 180.6 ($N\underline{C}OMe$), 164.1 (C^a), 164.0 ($\underline{C}O_2Et$), 163.0 ($\underline{C}O_2Et$), 158.7, 154.0 (C^e), 136.0 (C^j), 135.1, 132.6 (C^h), 132.2, 129.9, 129.0 (C^c), 127.9 (C^g), 127.8 (C^f), 127.5 (C^d), 127.1 (C^i), 67.7 (C^b), 60.8 ($CO_2\underline{C}H_2CH_3$ on C^2), 60.1 ($CO_2\underline{C}H_2CH_3$ on C^4), 24.5 ($NCO\underline{C}H_3$), 15.1 ($\underline{C}H_3$ on C^3), 14.2 ($CO_2CH_2\underline{C}H_3$ on C^2), 13.9 ($\underline{C}H_3$ on C^1), 13.5 ($CO_2CH_2\underline{C}H_3$ on C^4); HRMS (ESI) calcd for $C_{24}H_{25}NO_5Na$ [M + Na]$^+$ 430.1625, found 430.1623.

The ^{1}H and ^{13}C-NMR spectra of **5ca–5ce**, **6ca–6ce**, **6ca′–6cf′** and **7ca**, and crystal data and data collection parameters of **5ce** are available in Supplementary Materials.

4. Conclusions

In summary, we have established that an electron-deficient cationic Cp^E-rhodium(III) complex catalyzes the non-oxidative [2+2+2] annulation of *N*-(1-naphthyl)acetamide with two alkynoates via cleavage of the adjacent C–H and C–N bonds to give densely substituted phenanthrenes under mild conditions (at 40 °C under air). Importantly, a dearomatized spiro compound was isolated in this reaction, which may support the formation of a spiro rhodacycle intermediate in the catalytic cycle. The use of *N*-(1-naphthyl)acetamide in place of acetanilide switched the reaction pathway from the oxidative tandem [2+2+2] annulation-lactamization involving cleavage of adjacent two C–H bonds to the non-oxidative [2+2+2] annulation involving cleavage of the adjacent C–H and C–N bonds. This chemoselectivity switch may arise from stabilization of the carbocation in the above cationic spiro rhodacycle by the neighboring phenyl and acetylamino groups, resulting in nucleophilic attack of the dienylrhodium moiety to this carbocation followed by β-nitrogen elimination.

Author Contributions: Conceptualization, Y.S. and K.T.; Experiments and analyses, J.T. and M.F.; Writing-Original Draft Preparation, Y.S. and K.T.; Writing-Review & Editing, K.T.; Visualization, J.T. and Y.S.; Supervision, K.T.; Project Administration, K.T.; Funding Acquisition, Y.S. and K.T.

Acknowledgments: We are grateful to Umicore for generous support in supplying the rhodium complex.

References

1. Pia–Quintana, A.; Roglans, A. Chiral Induction in [2+2+2] Cycloaddition Reactions. *Asian J. Org. Chem.* **2018**, *7*, 1706–1718. [CrossRef]
2. Beeck, S.; Wegner, H.A. Thieme Chemistry Journals Awardees—Where Are They Now? Aromaticity as a Source for Strain Energy? Synthesis of Curved Polycyclic Aromatic Structures via [2+2+2] Cycloaddition Reactions. *Synlett* **2017**, *28*, 1018–1027. [CrossRef]
3. Domínguez, G.; Pérez–Castells, J. Alkens in [2+2+2] Cycloadditions. *Chem. Eur. J.* **2016**, *22*, 6720–6739. [CrossRef]
4. Amatore, M.; Aubert, C. Recent Advances in Stereoselective [2+2+2] Cycloadditions. *Eur. J. Org. Chem.* **2015**, *2015*, 265–286. [CrossRef]
5. Satoh, Y.; Obora, Y. Niobium Complexes in Organic Transformations: From Stoichiometric Reactions to Catalytic [2+2+2] Cycloaddition Reactions. *Eur. J. Org. Chem.* **2015**, *2015*, 5041–5054. [CrossRef]
6. Tanaka, K. (Ed.) *Transition-Metal-Mediated Aromatic Ring Construction*; John Wiley & Sons: Hoboken, NJ, USA, 2013; Part 1; ISBN 978-1-118-14892-1.
7. Broere, D.L.J.; Ruijter, E. Recent Advances in Transition-Metal-Catalyzed [2+2+2]-Cyclo(co)trimerization Reactions. *Synthesis* **2012**, *44*, 2639–2672. [CrossRef]
8. Dominguez, G.; Pérez-Castells, J. Recent advances in [2+2+2] cycloaddition reactions. *Chem. Soc. Rev.* **2011**, *40*, 3430–3444. [CrossRef]
9. Weding, N.; Hapke, M. Preparation and synthetic applications of alkene complexes of group 9 transition metals in [2+2+2] cycloaddition reactions. *Chem. Soc. Rev.* **2011**, *40*, 4525–4538. [CrossRef]
10. Kotha, S.; Lahiri, K.; Sreevani, G. Design and Synthesis of Aromatics through [2+2+2] Cyclotrimerization. *Synlett* **2018**, *29*, 2342–2361. [CrossRef]
11. Tanaka, K.; Kimura, Y.; Murayama, K. Enantioselective Helicene Synthesis by Rhodium-Catalyzed [2+2+2] Cycloadditions. *Bull. Chem. Soc. Jpn.* **2015**, *88*, 375–385. [CrossRef]
12. Shibata, Y.; Tanaka, K. Rhodium-Catalyzed [2+2+2] Cycloaddition of Alkynes for the Synthesis of Substituted Benzenes: Catalysts, Reaction Scope, and Synthetic Applications. *Synthesis* **2012**, *3*, 323–350. [CrossRef]
13. Li, S.; Zhou, L.; Kanno, K.; Takahashi, T. Recent development for enantioselective synthesis of aromatic compounds from alkynes via metallacyclopentadienes. *J. Heterocycl. Chem.* **2011**, *48*, 517–528. [CrossRef]
14. Galan, B.R.; Rovis, T. Beyond Reppe: Building Substituted Arenes by [2+2+2] Cycloadditions of Alkynes. *Angew. Chem. Int. Ed.* **2009**, *48*, 2830–2834. [CrossRef]
15. Hsieh, J.; Cheng, C. *O*-Dihaloarenes as aryne precursors for nickel-catalyzed [2+2+2] cycloaddition with alkynes and nitriles. *Chem. Commun.* **2008**, *0*, 2992–2994. [CrossRef]
16. Hsieh, J.; Cheng, C. Nickel-catalyzed cocyclotrimerization of arynes with diynes; a novel method for synthesis of naphthalene derivatives. *Chem. Commun.* **2005**, *0*, 2459–2461. [CrossRef]
17. Yoshikawa, E.; Radhakrishnan, K.V.; Yamamoto, Y. Palladium-Catalyzed Controlled Carbopalladation of Benzyne. *J. Am. Chem. Soc.* **2000**, *122*, 7280–7286. [CrossRef]
18. Zhang, X.; Yu, X.; Feng, X.; Yamamoto, Y.; Almansour, A.I.; Arumugam, N.; Kumar, R.S.; Bao, M. Rhodium-Catalyzed Oxidative Benzannulation of *N*-Pivaloylanilines with Internal Alkynes through Dual C–H Bond Activation: Synthesis of Highly Substituted Naphthalenes. *Chem. Asian J.* **2016**, *11*, 3241–3250. [CrossRef]
19. He, Z.; Huang, Y. Diverting C–H Annulation Pathways: Nickel-Catalyzed Dehydrogenative Homologation of Aromatic Amides. *ACS Catal.* **2016**, *6*, 7814–7823. [CrossRef]
20. Pham, M.V.; Cramer, N. Aromatic Homologation by Non-Chelate-Assisted Rh^{III}-Catalyzed C–H Functionalization of Arenes with Alkynes. *Angew. Chem. Int. Ed.* **2014**, *53*, 3484–3487. [CrossRef]
21. Gao, M.; Lam, J.W.Y.; Liu, Y.; Li, J.; Tang, B.Z. A new route to functional polymers: Atom-economical synthesis of poly(pyrazolylnaphthalene)s by rhodium-catalyzed oxidative polycoupling of phenylpyrazole and internal diynes. *Polym. Chem.* **2013**, *4*, 2841–2849. [CrossRef]

22. Song, G.; Gong, X.; Li, X. Synthesis of Quinolines via Rh(III)-Catalyzed Oxidative Annulation of Pyridines. *J. Org. Chem.* **2011**, *76*, 7583–7589. [CrossRef]
23. Fukui, M.; Shibata, Y.; Hoshino, Y.; Sugiyama, H.; Teraoka, K.; Uekusa, H.; Noguchi, K.; Tanaka, K. Rhodium(III)-Catalyzed Tandem [2+2+2] Annulation–Lactamization of Anilides with Two Alkynoates via Cleavage of Two Adjacent C–H or C–H/C–O bonds. *Chem. Asian J.* **2016**, *11*, 2260–2264. [CrossRef]
24. Honjo, Y.; Shibata, Y.; Kudo, E.; Namba, T.; Masutomi, K.; Tanaka, K. Room Temperature Decarboxylative and Oxidative [2+2+2] Annulation of Benzoic Acids with Alkynes Catalyzed by an Electron-Deficient Rhodium(III) Complex. *Chem. Eur. J.* **2018**, *24*, 317–321. [CrossRef]
25. Loginov, D.A.; Muratov, D.V.; Nelyubina, Y.V.; Laskova, J.; Kudinov, A.R. μ-Borole triple-decker complexes as catalysts for oxidative coupling of benzoic acid with alkynes. Structure of a hybrid rhodacyclopentadienyl/borole triple-decker complex. *J. Mol. Catal.* **2017**, *426*, 393–397. [CrossRef]
26. Loginov, D.A.; Belova, A.O.; Kudinov, M.R. Rhodacarboranes as catalysts for oxidative coupling of benzoic acid with diphenylacetylene. *Russ. Chem. Bull.* **2014**, *63*, 983–986. [CrossRef]
27. Ueura, K.; Satoh, T.; Miura, M. Rhodium- and Iridium-Catalyzed Oxidative Coupling of Benzoic Acids with Alkynes via Regioselective C–H Bond Cleavage. *J. Org. Chem.* **2007**, *72*, 5362–5367. [CrossRef]
28. Geng, K.; Fan, Z.; Zhang, A. Available Rhodium-catalyzed oxidative coupling of N-acyl anilines with alkynes using an acylamino moiety as the traceless directing group. *Org. Chem. Front.* **2016**, *3*, 349–353. [CrossRef]
29. Saxena, P.; Thirupathi, N.; Nethaji, M. Depalladation of Neutral Monoalkyne- and Dialkyne-Inserted Palladacycles and Alkyne Insertion/Depalladation Reactions of Cationic Palladacycles Derived from *N,N′,N″*-Triarylguanidines as Facile Routes for Guanidine-Containing Heterocycles/Carbocycles: Synthetic, Structural, and Mechanistic Aspects. *Organometallics* **2014**, *33*, 5554–5565. [CrossRef]
30. Aparna, P.S.; Prabha, B.; Prakash, P.; Jijy, E.; Varma, R.L.; Radhakrishnan, K.V. Ruthenium catalyzed desymmetrization of diazabicyclic olefins to access heteroaryl substituted cyclopentenes through C–H activation of phenylazoles. *Tetrahedron Lett.* **2014**, *55*, 865–868. [CrossRef]
31. Qi, Z.; Li, X. Rhodium(III)-Catalyzed Coupling of Arenes with 7-Oxa/Azabenzonorbornadienes by C–H Activation. *Angew. Chem. Int. Ed.* **2013**, *52*, 8995–9000. [CrossRef]
32. Ohtaka, J.; Sakamoto, T.; Kikugawa, Y. A one-pot procedure for trifluoroacetylation of arylamines using trifluoroacetic acid as a trifluoroacetylating reagent. *Tetrahedron Lett.* **2009**, *50*, 1681–1683. [CrossRef]
33. Li, X.; Gong, X.; Zhao, M.; Song, G.; Deng, J.; Li, X. Rh(III)-Catalyzed Oxidative Olefination of *N*-(1-Naphthyl)sulfonamides Using Activated and Unactivated Alkenes. *Org. Lett.* **2011**, *13*, 5808–5811. [CrossRef] [PubMed]
34. Mezaki, N.; Yagi, S.; Yoshigami, R.; Maeda, J.; Suzuki, T.; Ohsawa, S.; Tsukamoto, K.; Tanaka, T. Pd-Catalyzed Sulfinylzincation of Activated Alkynes with 1-Alkynyl Sulfoxides as a Sulfinyl Source. *J. Org. Chem.* **2003**, *68*, 5550–5558. [CrossRef] [PubMed]
35. Dieter, R.N.; Lu, K. α-(*N*-Carbamoyl)alkylcuprate Chemistry in the Synthesis of Nitrogen Heterocycles. *J. Org. Chem.* **2002**, *67*, 847–855. [CrossRef] [PubMed]
36. Patel, P.; Borah, G. Direct Access to Indoles by Ir^{III}-Catalyzed C–H Functionalization of Acetanilides with Diazo Compounds. *Eur. J. Org. Chem.* **2017**, 2271–2279. [CrossRef]

8

Amide Activation in Ground and Excited States

Ervin Kovács [1], Balázs Rózsa [2], Attila Csomos [3], Imre G. Csizmadia [4] and Zoltán Mucsi [4,5,*]

1 Institute of Materials and Environmental Chemistry, Research Centre for Natural Sciences, Hungarian Academy of Sciences, Budapest H-1117, Hungary; kovacs.ervin@ttk.mta.hu
2 Laboratory of 3D Functional Imaging of Neuronal Networks and Dendritic Integration, Institute of Experimental Medicine, Hungarian Academy of Sciences, Budapest H-1083, Hungary; rozsabal@koki.hu
3 Institute of Chemistry, Eötvös University, Budapest H-1117, Hungary; csomosattila@caesar.elte.hu
4 Department of Chemistry, University of Toronto, Toronto, ON M5S 3H6, Canada; icsizmad@hotmail.com
5 Department of Chemistry, Femtonics Inc., Budapest H-1094, Hungary
* Correspondence: zoltanmucsi@gmail.com or zmucsi@femtonics.eu

Academic Editor: Michal Szostak

Abstract: Not all amide bonds are created equally. The purpose of the present paper is the reinterpretation of the amide group by means of two concepts: amidicity and carbonylicity. These concepts are meant to provide a new viewpoint in defining the stability and reactivity of amides. With the help of simple quantum-chemical calculations, practicing chemists can easily predict the outcome of a desired process. The main benefit of the concepts is their simplicity. They provide intuitive, but quasi-thermodynamic data, making them a practical rule of thumb for routine use. In the current paper we demonstrate the performance of our methods to describe the chemical character of an amide bond strength and the way of its activation methods. Examples include transamidation, acyl transfer and amide reductions. Also, the method is highly capable for simple interpretation of mechanisms for biological processes, such as protein splicing and drug mechanisms. Finally, we demonstrate how these methods can provide information about photo-activation of amides, through the examples of two caged neurotransmitter derivatives.

Keywords: amide; activation; amidicity; carbonylicity; transamidation; acyl transfer; excited state

---

1. Introduction

The amide functional group is one of the most generally used moieties both in synthetic organic chemistry and bioorganic chemistry [1–5]. They have been known and studied for more than a century and can be commonly found in peptides/proteins and biologically active compounds, as well as in a broad range of synthetic drugs and toxins as a key functionality [1,6–9]. In general, when speaking of amides, the idea of a stable, unreactive group comes in our minds, however, they can play a significant role as reactants in a vast number of processes.

The wide variability in their reactivity can be attributed to the possibility of fine-tuning the bond strength. This can be controlled by attached substituent groups at the carbon and nitrogen atoms. The correlation between poor reactivity and planarity [10,11] of amides was proposed firstly by Pauling and later it was understood as being a result of the strength of the resonance-stabilized amide C–N bond [12]. Pauling predicted that a typical planar amide exhibits approximately 40% partial double-bond character, which renders the amide bond one of the most solid and least reactive functional groups (Figure 1) [13,14]. For example, the conjugation of an amide bond, which is traditionally specified as n(N) $\rightarrow$ π^* (C=O), in formamide results in resonance energy (RE) of 91.29 kcal mol^{-1} [15,16]. In contrast to that, embedding the amide bond into a rigid bicyclic frame with nitrogen at the ring fusion, significantly modifies the degree of the resonance, therefore increasing the instability of the strained amide bond, for example in β-lactam antibiotics [17–19]. More intense conjugation implies a larger contribution of resonance stabilization, which increases the system stability thus lowering the reactivity and vice versa. In other words, the extent of conjugation predetermines chemical reactivity of the amides [20].

Amides can readily be cleaved by enzymes such as protease [15,16], but synthetically it is difficult to selectively break the carbon–nitrogen bond [21–31]. How can nature solve a problem easily, that is so hard for us? The answer is hidden in the process of activation of the unreactive group.

Organic chemists and "Nature" have developed methods to activate stable amide bonds, to make possible various reactions. Numerous examples of these can be found in the field of synthetic organic chemistry and biochemistry. The amide activation is typically based on the lowering of the amidic character by forcing the amide group to lose its conjugation. This can be achieved by substituent groups, as well as by forcing them into an inflexible structure. Moreover, as Thakkar et al. have described, the presence of the cis or trans isomer in the planar character of the amide bond in stable geometries can also influence the overall reactivity [32]. In connection with special chromophores, amide bonds can also be activated under excitation, as was demonstrated in some earlier studies [33–35]. Some of these processes include amidation, transamidation, reduction, coupling reactions by transition metallic complexes, and spontaneous or enzymatic hydrolysis.

The aim of our paper is to characterize the activity of amides using the amidicity and carbonylicity concepts and to demonstrate possibilities to increase or decrease their activity based on the bond strength [14]. In this work, using quantum chemical computation by using Gaussian 09 [36] at B3LYP/6-31G(d,p) level of theory [37], we reevaluated the data reported in the literature for reactivity of amide bonds. Here we reinterpreted some selected well-known compounds and reactions using the simple rule of amidicity, which can estimate the direction and rate of similar reactions involving more complex reactants.

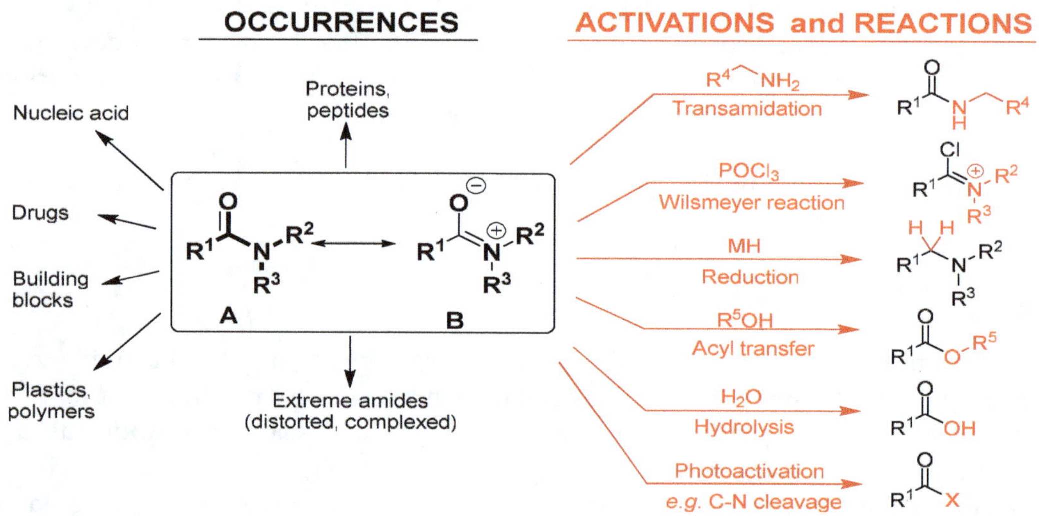

Figure 1. Contributing resonance structures of an amide bond (in the box). The occurrences (**left**) of the amide functionality and their typical reactions (**right**).

2. Discussion

The amide group can be described generally by its resonance structures (A and B in Figure 1), where the nitrogen lone electron pair is delocalized between the nitrogen and the neighboring carbonyl. The condition of this phenomenon is a planar structure of the amide group. For the resonance structure B (Figure 1), the nucleophilic attack to the group is far less possible, thus making a general amide a resistant and unreactive group. The chemical and biochemical stability and reactivity is determined by the contribution of the minor resonance structure (B in Figure 1), in which formally a double bond is formed between the carbon and nitrogen atoms. These characteristics can be represented on a wide spectrum.

Earlier, the strength of a given amide bond was successfully characterized and the reactivity of different amides was made comparable to each other by creating the so called linear "amidicity scale" (Figure 2B) [27]. Analogously, a linear "carbonylicity scale" (Figure 2A) was also constructed, extending the group of amides toward the entire world of carbonyl compounds, making all of them comparable [20]. Other methods, such as carbonyl substitution nitrogen atom replacement [6,38,39] and transamidation (TA) [14] which also aim to measure the amide stabilization energy, were also published in the literature. Using this scale, we are able not only to explain the reactivity of the amide bond, but also predict the effect of different substituents on the bond strength. Predicting amide bond strength effectively gives us the opportunity to activate or reactivate the desired amide bond. Manipulating the amidicity of a selected amide of a molecule has numerous synthetic applications. Also, calculating the amidicity—in other words, estimating the activity of the amide bond in biochemical compounds—may clarify the mechanism of action in which they are involved.

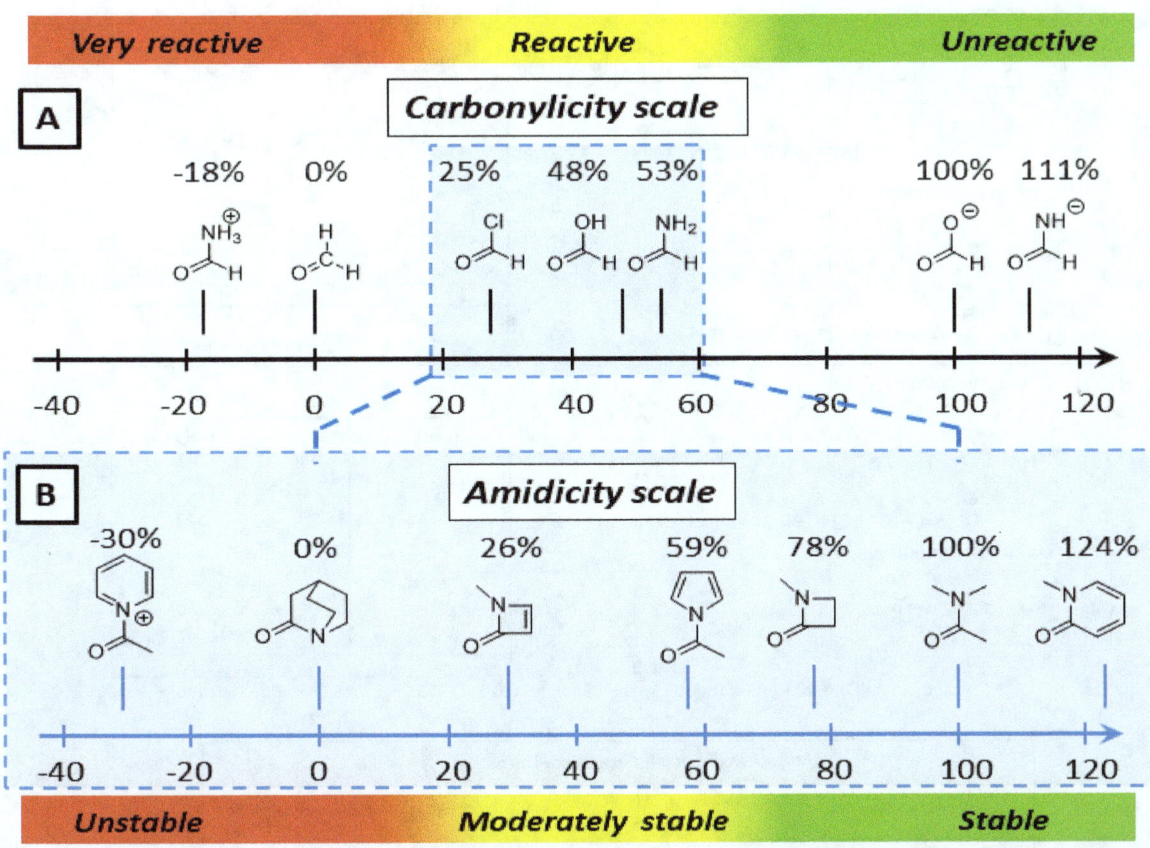

Figure 2. Typical carbonyl and amide compounds represented on the carbonylicity (**A**) and amidicity (**B**) scales. The direction of reactivity and stability increase are shown by red and green strips, respectively.

2.1. *Amidicity Scale (AM%) and Carbonylicity Scale (CA%) and Their Resonance Enthalpies (H_{RE})*

The "amidicity scale" [27] quantifies the amide bond strength on a linear scale (Figure 3A), based on the computed enthalpy of hydrogenation (ΔH_{H2}) of the examined compound. A few examples positioned on the scale demonstrate the wide variety of amides. The scale is based on two specifically chosen reference compounds: A1 and B1 (Figure 3A). Compound aza-adamantane-2-on (A1 in Figure 3A) represents the lack of amidic charater, while dimethylacetamide (B1 in Figure 3A) corresponds to the perfect amide bond. In the structure of A1, the planar geometry is missing, due to the forced arrangement of the bridged N atom. To create the amidicity scale (AM%), the enthalpy of hydrogenation (ΔH_{H2}) of compounds A1 and B1 was calculated Equations (1) and (2) and amidicity was defined using a linear fit to these points. The enthalpy of hydrogenation for A1 (ΔH_{H2}[A]) is considered to have a 0% AM% value, while that for B1 (ΔH_{H2}[B]) is considered to have a 100% AM% value.

$$\Delta H_{H2}[\mathrm{A}] = H[\mathrm{A2}] - \{H[\mathrm{A1}] + H[\mathrm{H_2}]\} = -44.62 \text{ kJ mol}^{-1} \quad (1)$$

$$\Delta H_{H2}[\mathrm{B}] = H[\mathrm{B2}] - \{H[\mathrm{B1}] + H(\mathrm{H_2})\} = 34.88 \text{ kJ mol}^{-1} \quad (2)$$

As specified before, the carbonylicity scale (CA%) [20] quantifies the conjugation of the carbonyl group on a linear scale (Figure 3B), based on computed enthalpy of hydrogenation (ΔH_{H2}) of the examined carbonyl group (Equations (3) and (4)). Analogously to amidicity, the carbonylicity scale is also based on two specifically chosen reference compounds, C1 (formaldehyde; CA% = 0%) and D1 (formate; CA% = 100%) and defined using a linear fit to these points. Also, note that the amidicity scale is a defined section of the carbonylicity scale as it is represented on Figure 2, and differs only in the reference compunds chosen.

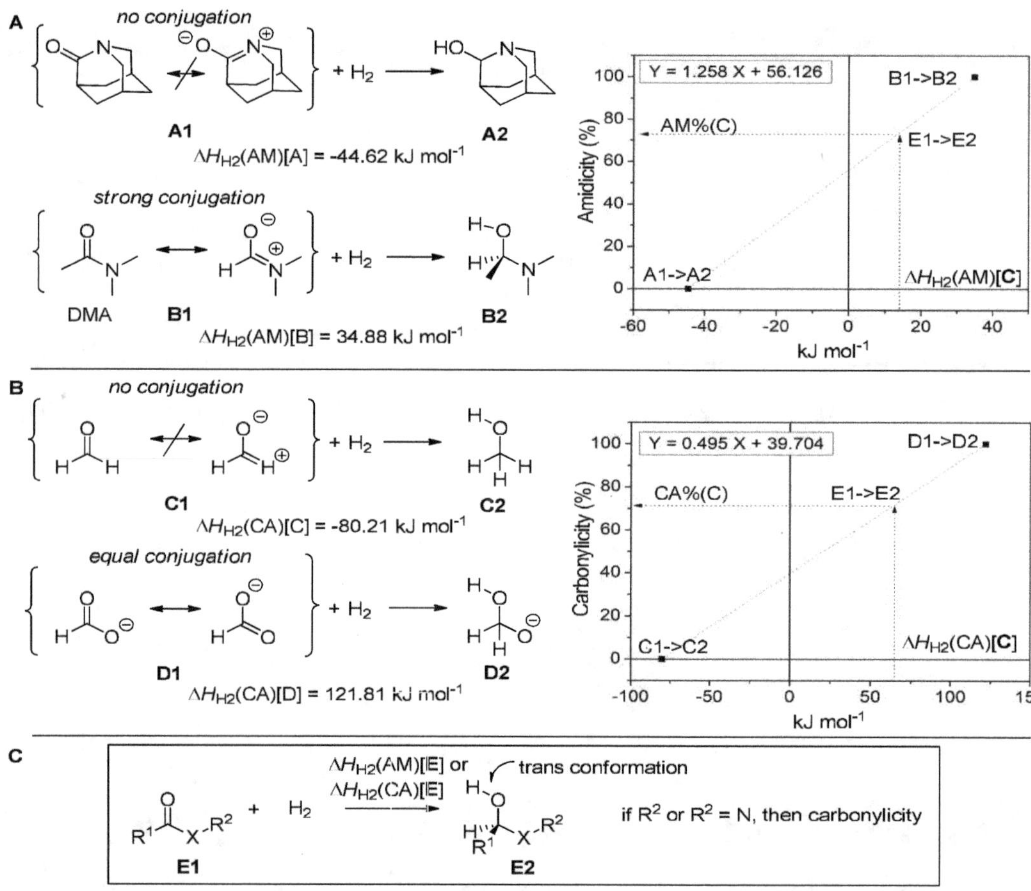

Figure 3. The definition of the amidicity (**A**) and carbonylicity (**B**) percentage based on the enthalpy of hydrogenation (ΔH_{H2}) of the carbonyl group of a general carbonyl compound E1 (**C**). The two specifically chosen reference structures and their hydrogenation are also illustrated as A1 and B1 for amidicity (AM%; top) as well as C1 and D1 for carbonylicity (CA%; bottom). Values were obtained from the B3LYP/6-31G(d,p) geometry-optimized structures. In structure E2, the O–C–X–R_2 and the H–O–C–X dihedral angles are chosen to be in the anti orientation.

$$\Delta H_{H2}[\mathrm{C}] = H[\mathrm{C2}] - \{H[\mathrm{C1}] + H(\mathrm{H_2})\} = -80.21 \text{ kJ mol}^{-1} \quad (3)$$

$$\Delta H_{H2}[\mathrm{D}] = H[\mathrm{D2}] - \{H[\mathrm{D1}] + H(\mathrm{H_2})\} = 121.81 \text{ kJ mol}^{-1} \quad (4)$$

The linear fits are described by two pairs of fitting parameters, the slope (m_{AM} and m_{CA}) and the intercept (b_{AM} and b_{CA}). The values (AM% and CA%) for a desired amide or carbonyl bond can be easily calculated with the help of Equations (6) and (8), for a known enthalpy of hydrogenation, ΔH_{H2}, from Equation (5). Both the amidicity and carbonylicity values can be transformed to the resonance enthalpy (H_{RE}[E]; Equations (7) and (9)), which helps to estimate the energy benefit coming from the resonance inside the group. Neither of these scales is limited to values between 0% and 100%; in extreme cases the bond is weaker than A1 or C1 and stronger than B1 or D1, so they may display values outside of the aformentioned range. The amidicity scale can be considered as a sub-section of the carbonylicity scale.

$$\Delta H_{H2}[E] = H[E2] - \{H[E1] + H(H_2)\} \quad (5)$$

$$AM\%[E] = m_{AM}\ \Delta H_{H2}[E] + b_{AM} \quad (6)$$

$$H_{RE}[E] = AM\%/m_{AM} \quad (7)$$

$$CA\%[E] = m_{CA}\ \Delta H_{H2}[E] + b_{CA} \quad (8)$$

$$H_{RE}[E] = CA\%/m_{CA} \quad (9)$$

2.2. Transamidation and Transacylation Reactions in Synthetic Organic Chemistry

The amidicity value is useful to predict the outcome of a general transamidation reaction (Figure 4, top). The amidicity change ΔAM in Equation (10) or the change in stabilization enthalpy (ΔH_{RE} in Equation 12) leads to a thermodynamic selection rule, allowing for the reactions to be categorized as being either thermodynamically favorable or unfavorable. Generally, ΔH_{RE} is proportional with the entalpy change of the transamidation reaction. The lower the amidicity value of the chosen group, the greater reactivity it has towards nucleophylic reagents, like amines. The direction of the transamidation reactions can be predicted by comparing the sum of amidicities on the product and reactant sides. If the amidicity value increases, the reaction is energetically favoured in the given direction. If the difference is a negative value, the reaction is not driven forward. If there are different possible reaction routes, the one with the greater increase in amidicity is favored. The amidicity value is also capable of predicting the most active amide group in a molecule. Such a thermodynamic selection rule may also be used to predict the selectivity of a reaction in the presence of competing amide functionalities. This principle is illustrated as being operative in cases of different reactions. However, it should be mentioned that reactions may have several other parameters that determine if a reaction can proceed or not, such as steric hinderance, kinetic consequences, side reactions and the role of the leaving group on the acyl group. These amidicity and carbonylicity values describe only the thermodynamic background and not the kinetic aspect of a reaction.

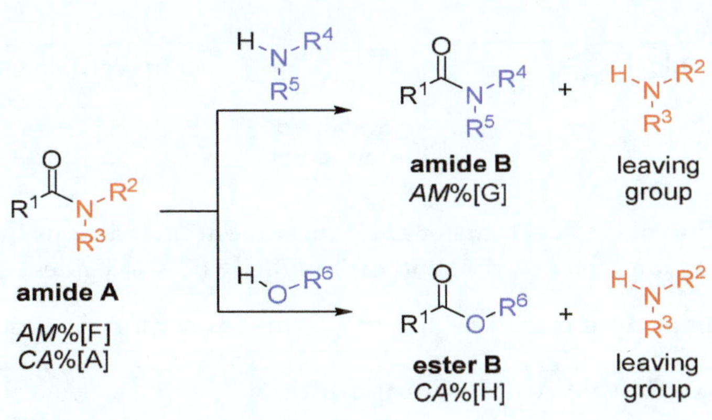

Figure 4. General transamidation reaction and the amidicity (AM%) and carbonylicity (CA%) percentages of the amides examined.

$$\Delta AM = AM\%[G] - AM\%[F] \quad (10)$$

$$\Delta CA = CA\%[H] - CA\%[F] \quad (11)$$

$$\Delta H_{TA} \approx \Delta H_{RE} = H_{RE}[B] - H_{RE}[A] \quad (12)$$

On the basis of the exploratory screening of the literature, four cases were defined (Figure 5) [40]. Cases I and II (one-step processes) are both thermodynamically favored yet differ in reaction barrier height. Reaction barriers in case I are low [41], while those of case II are high, with the latter requiring a suitably chosen Lewis acid catalyst ($AlCl_3$, BF_3, $TiCl_4$, HCl, etc.) to proceed normally [42,43]. This amine exchange reaction is sometimes carried out in a closed autoclave at high temperature by using the salt of the amine in the presence of BF_3 [44], often to convert simple urea to a substituted one. In contrast to that, both cases of III and IV (two-step processes) are thermodynamically unfavorable, requiring the activation of either the reacting amide (case III) or the reacting amine. In case III, the amide is effectively destabilized, so activated by nitrosation, nitration, or acylation step [*N*-activation by activating group (AG), case III, Figure 5]. This activation is manifested as an increase in the energy level of the starting state, wherein the activated amide intermediate provides a thermodynamically favorable and rapid transamidation process.

In case IV (*N*-deprotonation), a theoretically thermodynamically unfavorable transamidation, which may proceed in the presence of a strong base (NaH or NaOR). The product monosubstituted amide forms and stays in its deprotonated form (–CO–NHR^4) and the starting amide is disubstituted non-deprotanable (–CO–NR^{1}R^2, amide). Therein, the forming amide-anion intermediate is stabilized by the deprotonation. In this case, an originally thermodynamically unfavorable transamidation process involves a thermodynamically favorable reversible subprocess, which finally finishes irreversibly by protonation in the acidic work-up.

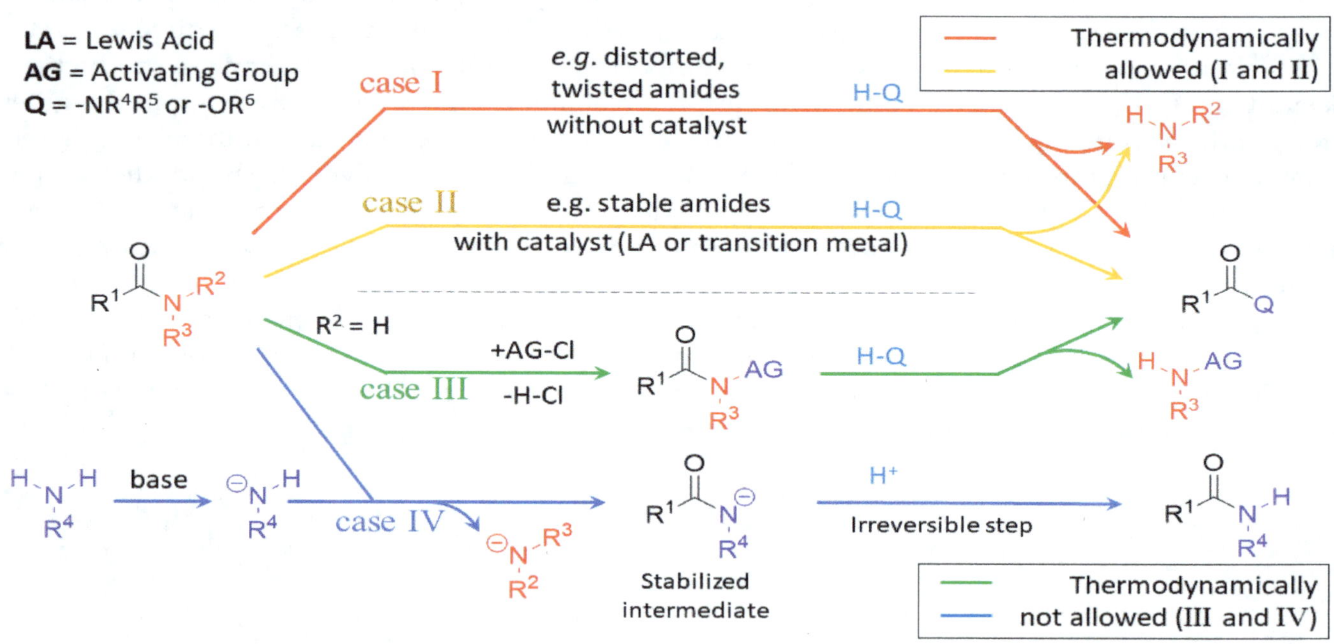

Figure 5. The classification of the acyl transfer and transamidation reactions into four cases (case I, II, III and IV), interpreted by amidicity (AM%) and carbonylicity (CA%) values.

2.2.1. Examples of the Thermodynamically Allowed Transamidation Reactions (case I and II)

Transamidation Processes of Simple Amide Compounds

In order to illustrate the essence of the transamidation reaction, we are demonstrating this concept by simple amides and amines. Unsubstituted amides, such as **1** and **2** in Figure 6, exhibit a moderately reduced value of amidicity (Figure 6) relative to mono-substituted or di-substituted amides, such as **3** and **4** (97–103%; Figure 6) [40], so one may therefore predict a transamidation process between them. Mono- and di-substituted amines (e.g., dimethylamine) are shown to react readily with formamide (**1**) at RT (25 °C) or above without any additional activation (case I). However, the transamidation reaction of acetamide (**2**) required elevated temperature and the presence of $AlCl_3$ as a Lewis acid activator in order to attain an acceptable reaction rate (case II). The harsh conditions are explained by the high activation energy of the sterically hindered reaction center [40].

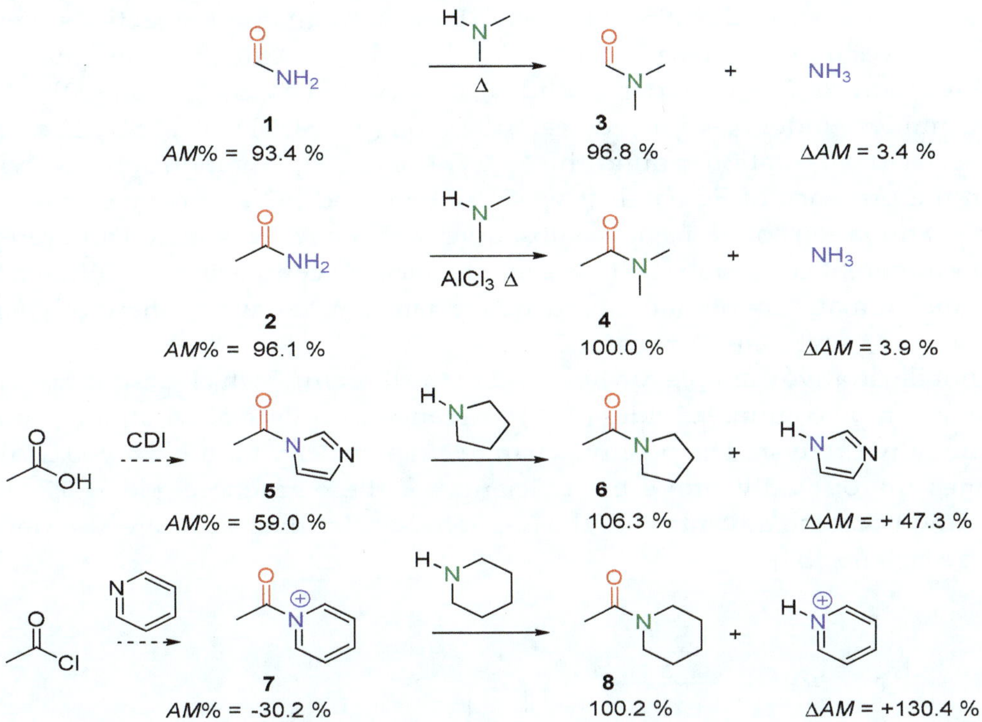

Figure 6. Examples for transamidation involving secondary amine, with the corresponding amidicity values computed at B3LYP/6-31G(d,p) level of theory.

Compound **5** represents mild acylating agents (Figure 6) taking part readily in transamidation reactions with amines (e.g., pyrrolidine), forming amide **6** beside imidazole as a side-product. The synthesis of **5** can be carried out simply by means of the corresponding acid and carbonyl diimidazole (CDI) reagent. The increased acylating potency (AM% = 59%) of **5** can be attributed to the competition for the N atom lone pair between the imidazole aromatic ring and the carbonyl group of the amide. This phenomenon decreases the amidicity percentage of **5** to a value as low as 59% compared to normal amides. During the acylation reaction **5** $\rightarrow$ **6**, the amidicity value increased significantly, providing the main driving force for this transamidation reaction. Compound **7** (Figure 6) exhibits an extremely low amidicity percentage (−30.2%), making this molecule an excellent acylating agent, which can be prepared in situ from AcCl and pyridine. Consequently, **7** readily furnishes the acylation with an amine (e.g., piperidine), resulting in product **8**. Here, the reaction is motivated by the extremely large change in amidicity (ΔAM, Figure 6) even at low temperature [24,31,40,45].

Amidicity value is also able to explain the inactivity and resistance of some commonly used organic amide solvents, For example, *N*,*N*-dimethylformamide (DMF, AM% = 100%) and *N*-methylpyrrolidinone (NMP, AM% = 136%) exhibit extremely high amidicity values.

When dissolving any amine in these species, no reaction occurs between them due to the negative ΔAM, which makes them unfavorable for taking part in any reaction even at high temperature [24,31,40,45].

Transamidation and Acyl Transfer Processes of *N*-Phenylamides.

In the literature, there are numerous examples where acylated *N*-alkyl anilines are reacted by amines or alcohols in the presence of transition metal catalysts (Pd, Ni, etc.). In these cases, the *N*-alky-anilids exhibit moderately low amidicity values, due to the dual effect of the phenyl ring. The phenyl ring suffers some steric hindrance and competes with the non-bonding electron-pair of the N atom against the acyl group at ones. In retrospect, it appears that the transamidation and acyl transfer reaction can occur without any catalyst, although, in the literature at different times, both of these types of reactions were reported as new approaches. Initially, these were reported to proceed without catalyst in the 1960's [46–49]. However, analogue reactions were highlighted later as a novel approach in the presence of Ni or Pd catalysts. Authors called these reactions new, cross-coupling reactions of *N*-Ar amides (anilides), as reported by Garg et al. [50]. The reactivity of theses aromatic amides, under Ni-catalyzed esterification (Ni/SIPr) was rationalized by the concept of so called "barrier-free" rotation around the N–CO bond [51]. Analogues transacylation reaction with alcohol in the presence of Pd catalyst was also reported [52]. However, the earlier metal or catalyst-free transamidation solutions have subsequently been reinvestigated and rereported as new results [53]. This apparent confusion can be easily resolved by comparing the difference of amidicity values between the aromatic amides (anilids) with aliphatic amides, which thermodynamically allows such transamidation or acyl transfer reaction.

A simple, but illustrative example can be found in the literature, which describes an intramolecular transamidation, where two amine-anilides (**9**,**11**) rearrange to aniline-alkyl-amide products (**10**,**12**; in Figure 7) without any catalyst. The positive changes in amidicity values as well as the exothermic enthalpy changes undoubtedly prove the efficiency of these reactions [46,54,55]. Other creative synthetic applications, taking advantage of the *N*-phenyl-acyl activation, were also developed for the synthesis of large lactams [53].

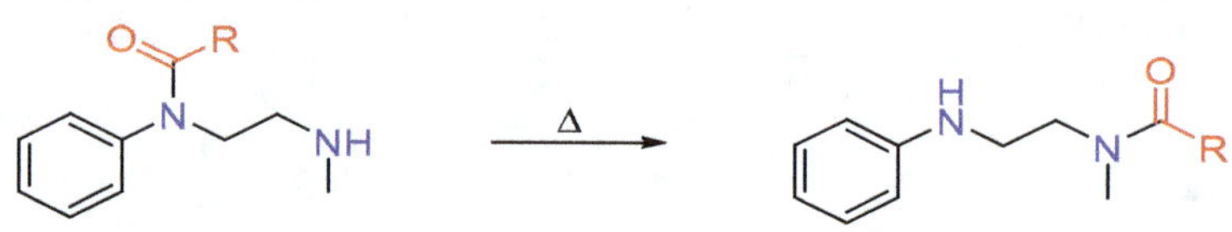

	R	AM%		R	AM%	ΔAM%	Δ*H*
9	Me	73.5%	**10**	Me	106.7%	33.2%	-32.2 kJ mol^{-1}
11	Ph	89.7%	**12**	Ph	93.9%	4.1%	-37.0 kJ mol^{-1}

Figure 7. Typical example of intramolecular transamidation rearrangement, with the corresponding amidicity values (AM%) as well as amidicity (ΔAM) and enthalpy change (ΔH), computed at B3LYP/6-31G(d,p) level of theory.

Acyl Transfer Reactions of Distorted or Twisted Amides

The structural distortion of an amide bond significantly modulates the degree of the amidic resonance, typically by lowering its amidicity/carbonylicity value [56,57]. The two classic twisted bridged lactams (aza-bicycloheptanone **13** [17], penicillin scaffold-penam **14** [58–60]) were synthetized more than 75 years ago. More than six decades later, this was followed by the preparation of **15** (in 1998) and its aromatic derivatives [61–64]. The aza-adamantane analogue **16** (in 2006 [65]), as well its substituted derivatives [66–68] made the series complete. These compounds furnish fully perpendicular arrangement within the amide bonds, resulting in high reactivity similar to that of ketones (Figure 8).

13 10%   **14** 22%   **15 and derivatives** 17.1%   **16** R = H, Me 0.0% if R = H (by definition)

Figure 8. Typical examples of synthetized twisted amides (**13**–**16**) with the corresponding amidicity computed at B3LYP/6-31G(d,p) level of theory.

At first, we should clarify the difference between the distorted and twisted amides (Figure 9). Updating a previous nomenclature [6,69], the distorted amides reflect some steric hindrance or ring strain in their entire structures, but they already possess significant conjugation, with mentionable amidicity values. Here the amides are almost planar. In contrast, twisted amides may be described by a near perpendicular amine plane to the carbonyl plane with pyramidalized N atom, where the whole structure itself is not distorted or hindered (Figure 9).

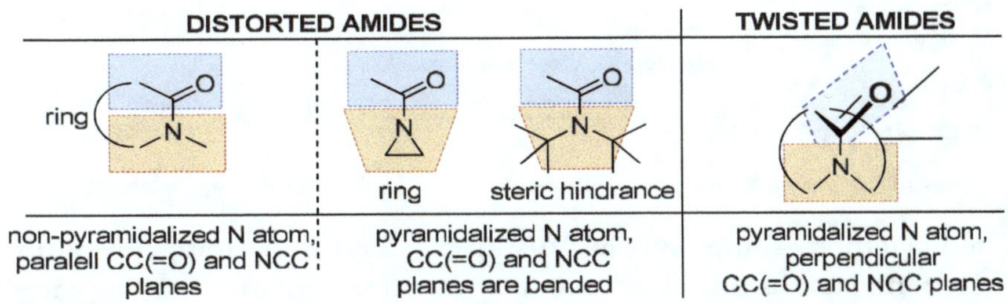

Figure 9. Redefined difference between the distorted and twisted amides. Distorted amides suffer steric hindrance or ring stress with pyramidalized or non-pyramidalized N atom. Twisted amides do not exhibit any strain, with perpendicular, non-overlapping amide planes. The carbonyl and amide planes are illustrated by the parallel- or perpendicularly-oriented blue and orange colors.

Recently, twisted and distorted amide bonds have slowly found real application as a potent reagent in the mainstream of general organic chemistry. This reagent advances the highly reactive N–C amide bond in metal-free as well as transition metal catalyzed cross-coupling reactions yielding various carbonyl derivatives.

Distorted ring-amide can be represented generally by the correlation between the cleavage of lactam rings and the ring size. In Figure 10 and the corresponding Table 1 we summarize the possible alcoholysis of four to seven-membered lactams (n = 1–4) to open chain amino ester in the light of the calculated carbonylicity values. It is important to clarify, that the CA% values are corrected with ring strain as described. The smallest ring-sized lactam (**17a**; n = 1 in Figure 10) has relatively low carbonylicity value (40.5%), which may be due to the insignificant conjugation and not only to the ring strain (Table 1). Thus, **17a** can transform to the corresponding open-chain amino ester **18a**, already having a normal value (ca. 56%; Figure 10, Method-A). Generally, five- and six-membered rings (**17b**, n = 2 and **17c**, n = 3) are the most stable among the small ring lactams, so one-step alcoholysis reactions are not allowed. More precisely, for five-member ring lactam **17b** this one-step process toward **18b** is forbidden due to the negative change of the carbonylicity value (ΔCA = −3%), while in the case of six-member lactam the ring opening process (**17c** → **18c**) exhibits an almost neutral change (ΔCA = +1%). In these cases, a more complicated multi-step processes (**17** → **19** → **20** → **18**; Method-B) should be used to cleave the lactam ring effectively. In this longer process, the hydrolysis with NaOH is allowed due to the increasing CA% (**19b**,**c**). The acid chloride formation (**20b**,**c**) activates their functionalities, which easily transform to the desired ester **18b**,**c**. Finally, the seven membered lactam (17d; n = 4) has a slightly lower carbonylicity value (53%), which allows a moderate potency to transform it to **18b** in alcohol in one step, with slightly increasing ΔCA. It should be mentioned that the longer procedure **17** → **19** → **20** → **18** is available for all ring sizes.

Table 1. The calculated carbonylicity values (CA%) of compounds **17a–d** and **18a–b** and the carbonylicity changes (ΔCA) for ring-opening alcoholysis depending on the ring size.

State	n = 1 (a)	n = 2 (a)	n = 3 (a)	n = 4 (a)
Start (CA%) 17	46.5%	60.3%	55.2%	53.4%
Product (CA%) 18	56.2%	56.1%	56.1%	56.2%
ΔCA	+10.7%	−4.2%	+0.9%	+2.8%
Method	A or B	only B	A [1] or B	A or B

[1] Extremely slow reaction.

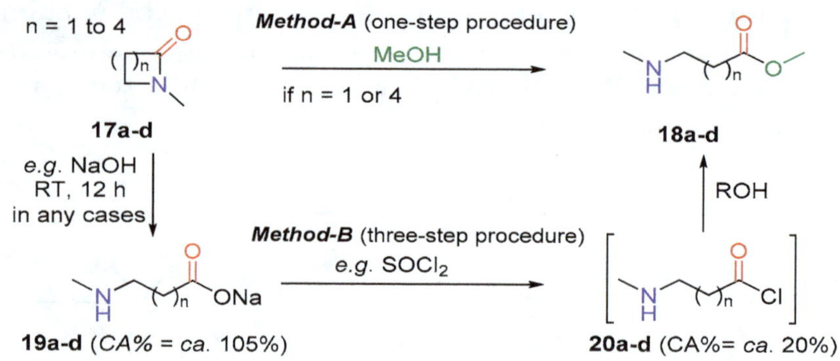

Figure 10. One step (Method-A; **top**) and multi-step (Method-B, **bottom**) procedures to transform cyclic amide (lactam) with various ring sizes to amino ester (for the corresponding carbonylicity values computed at B3LYP/6-31G(d,p) level of theory; see Table 1). The carbonylicity values for 19 and 20 are average values.

Some examples of sterically hindered amides were reported [70–72] to exhibit moderately low amidicity and carbonylicity values. One representative example is **21**, where the amide functionality is sterically hindered and, hence, distorted. Here, somewhat decreased carbonylicity makes the reaction possible with alcohol (for example MeOH) to yield an ester (**22**), as published earlier (Figure 11, top) [73]. Another example shows that a twisted amide (e.g., **23**), having an extremely low amidicity value, readily cleaves its CO–N bond by MeOH, resulting an open-chain amino-ester (**24**), as shown in Figure 11 (bottom). In both cases, the thermodynamic driving force can be easily explained by the positive change in carbonylicity [74,75]. The combination of the twisted amide and acyl-anilide concepts are also exemplified, where these amides furnish effective alcoholysis analogously to the previous cases [73,76].

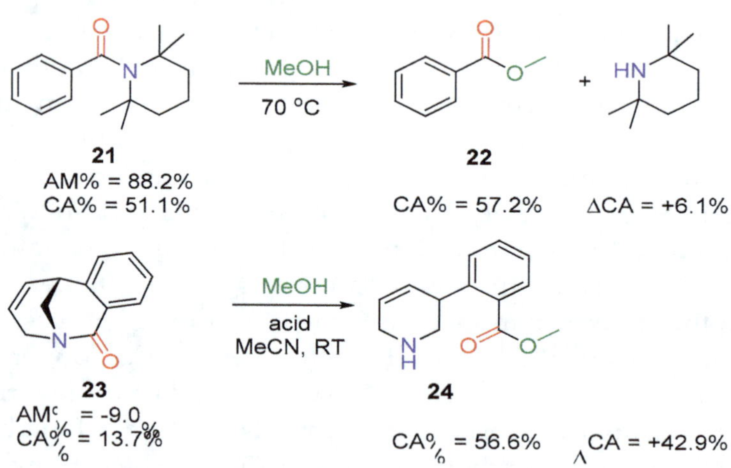

Figure 11. Relevant examples for acyl transfer reactions in the cases of distorted (**21**) and twisted amides (**23**) resulting esters (**22** and **24**), explained by the change in carbonylicity (CA%) with the corresponding carbonylicity and amidicity values computed at B3LYP/6-31G(d,p) level of theory.

2.2.2. Transamidation Reaction via Activated Amides (case III)

Non-strained secondary amides (e.g., *N*-metylacetamide; **25**) exhibit approximately 100% amidicity, so are typically not able to take part in transamidation reactions. For instance, the reaction between **25** and dimethylamine, yielding theoretically dimethyl acetamide (**26**), is not beneficial, due to the slightly smaller amidicity [40]. From the perspective of practicing chemists, there was a synthetic demand to activate and use these amides as reagents. To activate these resistant secondary amides, their high amidicities should be decreased by an activating group. Numerous methods were developed in recent decades and presented in the literature, summarized in Figure 12. All these solutions involve a preliminary activation step.

The introduction of *N*-nitroso (**27**) [77,78], *N*-nitro (**28**) [79], *N*-(tosyl) (**29**) [80], *N*-(trifluoromethanesulfonyl) (30) [80], or *N*-acyl (31) [81–83] amides are the most relevant examples. The following step describes how these activated amides (27–31) could react with, e.g., dimethylamine, to yield the desired and exemplified amide product 26. The byproducts of these reactions are typically good leaving groups. The most significant amidicity decrease can be measured for **28** and **30**, producing the most reactive acylating agents [79,80]. The *N*-BOC-*N*-methyl-acetamide **31** was prepared in a previous step from **25** by $(BOC)_2O$ reagent [82,83], a process known as *N*-Boc activation. Other, bisacyl and triacyl substituted amides (such as succinimide [84] and phthalimide derivatives [85]) may also behave as good transamidating agents in the presence of an amine. In all cases, the listed ΔAM values were large enough to provide enough driving force to convert to the desired product **26** [40].

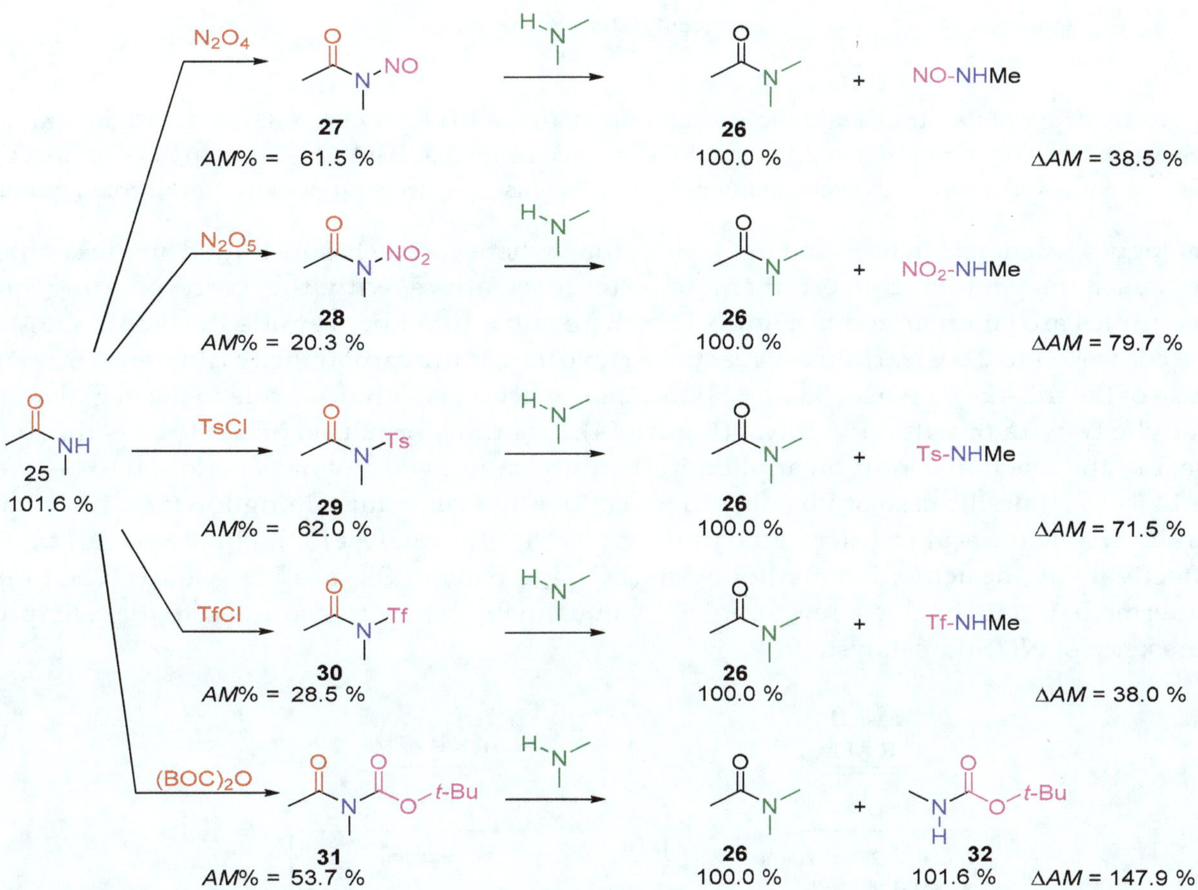

Figure 12. Some representative examples of amide activations (case III), using various reagents with the corresponding amidicity values computed at B3LYP/6-31G(d,p) level of theory. Ts = *p*-toluenesulfonyl; Tf = CF_3SO_2–; BOC = *t*–BuO–CO–.

Finally, a good example can be found for the transition metal catalyzed transamidation, where *N*-alkyl-benzamide (e.g., **33**) was transformed to dialkyl derivatives (e.g., **34a**). The authors previously reported that **33** could not be transformed to **34a** directly, even in the presence of an active transition metal catalyst (Ni/SIPr), without providing any theoretical interpretation. However, the amidicity concept is able to highlight the thermodynamic background of the process failure, due to the unfavorable amidicity change ($\Delta AM = -16.4\%$ in Figure 13). Nevertheless, with the activation of the amide by BOC group (**35**), the progress occurs smoothly to product **34a**, in the presence of Ni/SIPr catalyst. It seems reasonable to conclude that when the transamidation reaction exhibits positive amidicity change (as in **35** $\rightarrow$ **34**), the transition metal catalyst or any other catalyst may help.

However, when the reaction is substantially endothermic (as in **33** → **34a**), the reaction will not take place even if catalyst is used.

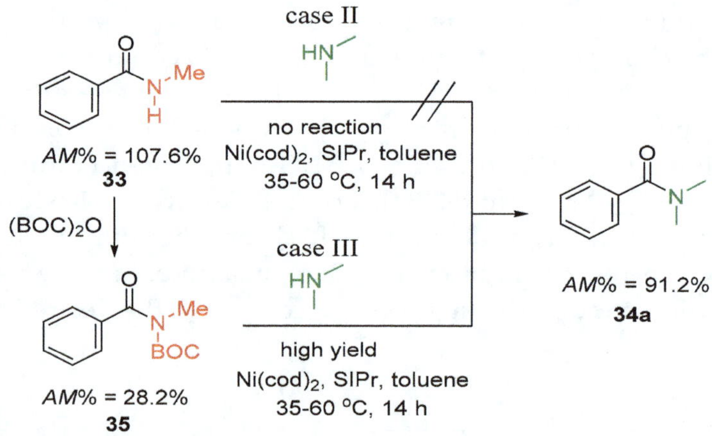

Figure 13. Exemplified transamidation reaction of **33** (case III) with (**33** → **35**) and without amide activation. The corresponding amidicity values were computed at B3LYP/6-31G(d,p) level of theory. BOC = *t*–BuO–CO–; cod = 1,5-cyclooctadiene; SIPr = 1,3-bis(2,6-diisopropylphenyl)imidazole-2-ylide.

Analogues reactions can be found in the literature, where variously substituted amide compounds are used as a reagent to convert them to ester derivatives with the corresponding alcohol. Some examples are summarized in Figure 14, where substituted benzamide derivatives (**33–34a,b**) could be converted to **22**. Clearly, these reactions can occur, if the carbonylicity difference is favorable. In the case of the transformation of **33** (R = H) to **22**, the reaction is forbidden due to the negative change in carbonylicity, with or without catalyst (Figure 14). The transformation of **34a** to **22** is theoretically possible, but the reaction condition applied in the publication was not harsh enough to be effective (Figure 14, case II). In the case of Ph substituents (**35**), however, maintaining low amidicity value is sufficient to enable the acyl transfer in the presence of Ni/SIPr catalyst [50]. Finally, **33** can transform to **22** effectively by the activation method with $(BOC)_2O$ reagent (Figure 14, case III). The prepared active intermediate **34b**, having a low amidicity value, finally can be transformed to the ester product, in the presence of Ni/SIPr catalyst.

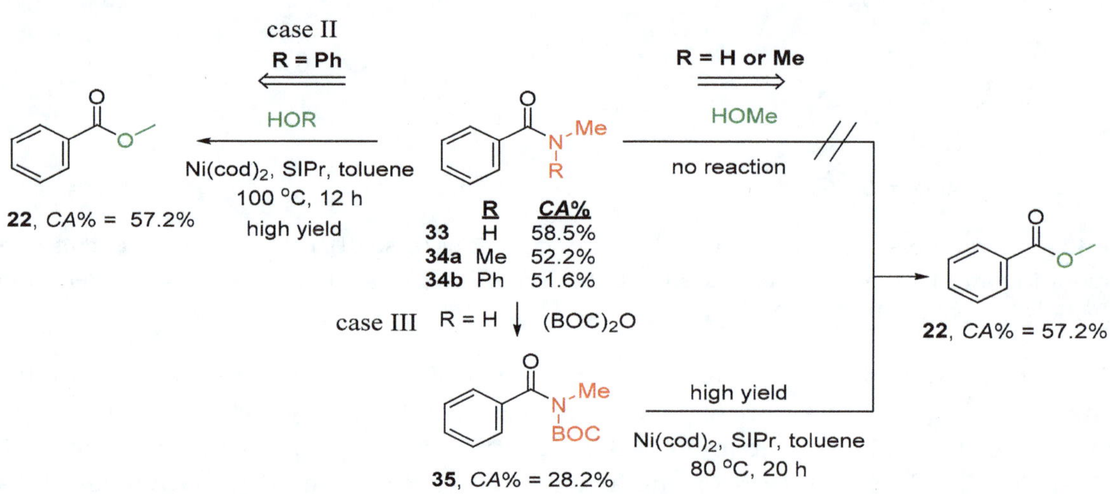

Figure 14. Exemplified acyl transfer reaction (case II and case III) with and without amide activation in the presence of catalyst. The corresponding carbonylicity values were computed at B3LYP/6-31G(d,p) level of theory. BOC = *t*–BuO–CO–; cod = 1,5-cyclooctadiene; SIPr = 1,3-bis(2,6-diisopropylphenyl)imidazole-2-ylide.

From these examples it becomes clear that there is no fundamental difference between transamidation reaction and other transacylation, because in both cases there is an acyl migration. However, for historical reasons, the transamidation and acyl transfer processes are still distinguished in the literature. Needless to say, all of these conclusions are true when the heteroatom is sulfur.

The reverse pathway with different substituent patterns, the transacylation from ester to amide, was also described in the literature in the presence of Ni/SIPr catalyst [86–88].

2.2.3. Activated Transamidation Reaction via Product Stabilization [case IV]

Because of its high amidicity value, *N,N*-dimethyl formamide (DMF; Figure 15) can be considered an unreactive amide. In contrast to this, deprotonated amides, such as **36**, possess even higher amidicity values (>150%) than tertiary amides. In strongly basic conditions, deprotonated amides (**36**) should, therefore, be more favorable than non-deprotonable DMF. A good example is found in the literature, where aniline derivatives **37a–c** can be deprotonated by NaH or NaOMe, generating their corresponding anion form of **38a–c**. The product amide anions **36a–c** were formed at elevated temperature (Figure 15) [36,89–91]. During the workup process, an aqueous acid was used to acidify and precipitate the neutral product **39a–c** in an irreversible way. The procedure as a whole is a thermodynamically forbidden transformation (**38** + DMF $\rightarrow$ **39**), due to the lower amidicity of the product **39a–c**. Nevertheless, only the reversible subprocess of **38** + DMF $\rightarrow$ **36** (in green box, Figure 15) is favorable. Overall, the driving force of this subprocess is controlled by the high ΔAM, so this formylation reaction is rapid, making this reaction practical [40].

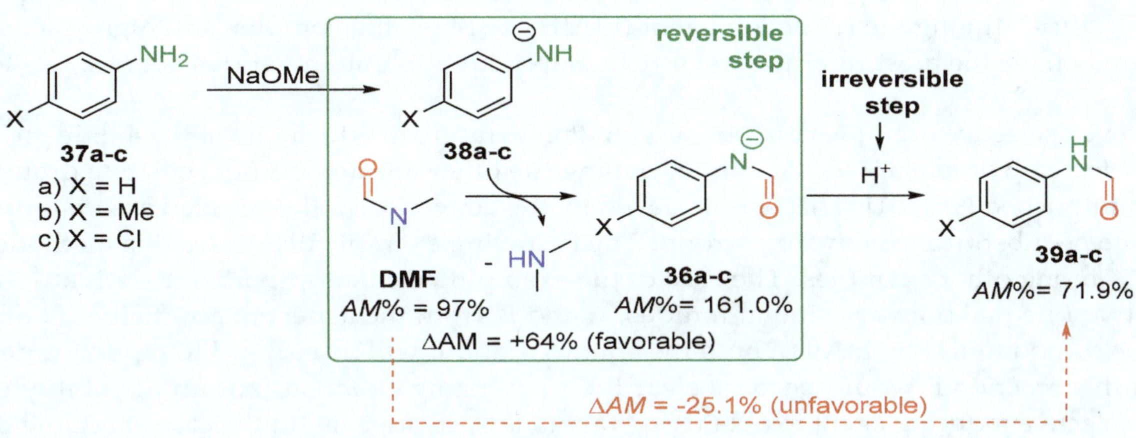

Figure 15. An example for case IV transamidation reaction with the corresponding amidicity values computed at B3LYP/6-31G(d,p) level of theory. The green box illustrates the thermodynamically favorable subprocess, while the red arrow shows the overall unfavorable reaction.

The use of the amidicity change (ΔAM) leads to a quasi-thermodynamic selection rule, allowing the reactions to be categorized as being either thermodynamically favorable or unfavorable. This simple rule of amidicity-change may be used to predict the selectivity of a reaction, where two competing functional groups are involved. Such an example is presented, where two types of amino groups are attached in one molecule (**40**). As one may predict by the amidicity change, only the alkyl amine group will react with formamide, yielding **41** exclusively with high yields in one step (Figure 16). The other isomer amide **42** and the diacylated **43**, including anilid type amide, could not be detected in the reaction mixture, as the amidicity change simply predicted [40].

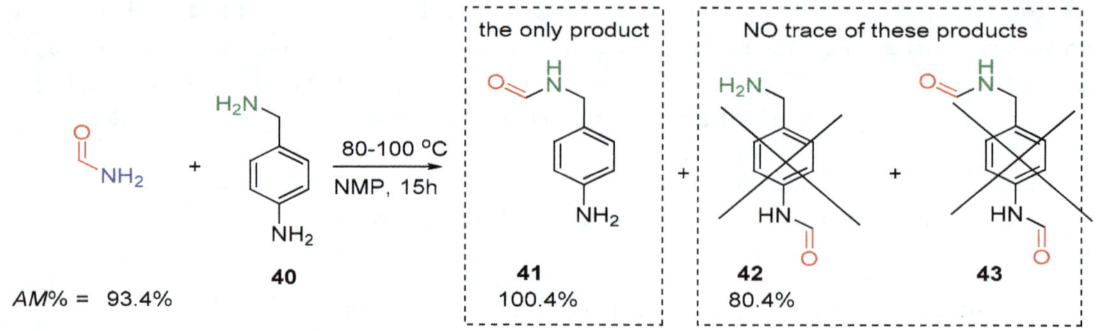

Figure 16. Selectivity of transamidation reaction of *N,N*-dimethylformamide (DMF) and compound **40**, having two types of amino groups. The three possible products **41–43** are shown. The corresponding amidicity values were computed at B3LYP/6-31G(d,p) level of theory.

2.3. *Amide Reduction via Amide Activation*

An unprecedented number of examples can be found in the literature to reduce amides to substituted amines, using numerous reduction agents (most generally metal hydrides). The situation is illustrated by a few selected citations [31,69,92–97]. Strong reduction agents are a common feature of these methods, since generally amides are very resistant against reduction. Consequently, these methods are typically not applicable to carry out selective reactions. This extreme resistance can be explained by the strong conjugative interaction within the structure of amide, exhibited by their high amidicity values. Among the reduction agent ($LiAlH_4$ or BH_3), the complex and highly active metal hydrides may have the most significant synthetic importance to obtain various amines.

However, selective examples can be found in the literature in which one of the amide groups was reduced selectively in one molecule, while keeping the other amide or other sensitive groups of the molecule untouched [98–101]. This was achieved by the activation of the selected amide group by an appropriate *N*-substituting activating group. The following example illustrates this methodology in Figure 17, among other examples. The case of the exemplified diketopiperazine derivative **44** [101] includes two unequal but very similar amides, A and B, from the aspect of amidicity. If one tries to reduce this compound **44** by $LiAlH_4$, both the amides A and B will be reduced to piperzine derivative **45**. From this reaction it would become clear for a practicing chemist, that no possibility exists of finding a selective reducing agent to obtain piperazidone derivative **46**. In the case of adding only one equivalent of reductive agent, a complex mixture of partially reduced forms can be observed.

However, analogously to case III reactions in the previous section, the secondary amide-A was acylated by BOC_2O reagent, which resulted in a triacylated intermediate **47**. In this structure, the amidicity of the acylated amide A is lowered to 34.6%, while that of amide B increased slightly. The carbonyl of the BOC group (amide-C) is as high as 109.1%, which exhibits large resistance to reduction. The lowered amidicity of the amide-A provides not only higher reactivity with the metal hydride, but allows the application of significantly weaker reducing agent, such as $NaBH_4$. This reagent is soft enough to avoid the reduction of amide-B and C. In the first reduction step, amide-A is reduced only to hydroxy amine intermediate **48**, protected by the BOC group, but with increased amidicity for amide-C. This intermediate **48** was transformed finally to the desired product **46** by a consecutive deprotection and reducing steps. This compound later proved to be an excellent building block in drug research. This method is not limited only to six-member diketopiperazines, but it can also be generalized for even open chain systems [101].

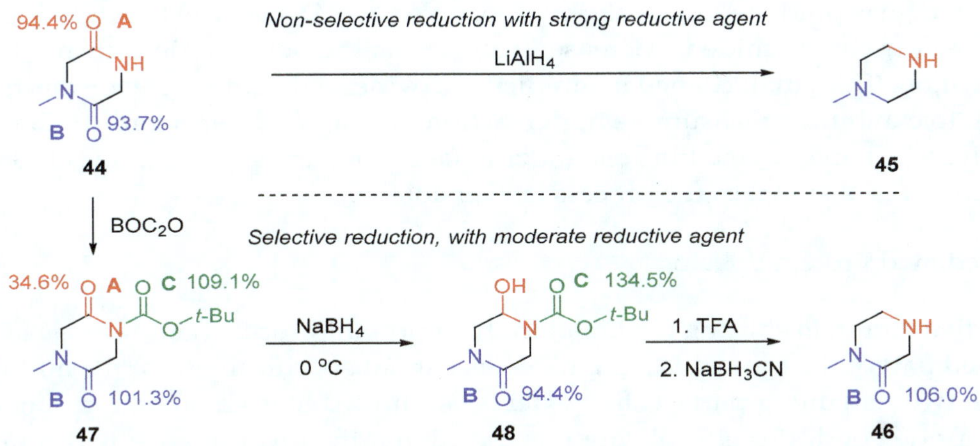

Figure 17. The non-selective and selective reduction of the amide bonds in a diketopiperazine derivative **44**, including amide-A and B, to yield **45** and **46**. The corresponding amidicity values were computed at B3LYP/6-31G(d,p) level of theory.

2.4. *Amide Reaction in the Biochemistry*

Natural products are very precisely assembled from selected molecular components in unique arrangements. They typically exhibit very effective and targeted mechanisms of actions in numerous biological processes. There is an observably large difference in complexity, activity and efficiency of human-designed compounds [102] and nature's biomolecules, such as dinucleotide coenzymes (NAD and FAD) [103], calicheamicin-γl [102,104,105], duocarmycin [106–108], syringolin A [18], aflatoxin B [18], and penicillin [109]. Biochemical processes also take advantage of the selectivity of acyl transfer reaction and amide activation, so three representative examples are presented in this section.

2.4.1. Cross-Linking in the Blood Clotting Process

The first example is taken from the multistep process of blood clotting [110–112]. In this case, in the final, thirteenth step (Figure 18) of the entire complex process, the two final protein intermediates **49** and **50** are linked to each other via a new side-chain amide bond [112].

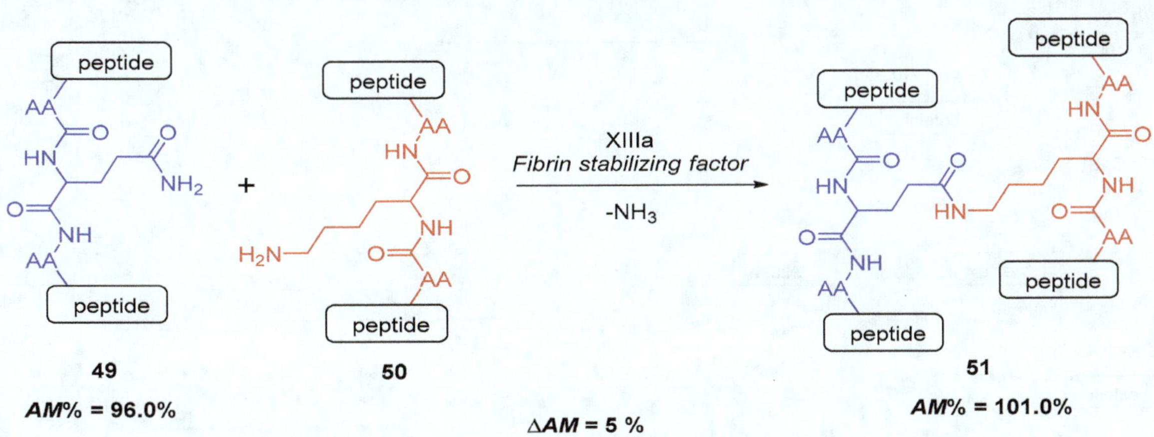

Figure 18. The final step of the blood clotting process from the aspect of amidicity change, catalyzed by the transamidinaze enzyme. The corresponding amidicity values were computed at B3LYP/6-31G(d,p) level of theory.

This creates a molecular polymeric net (**51**), which forms a barrier against bleeding. Analyzing the process from the aspect of amidicity values (Figure 18), this transamidation reaction is spontaneous. The process takes part between a glutamine side chain amide (blue) and a lysin side chain amine (red).

The amidicity change is predicted to be slightly positive (ΔAM), because the initial 96.0% amidicity of the glutamine side chain amide is increased to 101% as the new amide is formed, not requiring additional activation. This small change in amidicity provides a driving force for this bioreaction, but it is not enough to exhibit the measurable high reaction rate at body temperature. Consequently this process is catalyzed by an enzyme fibrin stabilizing factor or transglutaminase [112,113]. Moreover, the releasing ammonia also makes the process irreversible.

2.4.2. Intein-Mediated Protein Splicing

Intein-mediated protein splicing is a biologically important process (Figure 19) [114–117], where a small but defined part of the protein **52**, intein (**53**), is cut out specifically from the middle of a protein. Meanwhile, the two remaining parts, called exteins, are linked to each other, forming a new protein **54** via a new peptide bond. Protein splicing is so rapid that the precursor protein is rarely observed in native systems. It was shown that the process is auto-catalytic and folding-dependent [118–120]. The N-extein residues play unique and important roles in protein splicing [120]. The intein peptide sequence is supposed to contain no sufficient information originating from an external source [114–117].

From the chemical aspect, here two amides are involved in transamidation process rather than an amide and an amine shown previously [33]. The intein and the C-terminal extein residue contain sufficient information for splicing in proteins, which involves four basic chemical steps. In the first step of the mechanism [114–117], the serine or cysteine residue, at the N-terminal border of the intein, attacks the neighboring peptide bond forming an intermediate **55**, which in a subsequent step rearranges to **56** and **57**. In the final step, the intein leaves the native sequence resulting in the edited protein **54**. Taking into account only the starting and ending state, the overall process exhibits a large negative ΔAM value (−45 %), which may suppose an endothermic reaction. One may argue that the exothermic refolding of the instantly cut intein sequence provides additional driving force, covering the energy demand of the overall amidicically unfavorable process. Moreover, the N-extein-intein amide linkage is a typically distorted amide (blue in Figure 19.) due to the folding, resulting in lower amidicity, which could initiate the acyl migration (blue arrow).

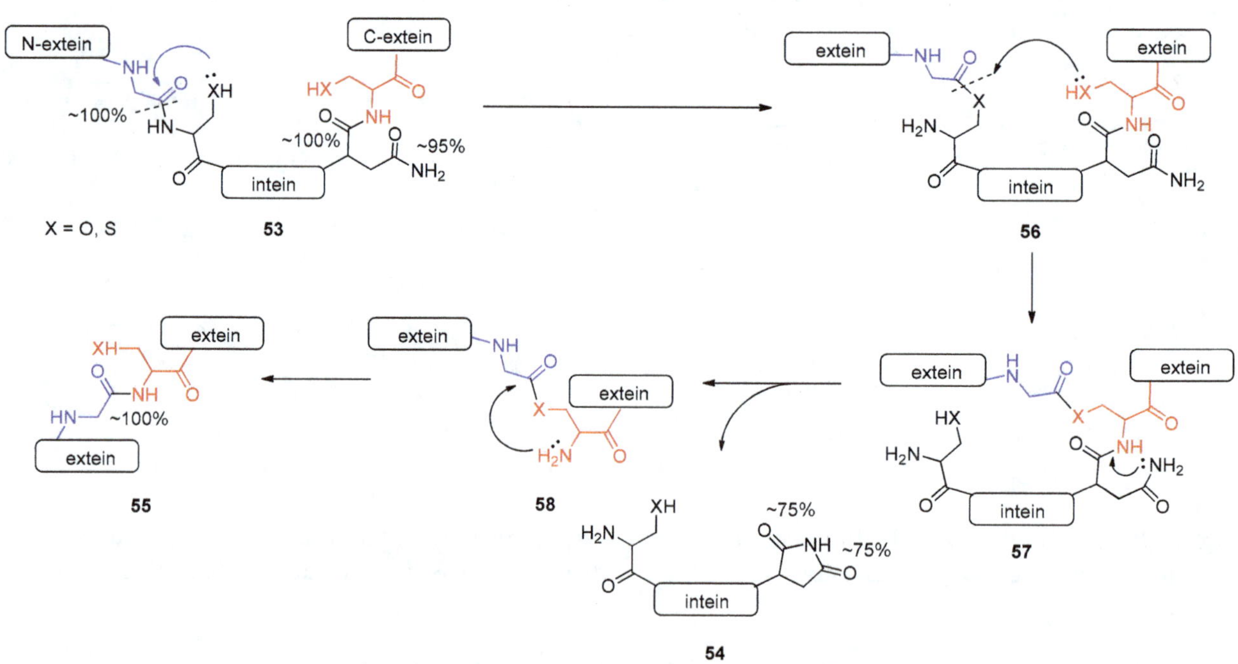

Figure 19. The mechanism of action of protein splicing as a special type of transamidation reaction.

2.4.3. Penicillin

One of the most important small biogen amides is penicillin (58), produced by the *Penicillium* genus, which is considered to be one of the great discoveries of the 20th century [121,122]. This antibiotic inhibits penicillin binding proteins, such as transpeptidase, by blocking their serine residue via acylation in gram-positive bacteria. In this way, bacterial cell wall synthesis stops, leading to deadly susceptibility to osmotic effects and cell bursting. This bacterial enzyme plays a crucial role in transferring the D-Ala-D-Ala dipeptide into the bacterial cell wall synthesis as illustrated by Figure 20. The perennial war between bacteria and fungus led to the evolution of β-lactam related antibiotics, and resulted in many variants, of which cephalosporine [123] and thienamycin [124,125] are just two examples.

During the last century, resistant bacteria strains have developed some defending mechanisms against intensively and overused β-lactam antibiotics. Consequently, antibiotic research is once again at the forefront of drug research. The β-lactamase enzyme is synthesized by resistant bacteria for deactivation of penicillin-type compounds in order to avoid their lethal effects [122,126–129]. Regardless, continued characterization of the β-lactam structure and mechanism aids in the understanding and design of novel antibiotics with desired effects and great precision [126].

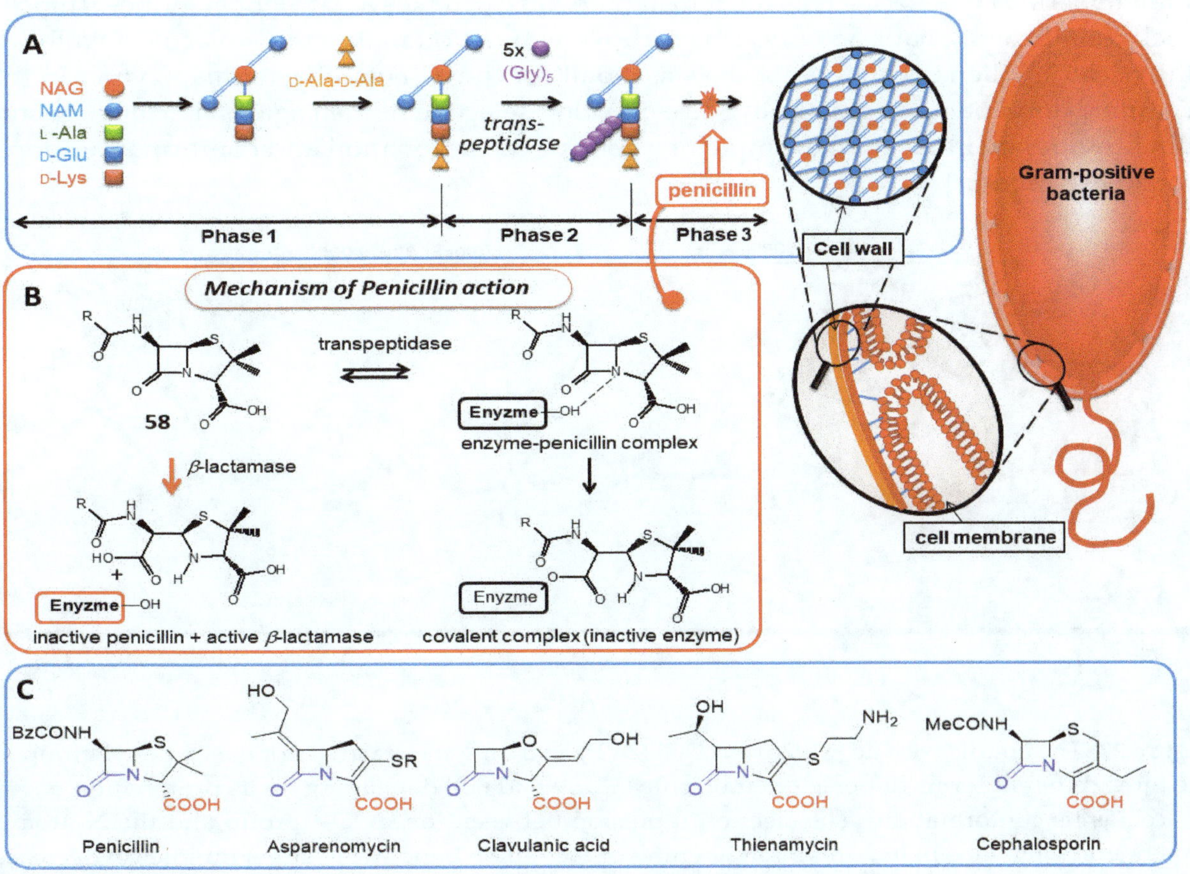

Figure 20. (**A**) The schematic mechanism of biological action of the penicillin antibiotic in the course of cell wall synthesis. (**B**) Penicillin inhibits transpeptidase enzyme by bonding covalently to the serine side-chain producing non-hydrolysable enzyme-substrate complex, blocking irreversibly the enzyme activity. (**C**) Some representative examples of β-lactam antibiotics. Their common feature is the strained β-lactam ring neighbouring a carboxylic functionate, which are both essential for the antibiotic activity. NAG = *N*-acetylglucoseamine; NAM = *N*-acetylmuramic acid.

Since the discovery of penicillin's chemical structure [18,129], the way this molecule avoids the aqueous hydrolytic effects in body fluid, while immediately awakening its hidden ability to acylate the transpeptidase's serine side-chain oxygen, has remained a mystery [130,131]. Recently, bacterial resistance has been investigated intensively, wherein penicillin hydrolysis was studied [131–135], and on the bases of this mechanism its activity was understood [132–134]. The "strange" fused β-lactam-type structure [132] was dedicated to explain its strength as an acylating agent completely, approximated earlier to that of acyl chlorides [133,134]. In organisms, however, such a strong acylating agent should decompose quickly before reaching its destination, e.g., the transpeptidase enzyme of the bacteria.

Previously, it was shown that the unexpected stability of penicillin can be attributed to its deprotonated form (**59** in Figure 21), which is the dominant component of the neutral aqueous environment [60]. According to the new, higher level quantum chemical calculations at MP2(full)/DGDZVP, the amidicity value [40] of the anion form was found to be higher (AM% = 71.7%) than the neutral form (**58**), referring to the more stable amide bond in the anionic form. However, when the penicillin reached its target enzyme, the transpeptidase, it gets protonated (**59** in Figure 21) by the enzyme. The forming neutral form of penicillin (**58**) immediately lowers its amidicity to 22.1%, so it has been triggered for the reaction. Finally, the H atom on the carboxylic acid group can also easily turn back to the pyramidaziled amid N atom, forming an H-bond. This conformational change turns on the final, superactive form of the molecule (**60** in Figure 21), reflected by the extremely low amidicity (AM% = −8.7%). This form instantaneously makes a covalent enzyme-drug complex, and knocks out the enzymatic activity. This carboxylic function makes the molecule a well-designed acylating bacteria killer, because it also acts as a bait for the essential bacterial enzyme. In this way, penam scaffold antibiotics are practically self-activating selective hitmen against bacteria. In summary, penicillin provides an interesting example of a controlled protonation and conformation-dependent amide bond activation.

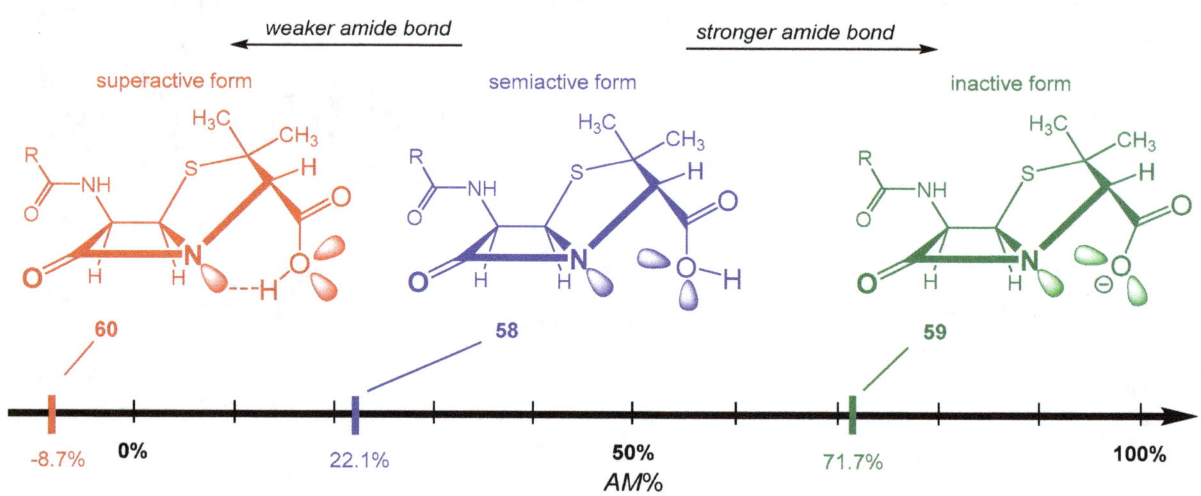

Figure 21. The amidicity scale [MP2(full)/DGDZVP], showing the reactivity of the carbonyl group for the three different forms of penicillin molecule (58, 59 and 60), depending on its protonation degree as well as its conformation. The electron repulsion between the COO− group and the N atom is illustrated by the lone pairs. This strengthening of the amide bond in 59 (larger amidicity) decreases reactivity with respect to 58. In form 60, the internal H-bond withdraws density from the amide bond, weakening it (extremely low amidicity) leading to extremely high reactivity toward nucleophiles.

2.5. Amide Activation in Excited State.

The electronic excitation of the chromophore functionality is able to transfer the energy toward a neighboring amide bond, resulting in an amide activation and in an increased reactivity of the amide bond, which is utilized for various processes [33]. Photocleavage (also known as uncaging) is based on this phenomenon, providing a methodology to release selectively a bioactive compound from its covalently bonded cage-form in a localized volume. Otherwise, the cage scaffold should effectively block the activity of the bonded biomolecule, but undergoes efficient removal after excitation of the moiety, through the scissile bond. Until now, numerous cage molecules have been developed (Figure 22); however, nitro-indoline and coumarine scaffolds are the most popular of these, and are usually linked with glutamate as excitatory neurotransmitter [136,137].

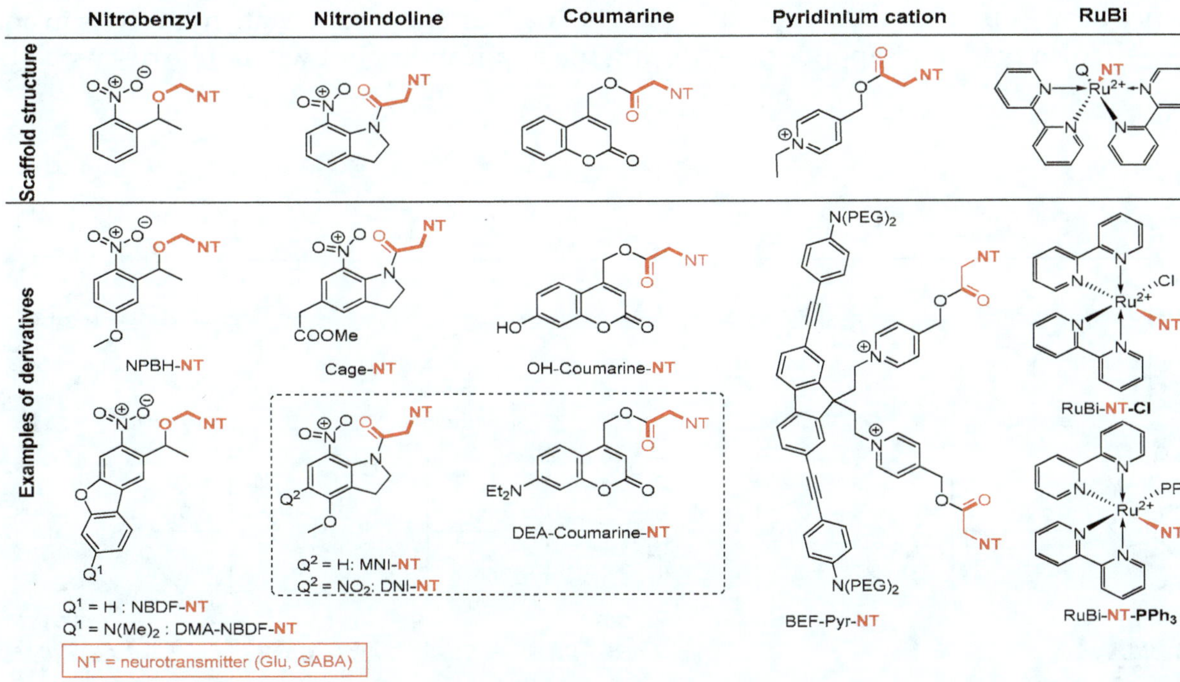

Figure 22. Collection of the most prominent cage-scaffolds (**top**) and some derivatives (**bottom**). NT = neurotransmitter. Compounds in the box are the subject of discussion.

2.5.1. Photochemical Release of Glutamate from MNI-Glu and DNI-Glu at Excited State

The most popular cage compounds, 5-mono-nitro-indoline (MNI) and 5,7-dinitro-indoline (DNI), have proved their effectiveness during the last decade and have been widely applied in numerous neuroscientific papers. Due to the large number of scientific results, in this paragraph we focus only on DNI-Glu and MNI-Glu [138]. MNI-Glu and DNI-Glu fulfill the main requirements for an ideal cage compound, including an efficient uncaging process with relatively high chemical yield, rapid two-photon induced release (quantum yield 8.5%) [139], low spontaneous hydrolysis rate, (Figure 23) and low biochemical side effects [140]. These properties were demonstrated to be useful in neurophysiological experiments.

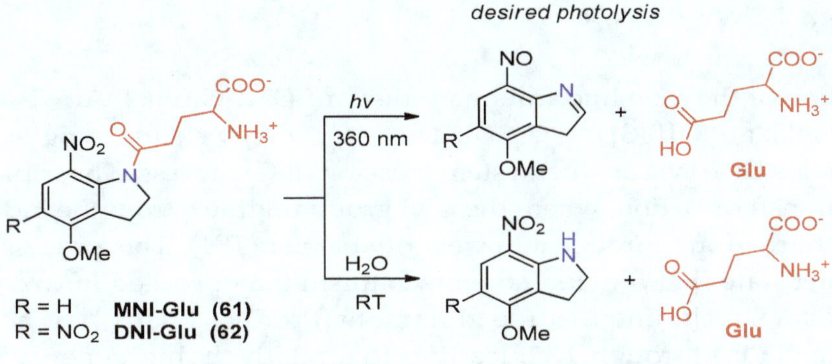

Figure 23. Chemical structures of MNI–Glu (R = H) and DNI-Glu (R = NO_2) and their photochemical reactivity (**top**) and hydrolytic reactivity (**bottom**).

The reaction mechanisms of the desired photochemical glutamate release of MNI-Glu (**61**) and DNI-Glu (**62**) were studied and compared on quantum chemical bases using GAUSSIAN 09 software [141] (Figures 24 and 25). Their reaction profile proved to be identical, but the energy values varied in a significant range (4–10 kJ/mol). Here, only the photochemical reaction of DNI-Glu (**62**) will be presented. The photocleavage of DNI-Glu (**62**) starts from the initial state A(S0) and follows the sequence of A → B → C → D → E → F → G. The inadvertent ground state hydrolysis in aqueous media (A → I) also plays an important role during the applications, but will not be discussed here.

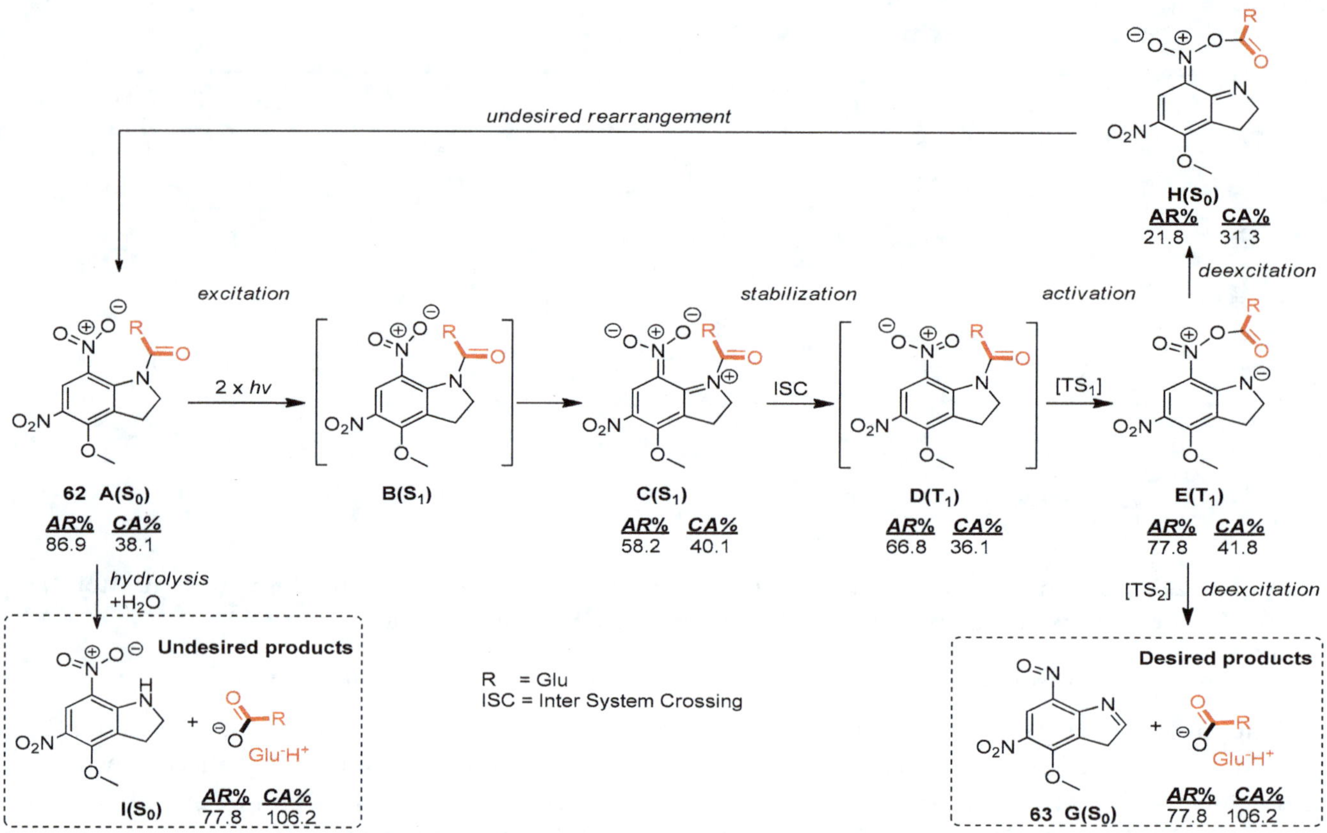

Figure 24. Quantum chemical modelling of the photochemical reactions as well as the simplified mechanisms of ground state hydrolysis for DNI-Glu (**62**). The aromaticity (AR%) of the benzene ring and carbonylicity of the carbonyl functionality along the mechanism are computed at B3LYP/6-31G(d,p)//PCM(water) level of theory [37].

After the excitation of the aromatic antenna moiety of DNI-Glu [**62**; $A(S_0) \rightarrow B(S_1)$] the structure follows a vibrational relaxation [$B(S_1) \rightarrow C(S_1)$]. The high energy excited singlet state $C(S_1)$ tends to transform to a triplet state $D(T_1)$ via an intersystem crossing (ISC) process. The triplet state intermediate takes part in an acyl transfer reaction, where the acyl group migrates from the indoline N atom to one of the O atoms of the nitro group through a low energy barrier (TS1). This process finally results in an unusually stable triplet state [$E(T_1)$]. Here, the mechanism may proceed in two different directions. The desired product state, including the free glutamate [Product 1; $G(S_0)$], is achieved by the N–O bond cleavage at stage $E(T_1)$ through TS2, leading to a relatively high energy state $F(T_1)$. Finally, the deexcitation to ground state [$G(S_0)$] results in the final products. However, due to the relative stability of the triplet state $E(T_1)$, it can undergo a deexcitation process to give a very reactive species $H(S_0)$, which can return to the starting state $A(S_0)$.

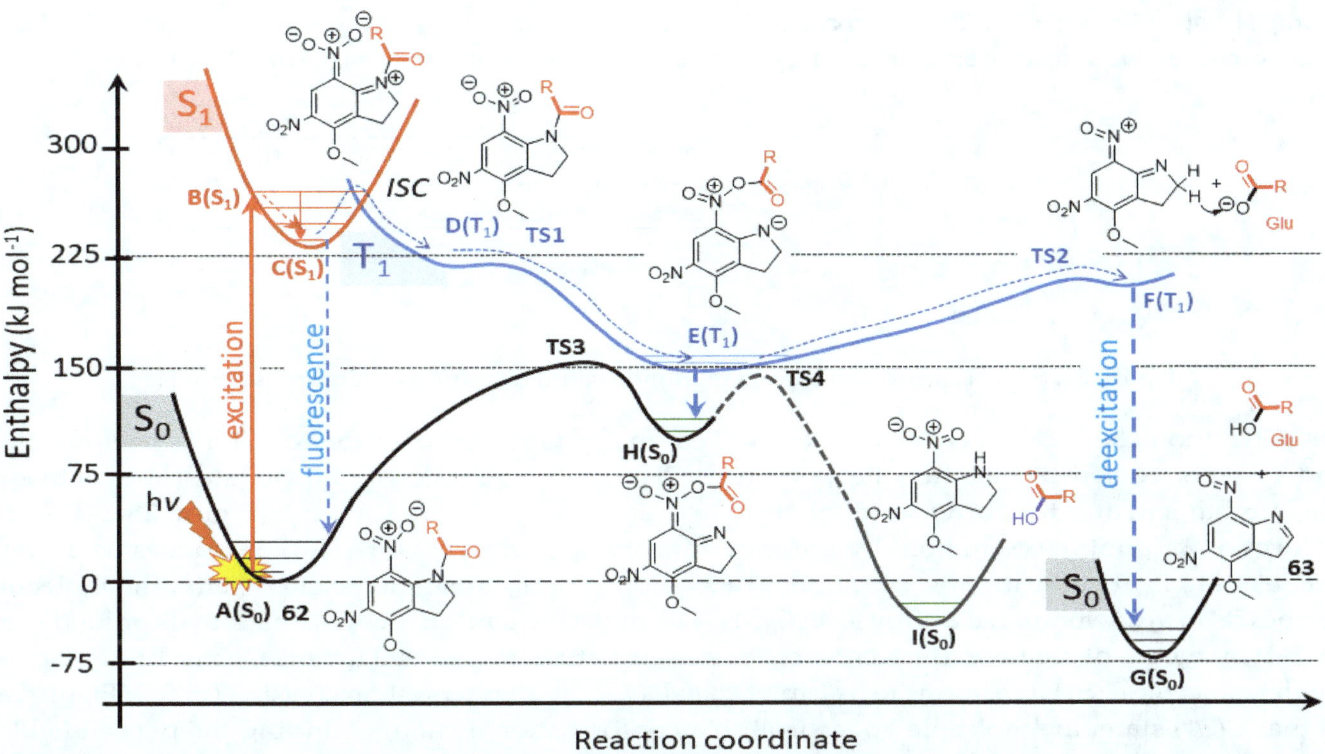

Figure 25. Quantum chemical modelling of the potential enthalpy profile of the singlet (red), triplet (blue) and ground state (black) mechanism for DNI-Glu (**62**) at B3LYP/6-31G(d,p)//PCM(water) level of theory. For detailed data see [138].

It is possible to explain the overall process by means of systems chemistry. Through the change in the aromaticity (AR%) [142] and carbonylicity (CA%) values [computed at B3LYP/6-31G(d,p)//PCM(water)] it is possible to understand how the electron distribution follows the energetic changes in the triplet excited as well as the ground state during the photochemical process. DNI-Glu (**62**) is composed of an aromatic ring and a carbonyl group (Figure 24), characterized by values of AR% and CA%. During the excitation [A(S_0)] → [C(S_1)] the aromaticity is significantly reduced (AR%: 86.9% → 58.2%), while the carbonylicity value remained relatively constant (CA%: 38.1% → 40.1%). After intersystem crossing (ISC) from singlet to triplet excited state [C(S_1)]→[D(T_1)], the molecule tries to regain its stability by slightly increasing its aromaticity (AR%: 58.2% → 66.8%) at the expense of carbonylicity (CA%: 40.1% → 36.1%). The reduced CA% in state D(T_1) indicates that the amid group is already triggered and favors the next intramolecular rearrangement. This results in a relatively stable triplet state molecule E(T_1) with an increased aromaticity (77.8%) as well as carbonylicity (41.8%). Subsequently, an endothermic cleavage of the N–O bond leads to stage F(T_1), then to the product **63** G(S_0), recovering finally both the AR% and CA% after a de-excitation. During the deexcitation from state E(T_1) to the ground state G(S_0), the rearranged molecule loses a lot of stabilization energy, represented by the sharp decreases of AR% (21.8%) and CA% (31.3%), summarized in Figure 24.

The photochemical uncaging process of DNI–Glu is an excellent example of how the electronic excitation of an aromatic moiety can activate a neighboring amide group, leading to an acyl transfer reaction.

2.5.2. Photochemical Release of Glutamate from Coumarin-Caged Transmitters at Excited State

The neurotransmitter derivatives of coumarin-4-yl-methyl have been developed as a new class of efficient caging scaffolds and were applied successfully to mask the biologically active amino, carbonyl [143] and hydroxyl compounds via carbamate [144–146] or carbonate [33] linkers.

Some phosphate [56,147–155], carboxylate [56,155,156], sulfate [154], sulfonate [154] and diol [157] functionalities were also been protected by various coumarin derivatives (Figure 26) [33].

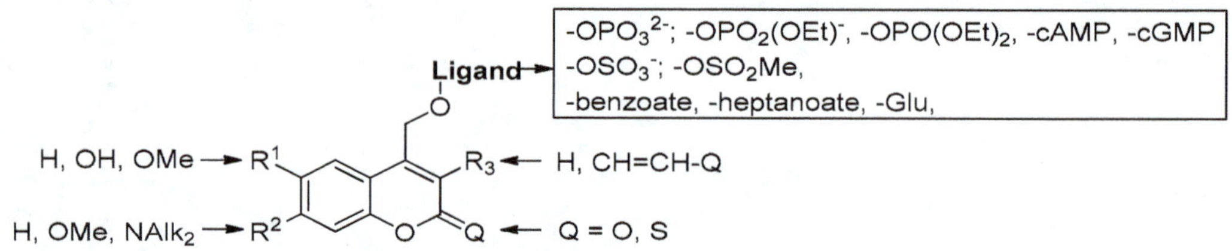

Figure 26. Some examples of cage compounds based on various coumarin scaffolds.

The photochemical mechanism of diethyl coumarin-Glu cage compound (**64**) has been studied and are reported here by quantum theory (Figure 27) at B3LYP/6-31G(d,p//PCM(water) level of theory. The mechanism and the corresponding enthalpy reaction profile are shown in Figure 28. After the absorption of photon (excitation) by the ground state of coumarin 64 [A(S_0)], it reaches its excited singlet state B(S1). During the vibrational relaxation process at singlet excited state, the molecule reaches its lowest vibrational energy level C(S1). The molecule tends to deexcite back to its ground state [A(S_0)] by means of emission (fluorescence) or non-radiative processes. However, some of the excited singlet molecule C(S1) transforms to its triplet state D(T1) via an intersystem crossing (ISC, in Figure 28). In that C(S1) state, the molecule goes through a conformational change, finding the perpendicular arrangement of the side chain carbamate to the aromatic ring as the energy optimum E(T_1). After a relatively high enthalpy TS [34.5 kJ mol^{-1}; F(T_1)], the CH_2–O bond is cleaved heterolytically, forming a triplet state coumarin cation (**65**, [G(T_1)]) and a ground state carbamate anion derivative (66, S_0). The formation of the real solvent-separated ions is an energy demanding process (estimated as high as +265.4 kJ mol^{-1}), so before the ion-pair separation, the positively charged coumarin scaffold deexcites to its ground state and reacts with a water molecule immediately, resulting in coumarin alcohol [66; S_0(H)]. At the same time, the carbamate derivative **66** yields the desired amino neurotransmitter (in the present cases, Me–NH_2), while it loses CO_2. Notably, the overall enthalpy change is endothermic with respect to the starting material in its S_0 state, however, the reaction is greatly exothermic with respect to the reactant first excited state S_1[B] or S_1[C].

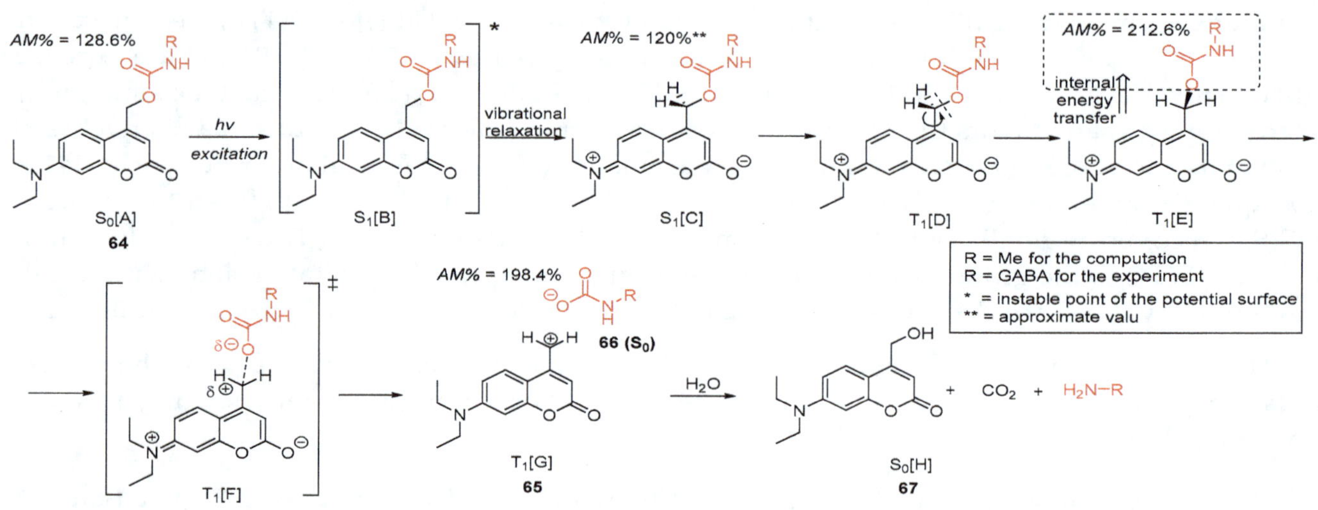

Figure 27. Mechanism of the photochemical initiated scission of coumarin-caged model compound **64** to products **66** + **67** by an internal energy transfer, computed at B3LYP/6-31G(d,p)//PCM(water) level of theory. The amidicity values (AM%) are also shown for some structures.

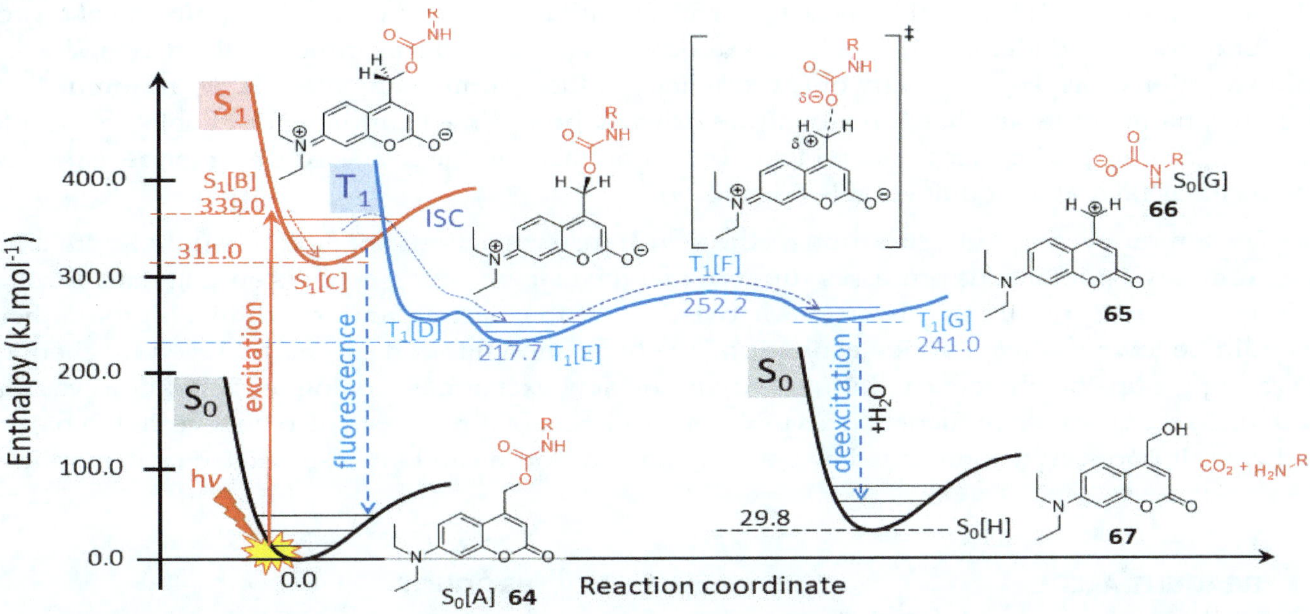

Figure 28. Reaction enthalpy (H in kJ mol^{-1}) profile of the photochemical initiated scission of coumarin-caged model compound **63** to **65** + **66**, computed at B3LYP/6-31G(d,p)//PCM(water) level of theory.

During the photochemical reaction, the original high amidicity value (AM% = 128.6%) remains nearly constant in its S_1 state, but in its subsequent triple state (T_1) its amidicity increased to an unprecedently high value (212.6%), which indicates that the carbamate functionality became a good leaving group, allowing the C–O dissociation. Finally, the forming of ground state carbamate **66** also exhibits a high value. It appears that there has been an internal energy transfer from the excited aromatic ring to the carbamate group. This means that amide, in fact, is stabilized and not activated, as might have been supposed from the study of DNI-Glu in the previous section.

3. Materials and Methods

All computations were carried out using the Gaussian09 program package (G09) [141]. Geometry optimizations and subsequent frequency analyses were carried out on selected amide-containing systems from which the values for the enthalpy of hydrogenation (ΔH_{H2}) were extracted. Computations were carried out at B3LYP/6-31G(d,p) level of theory [37]. Method and basis sets were chosen for their reliability in the characterization of amidicity [40] and carbonylicity [20], in agreement with works published earlier. The vibrational frequencies were computed at the same levels of theory as those used for geometry optimization, in order to properly confirm all structures as residing at minima on their potential energy hypersurfaces (PESs). In some cases, the acetonitrile or water solvent was considered by the default IEF-PCM (integral equation formalism polarizable continuum medium) method.

4. Conclusions

In general, amide groups have been considered to be very stable and functional for many years, due to their extensive conjugation of the π electron system consisting of the four π electrons. Due to this characteristic, amides can be found frequently in natural or human-made molecules as stable linkages between functioning building blocks, maintaining them for hundreds of years. During synthetic work, they are often and willingly used to inactivate a part of a given molecule as a protecting group. However, as was earlier revealed, the chemistry of the amide bond can be tricked, turning up their reactivity and forming a reactive functional group.

In this paper we showed that there is a semi-quantitative rule to predict the outcome of an acyl transfer or transamidation reaction. This thermodynamic selection rule indicates the driving force of amide reactions based on amidicity or carbonylicity values, numerically measuring the amide bond strengths, namely the stabilization enthalpies, toward providing a simple and reliable protocol for practicing chemists. The change of amidicity or carbonylicity in the course of a reaction revealed that the process is favorable or unfavorable (Figure 29).

The amidicity-based classification of the acyl transfer reactions was demonstrated with a few case examples. Spontaneous processes suppose that the reagent amide exhibits an originally lowered amidicity value, typically by distortion or twisting. If the starting amide exhibits high amidicity it should be lowered by an amide activation. We have shown that it can be achieved by chemical activating groups or photoexcitation, to initiate the acyl exchange. Analogues activation was also used for selective amide reduction. This ON/OFF function of amides can also be revealed in natural biochemical processes, demonstrated by penicillin, blood clotting and intein-mediated protein splicing.

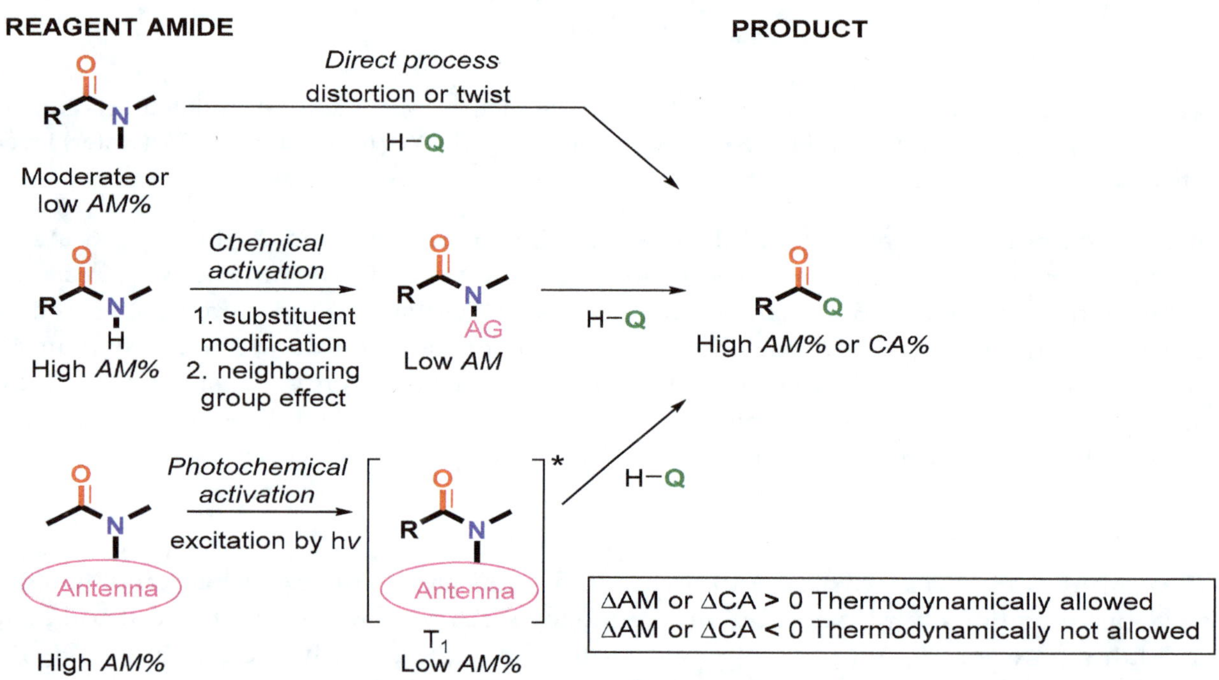

Figure 29. Summary of the amide activation processes. T_1 = triplet state; AM% = amidicity; CA% = carbonylicity.

Author Contributions: Author Contributions: conceptualization, E.K., A.C., I.G.C. and Z.M.; methodology, Z.M.; software, Z.M.; validation, E.K., A.C. and I.G.C.; formal analysis, E.K., A.C. and I.G.C.; investigation, Z.M.; writing—original draft preparation, Z.M.; writing—review and editing E.K., A.C., I.G.C. and Z.M.; visualization A.C. and Z.M.; supervision I.G.C. and B.R.; financing B.R.; project administration, Z.M.

References

1. Pattabiraman, V.R.; Bode, J.W. Rethinking amide bond synthesis. *Nature* **2011**, *480*, 471–479. [CrossRef] [PubMed]
2. Mahesh, S.; Tang, K.; Raj, M. Amide Bond Activation of Biological Molecules Sriram. *Molecules* **2018**, *23*, 2615. [CrossRef] [PubMed]
3. Figueiredo, R.M.; Suppo, J.; Campagne, J. Nonclassical Routes for Amide Bond Formation. *Chem. Rev.* **2016**, *116*, 12029–12122. [CrossRef] [PubMed]

4. Acosta-Guzmán, P.; Mateus-Gómez, A.; Gamba-Sánchez, D. Direct Transamidation Reactions: Mechanism and Recent Advances. *Molecules* **2018**, *23*, 2382. [CrossRef] [PubMed]
5. Valeur, E.; Bradley, M. Amide bond formation: Beyond the myth of coupling reagents. *Chem. Soc. Rev.* **2009**, *38*, 606–631. [CrossRef] [PubMed]
6. Greenberg, A.; Breneman, C.M.; Liebman, J.F. *The Amide Linkage: Structural Significance in Chemistry, Biochemistry, and Materials Science*; Wiley-VCH: New York, NY, USA, 2003; ISBN 978-0-471-35893-0.
7. Pauling, L.; Corey, R.B.; Branson, H.R. The structure of proteins: Two hydrogen-bonded helical configurations of the polypeptide chain. *Proc. Natl. Acad. Sci. USA* **1951**, *37*, 205–211. [CrossRef] [PubMed]
8. Hughes, A.B. (Ed.) *Amino Acids, Peptides and Proteins in Organic Chemistry*; Wiley-VCH: Weinheim, Germany, 2011; ISBN 978-3-527-32102-5.
9. Kaspar, A.A.; Reichert, J.M. Future directions for peptide therapeutics development. *Drug Discov. Today* **2013**, *18*, 807–817. [CrossRef] [PubMed]
10. Pauling, L. The nature of the chemical bond. Application of results obtained from the quantum mechanics and from a theory of paramagnetic susceptibility to the structure of molecules. *J. Am. Chem. Soc.* **1931**, *53*, 1367–1400. [CrossRef]
11. Pauling, L. *The Nature of the Chemical Bond*; Oxford University Press: London, UK, 1940.
12. Greenberg, A.; Breneman, C.M.; Liebman, J.F. (Eds.) *Amide Linkage: Structural Significance in Chemistry, Biochemistry, and Materials Science*; Wiley-VCH: New York, NY, USA, 2000.
13. Wiberg, K.B.; Breneman, C.M. Resonance Interactions in Acyclic Systems. 3. Formamide Internal Rotation Revisited. Charge and Energy Redistribution along the C-N Bond Rotational Pathway. *J. Am. Chem. Soc.* **1992**, *114*, 831–840. [CrossRef]
14. Glover, S.A.; Rosser, A.A. Reliable determination of amidicity in acyclic amides and lactams. *J. Org. Chem.* **2012**, *77*, 5492–5502. [CrossRef] [PubMed]
15. Van Vranken, D.; Weiss, G.A. *Introduction to Bioorganic Chemistry and Chemical Biology*; Garland Science (Taylor & Francis): Oxford, UK, 2013; ISBN 978-0815342144.
16. Brix, K.; Stöcker, W. *Proteases: Structure and Function*; Brix, K., Stöcker, W., Eds.; Springer: Berlin, Germany, 2013; ISBN 978-3-7091-0885-7.
17. Lukes, R. Sur une nouvelle application de la règle de bredt. *Collect. Czechslovak Chem. Commun.* **1938**, *10*, 148–152. [CrossRef]
18. Woodward, R.B.; Neuberger, A.; Trenner, N.R. *The Chemistry of Penicillin*; Clarke, H.T., Johnson, J.R., Robinson, R., Eds.; Princeton University Press: New Jersey, NJ, USA, 1949.
19. Wasserman, H.H. Chemistry: Synthesis with a twist. *Nature* **2006**, *441*, 699–700. [CrossRef] [PubMed]
20. Mucsi, Z.; Chass, G.A.; Viskolcz, B.; Csizmadia, I.G. Quantitative scale for the extent of conjugation of carbonyl groups: "Carbonylicity" percentage as a chemical driving force. *J. Phys. Chem. A* **2008**, *112*, 9153–9165. [CrossRef] [PubMed]
21. Kemnitz, C.R.; Loewen, M.J. "Amide resonance" correlates with a breadth of C-N rotation barriers. *J. Am. Chem. Soc.* **2007**, *129*, 2521–2528. [CrossRef] [PubMed]
22. Mujika, J.I.; Matxain, J.M.; Eriksson, L.A.; Lopez, X. Resonance structures of the amide bond: The advantages of planarity. *Chem.-Eur. J.* **2006**, *12*, 7215–7224. [CrossRef] [PubMed]
23. Bechara, W.S.; Pelletier, G.; Charette, A.B. Chemoselective synthesis of ketones and ketimines by addition of organometallic reagents to secondary amides. *Nat. Chem.* **2012**, *4*, 228–234. [CrossRef] [PubMed]
24. Mujika, J.I.; Mercero, J.M.; Lopez, X. Water-promoted hydrolysis of a highly twisted amide: Rate acceleration caused by the twist of the amide bond. *J. Am. Chem. Soc.* **2005**, *127*, 4445–4453. [CrossRef] [PubMed]
25. Wang, B.; Cao, Z. Acid-catalyzed reactions of twisted amides in water solution: Competition between hydration and hydrolysis. *Chem.-Eur. J.* **2011**, *17*, 11919–11929. [CrossRef] [PubMed]
26. Cho, S.J.; Cui, C.; Lee, J.Y.; Park, J.K.; Suh, S.B.; Park, J.; Kim, B.H.; Kim, K.S. N-Protonation vs O-Protonation in Strained Amides: *Ab Initio* Study. *J. Org. Chem.* **1997**, *62*, 4068–4071. [CrossRef]
27. Mucsi, Z.; Tsai, A.; Szori, M.; Chass, G.A.; Viskolcz, B.; Csizmadia, I.G. A Quantitative Scale for the Extent of Conjugation of the Amide Bond. Amidity Percentage as a Chemical Driving Force. *J. Phys. Chem. A* **2007**, *111*, 13245–13254. [CrossRef] [PubMed]
28. Morgan, J.; Greenberg, A.; Liebman, J.F. Paradigms and paradoxes: O- and N-protonated amides, stabilization energy, and resonance energy. *Struct. Chem.* **2012**, *23*, 197–199. [CrossRef]

29. Morgan, J.; Greenberg, A. Novel bridgehead bicyclic lactams: (A) Molecules predicted to have O-protonated and N-protonated tautomers of comparable stability; (B) hyperstable lactams and their O-protonated tautomers. *J. Chem. Thermodyn.* **2014**, *73*, 206–212. [CrossRef]
30. Tinnis, F.; Volkov, A.; Slagbrand, T.; Adolfsson, H. Chemoselective Reduction of Tertiary Amides under Thermal Control: Formation of either Aldehydes or Amines. *Angew. Chem. Int. Ed.* **2016**, *55*, 4562–4566. [CrossRef] [PubMed]
31. Szostak, M.; Spain, M. Highly Chemoselective Reduction of Amides (Primary, Secondary, Tertiary) to Alcohols using SmI2/Amine/H_2O under Mild Conditions. *J. Am. Chem. Soc.* **2014**, *136*, 2268–2271. [CrossRef] [PubMed]
32. Thakkar, B.; Svendsen, J.; Engh, R. Density Functional Studies on Secondary Amides: Role of Steric Factors in Cis/Trans Isomerization. *Molecules* **2018**, *23*, 2455. [CrossRef] [PubMed]
33. Klán, P.; Šolomek, T.; Bochet, C.G.; Blanc, A.; Givens, R.; Rubina, M.; Popik, V.; Kostikov, A.; Wirz, J. Photoremovable protecting groups in chemistry and biology: Reaction mechanisms and efficacy. *Chem. Rev.* **2013**, *113*, 119–191. [CrossRef] [PubMed]
34. Ellis-Davies, G.C.R. Caged compounds: Photorelease technology for control of cellular chemistry and physiology. *Nat. Methods* **2007**, *4*, 619–628. [CrossRef] [PubMed]
35. Mayer, G.; Hechel, A. Biologically active molecules with a "light switch". *Angew. Chem. Int. Ed.* **2006**, *45*, 4900–4921. [CrossRef] [PubMed]
36. Kotsuki, H.; Iwasaki, M.; Nishizawa, H. A new powerful method for the transformation of lactams into ω-amino-carboxamides under high pressure conditions. *Tetrahedron Lett.* **1992**, *33*, 4945–4948. [CrossRef]
37. Becke, A.D. Density-functional thermochemistry. III. The role of exact exchange. *J. Chem. Phys.* **1993**, *98*, 5648–5652. [CrossRef]
38. Greenberg, A.; Moore, D.T.; DuBois, T.D. Small and medium-sized bridgehead bicyclic lactams: A systematic ab initio molecular orbital study. *J. Am. Chem. Soc.* **1996**, *118*, 8658–8668. [CrossRef]
39. Greenberg, A.; Venanzi, C.A. Structures and Energetics of Two Bridgehead Lactams and Their N- and O-Protonated Forms: An ab Initio Molecular Orbital Study. *J. Am. Chem. Soc.* **1993**, *115*, 6951–6957. [CrossRef]
40. Mucsi, Z.; Chass, G.A.; Csizmadia, I.G. Amidicity change as a significant driving force and thermodynamic selection rule of transamidation reactions. A synergy between experiment and theory. *J. Phys. Chem. B* **2008**, *112*, 7885–7893. [CrossRef] [PubMed]
41. Lukács, A.; Szabó, L.; Hazai, L.; Szántay, C.; Mák, M.; Gorka, Á.; Szántay, C. Seven-membered ring formation through Grewe-cyclization. *Tetrahedron* **2001**, *57*, 5843–5850. [CrossRef]
42. Sibi, M.P.; Hasegawa, H.; Ghorpade, S.R. A convenient method for the conversion of N-acyloxazolidinones to hydroxamic acids. *Org. Lett.* **2002**, *4*, 3343–3346. [CrossRef] [PubMed]
43. Galat, A.; Elion, G. The Interaction of Amides with Amines: A General Method of Acylation. *J. Am. Chem. Soc.* **1943**, *65*, 1566–1567. [CrossRef]
44. Sowa, F.J.; Nieuwland, J.A. Organic Reactions with Boron Fluoride. XIV. The Reaction of Amides with Acids and Amines. *J. Am. Chem. Soc.* **1937**, *59*, 1202–1203. [CrossRef]
45. Varga, T.R.; Nemes, P.; Mucsi, Z.; Scheiber, P. A concise synthetic pathway towards 5-substituted indolizidines. *Tetrahedron Lett.* **2007**, *48*, 1159–1161. [CrossRef]
46. Wright, W.; Brabander, H.; Hardy, R. Notes- The Rearrangement of N-(Methylaminoalkyl)anilides. *J. Org. Chem.* **1961**, *26*, 2120–2123. [CrossRef]
47. Kano, S.; Ebata, T.; Shibuya, S. Intra- and Intermolecular Nucleophilic Cleavage of the Amide Bond of b-Laactams. *Chem. Pharm. Bull.* **1979**, *27*, 2450–2455. [CrossRef]
48. Wade, P.C.; Vogt, B.R.; Toeplitz, B.; Puar, M.S.; Gougoutas, J.Z. 1,2,4-Triazolo- and 1,2,5-triazino[4,3-d][1,4]benzodiazepinone ring systems: synthesis and barrier to ring inversion. *J. Org. Chem.* **1979**, *44*, 88–99. [CrossRef]
49. Kano, S.; Ebata, T.; Denta, Y.; Hibino, S.; Shibuya, S. Formtion of Some Heterocycles through Ring Transformation of 1-Arylazetidin-2-ones. *Heterocycles* **1977**, *8*, 411–416. [CrossRef]
50. Hie, L.; Fine Nathel, N.F.; Shah, T.K.; Baker, E.L.; Hong, X.; Yang, Y.F.; Liu, P.; Houk, K.N.; Garg, N.K. Conversion of amides to esters by the nickel-catalysed activation of amide C-N bonds. *Nature* **2015**, *524*, 79–83. [CrossRef] [PubMed]

51. Itai, A.; Toriumi, Y.; Tomioka, N.; Kagechika, H.; Azumaya, I.; Shudo, K. Stereochemistry of N-methylbenzanilide and benzanilide. *Tetrahedron Lett.* **1989**, *30*, 6177–6180. [CrossRef]
52. Jin, Y.; Chen, M.; Ge, S.; Hartwig, J.F. Palladium-Catalyzed, Enantioselective α-Arylation of α-Fluorooxindoles. *Org. Lett.* **2017**, *19*, 1390–1393. [CrossRef] [PubMed]
53. Di Gregorio, G.; Mari, M.; Bartolucci, S.; Bartoccini, F.; Piersanti, G. Divergent reactions of oxindoles with amino alcohols: Via the borrowing hydrogen process: Oxindole ring opening vs. C3 alkylation. *Org. Chem. Front.* **2018**, *5*, 1622–1627. [CrossRef]
54. Reverdito, A.M.; Orelli, L.; Dalmaso, M.; Perillo, I.; Fernández, B.M. Synthesis and hydrolysis of substituted tetrahydropyrimidinium salts. Behaviour of the degradation products on varying pH. *J. Heterocycl. Chem.* **1991**, *28*, 273–281. [CrossRef]
55. Milos, S.; Hejtmánková, L.; Hanusek, J.; Machác, V. Synthesis and Ring Transformation of Substituted S-(1-Phenylpyrrolidine-2-ones-3-yl) isothiuronium Salts to Substituted. *J. Heterocycl. Chem.* **2002**, *39*, 1105–1107. [CrossRef]
56. Schade, B.; Hagen, V.; Schmidt, R.; Herbrich, R.; Krause, E.; Eckardt, T.; Bendig, J. Deactivation Behavior and Excited-State Properties of (Coumarin-4-yl)methyl Derivatives. 1. Photocleavage of (7-Methoxycoumarin-4-yl)methyl-Caged Acids with Fluorescence Enhancement. *J. Org. Chem.* **1999**, *64*, 9109–9117. [CrossRef]
57. Liu, C.; Szostak, M. Twisted Amides: From Obscurity to Broadly Useful Transition-Metal-Catalyzed Reactions by N–C Amide Bond Activation. *Chem.-Eur. J.* **2017**, *23*, 7157–7173. [CrossRef] [PubMed]
58. Borthwick, A.D.; Davies, D.E.; Exall, A.M.; Livermore, D.G.; Sollis, S.L.; Nerozzi, F.; Allen, M.J.; Perren, M.; Shabbir, S.S.; Woollard, P.M.; et al. 2,5-Diketopiperazines as potent, selective, and orally bioavailable oxytocin antagonists. 2. Synthesis, chirality, and pharmacokinetics. *J. Med. Chem.* **2005**, *48*, 6956–6969. [CrossRef] [PubMed]
59. Matulenko, M.A.; Lee, C.H.; Jiang, M.; Frey, R.R.; Cowart, M.D.; Bayburt, E.K.; DiDomenico, S.; Gfesser, G.A.; Gomtsyan, A.; Guo, Z.Z.; et al. 5-(3-Bromophenyl)-7-(6-morpholin-4-ylpyridin-3-yl)pyrido[2,3-d] pyrimidin-4-ylamine: Structure-activity relationships of 7-substituted heteroaryl analogs as non-nucleoside adenosine kinase inhibitors. *Bioorg. Med. Chem.* **2005**, *13*, 3705–3720. [CrossRef] [PubMed]
60. Mucsi, Z.; Chass, G.A.; Ábrányi-Balogh, P.; Jójárt, B.; Fang, D.C.; Ramirez-Cuesta, A.J.; Viskolcz, B.; Csizmadia, I.G. Penicillin's catalytic mechanism revealed by inelastic neutrons and quantum chemical theory. *Phys. Chem. Chem. Phys.* **2013**, *15*, 20447–20455. [CrossRef] [PubMed]
61. Blackburn, G.M.; Skaife, C.J.; Kay, I.T. Strain effects in acyl transfer reactions. *J. Chem. Res. Synopses* **1980**, *9*, 294–295.
62. Wang, Q.P.; Bennet, A.J.; Brown, R.S.; Santarsiero, B.D. Distorted amides as models for activated peptide N-C=O units. 2. The synthesis, hydrolytic profile, and molecular structure of 3,4-dihydro-2-oxo-1,4-propanoquinoline. *Can. J. Chem.* **1990**, *68*, 1732–1739. [CrossRef]
63. Wang, Q.P.; Bennet, A.J.; Brown, R.S.; Santarsiero, B.D. Distorted Amides as Models for Activated Peptide N-C(O) Units. 3. Synthesis, Hydrolytic Profile, and Molecular Structure of 2,3,4,5-Tetrahydro-2-oxo-l,5-propanobenzazepine. *J. Am. Chem. Soc.* **1991**, *113*, 5757–5765. [CrossRef]
64. Somayaji, V.; Brown, R.S. Distorted Amides as Models for Activated Peptide N—C=0 Units Produced during Enzyme-Catalyzed Acyl Transfer Reactions. 1. The Mechanism of Hydrolysis of 3,4-Dihydro-2-oxo-l,4-ethanoquinoline and 2,3,4,5-Tetrahydro-2-oxo-l,5-ethanobenzazepine. *J. Org. Chem.* **1986**, *51*, 2676–2686. [CrossRef]
65. Tani, K.; Stoltz, B.M. Synthesis and structural analysis of 2-quinuclidonium tetrafluoroborate. *Nature* **2006**, *441*, 731–734. [CrossRef] [PubMed]
66. Kirby, A.J.; Komarov, I.V.; Feeder, N. Synthesis, structure and reactions of the most twisted amide. *J. Chem. Soc. Perkin Trans. 2* **2001**, 522–529. [CrossRef]
67. Kirby, A.J.; Komarov, I.V.; Feeder, N. Spontaneous, millisecond formation of a twisted amide from the amino acid, and the crystal structure of a tetrahedral intermediate [1]. *J. Am. Chem. Soc.* **1998**, *120*, 7101–7102. [CrossRef]
68. Kirby, A.J.; Komarov, I.V.; Wothers, P.D.; Feeder, N. The Most Twisted Amide: Structure and Reactions. *Angew. Chem. Int. Ed.* **1998**, *37*, 785–786. [CrossRef]
69. Andersson, P.G.; Munslow, I.J. (Eds.) *Modern Reduction Methods*; Wiley-VCH: New York, NY, USA, 2008; ISBN 978-3-527-31862-9.

70. Levkoeva, E.I.; Nikitskaya, E.S.; Yakhontov, L.N. Reaction of 2-quinuclidones with and without C-N bond breaking. *Dokl. Akad. Nauk* **1970**, *192*, 342.
71. Von Pracejus, H.; Kehlen, M.; Kehlen, H.; Matschiner, H. Neues zur sterischen mesomeriehinderung bei lactamen vom typ des α-chinuclidons. *Tetrahedron* **1965**, *21*, 2257–2270. [CrossRef]
72. Werstiuk, N.H.; Brown, R.S.; Wang, Q. An AM1 calculational study of the protonation and reactions of 3,4-dihydro-2-oxo-1,4-ethanoquinoline, 3,4-dihydro-2-oxo-1,4-propanoquinoline, 3,3,4,5-tetrahydro-2-oxo-1,5-ethanobenzazepine, 3,3,4,5-tetrahydro-2-oxo-1,5-propanobenzazepine, and N-methyl-4-b. *Can. J. Chem.* **1996**, *74*, 524–532. [CrossRef]
73. Hu, F.; Lalancette, R.; Szostak, M. Structural Characterization of N-Alkylated Twisted Amides: Consequences for Amide Bond Resonance and N-C Cleavage. *Angew. Chem. Int. Ed.* **2016**, *55*, 5062–5066. [CrossRef] [PubMed]
74. Hutchby, M.; Houlden, C.E.; Gair Ford, J.; Tyler, S.N.G.; Gagné, M.R.; Lloyd-Jones, G.C.; Booker-Milburn, K.I. Hindered ureas as masked isocyanates: Facile carbamoylation of nucleophiles under neutral conditions. *Angew. Chem. Int. Ed.* **2009**, *48*, 8721–8724. [CrossRef] [PubMed]
75. Hutchby, M.; Houlden, C.E.; Haddow, M.F.; Tyler, S.N.G.; Lloyd-Jones, G.C.; Booker-Milburn, K.I. Switching pathways: Room-temperature neutral solvolysis and substitution of amides. *Angew. Chem. Int. Ed.* **2012**, *51*, 548–551. [CrossRef] [PubMed]
76. Somayaji, V.; Skorey, K.I.; Brown, R.S. Molecular Structure of and Its Reaction with Alcohols as a Model for the Acylation Step of the Serine Proteases. *J. Org. Chem.* **1986**, *51*, 4866–4872. [CrossRef]
77. Garcia, J.; González, J.; Segura, R.; Urpí, F.; Vilarrasa, J. Reaction of N-Nitroso- and N-Nitro-N-alkylamides with Amines. *J. Org. Chem.* **1984**, *49*, 3322–3327. [CrossRef]
78. Garcia, J.; Vilarrasa, J. New synthetic "tricks" using old reagents. A mild method for the conversion of RCONHR′ to RCONHR″. *Tetrahedron Lett.* **1982**, *23*, 1127–1128. [CrossRef]
79. Hendrickson, J.B.; Bergeron, R. Trifumides: New Acylating Ard Triflating Reagents. *Tetrahedron Lett.* **1973**, *14*, 4607–4610. [CrossRef]
80. Bon, E.; Bigg, D.C.H.; Bertrand, G. Aluminum Chloride-Promoted Transamidation Reactions. *J. Org. Chem.* **1994**, *59*, 4035–4036. [CrossRef]
81. Shah, K.R.; Blanton, C.D.W. Reaction of Maleimides and Ethyl 3-Aminocrotonates. A Reinvestigation Leading to an Improved Synthesis of Pyrrolo[3,4-c]pyridines. *J. Org. Chem.* **1982**, *47*, 502–508. [CrossRef]
82. Davidsen, S.K.; May, P.D.; Summers, J.B. Di-tert-butyl N-Acylimidodicarbonates as Isolable Acylating Agents: Mild Conversion of Primary Carboxamides to Substituted Amides. *J. Org. Chem.* **1991**, *56*, 5482–5485. [CrossRef]
83. Shapiro, G.; Marzi, M. Facile and Selective O-Alkyl Transesterification of Primary Carbamates with Titanium(IV) Alkoxides. *J. Org. Chem.* **1997**, *62*, 7096–7097. [CrossRef] [PubMed]
84. Goodman, C.A.; Eagles, J.B.; Rudahindwa, L.; Hamaker, C.G.; Hitchcock, S.R. Synthesis, X-ray crystallography, and reactions of N-acyl and N-carbamoyl succinimides. *Synth. Commun.* **2013**, *43*, 2155–2164. [CrossRef]
85. Goodman, C.A.; Hamaker, C.G.; Hitchcock, S.R. Synthesis and evaluation of some variants of the Nefkens' reagent. *Tetrahedron Lett.* **2013**, *54*, 6012–6014. [CrossRef]
86. Ji, C.-L.; Xie, P.-P.; Hong, X. Computational Study of Mechanism and Thermodynamics of Ni/IPr-Catalyzed Amidation of Esters. *Molecules* **2018**, *23*, 2681. [CrossRef] [PubMed]
87. Li, Z.; Zhang, S.L.; Fu, Y.; Guo, Q.X.; Liu, L. Mechanism of Ni-catalyzed selective C-O bond activation in cross-coupling of aryl esters. *J. Am. Chem. Soc.* **2009**, *131*, 8815–8823. [CrossRef] [PubMed]
88. Yu, H.; Fu, Y. Mechanistic origin of cross-coupling selectivity in Ni-catalysed Tishchenko reactions. *Chem.-Eur. J.* **2012**, *18*, 16765–16773. [CrossRef] [PubMed]
89. Stach, H.; Hesse, M. Synthesis of macrocyclic compounds by ring enlargement. *Tetrahedron* **1988**, *44*, 1573–1590. [CrossRef]
90. Süsse, M.; Hájícek, J.; Hesse, M. Studies on the Carbon Zip Reaction of 2-Oxocycloalkane-l-carbonitriles. *Helv. Chim. Acta* **1985**, *68*, 1986–1997. [CrossRef]
91. Askitoglu, E.; Guggisberg, A.; Hesse, M. N-substituted 3-aminopropanenitriles and 2-aminoacetonitriles asSchiff-base equivalents. *Helv. Chim. Acta* **1985**, *68*, 750–759. [CrossRef]
92. Seyden-Penne, J. *Reductions by the Alumino and Borohydrides in Organic Synthesis*; Wiley-VCH: New York, NY, USA, 1997; ISBN 978-0-471-19036-3.

93. Hudlicky, M. (Ed.) *Reductions in Organic Chemistry*; John Wiley & Sons: New York, NY, USA, 1984.
94. Trost, B.M.; Fleming, I. (Eds.) *Comprehensive Organic Synthesis*; Pergamon: Oxford, UK, 1991.
95. Gribble, G.W. Sodium borohydride in carboxylic acid media: A phenomenal reduction system. *Chem. Soc. Rev.* **1998**, *27*, 395. [CrossRef]
96. Reeves, J.T.; Tan, Z.; Marsini, M.A.; Han, Z.S.; Xu, Y.; Reeves, D.C.; Lee, H.; Lu, B.Z.; Senanayake, C.H. A practical procedure for reduction of primary, secondary and tertiary amides to amines. *Adv. Synth. Catal.* **2013**, *355*, 47–52. [CrossRef]
97. Simmons, B.J.; Hoffmann, M.; Hwang, J.; Jackl, M.K.; Garg, N.K. Nickel-Catalyzed Reduction of Secondary and Tertiary Amides. *Org. Lett.* **2017**, *19*, 1910–1913. [CrossRef] [PubMed]
98. Ortín, I.; González, J.F.; de la Cuesta, E.; Avendaño, C. Synthesis of a novel tetrahydroisoquinoline pentacyclic framework. *Tetrahedron* **2009**, *65*, 2201–2211. [CrossRef]
99. Ortín, I.; González, J.F.; de la Cuesta, E.; Manguan-García, C.; Perona, R.; Avendaño, C. Pyrazino[1,2-b]isoquinolines: Synthesis and study of their cytostatic and cytotoxic properties. *Bioorganic Med. Chem.* **2008**, *16*, 9065–9078. [CrossRef] [PubMed]
100. Valente, M.W.N.; Williams, R.M. The Concise and Versatile Synthesis of Epi-malbrancheamide and Structurally Related Analogs. *Heterocycles* **2006**, *70*, 249–259. [CrossRef]
101. Lee, B.H.; Clothier, M.F. Selective reduction of secondary amides to amines in the presence of tertiary amides. *Tetrahedron Lett.* **1999**, *40*, 643–644. [CrossRef]
102. Zein, N.; Sinha, A.M.; McGahren, W.J.; Ellestad, G.A. Calicheamicin gamma 1I: An antitumor antibiotic that cleaves double-stranded DNA site specifically. *Science* **1988**, *240*, 1198. [CrossRef] [PubMed]
103. Zein, N.; Poncin, M.; Nilakantan, R.; Ellestad, G. Calicheamicin gamma 1I and DNA: Molecular recognition process responsible for site-specificity. *Science* **1989**, *244*, 697–699. [CrossRef] [PubMed]
104. Boger, D.L.; Garbaccio, R.M. Shape-dependent catalysis: Insights into the source of catalysis for the CC-1065 and duocarmycin DNA alkylation reaction. *Acc. Chem. Res.* **1999**, *32*, 1043–1052. [CrossRef]
105. Boger, D.L.; Johnson, D.S. CC-1065 and the duocarmycins: Understanding their biological function through mechanistic studies. *Angew. Chem. Int. Ed.* **1996**, *35*, 1438–1474. [CrossRef]
106. Boger, D.L.; Boyce, C.W.; Garbaccio, R.M.; Goldberg, J.A. CC-1065 and the duocarmycins: Synthetic studies. *Chem. Rev.* **1997**, *97*, 787–828. [CrossRef] [PubMed]
107. Groll, M.; Schellenberg, B.; Bachmann, A.S.; Archer, C.R.; Huber, R.; Powell, T.K.; Lindow, S.; Kaiser, M.; Dudler, R. A plant pathogen virulence factor inhibits the eukaryotic proteasome by a novel mechanism. *Nature* **2008**, *452*, 755–758. [CrossRef] [PubMed]
108. Crawford, J.M.; Thomas, P.M.; Scheerer, J.R.; Vagstad, A.L.; Kelleher, N.L.; Townsend, C.A. Deconstruction of iterative multidomain polyketide synthase function. *Science* **2008**, *320*, 243–246. [CrossRef] [PubMed]
109. Bud, R. The fall of a wonder drug. *Nature* **2007**, *446*, 981. [CrossRef]
110. Palta, S.; Saroa, R.; Palta, A. Overview of the coagulation system. *Indian J. Anaesth.* **2014**, *58*, 515–523. [CrossRef] [PubMed]
111. Adams, R.L.C.; Bird, R.J. Review article: Coagulation cascade and therapeutics update: Relevance to nephrology. Part 1: Overview of coagulation, thrombophilias and history of anticoagulants. *Nephrology* **2009**, *14*, 462–470. [CrossRef] [PubMed]
112. Hoffman, R.; Benz, E.; Shattil, S.; Furie, B.; Cohen, H. (Eds.) *Hematology: Basic Principles and Practice*, 4th ed.; Elsevier: Philadelphia, PA, USA, 2005; ISBN 978-0443079542.
113. Griffin, M.; Casadio, R.; Bergamini, C.M. Transglutaminases: Nature's biological glues. *Biochem. J.* **2002**, *368*, 377–396. [CrossRef] [PubMed]
114. Kane, P.; Yamashiro, C.; Wolczyk, D.; Neff, N.; Goebi, M.; Stevens, T. Protein splicing converts the yeast TFP1 gene product to the 69-kD subunit of the vacuolar H(+)-adenosine triphosphatase. *Science* **1990**, *250*, 651–657. [CrossRef] [PubMed]
115. Anraku, Y.; Mizutani, R.; Satow, Y. Protein splicing: Its discovery and structural insight into novel chemical mechanisms. *IUBMB Life* **2005**, *57*, 563–574. [CrossRef] [PubMed]
116. Hirata, R.; Ohsumi, Y.; Nakano, A.; Kawasaki, H.; Suzuki, K.; Anraku, Y. Molecular Structure of a Gene, VMA1, Encoding the Catalytic Subunit of H + -Translocating Adenosine Triphosphatase from Vacuolar Membranes of Saccharomyces cerevisiae. *J. Biol.Chem.* **1990**, *265*, 6726–6733. [PubMed]
117. Gimble, F.S.; Thorner, J. Homing of a DNA endonuclease gene by meiotic gene conversion in Saccharomyces cerevisiae. *Nature* **1992**, *357*, 301–306. [CrossRef] [PubMed]

118. Kawasaki, M.; Makino, S.I.; Matsuzawa, H.; Satow, Y.; Ohya, Y.; Anraku, Y. Folding-dependent in vitro protein splicing of the Saccharomyces cerevisiae VMA1 protozyme. *Biochem. Biophys. Res. Commun.* **1996**, *222*, 827–832. [CrossRef] [PubMed]
119. Kawasaki, M.; Nogami, S.; Satow, Y.; Ohya, Y.; Anraku, Y. Identification of three core regions essential for protein splicing of the yeast Vma1 protozyme. A random mutagenesis study of the entire VMA1- derived endonuclease sequence. *J. Biol. Chem.* **1997**, *272*, 15668–15674. [CrossRef] [PubMed]
120. Kawasaki, M.; Satow, Y.; Ohya, Y.; Anraku, Y. Protein splicing in the yeast Vma1 protozyme: Evidence for an intramolecular reaction. *FEBS Lett.* **1997**, *412*, 518–520. [CrossRef]
121. Walsh, C. Molecular mechanisms that confer antibacterial drug resistance. *Nature* **2000**, *406*, 775. [CrossRef] [PubMed]
122. Dennis, C. The bugs of war. *Nature* **2001**, *411*, 232. [CrossRef] [PubMed]
123. Theilgaard, H.B.A.; Kristiansen, K.N.; Henriksen, C.M.; Nielsen, J. Purification and characterization of delta-(L-alpha-aminoadipyl)-L-cysteinyl-D-valine synthetase from Penicillium chrysogenum. *Biochem. J.* **1997**, *327*, 185–191. [CrossRef] [PubMed]
124. Huffman, G.W.; Gesellchen, P.D.; Turner, J.R.; Rothenberger, R.B.; Osborne, H.E.; Miller, F.D.; Chapman, J.L.; Queener, S.W. Substrate Specificity of Isopenicillin N Synthase. *J. Med. Chem.* **1992**, *35*, 1897–1914. [CrossRef] [PubMed]
125. Roach, P.L.; Clifton, I.J.; Fülöp, V.; Harlos, K.; Barton, G.J.; Hajdu, J.; Andersson, I.; Schofield, C.J.; Baldwin, J.E. Crystal structure of isopenicillin N synthase is the first from a new structural family of enzymes. *Nature* **1995**, *375*, 700. [CrossRef] [PubMed]
126. Voet, D.; Voet, J.G. *Biochemistry*, 3rd ed.; John Wiley & Sons, Inc.: Hoboken, NJ, USA, 2004.
127. Cohen, M.L. Changing patterns of infectious disease. *Nature* **2000**, *406*, 762. [CrossRef] [PubMed]
128. Smaglik, P. Antibiotic resistance must be monitored, US Senate is told. *Nature* **2000**, *407*, 437. [CrossRef] [PubMed]
129. Nogrady, T. *Medicinal Chemistry A Biochemical Approach*; Oxford University Press: New York, NY, USA, 1985.
130. Sheehan, J.C. *The enchanted Ring: The Untold Story of Penicillin*; The MIT Press: Cambridge, MA, USA, 1982; ISBN 9780262192040.
131. Johnson, J.R.; Woodward, R.B.; Robinson, R. *The Chemistry of Penicillin*; Clarke, H.T., Ed.; Princton University Press: Princton, NJ, USA, 1949; ISBN 9780691627489.
132. Woodward, R.B. Recent advances in the chemistry of natural products. *Science* **1966**, *153*, 487–493. [CrossRef] [PubMed]
133. Newton, G.; Abraham, E.P. Cephalosporin C, a New Antibiotic containing Sulphur and D-α-Aminoadipic Acid. *Nature* **1955**, *175*, 548. [CrossRef] [PubMed]
134. Tally, F.P.; Jacobus, N.V.; Gorbach, S.L. In vitro activity of thienamycin. *Antimicrob. Agents Chemother.* **1978**, *14*, 436–438. [CrossRef] [PubMed]
135. Johnston, D.B.R.; Schmitt, S.M.; Bouffard, F.A.; Christensen, B.G. Total Synthesis of (±)-Thienamycin. *J. Am. Chem. Soc.* **1978**, *100*, 313–315. [CrossRef]
136. Araya, R. Input transformation by dendritic spines of pyramidal neurons. *Front. Neuroanat.* **2014**, *8*, 1–18. [CrossRef] [PubMed]
137. Granger, A.J.; Nicoll, R.A. Expression mechanisms underlying long-term potentiation: A postsynaptic view, 10 years on. *Philos. Trans. R. Soc. Lond. B Biol. Sci.* **2014**, *369*, 1–6. [CrossRef]
138. Pálfi, D.; Chiovini, B.; Szalay, G.; Kaszás, A.; Turi, G.F.; Katona, G.; Ábrányi-Balogh, P.; Szori, M.; Potor, A.; Frigyesi, O.; et al. High efficiency two-photon uncaging coupled by the correction of spontaneous hydrolysis. *Org. Biomol. Chem.* **2018**, *16*, 1958–1970. [CrossRef] [PubMed]
139. Papageorgiou, G.; Ogden, D.; Kelly, G.; Corrie, J.E.T. Synthetic and photochemical studies of substituted 1-acyl-7-nitroindolines. *Photochem. Photobiol. Sci.* **2005**, *4*, 887–896. [CrossRef] [PubMed]
140. Warther, D.; Gug, S.; Specht, A.; Bolze, F.; Nicoud, J.F.; Mourot, A.; Goeldner, M. Two-photon uncaging: New prospects in neuroscience and cellular biology. *Bioorg. Med. Chem.* **2010**, *18*, 7753–7758. [CrossRef] [PubMed]
141. Frisch, M.J.; Trucks, G.W.; Schlegel, H.B.; Scuseria, G.E.; Robb, M.A.; Cheeseman, J.R.; Scalmani, G.; Barone, V.; Mennucci, B.; Petersson, G.A.; et al. *Gaussian 09*, Revision D.01; Gaussian, Inc.: Wallingford, CT, USA, 2009.

142. Kovács, E.; Sághy, P.; Turczel, G.; Tóth, I.; Lendvay, G.; Domján, A.; Anastas, P.T.; Tuba, R. Metathesis of renewable polyene feedstocks–Indirect evidences of the formation of catalytically active ruthenium allylidene species. *J. Organomet. Chem.* **2017**, *847*, 213–217. [CrossRef]
143. Lu, M.; Fedoryak, O.D.; Moister, B.R.; Dore, T.M. Bhc-diol as a photolabile protecting group for aldehydes and ketones. *Org. Lett.* **2003**, *5*, 2119–2122. [CrossRef] [PubMed]
144. Montgomery, H.J.; Perdicakis, B.; Fishlock, D.; Lajoie, G.A.; Jervis, E.; Guy Guillemette, J. Photo-control of nitric oxide synthase activity using a caged isoform specific inhibitor. *Bioorg. Med. Chem.* **2002**, *10*, 1919–1927. [CrossRef]
145. Takaoka, K.; Tatsu, Y.; Yumoto, N.; Nakajima, T.; Shimamoto, K. Synthesis and photoreactivity of caged blockers for glutamate transporters. *Bioorg. Med. Chem. Lett.* **2003**, *13*, 965–970. [CrossRef]
146. Takaoka, K.; Tatsu, Y.; Yumoto, N.; Nakajima, T.; Shimamoto, K. Synthesis of carbamate-type caged derivatives of a novel glutamate transporter blocker. *Bioorg. Med. Chem.* **2004**, *12*, 3687–3694. [CrossRef] [PubMed]
147. Givens, R.S.; Matuszewski, B. Photochemistry of Phosphate Esters: An Efficient Method for the Generation of Electrophiles1. *J. Am. Chem. Soc.* **1984**, *106*, 6860–6861. [CrossRef]
148. Furata, T.; Torigai, H.; Sugimoto, M.; Iwamura, M. Photochemical Properties of New Photolabile cAMP Derivatives in a Physiological Saline Solution. *J. Org. Chem.* **1995**, *60*, 3953–3956. [CrossRef]
149. Hagen, V.; Bendig, J.; Frings, S.; Wiesner, B.; Schade, B.; Helm, S.; Lorenz, D.; Benjamin Kaupp, U. Synthesis, photochemistry and application of (7-methoxycoumarin-4-yl)methyl-caged 8-bromoadenosine cyclic 3′,5′-monophosphate and 8-bromoguanosine cyclic 3′,5′-monophosphate photolyzed in the nanosecond time region. *J. Photochem. Photobiol. B Biol.* **1999**, *53*, 91–102. [CrossRef]
150. Eckardt, T.; Hagen, V.; Schade, B.; Schmidt, R.; Schweitzer, C.; Bendig, J. Deactivation behavior and excited-state properties of (coumarin-4-yl)methyl derivatives. 2. Photocleavage of selected (coumarin-4-yl)methyl-caged adenosine cyclic 3′,5′-monophosphates with fluorescence enhancement. *J. Org. Chem.* **2002**, *67*, 703–710. [CrossRef] [PubMed]
151. Geißler, D.; Kresse, W.; Wiesner, B.; Bendig, J.; Kettenmann, H.; Hagen, V. DMACM-caged adenosine nucleotides: Ultrafast phototriggers for ATP, ADP, and AMP activated by long-wavelength irradiation. *ChemBioChem* **2003**, *4*, 162–170. [CrossRef] [PubMed]
152. Hagen, V.; Frings, S.; Wiesner, B.; Helm, S.; Kaupp, U.B.; Bendig, J. [7-(Dialkylamino)coumarin-4-yl]methyl-caged compounds as ultrafast and effective long-wavelength phototriggers of 8-bromo-substituted cyclic nucleotides. *ChemBioChem* **2003**, *4*, 434–442. [CrossRef] [PubMed]
153. Furuta, T.; Takeuchi, H.; Isozaki, M.; Takahashi, Y.; Kanehara, M.; Sugimoto, M.; Watanabe, T.; Noguchi, K.; Dore, T.M.; Kurahashi, T.; et al. Bhc-cNMPs as either water-soluble or membrane-permeant photoreleasable cyclic nucleotides for both one- and two-photon excitation. *ChemBioChem* **2004**, *5*, 1119–1128. [CrossRef] [PubMed]
154. Geißler, D.; Antonenko, Y.N.; Schmidt, R.; Keller, S.; Krylova, O.O.; Wiesner, B.; Bendig, J.; Pohl, P.; Hagen, V. (Coumarin-4-yl)methyl esters as highly efficient, ultrafast phototriggers for protons and their application to acidifying membrane surfaces. *Angew. Chem. Int. Ed.* **2005**, *44*, 1195–1198. [CrossRef] [PubMed]
155. Hagen, V.; Dekowski, B.; Nache, V.; Schmidt, R.; Geißler, D.; Lorenz, D.; Eichhorst, J.; Keller, S.; Kaneko, H.; Benndorf, K.; Wiesner, B. Coumarinylmethyl esters for ultrafast release of high concentrations of cyclic nucleotides upon one- and two-photon photolysis. *Angew. Chem. Int. Ed.* **2005**, *44*, 7887–7891. [CrossRef] [PubMed]
156. Furuta, T.; Wang, S.S.-H.; Dantzker, J.L.; Dore, T.M.; Bybee, W.J.; Callaway, E.M.; Denk, W.; Tsien, R.Y. Brominated 7-hydroxycoumarin-4-ylmethyls: Photolabile protecting groups with biologically useful cross-sections for two photon photolysis. *Proc. Natl. Acad. Sci. USA* **1999**, *96*, 1193–1200. [CrossRef] [PubMed]
157. Lin, W.; Lawrence, D.S. A strategy for the construction of caged diols using a photolabile protecting group. *J. Org. Chem.* **2002**, *67*, 2723–2726. [CrossRef] [PubMed]

9

Electronic Effect on the Molecular Motion of Aromatic Amides: Combined Studies using VT-NMR and Quantum Calculations

Sungsoo Kim [1], Jungyu Kim [2,3], Jieun Kim [1], Daeun Won [1], Suk-Kyu Chang [1], Wansik Cha [4], Keunhong Jeong [5,*], Sangdoo Ahn [1,*] and Kyungwon Kwak [2,3,*]

1 Department of Chemistry, Chung-Ang University, Seoul 156-756, Korea; hesperus86@gmail.com (S.K.); wldms01477@gmail.com (J.K.); panda881020@gmail.com (D.W.); skchang@cau.ca.kr (S.-K.C.)
2 Center for Molecular Spectroscopy and Dynamics, Institute for Basic Science (IBS), Seoul 02841, Korea; kimjoongyu@naver.com
3 Department of Chemistry, Korea University, Seoul 02841, Korea
4 Nuclear Chemistry Research Division, Korea Atomic Energy Research Institute, Daejeon 34057, Korea; wscha@kaeri.re.kr
5 Department of Chemistry, Korea Military Academy, Seoul 01805, Korea
* Correspondence: doas1mind@kma.ac.kr (K.J); sangdoo@cau.ac.kr (S.A.); kkwak@korea.ac.kr (K.K.)

Abstract: Rotational barrier energy studies to date have focused on the amide bond of aromatic compounds from a kinetic perspective using quantum calculations and nuclear magnetic resonance (NMR). These studies provide valuable information, not only regarding the basic conformational properties of amide bonds but also the molecular gear system, which has recently gained interest. Thus, we investigate the precise motion of the amide bonds of two aromatic compounds using an experimental rotational barrier energy estimation by NMR experiments and a theoretical evaluation of the density functional theory calculation. The theoretical potential energy surface scan method combined with the quadratic synchronous transit 3 method and consideration of additional functional group rotation with optimization and frequency calculations support the results of the variable temperature ^{1}H NMR, with deviations of less than 1 kcal/mol. This detailed experimental and theoretical research strongly supports molecular gear motion in the aromatic amide system, and the difference in kinetic energy indicates that the electronic effect from the aromatic structure has a key role in conformational movements at different temperatures. Our study provides an enhanced basis for future amide structural dynamics research.

Keywords: rotational barrier energy; amide bond; nuclear magnetic resonance; kinetic; density functional theory

---

1. Introduction

The amide bond, a basic unit of proteins, has unique steric and energetic characteristics [1,2]. Therefore, studying the dynamics of these bonds is important to understand protein dynamics. Conformational studies of aromatic amide bonds, in particular, have been extensively carried out owing to their potential applications in fields such as asymmetric synthesis, molecular gear systems, single-molecule motors, and single-molecule devices [3].

An aromatic amide has two rotational motions [4] around the central carbonyl group: the rotation around aryl-CO and the rotation around the C-N bond in the amide unit. Therefore, the basic structural change of an aromatic amide involves three bond rotational processes [4,5] that compose the interconversions of the conformers. The two independent rotations of the C-N bond or aryl-CO bond interconvert two conformers (exo and endo forms), and the concerted rotation of these bonds represents the molecular gear. Both rotations and their three rotational processes [6,7] have received intense interest because rotational dynamics can be applied to single molecule dynamic studies and experimental data on rotational dynamics can be achieved with variable temperature nuclear magnetic resonance (VT-NMR) spectroscopy [8–10] in order to investigate the Gibbs free energy of the concerted C-N/aryl-CO bond rotation and/or independent aryl-CO and C-N bond rotation around the carbonyl group on the aromatic amide. These rotations can be controlled through chemical modification of the structure or simple temperature variation [11].

The internal rotation [12] around the C-N bond in the amides provides information regarding the conformational characteristics of the backbone of peptides and proteins. There have been many reports concerning determination of the steric effect involving rotational dynamics [5,6], although few of these have discussed electronic effects owing to the difficulty of separately measuring the steric and electronic effects. The other rotational dynamic of aromatic amides, aryl-CO, has been successfully explained in the case of the C-C single bond rotational process in *N*-methylbenzamide [7]. However, the electron-rich aromatic ring in molecular gear studies is usually designed and connected to the *N* of amide groups. Thus, the aryl-CO bond is correlated to the C-N bond in many experimental studies [11,13,14]. It is therefore important to study the electronic effects of aromatic amides, including aryl-CO and C-N bonds, using both theoretical and experimental methods, which will clearly explain the molecular motion. Herein, we use quantum calculations and VT-NMR to investigate the rotational process of aryl-CO and C-N bonds in specifically designed *N,N*-diethylamide derivatives.

2. Results

2.1. Interpretation of VT-NMR: Coalescence

Characterization of *N,N*-diethylamide Derivatives (1), (2)

N,N-diethylamide derivatives (1), (2) were synthesized to measure the rotational barrier of the aryl-CO and C-N bonds (Figure 1). Before we evaluated the rotational barrier using ^{1}H VT-NMR, the H and Me peaks from two ethyl groups attached to N were assigned to understand which rotations were occurring. The peaks of chemical shifts in *N,N*-diethylpyrene-1-carboxamide (PCDEA) have a similar trend to those in *N,N*-diethyl-1-naphthamide (NCDEA) at 25 °C. As shown in Table 1, the ^{1}H NMR spectrum in NCDEA has two H chemical shift peaks (δ 3.51 ppm, δ 3.82 ppm), as does the ^{1}H NMR spectrum in PCDEA (δ 3.64 ppm, δ 3.89 ppm). These two peaks occur because two H atoms are located differently with respect to each aromatic group, resulting in a more de-shielded proton signal when the position is closer to the aromatic group. Similarly, Me peaks are better shielded than H peaks in PCDEA (δ 1.32 ppm, δ 1.77 ppm) and in NCDEA (δ 0.98 ppm, δ 1.34 ppm).

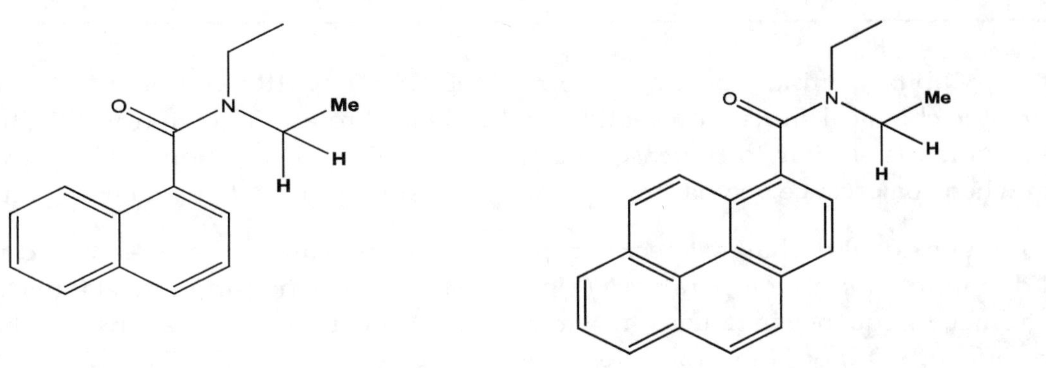

N,N-diethyl-1-naphthamide (**NCDEA, 1**) *N,N*-diethylpyrene-1-carboxamide (**PCDEA, 2**)

Figure 1. Structures of *N,N*-diethylamide derivatives.

Table 1. Experimental ^{1}H NMR for *N,N*-diethylamide derivatives (aliphatic part only).

Compound	Signal	Chemical Shifts (δ)/ppm
NCDEA	H	3.51/3.82
NCDEA	Me	0.98/1.34
PCDEA	H	3.64/3.89
PCDEA	Me	1.32/1.77

Line shape analysis of the ^{1}H VT-NMR experiment was applied to evaluate the relative Gibbs free energy difference and the enthalpy difference for aryl-CO and C-N rotation. Figure 2 shows that the experimental values are in good agreement with the theoretical analysis. The following (Eyring equation) estimates $\Delta G^{\ddagger}$ in ^{1}H VT-NMR using the coalescence temperature from the two-site exchange [9,15].

$$\Delta G^{\ddagger} = 0.01914 \times \mathrm{Tc} \times \left[10.319 + \log_{10}\left(\frac{Tc}{k}\right)\right],\ (k = \frac{\pi \Delta \upsilon}{\sqrt{2}}) \tag{1}$$

In the experiment, the split NMR signals are from two ground states. If their exchange rate constant by rotation is greater than their frequency separation, the peaks are merged into one. Therefore, if two peaks are merged into one peak, the rotational barriers of aryl-CO and C-N and the rate constant are detected by the coalescence temperature. The rate constant (k) has a fast exchange for aryl-CO and C-N rotations.

As the temperature increases, the merging of H-H into one relatively sharp peak corresponds to aryl-CO, and merging of Me-Me into one broader peak represents concerted aryl-CO/C-N rotation, which is well supported by a previous extensive theoretical study [16].

NCDEA has two aryl-CO and C-N bond rotations. Two different characteristics are explained by the peak coalescence of the ^{1}H VT-NMR experiment as the temperature increases. The spectra show that aryl-CO rotation represents the broadening signals of H-H peaks from δ 3.60 to δ 3.80 ppm (Figure 2b). As the temperature increases, the signals of Me-Me from δ 1.0 to δ 1.4 ppm are assigned to the C-N rotation (Figure 2c). At 65 °C, it is clear that two peaks coalesce in the aryl-CO bond rotation, as shown in Figure 2. On the other hand, the coalescence temperature of the C-N rotation is higher than that of the aryl-CO rotation (Table 2).

Table 2. Bond rotation in *N,N*-diethylamide derivatives measured by ^{1}H VT-NMR.

Compound	Coalescing Signals	$\Delta\upsilon$/Hz	T_c/°C	k/s^{-1}	$\Delta G^{\ddagger}$/kcal·mol^{-1}
NCDEA	H-H	175.43	60	602	15.40
NCDEA	Me-Me	273.93	115	389	18.00
PCDEA	H-H	219.62	57	478	15.62
PCDEA	Me-Me	220.28	115	478	17.65

PCDEA exhibits a similar spectral pattern in the ^{1}H VT-NMR experiment. At the coalescence temperature, the chemical shifts representing aryl-CO bond rotation range from δ 3.70 to δ 3.90 ppm (Figure 2e), which is similar to the case of NCDEA. C-N bond rotation is shown from δ 1.00 to δ 1.50 ppm when coalescence begins, and again, is very similar to that of NCDEA (Figure 2f).

A two-dimensional exchange spectroscopy (EXSY) experiment was performed to confirm the rotational dynamics at high temperatures (Figure S1). The mixing time was 200 ms, and $\omega_o \tau_c \ll 1$ (ω_o = spectrometer frequency in radians; τ_c = rotational correlation time) was satisfied; thus, the nuclear overhauser effect (NOE) of H in molecules is not expected to exist [17]. Therefore, the cross-peaks shown in Figure S1 can only be expressed by chemical exchange mechanisms. The 2D EXSY spectrum shows how each bond rotation progresses at three increasing temperatures.

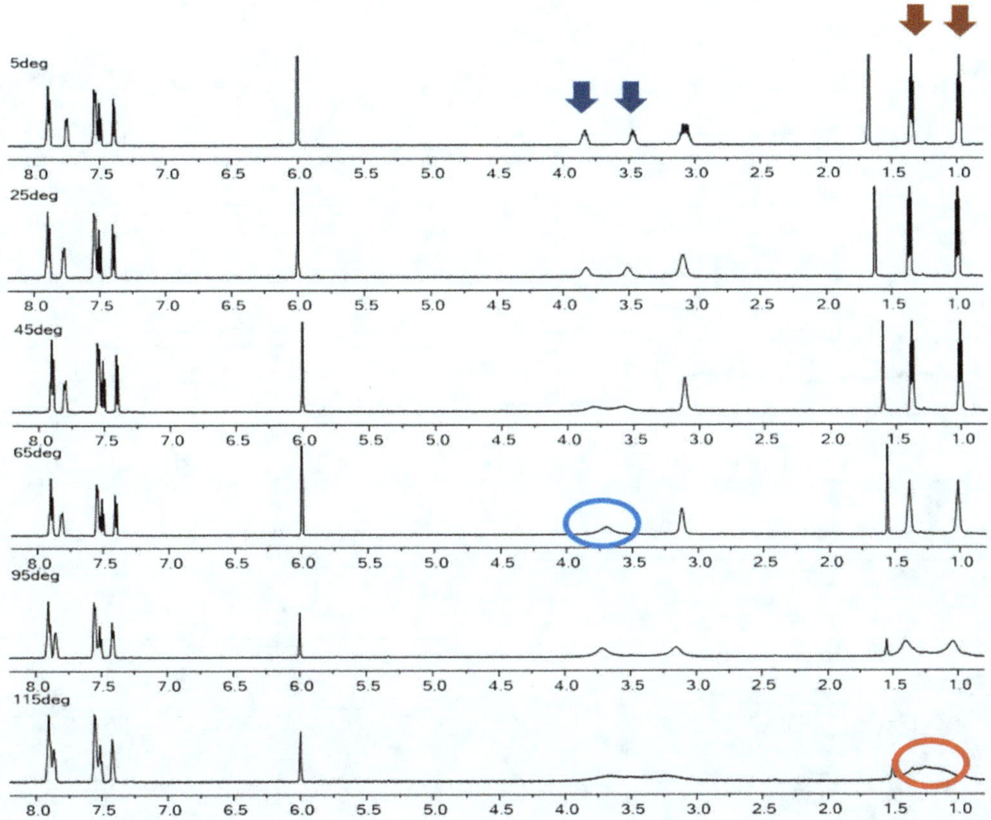

(a)

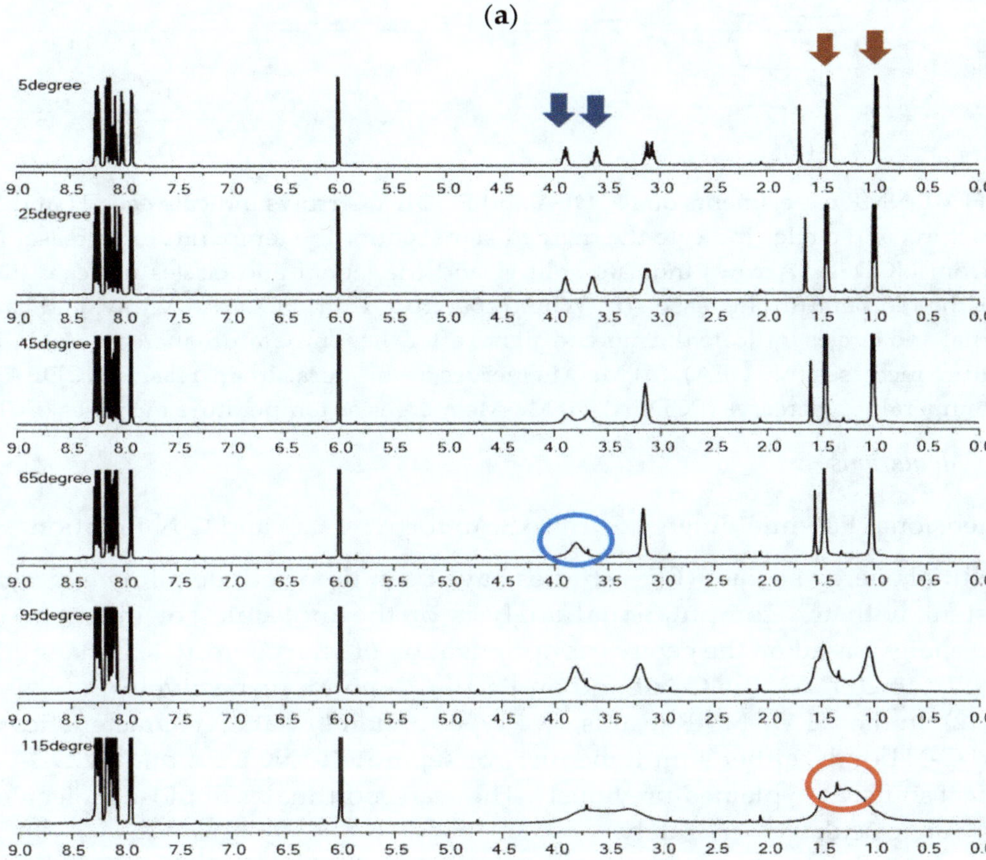

(b)

Figure 2. *Cont.*

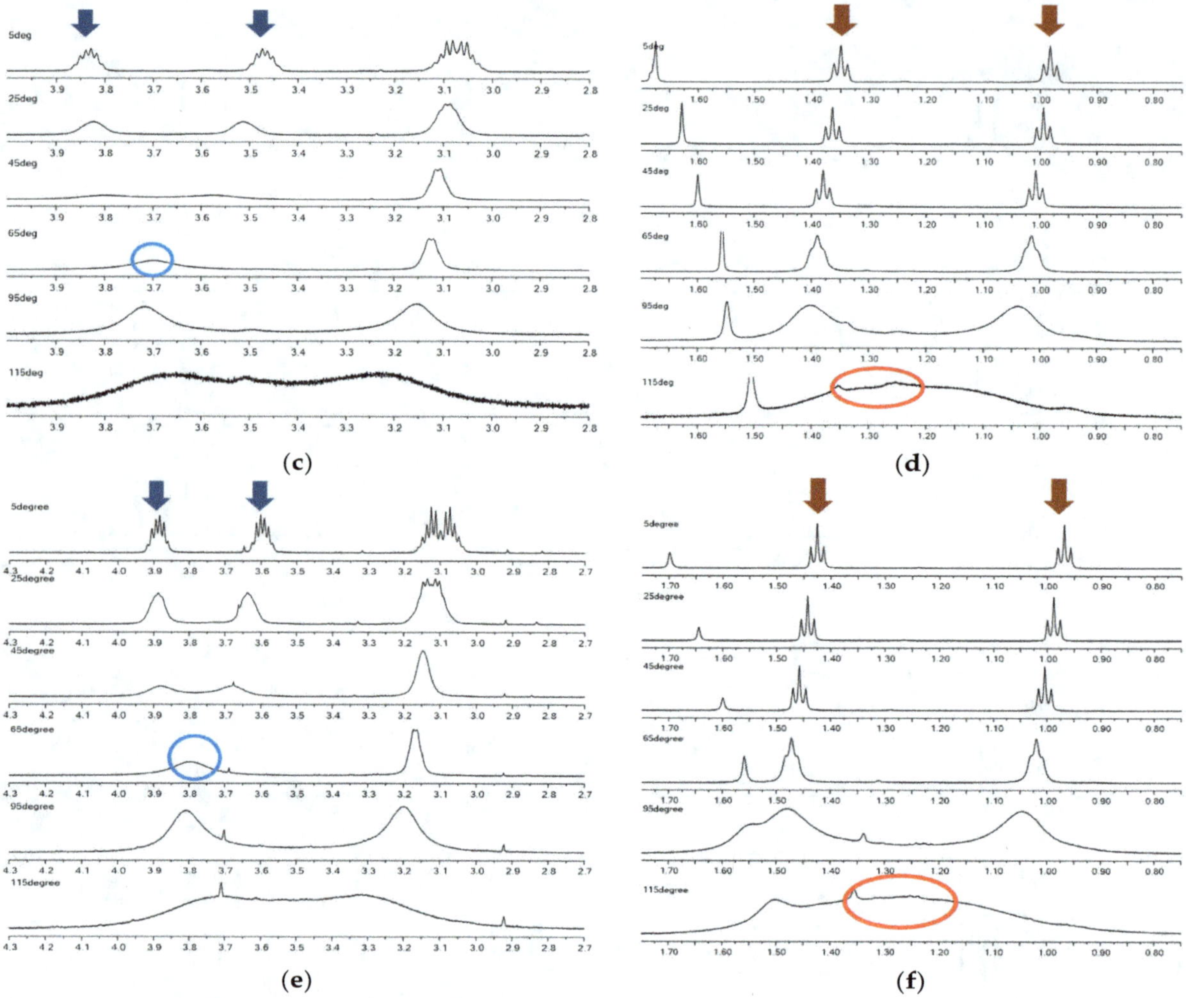

Figure 2. ^{1}H VT-NMR experiments on NCDEA and PCDEA. Arrows indicate each H and Me signal in each structure, and circles indicate the merged signals after the temperature increase. (**a**) Whole spectrum from NCDEA. Arrows indicate split H and Me signal and circles indicate the merged signal after the temperature increase. (**b**) Whole spectrum from PCDEA. Arrows indicate split H and Me signal and circles indicate the merged signal after the temperature increase. (**c**) H-H merges as temperature increases (NCDEA). (**d**) Me-Me merges as temperature increases (NCDEA). (**e**) H-H merges as temperature increases (PCDEA). (**f**) Me-Me merges as temperature increases (PCDEA).

2.2. Theoretical Calculations

2.2.1. One-Dimensional Potential Energy Surface Scan for Aryl-CO and C-N Rotation

Many potential energy surface (PES) studies have been reported, detailing experimental results that are consistent with the computational analysis of the molecules of interest. Likewise, we investigated the energy based on the geometric optimization of scan coordinates, changing the dihedral angles from $-170°$ to $200°$ for the C-N bond and aryl-CO bond, respectively, in *N*,*N*-diethylamide derivatives (1), (2). In the ^{1}H VT-NMR results, two types of bond rotation characteristics were detected on aryl-CO and C-N bonds, either simultaneously or separately. NCDEA and PCDEA have discrete values of chemical shifts, as explained previously. The scan coordinates of those molecules on C-N and aryl-CO bonds were calculated separately.

When the aryl-CO bond rotates, the diethyl group attached to the N is directly confronted with the steric effect from the naphthalene (1) and pyrene (2) rings, which increases the optimized energy representing repulsion (Figures S2a and S3a). On the other hand, when concerted aryl-CO and C-N bond rotation occurs, all the diethylamide derivatives manifest a similar tendency for molecular orientations and elucidate the concurrent rotations of aryl-CO and C-N bonds; thus, there are two similar transition states (Figures S2b and S3b).

We determined that the PES of the C-N bond rotation of NCDEA and PCDEA indicates that the signals of peak coalescence in VT-NMR show similar chemical shifts. This suggests that the PES of aryl-CO and C-N bond rotations have two ground states, which are enantiomers, and two transition states, indicating that they have two rotational barriers. The Gibbs free energy between the transition state and ground state approximately demonstrates the experimental NMR analysis (Table S1). However, the true transition states would enable a more accurate theoretical description of the actual Gibbs free energy difference.

2.2.2. Calculated Result for QST3 and Changing Dihedral Angles of Di-ethyl Groups

Scheme for Interconversion in *N,N*-Diethylamide Derivatives (1), (2)

Before the transition states were obtained through quadratic synchronous transit 3 (QST3) [18], we obtained the change in dihedral angle of the aryl-CO bond as the dihedral angle of the C-N bond increased, or vice versa, by 1D PES analysis. As indicated in Figures 3 and 4, whereas aryl-CO bond rotation is independent of C-N rotation, C-N bond rotation has an influence on aryl-CO bond rotation in that it experiences concerted rotation with the aryl-CO bond, consistent with previous NMR and calculation results.

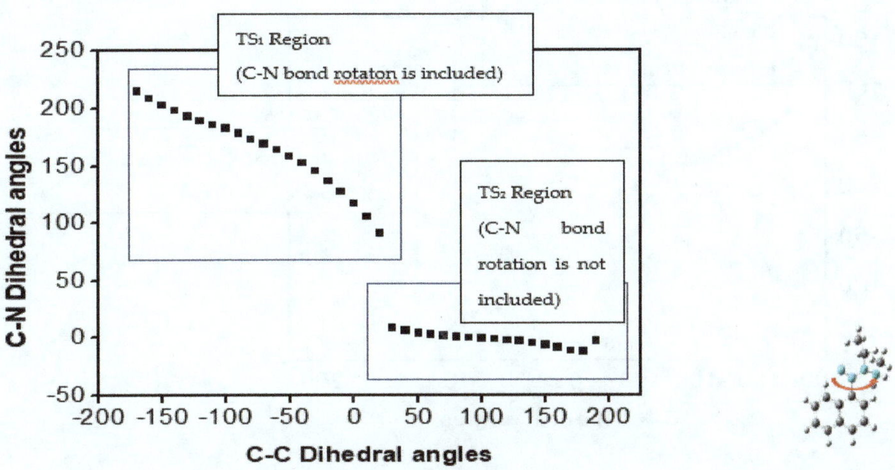

(**a**) NCDEA

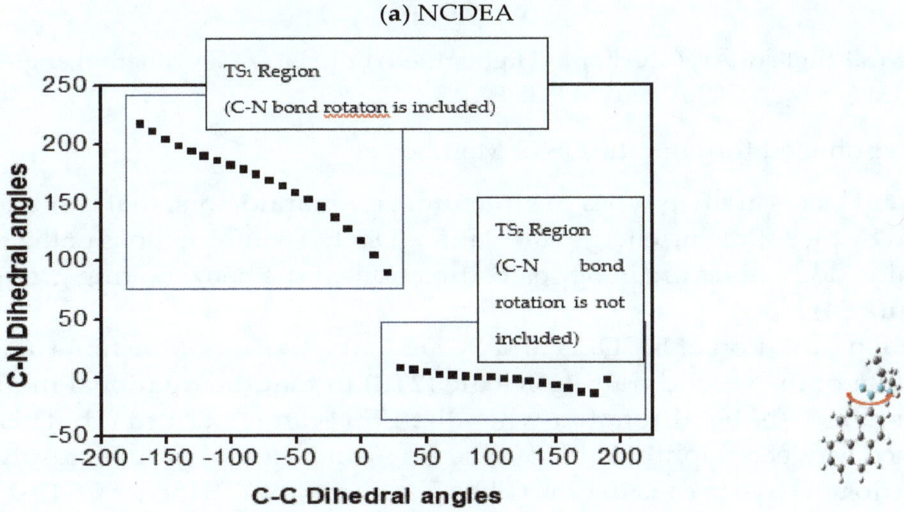

(**b**) PCDEA

Figure 3. Dihedral angle of C-N bond (highlighted) (Y axis) versus dihedral angle of Aryl-CO bond (Arrowed) (X axis) in (**a**) NCDEA and (**b**) PCDEA.

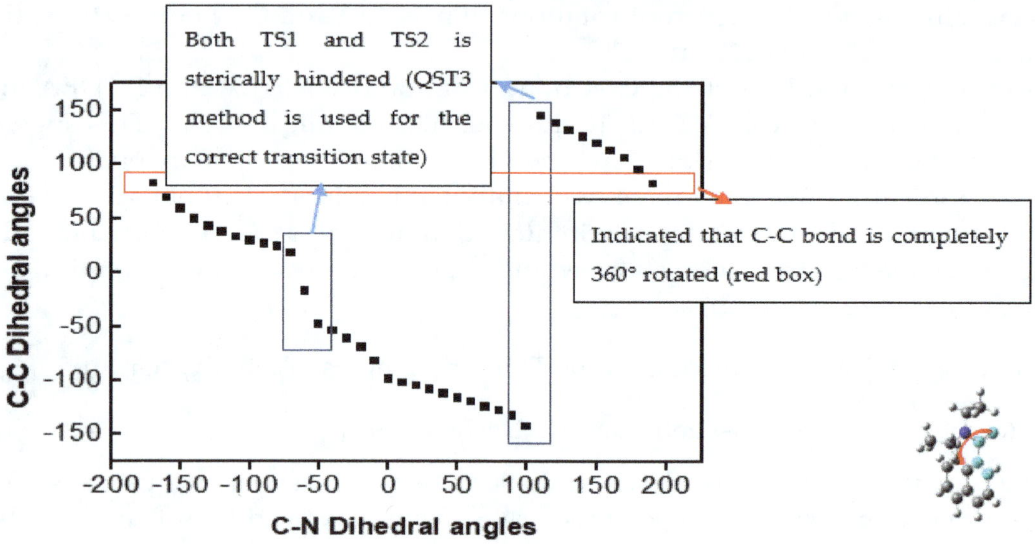

(**a**) NCDEA

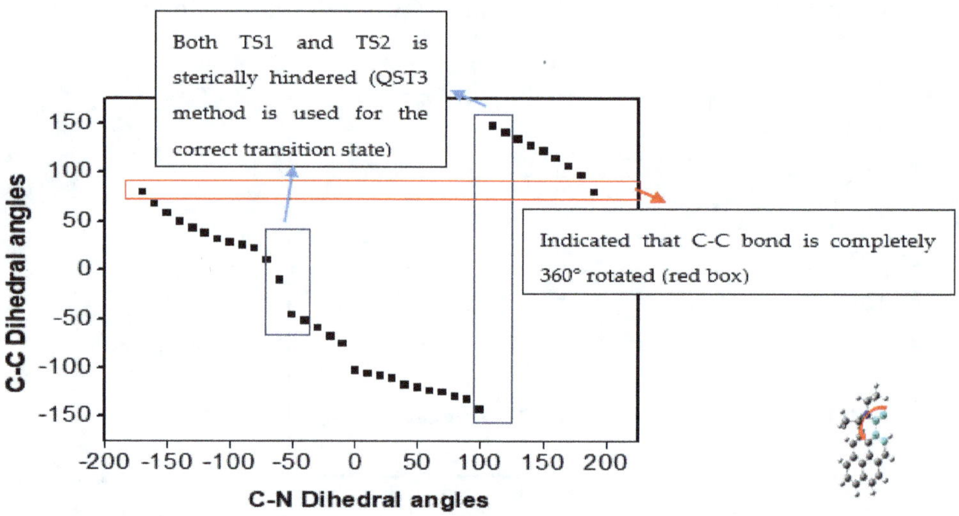

(**b**) PCDEA

Figure 4. Dihedral angle of Aryl-CO bond (Highlighted) (Y axis) versus dihedral angle of C-N bond (Arrowed) (X axis) in (**a**) NCDEA and (**b**) PCDEA.

Transition States Optimized through the QST3 Method

The two ground states are two types of structure, i.e., enantiomers, that exhibit the same energy, which is supported by the following energy calculation. Diethyl conformations of the two ground states of *N*,*N*-diethylamide derivatives are in opposite directions if the benzene ring group is considered a mirror plane (Figure S4).

For aryl-CO bond rotation of NCDEA and PCDEA, the barrier of the first transition state (1TS) is different from that of the second transition state (2TS) in that the rotational motion of 1TS in the aryl-CO bond contains C-N bond rotation when the PES scan is performed. This originates from the steric hindrance between naphthalene/pyrene rings and diethyl groups, resulting in C-N bond rotation. Thus, we do not have to consider the Gibbs free energy of 1TS (hence QST3-1TS of the aryl-CO bond is not included in Figure S4 or Table S2).

The scan coordinates do not confirm whether the two transition states found in this way are structurally true transition states, which are required for an adequate Gibbs energy difference, even though they have one negative imaginary frequency. Other optimized processes (QST3) for finding true transition states are required to subsequently characterize the structure and energy of each state (Figures S2 and S3); these optimized structures also have imaginary frequencies in which the mode corresponds to rotation of the aryl-CO bond and the concerted aryl-CO/C-N bond, respectively (Table S2). The structures of two GSs and one or two TSs for *N,N*-diethylamide derivatives (1), (2) are expressed as dihedral angles (Schemes S2 and S3) in order to more easily understand the energy difference (Figure S4). Furthermore, the study of intrinsic reaction coordinates analysis is added for confirming the transition state.

Dihedral Angle Change in the Diethyl Conformation of Transition States of the Aryl-CO Bond in NCDEA and PCDEA

Changing the dihedral angles of the diethyl moiety of transition states is necessary for explaining the routes to rotational barriers. As we described earlier, since the 1TS of aryl-CO bond rotation includes C-N rotation, and the increase of activation energy and energy of 1TS in aryl-CO/C-N bond rotation is greater than that of 2TS after the QST3 calculation, we only consider the 2TS of the aryl-CO bond and aryl-CO/C-N bond rotation and make the dihedral angle change for the proper theoretical value consistent with the experimental value. Results of calculations with various dihedral angles are described in detail, and these energy values are very close to the experimental data (Tables S3–S5).

In particular, for the energy value of aryl-CO bond rotation in NCDEA, the theoretical Gibbs free energy value can range from 14.24 to 15.19 kcal·mol^{-1}, a range of less than 1 kcal·mol^{-1}, which is highly consistent with the NMR experimental results; four of the di-ethyl rotation modes, however, do not correspond to aryl-CO bond rotation with more than an imaginary frequency (Table S3).

PCDEA shows a similar tendency in aryl-CO and C-N bond rotation to that of NCDEA. It is clear that the value estimated by di-ethyl rotation is similar to the experimental value. The importance of applying the dihedral angle change is thus confirmed; the value of the Gibbs free energy for QST3 optimization is 13.52 kcal·mol^{-1} and can be increased by di-ethyl rotation to 14.37 and 15.74 kcal·mol^{-1}, values that are only approximately 1 kcal lower than or the same as, respectively, the experimental values (Table S4).

Transition State of the 2D PES

As indicated in the 2D PES scan with frequency calculation in Figure 5, we also investigated independent aryl-CO and C-N dihedral angles for concerted aryl-CO/C-N bond rotation. As shown in Figure 4, as the C-N dihedral angle (X axis) decreases, the aryl-CO (Y axis) increases so that NCDEA and PCDEA have different Gibbs free energies for aryl-CO bond rotation (NCDEA $\Delta G^{\ddagger}$: 14.19 kcal and PCDEA $\Delta G^{\ddagger}$: 15.15 kcal) and C-N bond rotation (NCDEA $\Delta G^{\ddagger}$: 17.67 kcal and PCDEA $\Delta G^{\ddagger}$: 17.57 kcal). This means that C-N bond rotation affects the aryl-CO bond rotation, as it is a molecular gear model [3]. On the other hand, aryl-CO bond rotation does not affect C-N bond rotation; thus, it rotates independently. This means that the ethyl group connected to the tertiary amine could be sterically hindered by aromatic group when it rotates around. Those results robustly support our previous theoretical and experimental data. Thus, we confirmed via both of Gibbs free energy values (Table 3) and the transition state structures, that the transition state structure of C-N bond rotation is consistent with the structure of the concerted C-N/aryl-CO bond in 2D PES (Tables S6–S10).

Quantum Theory of Atoms in Molecules (QTAIM) Study

The information from the QTAIM study qualitatively supports all the results of this study. The electron density of the C-N/aryl-CO bond in the transition state is generally greater than that of the aryl-CO bond in each structure, demonstrating that the Gibbs free energy of C-N/aryl-CO rotation is greater than that of aryl-CO. Furthermore, the electronic effect from different aryl groups shows the different electron populations on the rotating bonds in the transition state, which reveal the Gibbs free energy difference between NCDEA and PCDEA (Table S11).

Table 3. Comparison of the relative experimental Gibbs free energy differences to computational results.

Compound	Bond	Variable Temperature ^{1}H NMR $\Delta G^{\ddagger}$/kcal·mol^{-1}	Theoretical Calculation (QST3//Di-ethyl Rotation *) $\Delta G^{\ddagger}$/kcal·mol^{-1}
NCDEA	C-N/aryl-CO	18.00	1TS: 20.61 ** 2TS: 17.13//16.09~20.04 2D TS: 17.67
	aryl-CO	15.40	1TS: C-N bond rotation is included 2TS: 14.24//14.24~15.19 2D TS: 14.19
PCDEA	C-N/aryl-CO	17.65	1TS: 19.94 ** 2TS: 17.04//15.97~19.92 2D TS: 17.57
	aryl-CO	15.62	1TS: C-N bond rotation is included 2TS: 13.52//13.38~15.74 2D TS: 15.15

Di-ethyl rotation * represents energy values between minimum and maximum. ** 1TS showed higher Gibbs free energy than 2TS in QST3, which was not considered for di-ethyl rotation calculation.

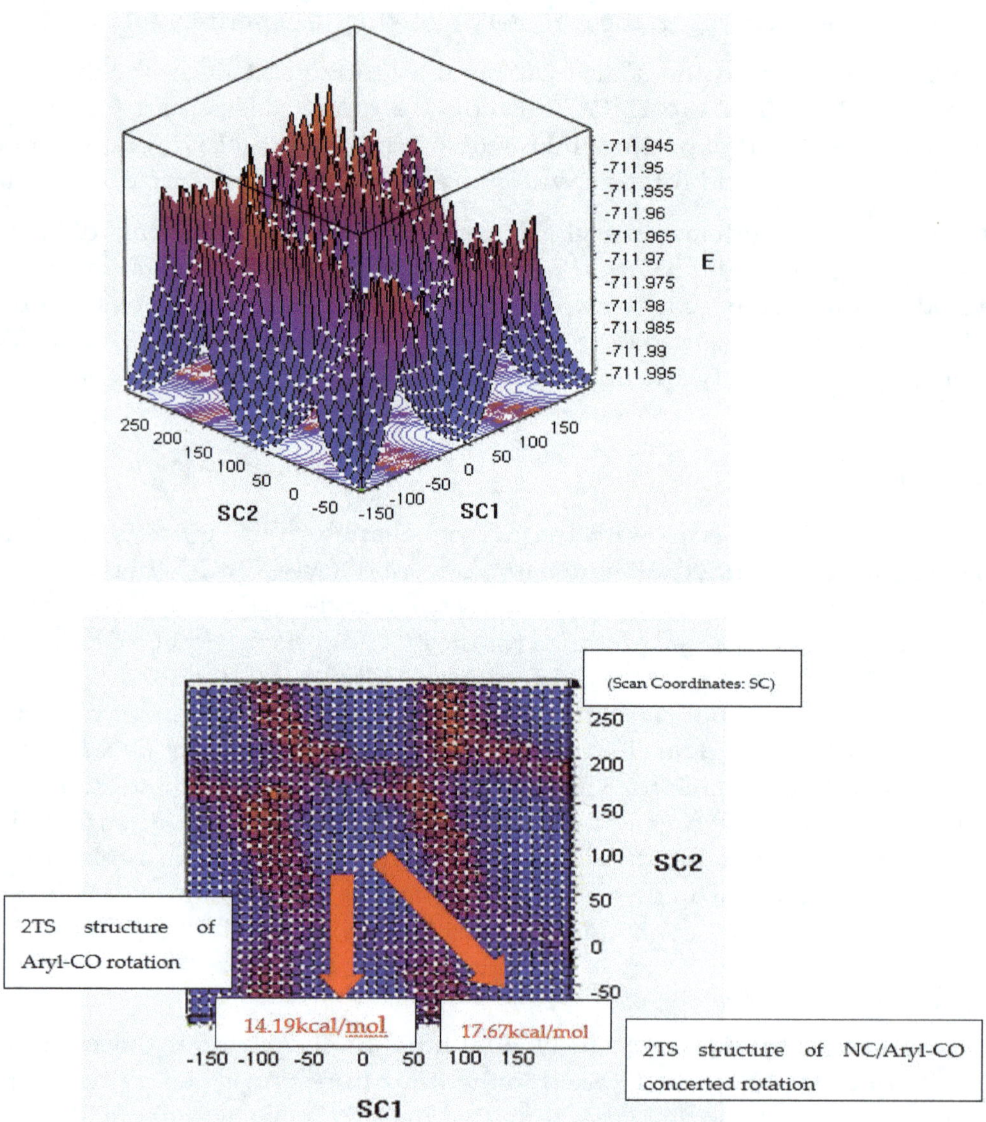

(a) NCDEA (SC1: Ayl-CO rotation, SC2: C-N rotation)

Figure 5. *Cont.*

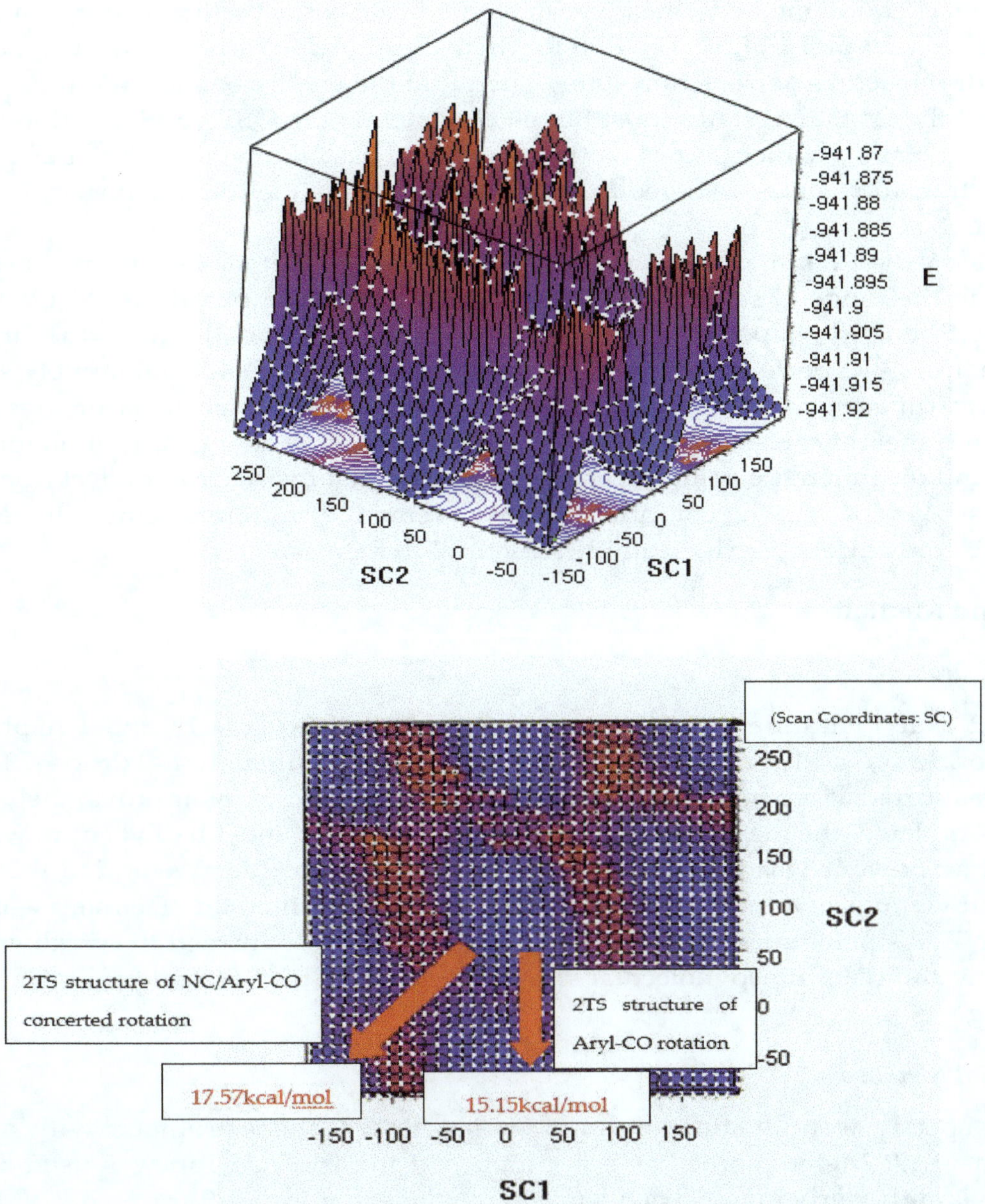

Figure 5. 2D-Potential Energy Surface on Aryl-C = O Dihedral angle and C-N Dihedral angle change of NCDEA (**a**) and PCDEA (**b**).

3. Discussion

Rotational barrier energy studies of amide bonds of aromatic compounds from a kinetic perspective provide valuable information regarding the basic conformational properties of amide bonds and their molecular gear system, which have been of recent scientific interest. To that end, we investigated rotational barrier energy using two methods to determine the precise motion and theoretical quantum chemical effects using an NMR experiment and density functional theory [19,20]. The theoretical PES scan method, with optimization and frequency calculations suggested by the theoretical study, supports the effect of ^{1}H VT-NMR, which assumes that if the bond rotation is fast enough to break out the double bond characteristic of interest, the two peaks are merged and cannot be distinguished from chemical shifts. Thus, we can predict and evaluate the energy barrier using the Eyring equation [9,15].

By quantum chemical methods, the PES can be predicted correctly for rotation in a simple system such as ethane (~3 kcal·mol^{-1}) [21]. The PES for diethylamide derivatives, however, has a limitation in explaining the structure of transition states, such that the energy barrier might be miscalculated. This means that the transition states must be re-calculated using QST3 optimization [18], followed by frequency calculation, which shows us that each state has an imaginary frequency. This clearly shows that the transition states assumed to be true demonstrate the rotational energy barriers in the NMR experiment.

The final step, di-ethyl rotations, is required to determine whether each transition states' energy can be varied in the range of several kcals, revealing that the computational result from this step is consistent with energy values from the NMR experiment (Table 3). To further investigate the rotational dynamics, we performed QTAIM studies [22]. The analysis qualitatively supported the Gibbs free energy difference and electronic effect from the different molecular structures. This detailed experimental and theoretical research strongly supports the molecular gear motion for the aromatic amide system and the difference in kinetic energy, provided that the electronic effect from the aromatic structure plays a key part in the conformational movement at different temperatures. Our study provides a basis for future research on amide structural dynamics.

4. Materials and Methods

4.1. Sample Preparation

We added oxalyl chloride to a mixture of pyrene-1-carboxylic acid and 1-naphthoic acid in methylene chloride and added dimethyl formamide after approximately 1–2 drops [23] (Scheme S1). The mixture was stirred at room temperature to maintain the reaction for approximately 4 h. Once the reaction was completed, the methylene chloride was evaporated, and 1 h of vacuum was performed. The dried yellow powder was then dissolved in methylene chloride. We added this mixture to a solution of diethylamine and methylene chloride by a dropwise method. This step was followed by extraction with D.W. three times. After sample preparation, we confirmed the product by thin layer chromatography and determined molecular structures with NMR, the results of which were used for data processing.

4.2. Experimental Methods

^{1}H NMR spectra were obtained using a Varian VNS 600 spectrometer with a temperature control system. 1,1,2,2-tetrachloroethane-d$_2$ purchased from Euriso-Top was used as the solvent for *N*,*N*-diethylamide derivatives. Each sample was injected into a 5-mm o.d. NMR tube to be analyzed. The chemical shifts of the spectra were obtained using the solvent peak as a reference.

In the VT-NMR experiment, prior to detection, the samples were left long enough for all samples to be in equilibrium at the designated temperature, so as to increase the reproducibility and accuracy of the results. The experiments were performed in the range of 273–388 K, considering the boiling point of the solvent and the limiting temperature of proton detection. The spectra were typically obtained at intervals of 10°. The interval was 2° beginning with the coalescence temperature. The possible error range was $\pm$0.5 K.

Two-dimensional exchange spectroscopy (EXSY) [24] was performed at different temperatures for the two *N*,*N*-diethylamide derivatives (Figure 1) using a normal nuclear Overhauser effect spectroscopy (NOESY) pulse sequence [25].

4.3. Computational Methodss

Calculations were performed at the B3LYP/6-311G(d) level by means of the Gaussian 09 software package (Gaussian, Inc., Wallingford, CT, USA) [26]. To find the Gibbs free energy difference between the ground state and transition state, we used density functional theory (DFT) [19,20,27] at the B3LYP/6-311G(d) level [28] with the PCM model [29] (dielectric constant

ε = 8.2 of 1,1,2,2-tetrachloroethane). Gibbs free energy difference was calculated under standard conditions (a temperature of 298.15K and a pressure of 1 atm).

4.3.1. 1D PES Method

First, the calculations in redundant coordinates were performed at 38 steps of every 10° increment of the aryl-CO bond and C-N bond in NCDEA and PCDEA. Harmonic vibrational frequencies were calculated for all stationary points. For each optimized state, the frequency analysis showed the absence of imaginary frequencies, whereas each transition state showed a single imaginary frequency [30]. Visual inspection of the corresponding normal mode was used to confirm that the correct transition state had been found.

4.3.2. QST3 Method and Changing Dihedral Angles of Diethyl Groups Method

After 1D PES calculations, two assumed transition states of each aryl-CO and C-N bond rotation were re-optimized in QST3 algorithms [18] to confirm transition states. For this calculation, two lowest energy states and an estimated transition state were required. After using the QST3 method, one transition state was calculated. Frequency calculations were also performed to find the imaginary frequency corresponding to bond rotations, meaning that one transition state of those bond rotations could be found.

Diethyl rotations result in a different Gibbs energy barrier; therefore, before we optimized the confirmed transition states for the aryl-CO bond and C-N bond in NCDEA and PCDEA, we prepared for the conformers, of which all the atoms except ethyl were fixed, and the dihedral angles (α, β) of C (one ethyl)-*N*-C (the other ethyl)-C (the other ethyl) were −60°, 60°, and 180°, respectively (Scheme S2). Thus, each conformer had nine different combinations (Scheme S2) and was optimized to have stable energy states. Frequency calculations were employed in these nine conformers to find the transition states of the aryl-CO and N-C bonds for selecting the optimum Gibbs energy barrier. We also performed intrinsic reaction coordinates of two transition states from QST3 algorithms and the nine conformers from 2nd transition states in aryl-CO and C-N bond rotations to find actual transition states including one imaginary frequency.

4.3.3. 2D PES Method

The activation energy of aryl-CO and C-N bond rotation can be specifically determined by independently calculating aryl-CO as the X axis and C-N bond as the Y axis. 2D PESs were calculated for 38 steps of every 10° increment of aryl-CO and C-N bond rotations. Frequency calculations were performed in an estimated transition state in both NCDEA and PCDEA.

4.3.4. QTAIM Study

In order to estimate the electron density on the rotating bonds (aryl-CO and C-N), we performed a QTAIM study using the Multiwfn program [22]. The electron density was calculated at the bond critical point (CP(3,−1)).

Supplementary Materials: Supplementary materials include 3 schemes, 4 figures, and 11 tables. Scheme S1: *N,N*-diethylamide derivatives (1)–(4) synthesis by Oxalyl Chloride, Scheme S2: Varying the Dihedral angle of C-C-N-C (α, β) on Aryl-CO bond and N-C bond, Scheme S3: Angles of Amide Structure, Figure S1: 2D-EXSY spectra of *N,N*-diethylamide derivatives (1),(2), Figure S2: The Potential Energy Surface Graph of Aryl-CO (a) and C-N (b) rotation on NCDEA, Figure S3: The Potential Energy Surface Graph of Aryl-CO (a) and C-N (b) rotation on PCDEA, Figure S4: Structure of ground state and transition states (QST3) of Aryl-CO bond and C-N bond in NCDEA and PCDEA, Table S1: Comparisons of Gibbs Free energy of ^{1}H VT-NMR with those of scan coordinates calculated with the gaussian09 program, Table S2: The imaginary frequency on NCDEA and PCDEA, Table S3: Experimental and Theoretical Gibbs free energy of Aryl-CO bond in NCDEA, Table S4: Experimental and Theoretical Gibbs free energy of Aryl-CO bond in PCDEA, Table S5: Structure data of 2TS (diethyl-rotated conformers) of Aryl-CO bond in NCDEA and PCDEA, Table S6: Structure data of 2TS (diethyl-rotated conformers) of C-N bond in NCDEA, Table S7: Structure data of 2TS (diethyl-rotated conformers) of C-N bond in PCDEA, Table S8: Structure data of 2TS (diethyl-rotated conformers) of C-N bond in NCDEA and

PCDEA, Table S9: Structure data of TS of the concerted Aryl-CO/C-N bond in NCDEA and PCDEA, Table S10: Structure data of TS of the concerted Aryl-CO/C-N bond of 2D PES in NCDEA and PCDEA, Table S11: Electron density of rotating bonds.

Author Contributions: Conceptualization, S.A. and K.K.; Methodology, X.X.; Software, S.K. and K.J.; Validation, S.K, J.K. (Jieun Kim), D.W., S.K.C., W.C., K.J., S.A. and K.K.; Resources, X.X.; Data Curation, S.K., J.K. (Jieun Kim), D.W., S.-K.C. and W.C., Y.Y. and Z.Z.; Formal Analysis, S.K. and K.J.; Investigation, S.K., J.K. (Jieun Kim)., D.W., S.K.C. and W.C.; Resources, X.X.; Writing-Original Draft Preparation, S.K., J.K. (Junkyu Kim)., D.W. and S.K.C.; Writing-Review & Editing, S.K., J.K. (Junkyu Kim) and K.J.; Visualization, S.K., J.K. (Junkyu Kim) and K.J.; Supervision, K.J., S.A. and K.K.; Project Administration, S.A. and K.K.; Funding Acquisition, W.C. and K.K.

References

1. Chakrabarti, P.; Dunitz, J.D. Structural characteristics of the carboxylic amide group. *Helv. Chim. Acta* **1982**, *65*, 1555–1562. [CrossRef]
2. Hirota, E.; Sugisaki, R.; Nielsen, C.J.; Sørensen, G.O. Molecular structure and internal motion of formamide from microwave spectrum. *J. Mol. Spectrosc.* **1974**, *49*, 251–267. [CrossRef]
3. Kottas, G.S.; Clarke, L.I.; Horinek, D.; Michl, J. Artificial molecular rotors. *Chem. Rev.* **2005**, *105*, 1281–1376. [CrossRef] [PubMed]
4. Bragg, R.A.; Clayden, J.; Morris, G.A.; Pink, J.H. Stereodynamics of bond rotation in tertiary aromatic amides. *Chem. Eur. J.* **2002**, *8*, 1279–1289. [CrossRef]
5. Clayden, J.; Pink, J.H. Concerted rotation in a tertiary aromatic amide: Towards a simple molecular gear. *Angew. Chem. Int. Ed.* **1998**, *37*, 1937–1939. [CrossRef]
6. Bragg, R.A.; Clayden, J. Using symmetry to monitor geared bond rotation in aromatic amides by dynamic NMR. *Org. Lett.* **2000**, *2*, 3351–3354. [CrossRef] [PubMed]
7. Vargas, R.; Garza, J.; Dixon, D.; Hay, B.P. C(sp^2)−C(aryl) bond rotation barrier in *N*-methylbenzamide. *J. Phys. Chem.* **2001**, *105*, 774–778. [CrossRef]
8. Hammaker, R.M.; Gugler, B.A. An NMR study of hindered internal rotation in *N*,*N*-dialkyl amides. *J. Mol. Spectrosc.* **1965**, *17*, 356–364. [CrossRef]
9. Olsen, R.A.; Liu, L.; Ghaderi, N.; Johns, A.; Hatcher, M.E.; Mueller, L.J. The amide rotational barriers in picolinamide and nicotinamide: NMR and ab initio studies. *J. Am. Chem. Soc.* **2003**, *125*, 10125–10132. [CrossRef] [PubMed]
10. Skorupska, E.A.; Nazarski, R.B.; Ciechańska, M.; Jóźwiak, A.; Kłys, A. Dynamic 1 h NMR spectroscopic study of hindered internal rotation in selected *N*,*N*-dialkyl isonicotinamides: An experimental and DFT analysis. *Tetrahedron* **2013**, *69*, 8147–8154. [CrossRef]
11. Ahmed, A.; Bragg, R.A.; Clayden, J.; Lai, L.W.; McCarthy, C.; Pink, J.H.; Westlund, N.; Yasin, S.A. Barriers to rotation about the chiral axis of tertiary aromatic amides. *Tetrahedron* **1998**, *54*, 13277–13294. [CrossRef]
12. Kang, Y.K.; Park, H.S. Internal rotation about the C–N bond of amides. *J. Mol. Struct.* **2004**, *676*, 171–176. [CrossRef]
13. Clayden, J.; Pink, J.H.; Yasin, S.A. Conformationally interlocked amides: Remote asymmetric induction by mechanical transfer of stereochemical information. *Tetrahedron Lett.* **1998**, *39*, 105–108. [CrossRef]
14. Clayden, J.; Johnson, P.; Pink, J.H.; Helliwell, M. Atropisomeric amides as chiral ligands: Using (−)-sparteine-directed enantioselective silylation to control the conformation of a stereogenic axis. *J. Org. Chem.* **2000**, *65*, 7033–7040. [CrossRef] [PubMed]
15. Reeves, L.W.; Shaddick, R.C.; Shaw, K.N. Nuclear magnetic resonance studies of multi-site chemical exchange. III. Hindered rotation in dimethylacetamide, dimethyl trifluoro-acetamide, and dimethyl benzamide. *Can. J. Chem.* **1971**, *49*, 3683–3691. [CrossRef]
16. Campomanes, P.; Menéndez, M.I.; López, R.; Sordo, T.L. Stereodynamics of bond rotation in tertiary 1-naphthoic acid amides: A computational study. *J. Comput. Chem.* **2005**, *26*, 365–373. [CrossRef] [PubMed]
17. Krishnan, V.V.; Thompson, W.B.; Maitra, K.; Maitra, S. Modulations in restricted amide rotation by steric induced conformational trapping. *Chem. Phys. Lett.* **2012**, *523*, 124–127. [CrossRef] [PubMed]
18. Peng, C.; Bernhard Schlegel, H. Combining synchronous transit and quasi-newton methods to find transition states. *Isr. J. Chem.* **1993**, *33*, 449–454. [CrossRef]
19. Becke, A.D. Density-functional thermochemistry. III. The role of exact exchange. *J. Chem. Phys.* **1993**, *98*, 5648–5652. [CrossRef]

20. Becke, A.D. Density-functional exchange-energy approximation with correct asymptotic behavior. *Phys. Rev. A* **1988**, *38*, 3098–3100. [CrossRef]
21. Ercolani, G. Determination of the rotational barrier in ethane by vibrational spectroscopy and statistical thermodynamics. *J. Chem. Educ.* **2005**, *82*, 1703. [CrossRef]
22. Lu, T.; Chen, F. Multiwfn: A multifunctional wavefunction analyzer. *J. Comput. Chem.* **2012**, *33*, 580–592. [CrossRef] [PubMed]
23. Eor, S.; Hwang, J.; Choi, M.G.; Chang, S.-K. Fluorescent signaling of oxone by desulfurization of thioamide. *Org. Lett.* **2011**, *13*, 370–373. [CrossRef] [PubMed]
24. Perrin, C.L.; Dwyer, T.J. Application of two-dimensional NMR to kinetics of chemical exchange. *Chem. Rev.* **1990**, *90*, 935–967. [CrossRef]
25. Kumar, A.; Ernst, R.R.; Wüthrich, K. A two-dimensional nuclear overhauser enhancement (2D NOE) experiment for the elucidation of complete proton-proton cross-relaxation networks in biological macromolecules. *Biochem. Biophys. Res. Commun.* **1980**, *95*, 1–6. [CrossRef]
26. Frisch, M.J.; Trucks, G.W.; Schlegel, H.B.; Scuseria, G.E.; Robb, M.A.; Cheeseman, J.R.; Scalmani, G.; Barone, V.; Petersson, G.A.; Nakatsuji, H.; et al. *Gaussian 09, Revision D.01*; Gaussian, Inc.: Wallingford, CT, USA, 2016.
27. Lee, C.; Yang, W.; Parr, R.G. Development of the colle-salvetti correlation-energy formula into a functional of the electron density. *Phys. Rev. B* **1988**, *37*, 785–789. [CrossRef]
28. Hehre, W.J.; Ditchfield, R.; Pople, J.A. Self—Consistent molecular orbital methods. XII. Further extensions of gaussian—Type basis sets for use in molecular orbital studies of organic molecules. *J. Chem. Phys.* **1972**, *56*, 2257–2261. [CrossRef]
29. Tomasi, J.; Mennucci, B.; Cammi, R. Quantum mechanical continuum solvation models. *Chem. Rev.* **2005**, *105*, 2999–3094. [CrossRef] [PubMed]
30. Wigner, E. The transition state method. *Trans. Faraday Soc.* **1938**, *34*, 29–41. [CrossRef]

10

Selective C-N σ Bond Cleavage in Azetidinyl Amides under Transition Metal-Free Conditions

Hengzhao Li, Zemin Lai, Adila Adijiang, Hongye Zhao and Jie An *

College of Science, China Agricultural University, No. 2 Yuanmingyuan West Road, Beijing 100193, China; lihengzhao@cau.edu.cn (H.L.); 2016310060420@cau.edu.cn (Z.L.); 2015310060229@cau.edu.cn (A.A.); hongyezhao@cau.edu.cn (H.Z.)

* Correspondence: jie_an@cau.edu.cn

Abstract: Functionalization of amide bond via the cleavage of a non-carbonyl, C-N σ bond remains under-investigated. In this work, a transition-metal-free single-electron transfer reaction has been developed for the C-N σ bond cleavage of *N*-acylazetidines using the electride derived from sodium dispersions and 15-crown-5. Of note, less strained cyclic amides and acyclic amides are stable under the reaction conditions, which features the excellent chemoselectivity of the reaction. This method is amenable to a range of unhindered and sterically encumbered azetidinyl amides.

Keywords: amides; C-N σ bond cleavage; sodium; crown ether

1. Introduction

Amide is among the most ubiquitous functional groups [1]. Although the reductive functionalization of amides has been studied extensively, the majority of strategies have focused on the amide reductions via C-O or C-N cleavage, to afford the corresponding amines or alcohols (Scheme 1) [2–11]. Only a few examples were reported for the amide bond functionalization via the selective activation of the non-carbonyl, C-N σ bond, despite its considerable potential in the synthesis of amide linkage in both chemistry and biology (Scheme 1) [12,13]. In 2005, Aube and co-workers reported a highly unusual C-N σ bond cleavage in a class of specialized bridged lactams under catalytic hydrogenation conditions [14]. The twisted amide bond is the possible reason for the high activities of the C-N σ bonds in those substrates. Recently, Szostak and co-workers have developed a more general single electron transfer (SET) method for the reductive cleavage of C-N σ bonds in both planar and pyramidalized amides, using TmI_2-ROH reagent, which forms from a nonclassical lanthanide (II) iodide [15–18]. Given the high price of thulium, a corresponding SET protocol mediated by cheap electron donor reagents will be more desirable, which is the subject of this work.

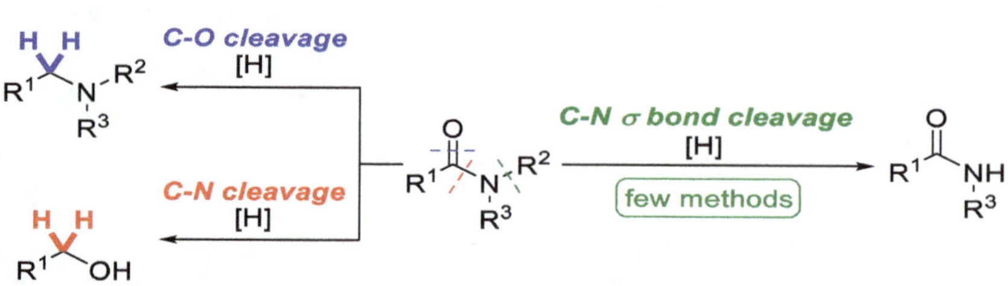

Scheme 1. Reductive functionalization of amides.

Electrides, in which anions are electrons, are a class of useful single electron-donor reagents. Solutions of alkali metal in liquid metals [19], first described by Sir Humphry Davy in 1803 [20], are among the most common electride systems, which have found wide applications in single-electron transfer reductions, including the venerable Birch reduction [21–23]. To avoid the hazards that are associated with the usage of liquid ammonia, new methods for the generation of electride salts using alkali metal and crown ethers were developed. However, freshly distilled sodium, potassium mirror, or highly pyrophoric potassium–sodium alloys were required to accelerate the reaction between the alkali metal and crown ethers [24–27]. Previously, our group developed a more practical protocol for the synthesis of electride salts, using sodium dispersions and 15-crown-5 [28]. Sodium dispersion in oil is a bench-stable and commercially available reagent with a high specific surface area [28–32]. The derived electride has already been successfully applied in a chemoselective ammonia-free Birch reduction [28]. However, the application of such an electride in other SET reactions remains under-investigated. Herein, we report the first electride-mediated C-N σ bond cleavage reaction in pyramidalized azetidinyl amides ($\tau = 3.3°$; $\chi_N = 32.5°$; 4-TolC(*O*)-azetidine, Winkler−Dunitz parameters [33]), using a cheap sodium dispersion/15-crown-5 reagent system under practical conditions.

2. Results and Discussion

Our study began with the optimization of the reaction conditions for the C-N σ bond cleavage process in azetidinyl amides, using **1a** as a model substrate. In the previous work, we have demonstrated that **1a** can be converted into the corresponding alcohol via C-N cleavage, using Na/EtOH [5]. We hypothesized that the absence of the proton donor would suppress the amide reduction pathway, and lead to the formation of **2a** via the C-N σ bond cleavage. The initial trial of the reaction, using 5.0 equiv. of sodium dispersions in Et_2O afforded **2a** in a moderate yield of 50% with the recovered starting material, accounting for the majority of the remaining mass balance (entry 1, Table 1). By-products derived from the amide reduction were not observed. The yield could be significantly improved by replacing Et_2O with tetrahydrofuran (THF), a solvent with higher dielectric constant, which indicated that the reaction might go through an outer-sphere electron transfer mechanism (entry 2, Table 1). As electrides are promising electron donors for the outer electron transfer processes, the feasibility of electride derived from sodium dispersions and 15-crown-5 was investigated. When 5.0 equiv. of Na/15-crown-5 was employed, satisfactory yields of **2a** were obtained in both Et_2O and THF (Entries 3 and 4, Table 1). Although Na/15-crown-5/*i*-PrOH is an effective system for Birch-type reductions, dearomatization was well-suppressed in the absence of a proton donor, and the reduction of the phenyl moiety in **1a** was not observed under the conditions using Na/15-crown-5 (Entries 3 and 4, Table 1). The reductive C-N σ bond cleavage is a two-electron process. However, shortening the amount of Na/15-crown-5 to 3.0 equiv. resulted in a much lower yield (entries 5 and 6, Table 1). Also, shortening the reaction time gave decreased yields (Entries 7 and 8, Table 1).

Table 1. Optimization of selective C-N σ bond cleavage in azetidinyl amides by Na/15-crown-5 [1].

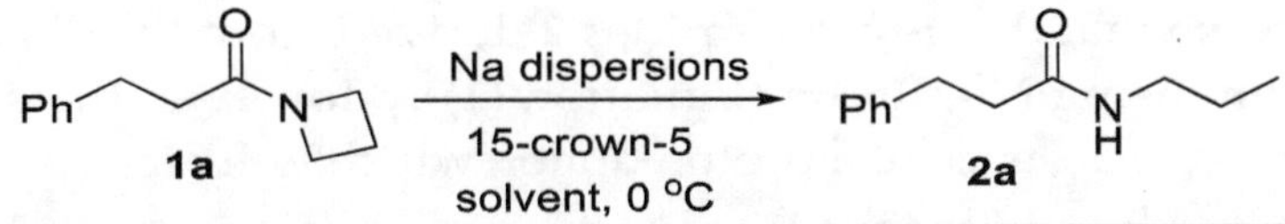

Entry	Sodium Dispersion [1] (Equiv)	15-crown-5 (Equiv)	Time	Solvent	Yield (%) [2]
1	5.0	0	2.0 h	Et_2O	50
2	5.0	0	2.0 h	THF	67
3	5.0	5.0	2.0 h	Et_2O	79
4	5.0	5.0	2.0 h	THF	80
5	4.5	4.5	2.0 h	Et_2O	65
6	3.0	3.0	2.0 h	Et_2O	56
7	5.0	5.0	1.0 h	THF	71
8	5.0	5.0	10 min	THF	47

THF: tetrahydrofuran. [1] Sodium dispersions in oil, 33.9 wt %, particle size 5–10 μm. [2] Determined by ^{1}H-NMR.

Next, the optimized conditions (Entry 4, Table 1) were applied to the selective C-N σ bond cleavage reactions. A broad range of aliphatic and aromatic azetidinyl amides were converted into the corresponding secondary *n*-butyl amides at high yields (Figure 1). Both the unhindered (e.g., **1a**, **1g**, and **1m**) and sterically encumbered (**1h**, **1i**, and **1j**) azetidinyl amides were viable substrates for this reaction. Aromatic rings were stable under the reaction conditions. By-products derived from the Birch-type dearomatization were not detected in any of the tested substrates (**1a–1h**). Substrate-bearing functional groups, such as methoxy group (**1d**) and alkene group (**1k**), were also readily converted into the corresponding *n*-butyl amides without the demethylation of the methoxy group or the reduction of the alkene group. In contrast, chloride (**1f**) were fully reduced when 8.0 equiv. of Na/15-crown-5 was used, which suggested that the potential application of this protocol in dehalogenation reactions. In addition, if the reaction with **1a** was quenched by D_2O, the corresponding deuterium labeled product, 3-phenyl-*N*-(propyl-3-*d*)propanamide, was detected, albeit in a low deuterium incorporation. Remarkably, this single electron transfer process is highly selective for azetidinyl amides. Less strained cyclic amides, such as pyrrolidinyl amide **1o** and piperidinyl amide **1p**, were very stable under the reaction conditions. Acyclic tertiary amide **1r** and secondary amide **1q** also did not undergo the cleavage reaction. Those observations suggested that the large ring strain in a four-membered ring of 25.4 kcal/mol (cf. aziridines, 27.5 kcal/mol) [33] is the possible driving force for the C-N σ bond cleavage process in azetidinyl amides.

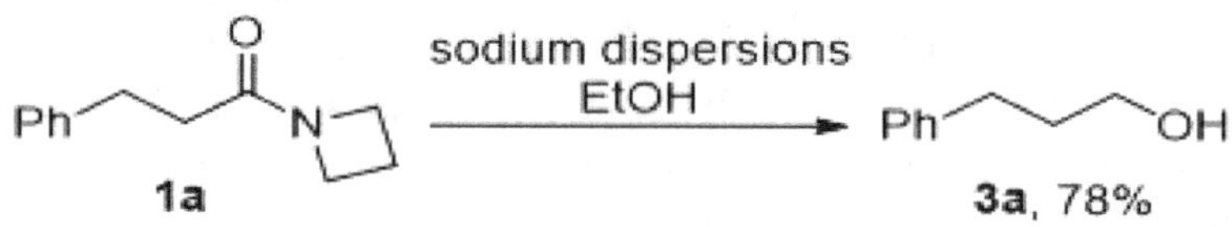

(1)

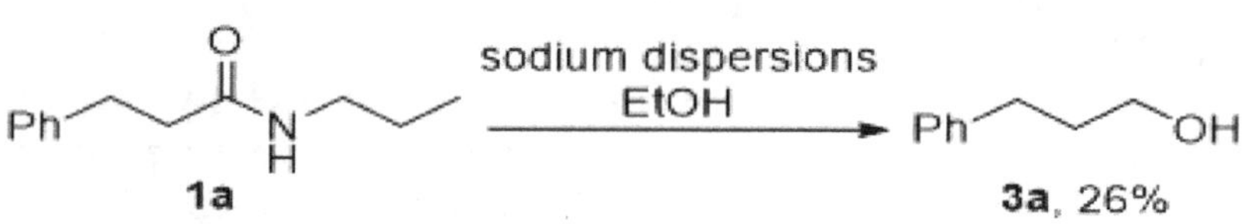

(2)

The key processes in this reaction involve the generation of ketyl-type radicals from a reversible electron transfer (**1**→**4**, Scheme 2) and the sequential C-N σ bond cleavage (**4**→**5**). The control reactions (Equations (1) and (2)) demonstrated that (a) in the presence of EtOH, alcohol **3** was formed as the major product; (b) it was difficult to convert secondary amide **2** to the corresponding alcohol **3** using Na/EtOH. These observations indicated that, in the presence of the proton donor, amide reduction via the C-N cleavage (**4**→**6**→**7**→**3**) was the dominant pathway. In addition, the ring-opening step (**4**→**5**) was relatively slow, so that the *n*-butyl amides **2** derived from the C-N σ bond cleavage were not detected in the reaction using Na/EtOH (Equation (1)). However, the second electron transfer (**6**→**7**) will be suppressed in the absence of a proton donor, which will alternately lead to the formation of a C-N σ bond cleavage product **2** (**4**→**5**→**2**).

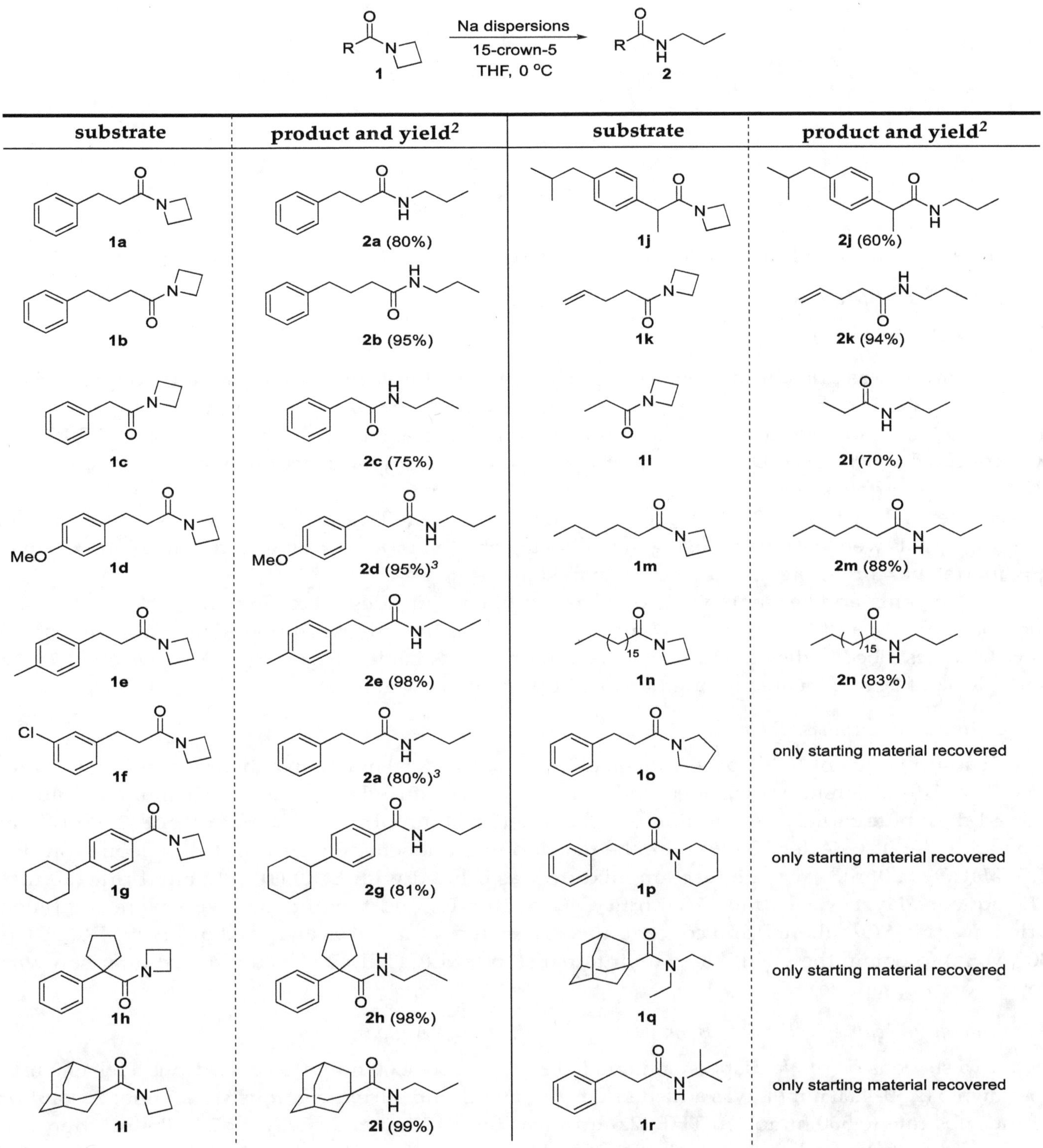

Figure 1. Selective C-N σ bond cleavage in azetidinyl amides by Na/15-crown-5 [1]. [1] Conditions: **1** (0.50 mmol, 1.0 equiv.), Na dispersions (5.0 equiv.), 15-crown-5 (5.0 equiv.), THF (3.0 mL), 0 °C. [2] Isolated yield. [3] Na dispersions (8.0 equiv.) and 15-crown-5 (8.0 equiv.) were used.

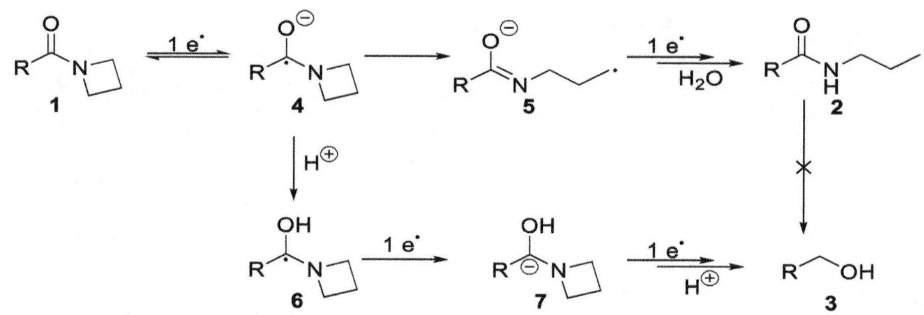

Scheme 2. Proposed mechanism for the C-N σ bond cleavage in azetidinyl amides with Na/15-crown-5.

3. Materials and Methods

3.1. General Information

Glassware was dried in an oven overnight before use. Thin-layer chromatography was carried out on SIL G/UV254 silica–aluminum plates, and plates were visualized using ultraviolet light (254 nm) and $KMnO_4$ solution. For flash column chromatography, silica gel 60, 35–70 μm was used. NMR data was collected at 300 MHz. Data was manipulated directly from the spectrometer or via a networked PC with appropriate software. All samples were analyzed in $CDCl_3$. Reference values for the residual solvent were taken as δ = 7.27 ($CDCl_3$) for ^{1}H-NMR; δ = 77.1 ($CDCl_3$) for ^{13}C-NMR. Multiplicities for coupled signals were designated using the following abbreviations: s = singlet, d = doublet, t = triplet, q = quartet, br = broad signal, and are given in Hz.

All solvents and reagents were used as supplied. Amides were prepared by the standard method [33]. **2g** and **2n** are novel compounds, and all the other compounds used in this study have been described in the literature or are commercially available. ^{1}H and ^{13}C NMR spectra of **2a–2n**, and HRMS of **2g** and **2n** are provided in the Supplementary Materials.

3.2. Optimization Studies (Table 1)

To a suspension of Na dispersion in oil (33.9 wt %, 1.50–2.50 mmol) in anhydrous solvent (0.5 mL), 15-crown-5 (0–2.50 mmol) was added under Ar at 0 °C and stirred vigorously for 5 min. The solution turned dark blue rapidly. A solution of substrate (0.500 mmol) in the same solvent (2.0 mL) was then added at 0 °C. After 10–120 min, the reaction was quenched by a saturated aqueous solution of $NaHCO_3$ (2.0 mL), and the reaction mixture was diluted with Et_2O (10 mL) and brine (20 mL). The aqueous layer was extracted with Et_2O (2 × 10 mL), and the organic layers were combined, dried over Na_2SO_4, filtered, and concentrated. Then, the sample was analyzed by ^{1}H-NMR ($CDCl_3$, 300 MHz) to obtain the yield, using an internal standard ($CHCl_2CHCl_2$) and by comparison with corresponding samples.

3.3. General Procedure for the C-N Bond Cleavage in Azetidinyl Amides

To a suspension of Na dispersion in oil (33.9 wt %, 2.50–4.00 mmol) in anhydrous THF (0.5 mL), 15-crown-5 (2.50–4.00 mmol) was added under Ar at 0 °C, and stirred vigorously for 5 min. A solution of the substrate (0.500 mmol) in THF (2.0 mL) was then added at 0 °C. After 2 h, the reaction was quenched by a saturated aqueous solution of $NaHCO_3$ (2.0 mL) and the reaction mixture was diluted with Et_2O (10 mL) and brine (20 mL). The aqueous layer was extracted with Et_2O (2 × 10 mL), and the organic layers were combined, dried over Na_2SO_4, filtered and concentrated. The crude product was purified by flash chromatography (silica, 0–50% hexane/EtOAc).

3-Phenyl-*N*-propylpropanamide (**2a**) [34]: white solid (76.5 mg, 80%). ^{1}H-NMR (300 MHz, $CDCl_3$) δ 7.32–7.24 (m, 2H), 7.24–7.15 (m, 3H), 5.56 (s, 1H), 3.16 (td, *J* = 7.1, 6.9 Hz, 2H), 2.96 (t, *J* = 7.7 Hz, 2H), 2.46 (t, *J* = 7.7 Hz, 2H), 1.44 (m, 2H), 0.84 (t, *J* = 7.4 Hz, 3H); ^{13}C-NMR (75 MHz, $CDCl_3$) δ 172.2, 140.9, 128.4, 128.3, 126.1, 41.2, 38.4, 31.8, 22.7, 11.3.

4-Phenyl-*N*-propylbutanamide (**2b**): colorless oil (97.5 mg, 95%). ^{1}H-NMR (300 MHz, $CDCl_3$) δ 7.33–7.23 (m, 2H), 7.23–7.13 (m, 3H), 5.49 (s, 1H), δ 3.20 (td, *J* = 7.1, 6.5 Hz, 2H), 2.65 (t, *J* = 7.5 Hz, 2H), 2.17 (t, *J* = 7.5 Hz, 2H), 1.97 (m, 2H), 1.51 (m, 2H), 0.91 (t, *J* = 7.4 Hz, 3H); ^{13}C-NMR (75 MHz, $CDCl_3$) δ 172.7, 141.6, 128.6, 128.4, 126. 0, 41.3, 36.0, 35.3, 27.2, 23.0, 11.4.

2-Phenyl-*N*-propylacetamide (**2c**) [35]: white solid (66.5 mg, 75%). ^{1}H-NMR (300 MHz, $CDCl_3$) δ 7.40–7.23 (m, 5H), 5.41 (s, 1H), 3.57 (s, 2H), δ 3.17 (td, *J* = 7.2, 6.2 Hz, 2H), 1.44 (m, 2H), 0.83 (t, *J* = 7.4 Hz, 3H); ^{13}C-NMR (75 MHz, $CDCl_3$) δ 171.0, 135.2, 129.5, 129.1, 127.4, 44.0, 41.4, 22.8, 11.3.

3-(4-Methoxyphenyl)-*N*-propylpropanamide (**2d**) [34]: white solid (105.1 mg, 95%). ^{1}H-NMR (300 MHz, $CDCl_3$) δ 7.11 (d, *J* = 8.1 Hz, 2H), 6.82 (d, *J* = 8.1 Hz, 2H), 5.53 (s, 1H), 3.77 (s, 3H), 3.16 (td, *J* = 6.7 × 2 Hz, 2H), 2.90 (t, *J* = 7.5 Hz, 2H), 2.42 (t, *J* = 7.5 Hz, 2H), 1.45 (m, 2H), 0.85 (t, *J* = 7.4 Hz, 3H); ^{13}C-NMR (75 MHz, $CDCl_3$) δ 172.2, 158.1, 133.0, 129.3, 114.0, 55.3, 41.2, 38.9, 31.0, 22.9, 11.3.

N-Propyl-3-(p-tolyl)propanamide (**2e**): white solid (100.6 mg, 98%). ^{1}H-NMR (300 MHz, $CDCl_3$) δ 7.18–6.99 (m, 4H), 5.72 (s, 1H), 3.16 (td, *J* = 6.5, 6.4 Hz, 2H), 2.91 (t, *J* = 7.8 Hz, 2H), 2.44 (t, *J* = 7.8 Hz, 2H), 2.30 (s, 3H), 1.45 (m, 2H), 0.85 (t, *J* = 7.4 Hz, 3H); ^{13}C-NMR (75 MHz, $CDCl_3$) δ 172.2, 137.9, 135.6, 129.1, 128.2, 41.2, 38.6, 31.4, 22.8, 21.0, 11.3.

3-Phenyl-*N*-propylpropanamide (**2a**) (derived from **1f**) [34]: white solid (76.5 mg, 80%).^{1}H-NMR (300 MHz, $CDCl_3$) δ 7.32–7.24 (m, 2H), 7.23–7.15 (m, 3H), 5.50 (s, 1H), 3.16 (td, *J* = 6.9, 6.5 Hz, 2H), 2.96 (t, *J* = 7.7 Hz, 2H), 2.46 (t, *J* = 7.7 Hz, 2H), 1.44 (m, 2H), 0.84 (t, *J* = 7.4 Hz, 3H); ^{13}C-NMR (75 MHz, $CDCl_3$) δ 172.1, 141.0, 128.6, 128.4, 126.3, 41.3, 38.6, 31.9, 22.9, 11.3.

4-Cyclohexyl-*N*-propylbenzamide (**2g**): white solid (99.4 mg, 81%). ^{1}H-NMR (300 MHz, $CDCl_3$) δ 7.73–7.66 (m, 2H), 7.26–7.21 (m, 2H), 6.32 (s, 1H), 3.39 (td, *J* = 7.4, 6.5 Hz, 2H), 2.53 (m, 1H), 1.94–1.73 (m, 5H), 1.61 (m, 2H), 1.43 – 1.22 (m, 5H), 0.96 (t, *J* = 7.4 Hz, 3H); ^{13}C-NMR (75 MHz, $CDCl_3$) δ 167.6, 151.6, 132.4, 127.0 (× 2), 44.5, 41.7, 34.3, 26.8, 26.1, 23.0, 11.5; HRMS (FTMS-ESI) m/z: $[M + 1]^+$ calc for $C_{16}H_{23}NO$ 246.1852, found 246.1849.

1-Phenyl-*N*-propylcyclopentane-1-carboxamide (**2h**): white solid (113.4 mg, 98%). ^{1}H-NMR (300 MHz, $CDCl_3$) δ 7.38–7.21 (m, 5H), 5.22 (s, 1H), 3.09 (td, *J* = 7.0, 5.9 Hz, 2H), 2.52–2.41 (m, 2H), 2.06–1.96 (m, 2H), 1.89–1.75 (m, 2H), 1.75–1.60 (m, 2H), 1.36 (m, 2H), 0.75 (t, *J* = 7.4 Hz, 3H); ^{13}C-NMR (75 MHz, $CDCl_3$) δ 176.4, 144.4, 128.6, 126.8 (× 2), 59.3, 41.4, 36.9, 24.0, 22.7, 11.1.

(3r,5r,7r)-*N*-Propyladamantane-1-carboxamide (**2i**): colorless oil (109.6 mg, 99%). ^{1}H-NMR (300 MHz, $CDCl_3$) δ 5.63 (s, 1H), 3.21 (td, *J* = 6.7, 6.3 Hz, 2H), 2.08–2.01 (m, 3H), 1.88–1.82 (m, 6H), 1.76–1.68 (m, 6H), 1.51 (m,2H), 0.91 (t, *J* = 7.4 Hz, 3H); ^{13}C-NMR (75 MHz, $CDCl_3$) δ 177.7, 41.0, 40.7, 39.4, 36.6, 28.3, 23.0, 11.4.

2-(4-Isobutylphenyl)-*N*-propylpropanamide (**2j**): colorless oil (74.2 mg, 60%). ^{1}H-NMR (300 MHz, $CDCl_3$) δ 7.22–7.16 (m, 2H), 7.14–7.08 (m, 2H), 5.41 (s, 1H), 3.53 (q, *J* = 7.2 Hz, 1H), 3.14 (td, *J* = 7.1, 5.9 Hz, 2H), 2.45 (d, *J* = 7.2 Hz, 2H), 1.84 (m, 1H), 1.51 (d, *J* = 7.2 Hz, 3H), 1.41 (m, 2H), 0.90 (d, *J* = 6.6 Hz, 6H), 0.80 (t, *J* = 7.4 Hz, 3H); ^{13}C-NMR (75 MHz, $CDCl_3$) δ 174.5, 140.7, 138.8, 129.6, 127.4, 46.8, 45.1, 41.3, 30.2, 22.8, 22.4, 18.5, 11.2.

N-Propylpent-4-enamide (**2k**): colorless oil (66.4 mg, 94%). ^{1}H-NMR (300 MHz, $CDCl_3$) δ 5.82 (t, *J* = 8.7 Hz, 1H), 5.69 (s, 1H), 5.04 (dd, *J* = 18.9, 13.6 Hz, 2H), 3.22 (d, *J* = 6.8 Hz, 2H), 2.33 (dt, *J* = 35.4, 7.5 Hz, 4H), 1.52 (dd, *J* = 14.5, 7.3 Hz, 2H), 0.92 (t, *J* = 7.2 Hz, 3H); ^{13}C-NMR (75 MHz, $CDCl_3$) δ 172.3, 137.2, 115.5, 41.3 36.0, 29.8, 22.9, 11.4.

N-Propylpropionamide (**2l**) [36]: colorless oil (40.3 mg, 70%). ^{1}H-NMR (300 MHz, $CDCl_3$) δ 5.44 (s, 1H), 3.22 (td, *J* = 7.2, 6.5 Hz, 2H), 2.20 (q, *J* = 7.6 Hz, 2H), 1.52 (m, 2H), 1.16 (t, *J* = 7.6 Hz, 3H), 0.93 (t, *J* = 7.6 Hz, 3H); ^{13}C-NMR (75 MHz, $CDCl_3$) δ 173.7, 41.3, 29.9, 23.0, 11.4, 10.0.

N-Propylhexanamide (**2m**): colorless oil (69.2 mg, 88%). ^{1}H-NMR (300 MHz, $CDCl_3$) δ 5.64 (s, 1H), 3.21 (td, *J* = 7.2, 6.1 Hz, 2H), 2.16 (t, *J* = 7.8 Hz, 2H), 1.63 (m, 2H), 1.52 (m, 2H), 1.36–1.26 (m, 4H), 0.96–0.85 (m, 6H); ^{13}C-NMR (75 MHz, $CDCl_3$) δ 173.2, 41.2, 36.9, 31.5, 25.6, 23.0, 22.5, 14.0, 11.4.

N-Propylstearamide (**2n**): white solid (135.1 mg, 83%). ^{1}H-NMR (300 MHz, $CDCl_3$) δ 5.63 (s, 1H), 3.21 (td, *J* = 6.9, 6.5 Hz, 2H), 2.16 (t, *J* = 7.6 Hz, 2H), 1.62 (m, 2H), 1.52 (m, 2H), 1.37–1.20 (m, 28H), 0.96–0.82 (m, 6H); ^{13}C-NMR (75 MHz, $CDCl_3$) δ 173.2, 41.2, 37.0, 32.0, 29.8 (× 5), 29.7 (× 3), 29.6, 29.4 (× 3), 25.9, 23.0, 22.7, 14.2, 11.4; HRMS (FTMS-ESI) m/z: $[M + 1]^+$ calc for $C_{21}H_{43}NO$ 326.3417, found 326.3408.

4. Conclusions

In summary, a transition metal-free method for the challenging C-N σ bond cleavage in azetidinyl amides has been developed, using sodium dispersions and 15-crown-5. This practical reaction requires only inexpensive air- and moisture-stable reagents. High yields were obtained across a broad range of aliphatic and aromatic azetidinyl amides. More importantly, full chemoselectivity over the reductive C-N σ bond cleavage of less strained cyclic amides and acyclic amides was achieved. This work represents the first application of an electride in the C-N σ bond cleavage in pyramidalized amides. The further application of the electride derived from sodium dispersions and crown ethers in new SET reactions will be the subject of our future research.

Author Contributions: Conceptualization, J.A. and H.L.; investigation, H.L., Z.L., A.A., and H.Z.; writing, J.A., H.L., Z.L., and H.Z.; supervision, J.A.

References

1. Greenberg, A.; Breneman, C.M.; Liebman, J.F. *The Amide Linkage: Structural Significance in Chemistry, Biochemistry and Materials Science*; Wiley: Hoboken, NJ, USA, 2000.
2. Volkov, A.; Tinnis, F.; Slagbrand, T.; Trillo, P.; Adolfsson, H. Chemoselective reduction of carboxamides. *Chem. Soc. Rev.* **2016**, *45*, 6685–6697. [CrossRef]
3. Dub, P.A.; Ikariya, T. Catalytic reductive transformations of carboxylic and carbonic acid derivatives using molecular hydrogen. *ACS Catal.* **2012**, *2*, 1718–1741. [CrossRef]
4. Addis, D.; Das, S.; Junge, K.; Beller, M. Selective reduction of carboxylic acid derivatives by catalytic hydrosilylation. *Angew. Chemie Int. Ed.* **2011**, *50*, 6004–6011. [CrossRef]
5. Zhang, B.; Li, H.; Ding, Y.; Yan, Y.; An, J. Reduction and reductive deuteration of tertiary amides mediated by sodium dispersions with distinct proton donor-dependent chemoselectivity. *J. Org. Chem.* **2018**, *83*, 6006–6014. [CrossRef] [PubMed]
6. Rasu, L.; John, J.M.; Stephenson, E.; Endean, R.; Kalapugama, S.; Clément, R.; Bergens, S.H. Highly enantioselective hydrogenation of amides via dynamic kinetic resolution under low pressure and room temperature. *J. Am. Chem. Soc.* **2017**, *139*, 3065–3071. [CrossRef] [PubMed]
7. Tinnis, F.; Volkov, A.; Slagbrand, T.; Adolfsson, H. Chemoselective reduction of tertiary amides under thermal control: formation of either aldehydes or amines. *Angew. Chemie Int. Ed.* **2016**, *55*, 4562–4566. [CrossRef] [PubMed]
8. Mukherjee, D.; Shirase, S.; Mashima, K.; Okuda, J. Chemoselective reduction of tertiary amides to amines catalyzed by triphenylborane. *Angew. Chemie Int. Ed.* **2016**, *55*, 13326–13329. [CrossRef]
9. Cabrero-Antonino, J.R.; Alberico, E.; Drexler, H.-J.; Baumann, W.; Junge, K.; Junge, H.; Beller, M. Efficient base-free hydrogenation of amides to alcohols and amines catalyzed by well-defined pincer imidazolyl–ruthenium complexes. *ACS Catal.* **2016**, *6*, 47–54. [CrossRef]

10. Szostak, M.; Spain, M.; Eberhart, A.J.; Procter, D.J. Highly chemoselective reduction of amides (primary, secondary, tertiary) to alcohols using SmI_2 /amine/H_2O under mild conditions. *J. Am. Chem. Soc.* **2014**, *136*, 2268–2271. [CrossRef]
11. Das, S.; Wendt, B.; Möller, K.; Junge, K.; Beller, M. Two iron catalysts are better than one: A general and convenient reduction of aromatic and aliphatic primary amides. *Angew. Chemie Int. Ed.* **2012**, *51*, 1662–1666. [CrossRef]
12. Hu, F.; Nareddy, P.; Lalancette, R.; Jordan, F.; Szostak, M. σ N-C Bond Difunctionalization in Bridged Twisted Amides: Sew-and-Cut Activation Approach to Functionalized Isoquinolines. *Org. Lett.* **2017**, *19*, 2386–2389. [CrossRef] [PubMed]
13. Hu, F.; Lalancette, R.; Szostak, M. Structural Characterization of N-Alkylated Twisted Amides: Consequences for Amide Bond Resonance and N-C Cleavage. *Angew. Chemie Int. Ed.* **2016**, *55*, 5062–5066. [CrossRef]
14. Lei, Y.; Wrobleski, A.D.; Golden, J.E.; Powell, D.R.; Aubé, J. Facile C-N cleavage in a series of bridged lactams. *J. Am. Chem. Soc.* **2005**, *127*, 4552–4553. [CrossRef] [PubMed]
15. Szostak, M.; Spain, M.; Procter, D.J. Uncovering the importance of proton donors in TmI_2-promoted electron transfer: facile C−N bond cleavage in unactivated amides. *Angew. Chemie Int. Ed.* **2013**, *52*, 7237–7241. [CrossRef] [PubMed]
16. Shi, S.; Szostak, M. Synthesis of nitrogen heterocycles using samarium (II) iodide. *Molecules* **2017**, *22*, 2018. [CrossRef] [PubMed]
17. Shi, S.; Szostak, R.; Szostak, M. Proton-coupled electron transfer in the reduction of carbonyls using SmI_2–H_2O: implications for the reductive coupling of acyl-type ketyl radicals with SmI_2–H_2O. *Org. Biomol. Chem.* **2016**, *14*, 9151–9157. [CrossRef] [PubMed]
18. Shi, S.; Szostak, M. Aminoketyl radicals in organic synthesis: Stereoselective cyclization of five- and six-membered cyclic imides to 2-azabicycles using SmI_2–H_2O. *Org. Lett.* **2015**, *17*, 5144–5147. [CrossRef] [PubMed]
19. Thompson, J.C. *Electrons in Liquid Ammonia*; Oxford University Press: Oxford, UK, 1976.
20. Thomas, S.J.M.; Edwards, P.P.; Kuznetsov, V.L. Sir Humphry Davy: boundless Chemist, physicist, poet and man of action. *Chem. Phys. Chem.* **2008**, *9*, 59–66. [CrossRef]
21. Rabideau, P.W.; Marcinow, Z. The Birch reduction of aromatic compounds. In *Organic Reactions*; John Wiley & Sons, Inc.: Hoboken, NJ, USA, 1992; pp. 1–334.
22. Zimmerman, H.E. A mechanistic analysis of the Birch reduction. *Acc. Chem. Res.* **2012**, *45*, 164–170. [CrossRef]
23. Cossy, J.; Gille, B.; Bellosta, V. Facile synthesis of spirocyclic systems through the intramolecular addition of ketyl radicals via the sodium/ammonia reduction of δ,ε-unsaturated carboxylic esters. *J. Org. Chem.* **1998**, *63*, 3141–3146. [CrossRef]
24. Dye, J.L. Electrons as anions. *Science* **2003**, *301*, 607–608. [CrossRef] [PubMed]
25. Dye, J.L. Electrides: ionic salts with electrons as the anions. *Science* **1990**, *247*, 663–668. [CrossRef] [PubMed]
26. Dye, J.L. Compounds of alkali metal snions. *Angew. Chemie Int. Ed. English* **1979**, *18*, 587–598. [CrossRef]
27. Jedliński, Z. Novel electron-transfer reactions mediated by alkali metals complexed by macrocyclic ligand. *Acc. Chem. Res.* **1998**, *31*, 55–61. [CrossRef]
28. Lei, P.; Ding, Y.; Zhang, X.; Adijiang, A.; Li, H.; Ling, Y.; An, J. A practical and chemoselective ammonia-free Birch reduction. *Org. Lett.* **2018**, *20*, 3439–3442. [CrossRef] [PubMed]
29. Ding, Y.; Luo, S.; Adijiang, A.; Zhao, H.; An, J. Reductive deuteration of nitriles: the synthesis of α,α-dideuterio amines by sodium mediated electron transfer reactions. *J. Org. Chem.* **2018**, *83*, 12269–12274. [CrossRef] [PubMed]
30. Han, M.; Ding, Y.; Yan, Y.; Li, H.; Luo, S.; Adijiang, A.; Ling, Y.; An, J. Transition-metal-free, selective reductive deuteration of terminal alkynes with sodium dispersions and EtOD-d_1. *Org. Lett.* **2018**, *20*, 3010–3013. [CrossRef]
31. Li, H.; Zhang, B.; Dong, Y.; Liu, T.; Zhang, Y.; Nie, H.; Yang, R.; Ma, X.; Ling, Y.; An, J. A selective and cost-effective method for the reductive deuteration of activated alkenes. *Tetrahedron Lett.* **2017**, *58*, 2757–2760. [CrossRef]
32. Han, M.; Ma, X.; Yao, S.; Ding, Y.; Yan, Z.; Adijiang, A.; Wu, Y.; Li, H.; Zhang, Y.; Lei, P.; et al. Development of a modified Bouveault–Blanc reduction for the selective synthesis of α,α-dideuterio alcohols. *J. Org. Chem.* **2017**, *82*, 1285–1290. [CrossRef]

33. Liu, C.; Achtenhagen, M.; Szostak, M. Chemoselective ketone synthesis by the addition of organometallics to *N*-Acylazetidines. *Org. Lett.* **2016**, *18*, 2375–2378. [CrossRef]
34. Štefane, B.; Polanc, S. hydrogenation of BF_2 complexes with 1,3-dicarbonyl ligands. *Tetrahedron* **2009**, *65*, 2339–2343. [CrossRef]
35. Ignatenko, V.A.; Deligonul, N.; Viswanathan, R. Branch-Selective Synthesis of Oxindole and Indene Scaffolds: Transition Metal-Controlled Intramolecular Aryl Amidation Leading to C3 Reverse-Prenylated Oxindoles. *Org. Lett.* **2010**, *12*, 3594–3597. [CrossRef] [PubMed]
36. Gajda, T.; Zwierzak, A. Phase-transfer-catalysed *N*-alkylation of carboxamides and sulfonamides. *Synthesis* **1981**, *1981*, 1005–1008. [CrossRef]

11

Chemistry of Bridged Lactams: Recent Developments

Roman Szostak [1] and Michal Szostak [2,*]

1 Department of Chemistry, Wroclaw University, F. Joliot-Curie 14, 50-383 Wroclaw, Poland; roman.szostak@chem.uni.wroc.pl
2 Department of Chemistry, Rutgers University, 73 Warren Street, Newark, NJ 07102, USA
* Correspondence: michal.szostak@rutgers.edu

Academic Editor: Michal Szostak

Abstract: Bridged lactams represent the most effective and wide-ranging method of constraining the amide bond in a non-planar conformation. A previous comprehensive review on this topic was published in 2013 (*Chem. Rev.* **2013**, *113*, 5701–5765). In the present review, which is published as a part of the *Special Issue on Amide Bond Activation*, we present an overview of the recent developments in the field of bridged lactams that have taken place in the last five years and present a critical assessment of the current status of bridged lactams in synthetic and physical organic chemistry. This review covers the period from 2014 until the end of 2018 and is intended as an update to *Chem. Rev.* **2013**, *113*, 5701–5765. In addition to bridged lactams, the review covers recent advances in the chemistry of bridged sultams, bridged enamines and related non-planar structures.

Keywords: amide bond; bridged lactams; twisted amides; amides; Winkler-Dunitz parameters; N–C activation; hypersensitivity; nitrogen heterocycles; distortion; bridged sultams

1. Introduction

The amide bond is arguably the most important linkage in chemistry and biology [1]. Typical amide bonds are planar as a result of amidic resonance ($n_N \rightarrow \pi^*_{C=O}$ conjugation, 15–20 kcal/mol) (Figure 1A) [2].

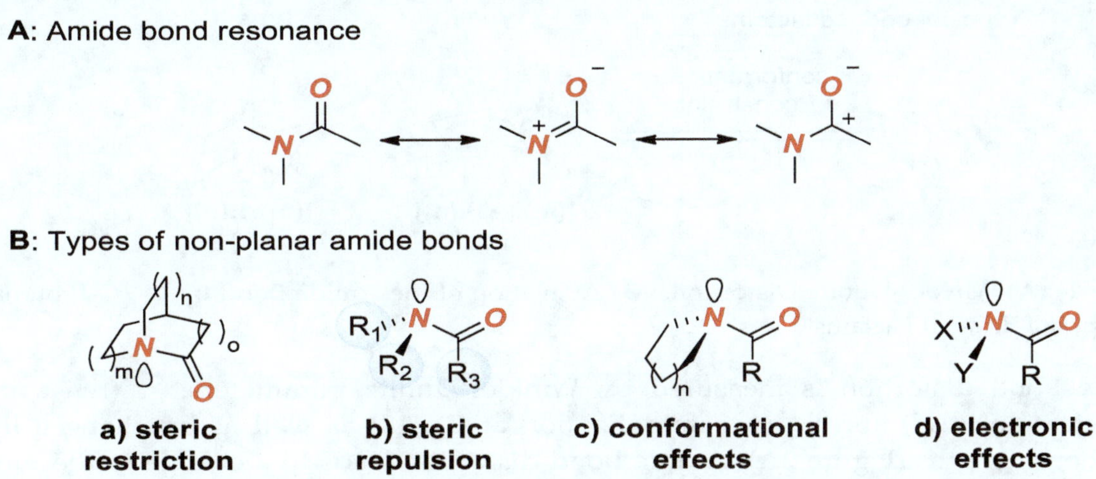

Figure 1. (**A**) Amide Bond Resonance. (**B**) Types of Distorted Amide Bonds.

The redesign of the amide bond geometry through structural and electronic changes of substituents comprising the amide bond has had a profound impact on the physico-chemical properties of amides [3–6]. The alteration of the amide bond geometry generally leads to a reversal of traditional properties of amides, such as lower barrier to cis-trans rotation, increased length of the N-C(O) bond, favored protonation at the nitrogen atom, and increased reactivity in nucleophilic addition and hydrolysis [3–6]. The geometric and structural changes of the amide bond are an established technique to affect properties of amide bonds in biology and medicinal chemistry [7–10], while recent advances in selective metal insertion into the amide bond driven by its distortion represent a thriving and general concept in organic synthesis [11,12]. In general, amide bond distortion can be achieved by four methods (Figure 1B): (1) steric restriction, (2) steric repulsion, (3) conformation effects, and (4) electronic effects. Out of these methods, the most effective one by far is steric restriction. Typically, steric restriction involves constraining the amide bond in a rigid bicyclic ring system with a nitrogen atom positioned at a bridgehead position. This allows one to constrain the typical planar amide bond in a non-planar conformation with the magnitude of distortion principally controlled by the type of ring system (Figure 2). To date, bridged lactams represent the only method that has allowed for a substantial distortion, exceeding 60% of the maximum theoretical value of the amide bond [3–6,11,12].

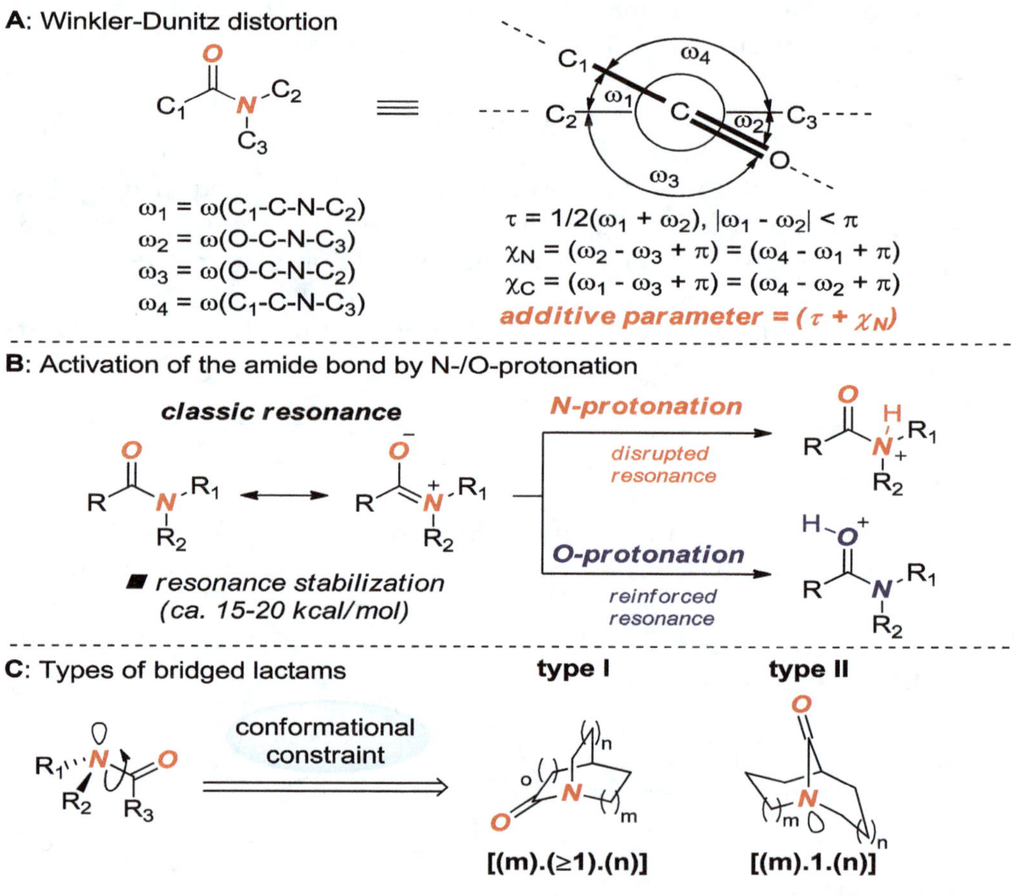

Figure 2. (**A**) Winkler-Dunitz Distortion. (**B**) Activation of the Amide Bond by *N*-/*O*-Protonation. (**C**) Types of Bridged Lactams.

Amide bond distortion is measured by Winkler-Dunitz parameters: τ (twist angle), χ_N (pyramidalization at N) and χ_C (pyramidalization at C) [13] as well as by changes in N–C(O) and C=O bond lengths (Figure 2A). Amide bond distortion leads to a change of thermodynamic *N*-/*O*-protonation aptitude, which is a key effect that controls the reactivity of non-planar amide bonds (Figure 2B) [11]. The properties of amide bonds in bridged lactams are further amplified by a type of bridged lactam scaffold (Figure 2C). In general, bridged lactams are classified into amides in which the N–C(O) bond is placed on a one-carbon bridge or on a larger bridge, with the former enjoying additional stabilization through transannular scaffolding effects.

In this review, published as a part of the *Special Issue on Amide Bond Activation*, we present an overview of the recent developments in the field of bridged lactams and present a critical assessment of the current status of bridged lactams. This review covers the period from 2014 until the end of 2018 and is intended as an update to the previous comprehensive review on topic, *Chem. Rev.* **2013**, *113*, 5701–5765 [3]. In addition to bridged lactams, the review covers recent advances in the chemistry of bridged sultams, bridged enamines and related non-planar structures. For additional coverage, the reader is referred to previous reviews on bridged lactams [4–6]. It is our hope that the review will serve as a useful reference for chemists involved in various aspects of activating the amide bond and stimulate further research in this area.

2. Synthesis, Properties and Reactivity of Bridged Lactams

Recent advances in the field of bridged lactams include: (1) identification of the additive Winkler-Dunitz parameter, (2) synthesis of extremely twisted non-stabilized amides, (3) synthesis of novel bridged lactams, and (4) new examples of reactivity of non-planar amides.

In 2015, we have identified the additive Winkler-Dunitz distortion parameter ($\Sigma\tau + \chi_N$), sum of twist and pyramidalization at nitrogen angles, as a more accurate prediction of the structural and energetic properties of non-planar amides than either twist or pyramidalization alone (Figures 3 and 4) [14,15]. A computational study to determine the effect of amide distortion on *N*-/*O*-protonation using a set of lactams comprehensively covering the entire distortion range (Figure 3) revealed a linear correlation between the composite Winkler-Dunitz parameter ($\Sigma\tau + \chi_N$) and *N*-/*O*-protonation aptitude (Figure 4) [14]. Our subsequent study demonstrated that the additive Winkler-Dunitz parameter ($\Sigma\tau + \chi_N$) gives linear correlations vs. structural and other energetic parameters (resonance energies, atomic charges, frontier molecular orbitals, infrared frequencies) [15].

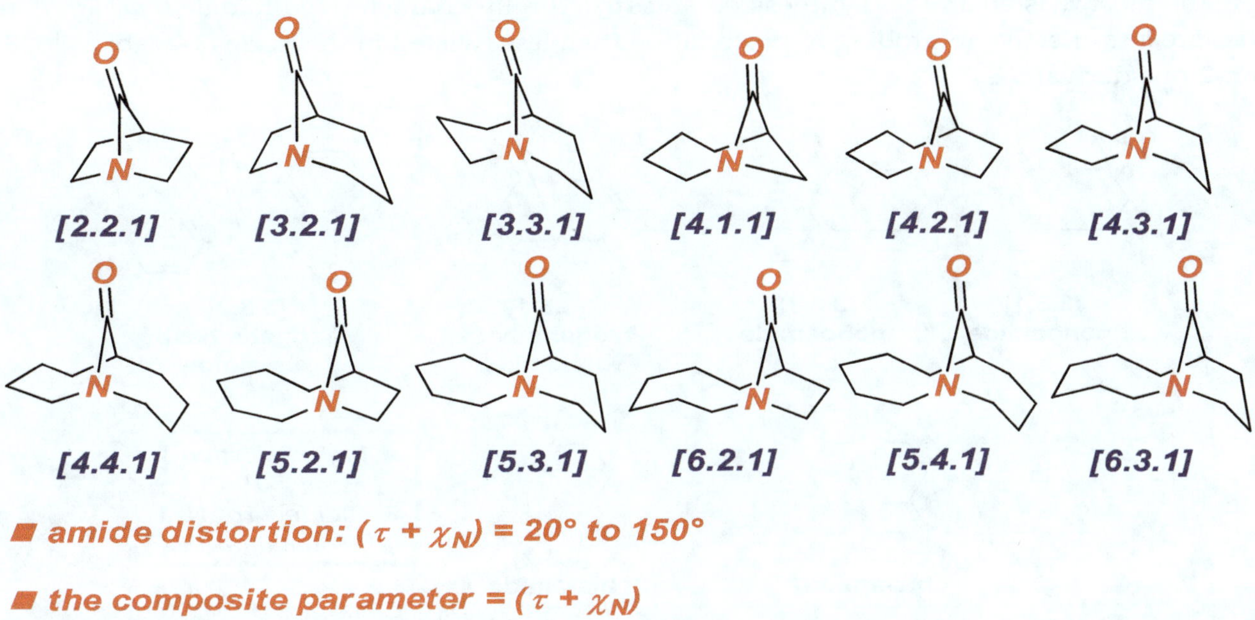

Figure 3. Additive Winkler-Dunitz Distortion Parameter: Sum of Twist Angle and Nitrogen Pyramidalization ($\Sigma\tau + \chi_N$).

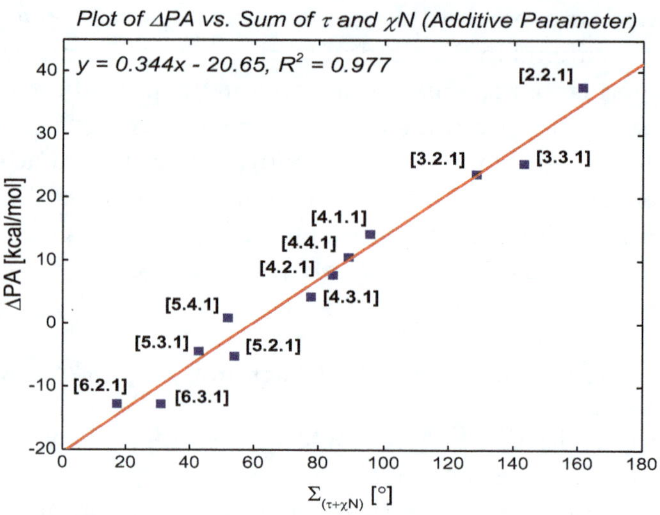

Figure 4. Plot of ΔPA (PA = Experimental Protonation Affinity) to the Sum of Twist and Pyramidalization at Nitrogen Angles (Additive Winkler-Dunitz Parameter: $\Sigma\tau + \chi_N$). Note that ($\Sigma\tau + \chi_N$) gives linear correlations vs. structural and other energetic parameters. See [15]. Also note comprehensive studies [16–18].

Since (1) amide bond distortion typically hinges upon both twist and pyramidalization, and (2) the additive Winkler-Dunitz parameter gives a more accurate prediction of geometric changes of the amide bond, this parameter should be routinely reported to describe structural variations of all non-planar amide bonds. Recently, we have utilized the additive Winkler-Dunitz parameter ($\Sigma\tau + \chi_N$) to determine the origin of high twist and *N*-/*O*-protonation aptitude of Tröger's base twisted amides (Figure 5) [19]. Perhaps surprisingly, we found that although Tröger's base twisted bis-amides are among the most twisted amides synthesized and structurally-characterized to date (*vide infra*), these amides are less effective in probing *N*-protonation than less twisted in this series 1-azabicyclo [3.3.1] nonan-2-one derivatives.

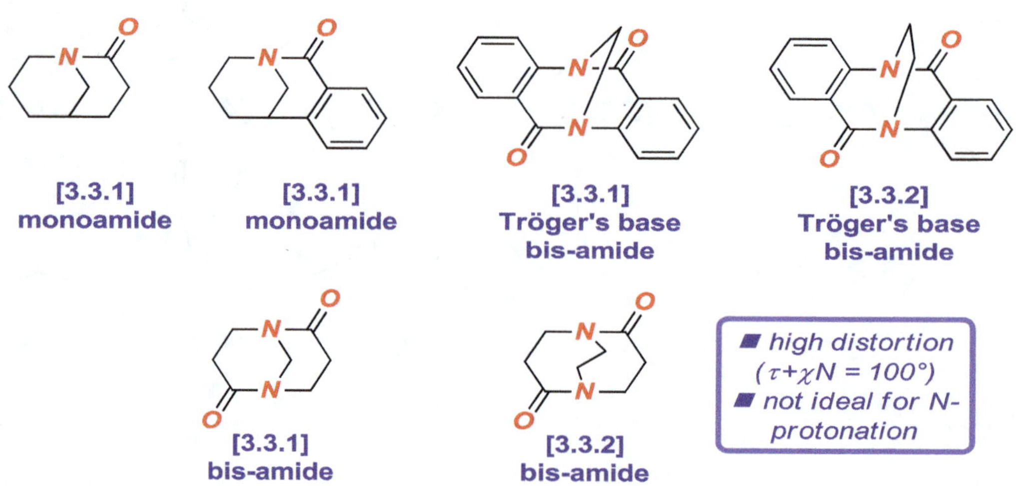

Figure 5. Additive Winkler-Dunitz Distortion Parameter in Tröger's Base Twisted Amides.

In 2016, the extremely twisted 7-hypoquinuclidonium tetrafluoroborate was reported by Stoltz and co-workers (Figure 6) [20]. The group has impressively exploited the intramolecular Aubé-Schmidt reaction to access the [2.2.1] bridged scaffold. This unconventional amide bond forming strategy represents a general approach to this and another extremely twisted amide, 2-quinuclidonium tetrafluoroborate (see Figure 6, box) [21], in the absence of nucleophiles that would likely decompose both compounds. It should be noted that the target twisted amide was isolated as a HBF_4 salt or BF_3

complex. The latter compound was fully characterized by X-ray crystallography, revealing one of the most twisted amide bonds isolated to date (τ = 90.0°, χ_N = 69.8°; N–C(O) = 1.526 Å, C=O = 1.186 Å).

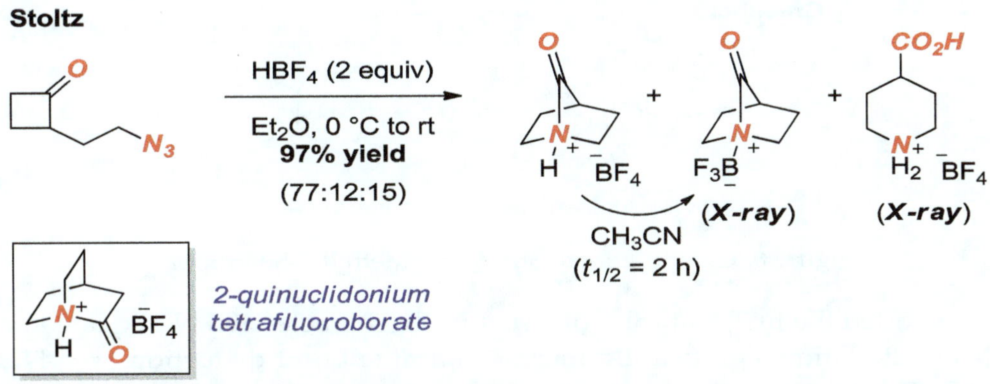

Figure 6. Synthesis of 7-Hypoquinuclidonium Tetrafluoroborate.

It is now well-established that *N*-coordination of the amide bond increases twist and pyramidalization (*vide infra*). Thus, with the exception of a structurally-unique 1-adamantan-2-one derivatives (see below), there are no examples of unconstrained structurally-characterized amides with a combined Winkler-Dunitz parameter ($\Sigma\tau + \chi_N$) exceeding 100° in the neutral form.

In 2016, the synthesis of another extremely twisted amide was reported by Komarov, Kirby et al. (Figure 7) [22]. These researchers achieved the synthesis of the parent 1-aza-2-adamantanone via a route consisting of thermal amidation of the *N*-Boc protected amino acid. Previous calculations showed a significant stabilizing effect of the methyl groups in the trimethyl twisted amide derivative (see Figure 7, box) [23]. The parent 1-aza-2-adamantanone was fully characterized after protonation as HBF_4 salt (τ = 88.1°, χ_N = 58.0°; N-C = 1.508 Å, C=O = 1.186 Å). Furthermore, the authors obtained the X-ray structure of the α-monomethylated 1-aza-2-adamantanone in a neutral form (τ = 90.0°, χ_N = 61.8°; N-C = 1.448 Å, C=O = 1.201 Å). Similar to the [2.2.1] amide synthesized by the Stoltz group (cf. [2.2.2] amide), the parent 1-aza-2-adamantanone was found to be more reactive in reactions with nucleophiles than the previous "most twisted amide" 3,5,7-trimethyl-1-azaadamantan-2-one [24].

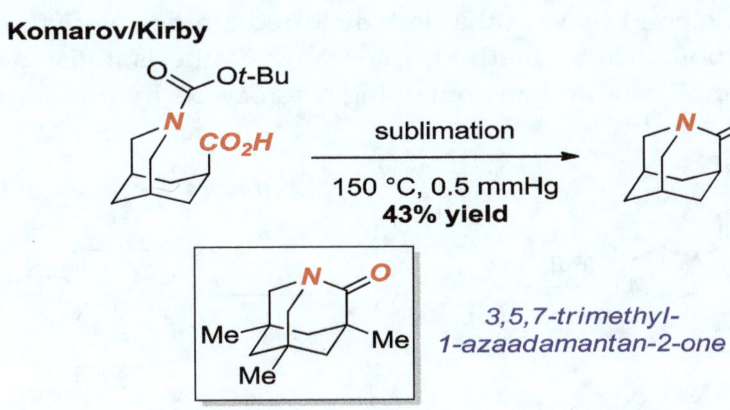

Figure 7. Synthesis of Parent 1-Azaadamantan-2-one.

In 2017, Greenberg and co-workers reported an interesting study of a silicon-containing twisted amide in a [3.3.3] scaffold (Figure 8) [25]. Heteroatom-containing derivatives of bridged lactams have received considerable attention as a means of facilitating the synthesis and tuning properties of the twisted amide bond. Through computations, the authors demonstrated that the nitrogen atom in 1-methyl-4-silatranone would be more similar to a lactam rather than a silatrane with a long intramolecular N-Si bond (N-Si = 2.902 Å vs. 1-methylsilatrane, N-Si = 2.466 Å). Although the attempted synthesis via condensation of $(HOCH_2CH_2)_2N(COCH_2OH)$ with various silanes was unsuccessful, the study lays a foundation for the synthesis of silicon-containing twisted amides.

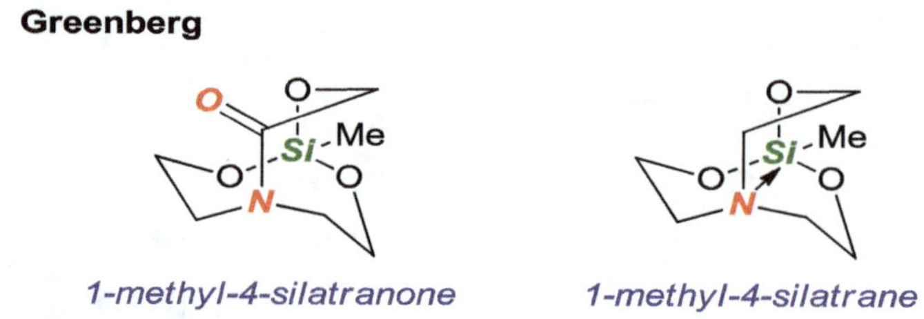

Figure 8. Computational Study of 1-Methyl-4-Silatranone.

In 2016, we reported the first examples of structurally-characterized *N*-alkylated bridged lactams (Figure 9) [26]. *N*-Alkylation significantly increased amide bond distortion (τ = 44.0°, χ_N = 58.3°; N-C = 1.554 Å, C=O = 1.192 Å). Furthermore, we demonstrated that *N*-coordination activated the twisted amide bond towards σ N–C bond activation by Pd-catalyzed hydrogenation.

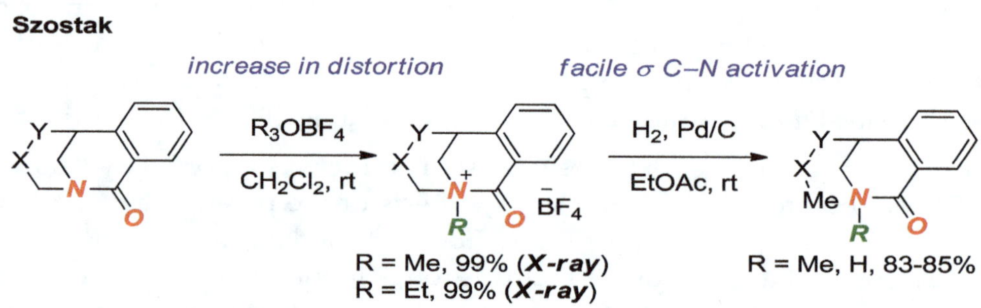

Figure 9. *N*-Alkylation of Bridged Lactams as a Trigger for σ N-C Bond Activation.

Transition-metal-free σ N-C bond activation in bridged lactams was reported by our group in 2017 (Figure 10) [27]. Facile assembly of the twisted amide scaffold by intramolecular Heck reaction, followed by *N*-alkylation and selective σ bond cleavage established a "sew-and-cut" approach to complex isoquinoline-2-ones by a formal di-functionalization of the N-C amide bond. The reactivity was correlated with amide bond twist in that less distorted amides were found unreactive. Given the utility of mild difunctionalization methods in organic synthesis, twisted lactams are attractive intermediates for the synthesis of nitrogen-containing heterocycles by this approach.

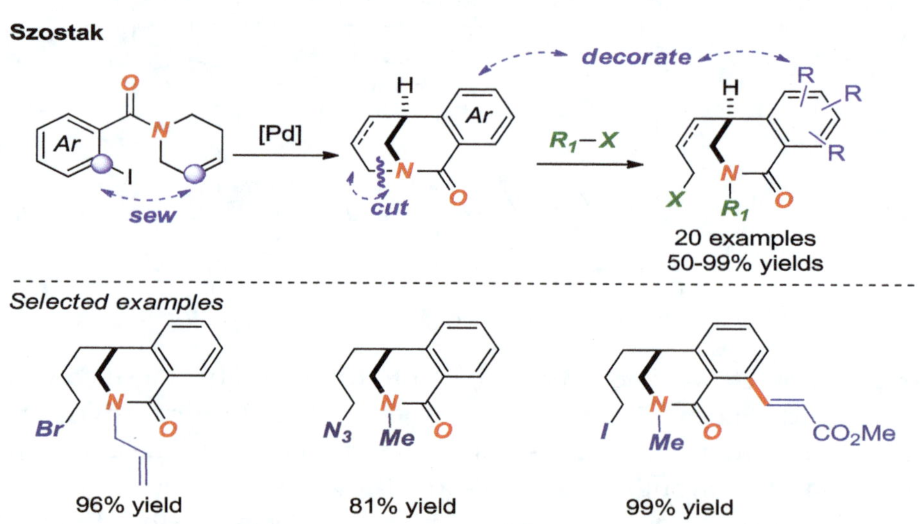

Figure 10. Sew-And-Cut of Bridged Lactams by a Transition-Metal-Free σ N–C Bond Activation.

In 2018, Marsden, Nelson and co-workers reported the synthesis of a set of bridged lactams with [3.3.1] and [4.3.1] scaffolds as part of their research on fragment-based drug discovery (Figure 11) [28]. This approach nicely utilizes the presence of an additional heteroatom in the twisted amide structure to facilitate the synthesis of starting materials. The cyclization was carried out according to the established lactamization protocol mediated by Bu_2SnO [29]. Due to the unique shape diversity, bridged lactams hold a significant potential as unexplored scaffolds in drug discovery.

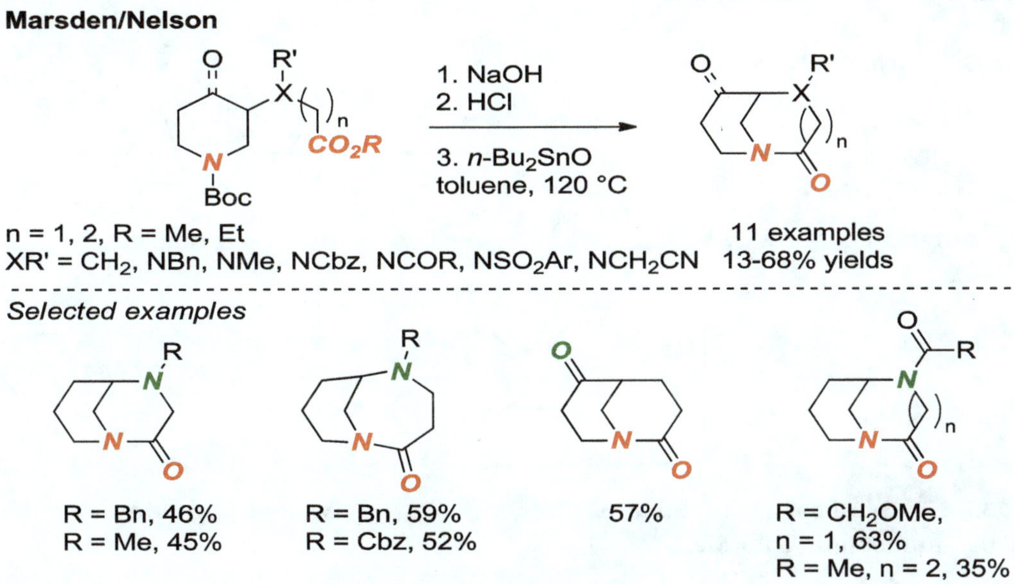

Figure 11. Fragment-Based Drug Discovery using Bridged Lactams.

In 2017, Stoltz and co-workers reported the synthesis of a bridged hydantoin by asymmetric α-allylation, Ru-catalyzed olefin isomerization, oxidative cleavage, Curtius rearrangement and *N*-cyclization onto the isocyanate (Figure 12) [30]. The cyclization is performed by the same mechanism as reported previously by Brouillette [31]; however, the method avoids the use of toxic lead acetate. The bridged hydantoin was fully characterized by X-ray crystallography (τ = 36.4°, χ_N = 50.2°; N-C(O) = 1.404 Å; C=O = 1.210 Å). This research illustrates one of the few methods for the synthesis of an enantioenriched twisted amide bond.

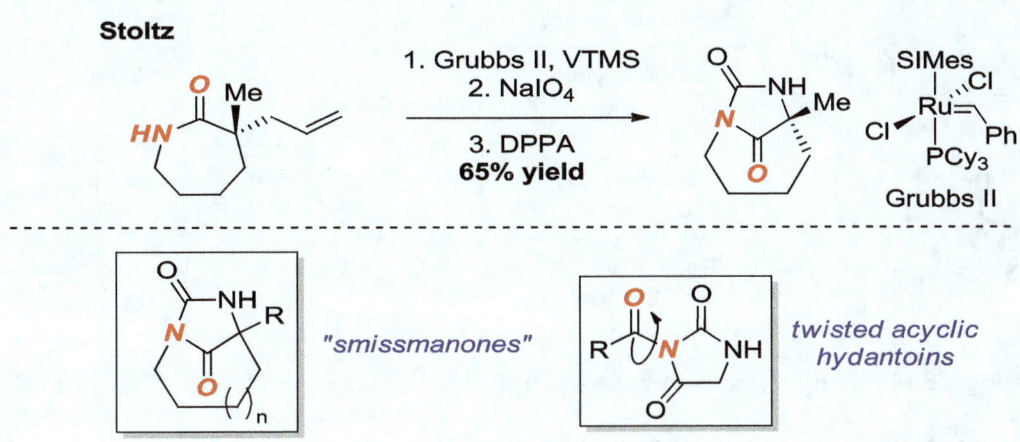

Figure 12. Synthesis of a Bridged Hydantoin. VTMS = Vinyloxytrimethylsilane. DPPA = diphenyl phosphoryl azide. SIMes = 1,3-Bis(2,4,6-trimethylphenyl)-2-imidazolidinylidene.

Bridged hydantoins of this type were originally proposed by Smissman as potential anticonvulsants (Figure 12, box) [32]. Our group has recently reported the structural characterization of related acyclic twisted *N*-acyl-hydantoins (Figure 12, box) [33].

In 2017, Gouverneur, Cvengros and co-workers reported the synthesis of ethano Tröger's base twisted bis-amides (Figure 13) [34]. Their approach involves one-step oxidation of the ethano-Tröger's base precursors under the conditions reported earlier by Wärnmark (Figure 13, box) [35]. The ethano Tröger's base twisted bis-amide is significantly twisted (τ = 29.8°, χ_N = 45.5°; N–C(O) = 1.401 Å, C=O = 1.215 Å). The advantage of this method is rapid access to a twisted amide bond; however, our study suggests that the presence of the fused aromatic ring is detrimental to the *N*-protonation reactivity in this class of twisted amides [19].

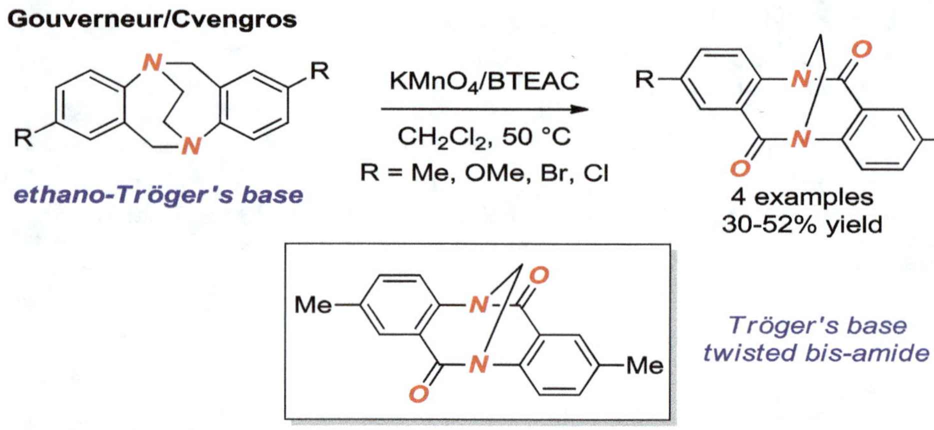

Figure 13. Synthesis of Ethano Tröger's Base Twisted Amides. BTEAC = Benzyltriethylammonium Chloride.

Satyanarayana and Helmchen reported asymmetric synthesis of bridged lactams in [3.3.1] and [4.3.1] scaffolds using Ir-catalyzed allylic amination as an enantioselectivity determining key step (Figure 14) [36]. With allylic amines in hand, the synthesis was completed by well-established amidation/Heck cyclization sequence. The structure of one of the lactams in the [3.3.1] series was confirmed by X-ray crystallography (τ = 30.8°, χ_N = 52.7°; N–C(O) = 1.393 Å, C=O = 1.218 Å).

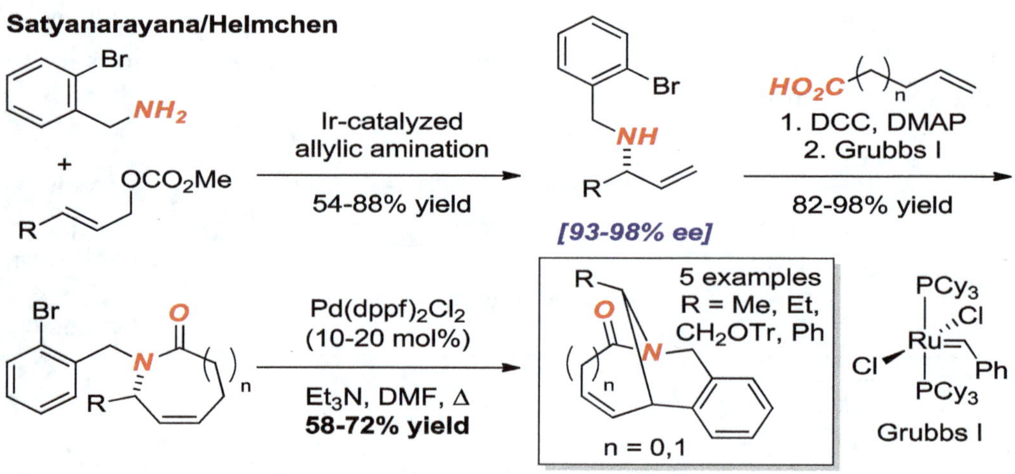

Figure 14. Enantioselective Synthesis of Bridged Lactams by Allylic Amination.

An inventive strategy for the synthesis of bridged lactams in the [5.3.1] and larger scaffolds was reported by Liu and co-workers (Figure 15) [37]. The authors developed a new process for radical aryl migration with chirality transfer to form macrocyclic ketones. As an application of this method, they subjected several azido-ketones to the transannular Aubé-Schmidt rearrangement, resulting in the formation of bridged lactams in 40–63% yields. Nevertheless, bridged lactams with the overall sum of carbon atoms forming the bridged structure of ten or more, are similar in properties to planar amides.

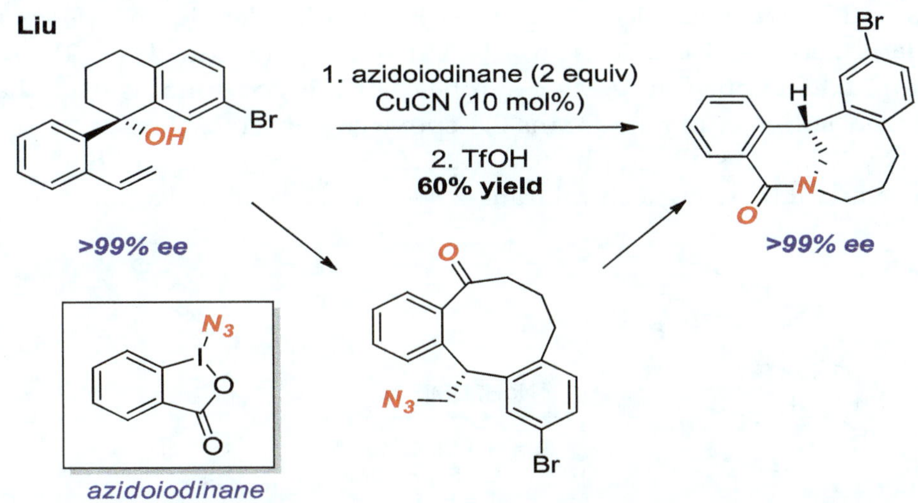

Figure 15. Enantioselective Synthesis of Bridged Lactams by Radical Aryl Migration.

3. Bridged Sultams

Bridged sultams (bridged sulfonamides) have attracted significant attention due to a wide range of biological activities of the sulfonamide bond [38]. In contrast to bridged lactams, constraining a sulfonamide bond in rigid bicyclic ring systems is easily possible due to the lack of Nlp to SO_2 conjugation (Nlp = nitrogen lone pair) [39]. Such bridged sultams are not hyper-reactive to hydrolysis and besides applications in medicinal chemistry have been used as template in stereoselective synthesis enabled by rapid scission of the N-SO_2 bond [3].

In 2017, Evans and co-workers in the continuation of their studies on bridged sultams reported an improved method for the synthesis of saturated sultams via intramolecular reductive Heck reaction (Figure 16) [40]. The use of a single Pd catalyst and a broad substrate scope are noteworthy features of this method. One of the saturated sultams in a [3.2.1] scaffold was characterized by X-ray crystallography (θ = 328.7; N-S = 1.643). The same group reported a bromonium-triggered 1,2-Wagner-Meerwein rearrangement of benzofused bridged sultams (not shown) [41].

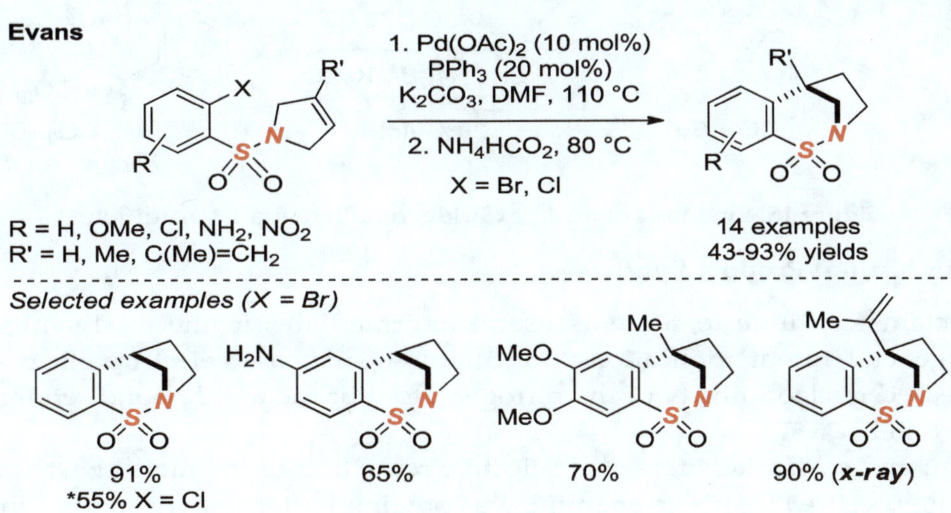

Figure 16. Synthesis of Saturated Bridged Sultams via Intramolecular Heck Reaction/Reduction.

In 2017, Das and co-workers nicely exploited intramolecular S_NAr-type cyclization to form bridged benzothiaoxazepine-1,1-dioxides in a [4.3.1] scaffold (Figure 17) [42]. An advantage of this method is a rapid, telescoped, three-step synthesis of bridged sultams from the corresponding *N*-aryl-2-fluorobenzenesulfonamides and trans-2,3-epoxy cinnamyl tosylates. One of the sultams was fully characterized by X-ray crystallography ($\theta = 342.0$; N-S = 1.668). The authors have also developed an enantioselective variant by using a chiral trans-2,3-epoxy cinnamyl alcohol.

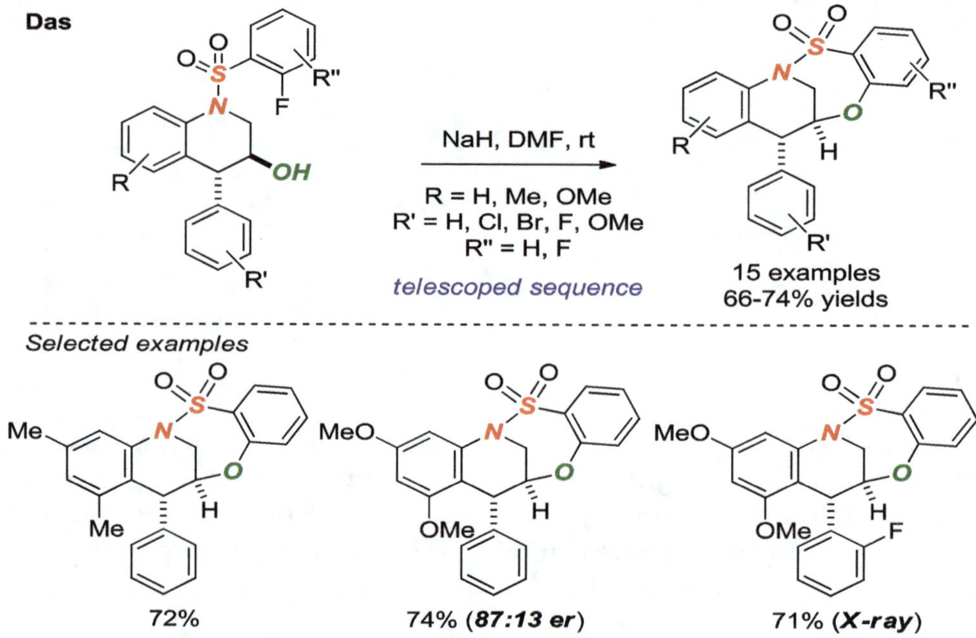

Figure 17. Synthesis of Bridged Benzothiaoxazepine-1,1-dioxides via S_NAr Cyclization.

One example of an "apex-type" bridged sultam in a [3.2.1] scaffold was reported by Sokolov and co-workers (Figure 18) [43]. The X-ray structure demonstrated significant pyramidalization of the nitrogen atom ($\theta = 325.5$; N–S = 1.668).

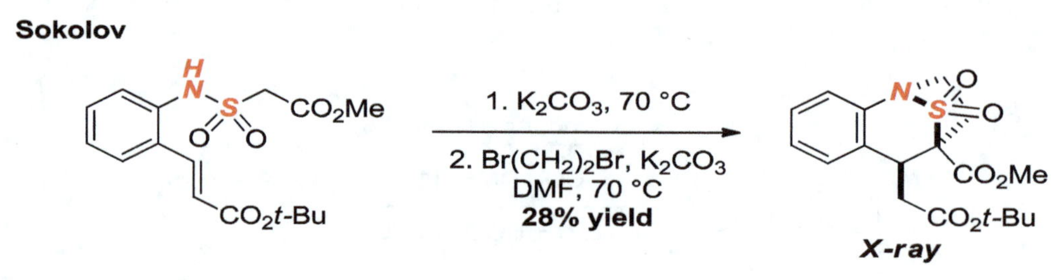

Figure 18. Synthesis of an Apex Bridged Sultam via 1,4-Addition.

4. Application in Natural Product Synthesis

Bridged lactams continue to serve as useful intermediates in the total synthesis of natural products [3]. In general, recent applications hinge upon the increased electrophilicity of the carbonyl group and increased nucleophilicity of the nitrogen atom of the amide bond rendered possible by geometric distortion.

In 2015, Zhu and co-workers reported selective reduction of the more twisted amide bond in scholarisine G (Figure 19A) [44]. The enamine was obtained after dehydration of the intermediate hemiaminal. The X-ray structure of scholarisine G showed a significantly distorted *N*-aryl amide bond ($\tau = 21.9°$, $\chi_N = 32.0°$; N-C(O) = 1.373 Å, C=O = 1.223 Å) vs. the aliphatic amide bond ($\tau = 2.9°$, $\chi_N = 19.9°$; N-C(O) = 1.352 Å, C=O = 1.227 Å). It is also possible that the selective reduction in this case can be explained by the exclusive stability of the five-membered ring lactams and by the fact that the six-membered lactam is an anilide. In contrast, Dai exploited the higher basicity of the oxygen atom in the aliphatic amide bond in a structurally-related leuconoxine to selectively maneuver reduction of the more electron-rich amide bond through electrophilic pathway (Figure 19B) [45].

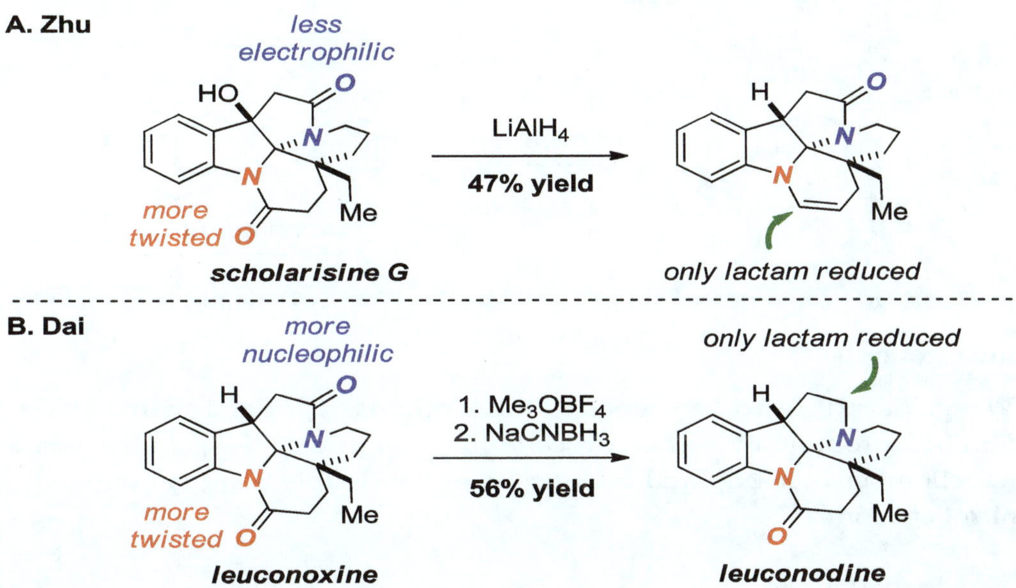

Figure 19. Reduction of a Bridged Lactam in Scholarisine G (**A**) and Leuconoxine Alkaloids (**B**).

A total synthesis of 3-*O*-demethylmacronine, an Amaryllidaceae alkaloid, utilizing a lactam-to-lactone rearrangement of a twisted amide was reported by Banwell and co-workers (Figure 20) [46]. This elegant method capitalized on the high basicity of the twisted amide nitrogen atom to form the acylium ion, which underwent trapping with a pendant hydroxyl group under mild conditions.

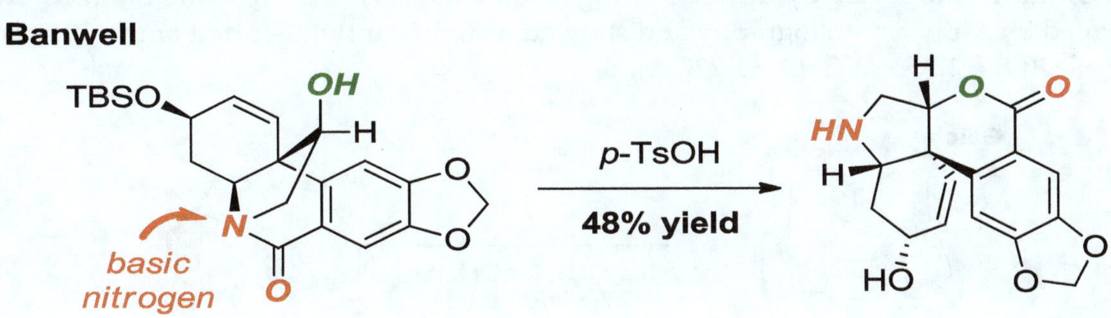

Figure 20. Lactam-to-Lactone Rearrangement of a Bridged Lactam in Haemanthidine Alkaloids.

Landais and co-workers reported a total synthesis of eucophylline, a dimeric terpene indole alkaloid, taking advantage of a high electrophilicity of the carbonyl group in a [3.3.1] bridged lactam scaffold (Figure 21) [47]. Bridged lactams are known to readily condense with amines to form amidines. In this approach, condensation of the twisted amide bond with aniline afforded a bridged amidine, which provided the key disconnection to the eucophylline core.

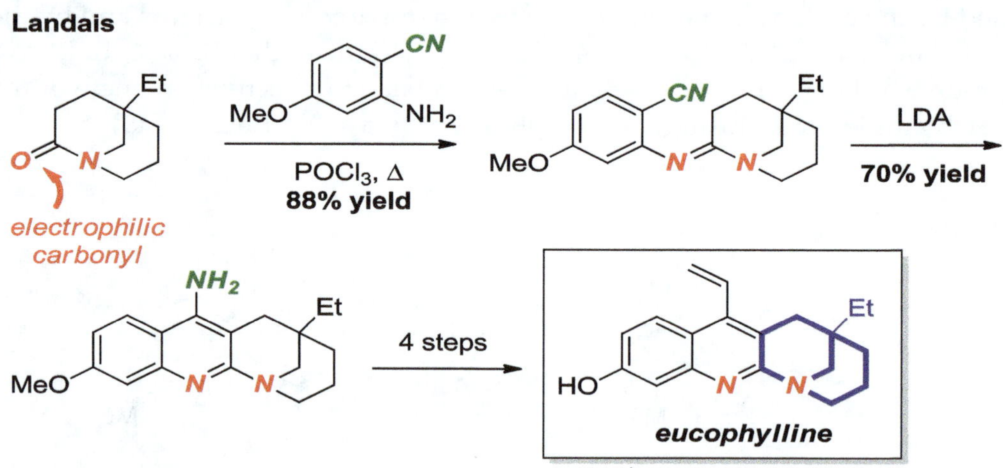

Figure 21. Bridged Amidine from a [3.3.1] Bridged Lactam in the Synthesis of Eucophylline.

5. Miscellaneous Examples

In 2015, Wang, Yu and co-workers reported the synthesis of bridged enamines via Au-catalyzed spiro-cyclization of 2-propargyl-β-tetrahydrocarbolines (Figure 22) [48]. In these heterocycles, the resonance interaction between Nlp and π electrons of the double bond is inhibited, resulting in a nucleophilic nitrogen atom.

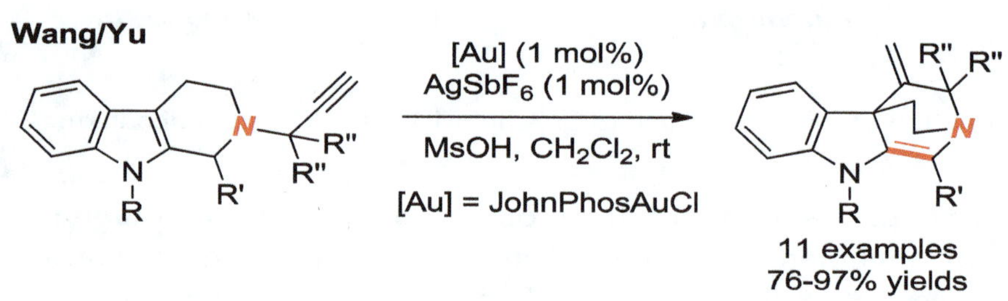

Figure 22. Synthesis of Bridged Enamines via Gold-Catalyzed Spriocyclization.

In their studies on the Witkop cyclization, Gaich and co-workers reported the synthesis of macrocyclic amides supported by the indole ring (Figure 23) [49]. The structure of one of the amides was confirmed by X-ray crystallography and showed a significantly distorted amide bond ($\tau = 32.3°$, $\chi_N = 0.0°$; N–C(O) = 1.353 Å, C=O = 1.227 Å).

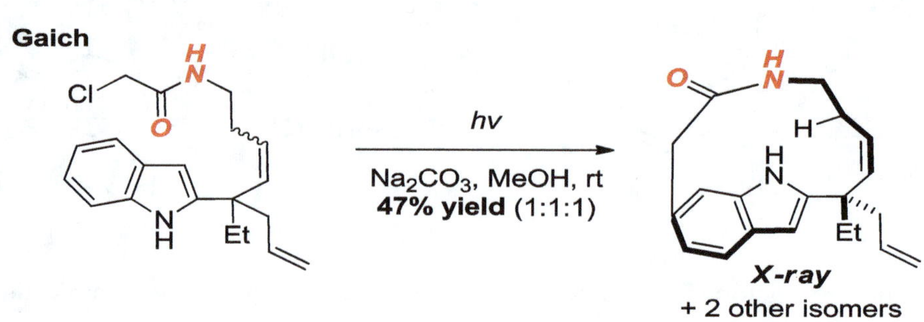

Figure 23. Synthesis of a Macrocyclic Lactam via Witkop Cyclization.

Yudin and co-workers developed several elegant methods for site-specific incorporation of amino acids [50], peptide sequencing [51] and conformational control [52] of cyclic peptides based on the twisted amide electrophilic sites (Figure 24). In their approach, the integration of a highly strained and *N*-pyramidalized aziridinyl ring allows for selective N–C(O) cleavage and amino acid incorporation, while the strained aziridinyl ring provides a handle for further functionalization by aziridine-ring opening with nucleophiles.

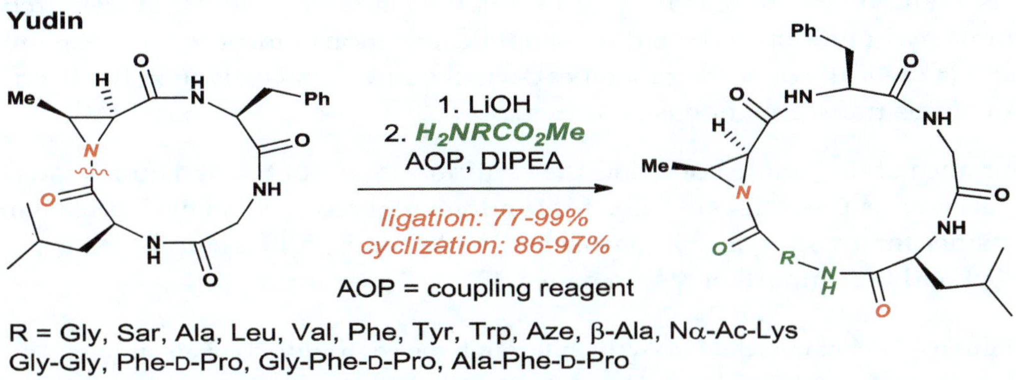

Figure 24. Cyclic Twisted Amide-Containing Tetrapeptides.

Several additional studies on non-planar amides have been reported. Computational studies on ion-pair interactions in Kirby's amide HBF_4 salts [53] and *N*-protonation of twisted amides by HF and HCl [54] were reported by Panday. Mykhailiuk and co-workers reported selective σ N-C bond scission in strained acyclic amides [55]. Stereoselective C-O bond cleavage in pyramidalized diketopiperazines induced by amide bond distortion was reported by Jahn and co-workers [56]. Intramolecular hydrogen bonding to electron-deficient oxygen in *N*-pyramidalized bicyclic amides was reported by Otani, Ohwada and co-workers [57]. Excellent reviews on the role of amide bond activation in biological molecules [58], amidicity [59], and heteroatom substitution at amide nitrogen [60] have been published. Additional studies on the properties of amides have been reported [61,62]. Recent relevant studies on acyclic non-planar amides should also be noted [11,12,63–66].

6. Conclusions

In summary, we have reviewed recent advances in the area of bridged lactams. This field continues to provide the most effective and wide-ranging method to achieve non-planarity of the amide bond. The main progress in the last five years includes: (1) identification of the additive Winkler-Dunitz parameter as a straightforward and accurate descriptor of the structural and energetic properties of the non-planar amide bond, and (2) the synthesis of extremely twisted bridged lactams in both quinuclidone and adamantanone series. Recently reported examples of novel bridged lactam scaffolds, divergent N–C cleavage reactivity and applications in the total synthesis of natural products are also worth noting.

Despite the considerable progress that has been made, the area is far from being mature. The recent remarkable progress in the chemistry of acyclic twisted amides has underlined the benefits of using conformational restriction (i.e., bridged lactams) to achieve non-planarity, but also brought to light the deficiencies of bridged lactams, and non-planar amide bonds in general. Most important is that except for the structurally-limited 1-aza-2-adamantanones there is a complete lack of isolated and fully characterized unhindered bridged lactams with a combined Winkler-Dunitz parameter ($\Sigma\tau + \chi_N$) exceeding 100° in the neutral form (or in other words merely 60% of the maximum theoretical distortion). It is beyond belief that the area has progressed so little since the first conception of a bridged lactam bond by Lukeš in 1938.

Further, very little is known about the stability of bridged lactams and, by extension, of non-planar amides. It is now safe to assume that a range of distortion of ($\Sigma\tau + \chi_N$) ca. 70–80° gives isolable lactams. However, this stability range has only been tested in a very limited set of lactams.

Even less is known about the generality of amide bond distortion, properties and structures across different sets of non-planar amides. This research is hindered by the very few types of non-planar bridged lactams that have been described to date. The recent example of Tröger's base twisted amides is as a good case in point [19]. Despite significant amide bond distortion, these lactams do not protonate readily as a consequence of Nlp conjugation onto the aromatic ring.

Another issue involves the reactivity of non-planar amides. Almost all research thus far has focused on simple hydrolytic stability and nucleophilic addition studies. At present, only one type of potentially very synthetically appealing metal-catalyzed σ N-C bond activation has been reported, and the mechanism of this transformation is unknown.

The importance of the amide bond and the continuous role of twisted amides in chemistry and biology have served as a powerful stimulus for fundamental research on bridged lactams. Given the obvious possibilities for applications of non-planar amide bonds, future studies are expected to have a broad impact beyond this important research area.

Author Contributions: The manuscript was written through the contributions of all authors.

References

1. Greenberg, A.; Breneman, C.M.; Liebman, J.F. *The Amide Linkage: Structural Significance in Chemistry, Biochemistry, and Materials Science*; Wiley: New York, NY, USA, 2000.
2. Pauling, L. *The Nature of the Chemical Bond*; Oxford University Press: London, UK, 1940.
3. Szostak, M.; Aubé, J. Chemistry of Bridged Lactams and Related Heterocycles. *Chem. Rev.* **2013**, *113*, 5701–5765. [CrossRef] [PubMed]
4. Szostak, M.; Aubé, J. Medium-Bridged Lactams: A New Class of Non-Planar Amides. *Org. Biomol. Chem.* **2011**, *9*, 27–35. [CrossRef] [PubMed]
5. Lease, T.G.; Shea, K.J. A Compilation and Analysis of Structural Data of Distorted Bridgehead Olefins and Amides. In *Advances in Theoretically Interesting Molecules*; JAI Press Inc.: Greenwich, CT, USA, 1992.
6. Hall, H.K., Jr.; El-Shekeil, A. Anti-Bredt bridgehead nitrogen compounds in ring-opening polymerization. *Chem. Rev.* **1983**, *83*, 549–555.
7. Poland, B.W.; Xu, M.Q.; Quiocho, F.A. Structural Insights into the Protein Splicing Mechanism of PI-SceI. *J. Biol. Chem.* **2000**, *275*, 16408–16413. [CrossRef] [PubMed]
8. Romanelli, A.; Shekhtman, A.; Cowburn, D.; Muir, T.W. Semisynthesis of a segmental isotopically labeled protein splicing precursor: NMR evidence for an unusual peptide bond at the N-extein–intein junction. *Proc. Natl. Acad. Sci. USA* **2004**, *101*, 6397–6402. [CrossRef] [PubMed]
9. Shemella, P.; Pereira, B.; Zhang, Y.M.; Van Roey, P.; Belfort, G.; Garde, S.; Nayak, S.K. Mechanism for Intein C-Terminal Cleavage: A Proposal from Quantum Mechanical Calculations. *Biophys. J.* **2007**, *92*, 847–853. [CrossRef] [PubMed]
10. Lizak, C.; Gerber, S.; Michaud, G.; Schubert, M.; Fan, Y.Y.; Bucher, M.; Darbare, T.; Aebi, M.; Reymond, J.L.; Locher, K.P. Unexpected reactivity and mechanism of carboxamide activation in bacterial N-linked protein glycosylation. *Nat. Commun.* **2013**, *4*, 2627. [CrossRef] [PubMed]
11. Pace, V.; Holzer, W.; Meng, G.; Shi, S.; Lalancette, R.; Szostak, R.; Szostak, M. Structures of Highly Twisted Amides Relevant to Amide N–C Cross-Coupling: Evidence for Ground-State Amide Destabilization. *Chem. Eur. J.* **2016**, *22*, 14494–14498. [CrossRef] [PubMed]
12. Shi, S.; Nolan, S.P.; Szostak, M. Well-Defined Palladium(II)-NHC (NHC = N-Heterocyclic Carbene) Precatalysts for Cross-Coupling Reactions of Amides and Esters by Selective Acyl CO–X (X = N, O) Cleavage. *Acc. Chem. Res.* **2018**, *51*, 2589–2599. [CrossRef] [PubMed]
13. Winkler, F.K.; Dunitz, J.D. The non-planar amide group. *J. Mol. Biol.* **1971**, *59*, 169–182. [CrossRef]
14. Szostak, R.; Aubé, J.; Szostak, M. An Efficient Computational Model to Predict Protonation at the Amide Nitrogen and Reactivity along the C–N Rotational Pathway. *Chem. Commun.* **2015**, *51*, 6395–6398. [CrossRef] [PubMed]
15. Szostak, R.; Aubé, J.; Szostak, M. Determination of Structures and Energetics of Small- and Medium-Sized One-Carbon Bridged Twisted Amides using ab Initio Molecular Orbital Methods. Implications for Amidic Resonance along the C–N Rotational Pathway. *J. Org. Chem.* **2015**, *80*, 7905–7927. [CrossRef] [PubMed]

16. Greenberg, A.; Venanzi, C.A. Structures and energetics of two bridgehead lactams and their N- and O-protonated forms: An ab initio molecular orbital study. *J. Am. Chem. Soc.* **1993**, *115*, 6951–6957. [CrossRef]
17. Greenberg, A.; Moore, D.T.; DuBois, T.D. Small and Medium-Sized Bridgehead Bicyclic Lactams: A Systematic ab Initio Molecular Orbital Study. *J. Am. Chem. Soc.* **1996**, *118*, 8658–8668. [CrossRef]
18. Morgan, J.; Greenberg, A. Novel bridgehead bicyclic lactams: Molecules predicted to have O-protonated and N-protonated tautomers of comparable stability; hyperstable lactams and their O-protonated tautomers. *J. Chem. Thermodyn.* **2014**, *73*, 206–212. [CrossRef]
19. Szostak, R.; Szostak, M. Tröger's Base Twisted Amides: High Amide Bond Twist and N-/O-Protonation Aptitude. *J. Org. Chem.* **2019**, in press. [CrossRef] [PubMed]
20. Liniger, M.; VanderVelde, D.G.; Takase, M.K.; Shahgholi, M.; Stoltz, B.M. Total Synthesis and Characterization of 7-Hypoquinuclidonium Tetrafluoroborate and 7-Hypoquinuclidone BF_3 Complex. *J. Am. Chem. Soc.* **2016**, *138*, 969–974. [CrossRef] [PubMed]
21. Tani, K.; Stoltz, B.M. Synthesis and structural analysis of 2-quinuclidonium tetrafluoroborate. *Nature* **2006**, *441*, 731–734. [CrossRef] [PubMed]
22. Komarov, I.V.; Yanik, S.; Ishchenko, A.Y.; Davies, J.E.; Goodman, J.M.; Kirby, A.J. The Most Reactive Amide As a Transition-State Mimic For cis–trans Interconversion. *J. Am. Chem. Soc.* **2015**, *137*, 926–930. [CrossRef]
23. Morgan, K.M.; Rawlins, M.L.; Montgomery, M.N. Influence of methyl substituents on the stability of 1-aza-2-adamantanone, Kirby's most twisted amide. *J. Phys. Org. Chem.* **2005**, *18*, 310–314. [CrossRef]
24. Kirby, A.J.; Komarov, I.V.; Wothers, P.D.; Feeder, N. The Most Twisted Amide: Structure and Reactions. *Angew. Chem. Int. Ed.* **1998**, *37*, 785–786. [CrossRef]
25. Morgan, J.P.; Weaver-Guevara, H.M.; Fitzgerald, R.W.; Dunlap-Smith, A.; Greenberg, A. Ab initio computational study of 1-methyl-4-silatranone and attempts at its conventional synthesis. *Struct. Chem.* **2017**, *28*, 327–331. [CrossRef]
26. Hu, F.; Lalancette, R.; Szostak, M. Structural Characterization of N-Alkylated Twisted Amides: Consequences for Amide Bond Resonance and N–C Cleavage. *Angew. Chem. Int. Ed.* **2016**, *55*, 5062–5066. [CrossRef] [PubMed]
27. Hu, F.; Nareddy, P.; Lalancette, R.; Jordan, F.; Szostak, M. σ N−C Bond Difunctionalization in Bridged Twisted Amides: Sew-and-Cut Activation Approach to Functionalized Isoquinolines. *Org. Lett.* **2017**, *19*, 2386–2389. [CrossRef] [PubMed]
28. Hassan, H.; Marsden, S.P.; Nelson, A. Design and synthesis of a fragment set based on twisted bicyclic lactams. *Bioorg. Med. Chem.* **2018**, *26*, 3030–3033. [CrossRef] [PubMed]
29. Steliou, K.; Poupart, M.A. Reagents for organic synthesis. Part 3. Tin-mediated esterification in macrolide synthesis. *J. Am. Chem. Soc.* **1983**, *105*, 7130–7138. [CrossRef]
30. Liniger, M.; Liu, Y.; Stoltz, B. Sequential Ruthenium Catalysis for Olefin Isomerization and Oxidation: Application to the Synthesis of Unusual Amino Acids. *J. Am. Chem. Soc.* **2017**, *139*, 13944–13949. [CrossRef]
31. Brouillette, W.J.; Jestkov, V.P.; Brown, M.L.; Akhtar, M.S.; DeLorey, T.M.; Brown, G.B. Bicyclic Hydantoins with a Bridgehead Nitrogen. Comparison of Anticonvulsant Activities with Binding to the Neuronal Voltage-Dependent Sodium Channel. *J. Med. Chem.* **1994**, *37*, 3289–3293. [CrossRef]
32. Smissman, E.E.; Matuszak, A.J.; Corder, C.N. Reduction of Barbiturates Under Hydroboration Conditions. *J. Pharm. Sci.* **1964**, *53*, 1541–1542. [CrossRef]
33. Szostak, R.; Liu, C.; Lalancette, R.; Szostak, M. Twisted N-Acyl-hydantoins: Rotationally Inverted Urea-Imides of Relevance in N–C(O) Cross-Coupling. *J. Org. Chem.* **2018**, *83*, 14676–14682. [CrossRef]
34. Pereira, R.; Pfeifer, L.; Fournier, J.; Gouverneur, V.; Cvengroš, J. Twisting the ethano-Tröger's base: The bisamide. *Org. Biomol. Chem.* **2017**, *15*, 628–633. [CrossRef] [PubMed]
35. Artacho, J.; Ascic, E.; Rantanen, T.; Karlsson, J.; Wallentin, C.J.; Wang, R.; Wendt, O.F.; Harmata, M.; Snieckus, V.; Wärnmark, K. Twisted Amide Analogues of Tröger's Base. *Chem. Eur. J.* **2012**, *18*, 1038–1042. [CrossRef] [PubMed]
36. Satyanarayana, G.; Helmchen, G. Enantioselective Syntheses of Bicyclic Lactams Based on Iridium-Catalyzed Asymmetric Allylic Substitution and Heck Cyclization. *Eur. J. Org. Chem.* **2014**, 2242–2252. [CrossRef]
37. Li, L.; Li, Z.L.; Wang, F.L.; Guo, Z.; Cheng, Y.F.; Wang, N.; Dong, X.W.; Fang, C.; Liu, J.; Hou, C.; et al. Radical aryl migration enables diversity-oriented synthesis of structurally diverse medium/macro- or bridged-rings. *Nat. Commun.* **2016**, *7*, 13852. [CrossRef] [PubMed]
38. Drews, J. Drug Discovery: A Historical Perspective. *Science* **2000**, *287*, 1960–1964. [CrossRef] [PubMed]

39. Paquette, L.A.; Leit, S.M. The First Examples of Bridgehead Bicyclic Sultams. *J. Am. Chem. Soc.* **1999**, *121*, 8126–8127. [CrossRef]
40. Khalifa, A.; Conway, L.; Geoghegan, K.; Evans, P. Ammonium formate-based one-pot reductive Heck reactions for the construction of cyclic sulfonamides. *Tetrahedron Lett.* **2017**, *58*, 4559–4562. [CrossRef]
41. Geoghegan, K.; Smullen, S.; Evans, P. Halonium Ion Triggered Rearrangement of Unsaturated Benzo-Annulated Bi- and Tricyclic Sulfonamides. *J. Org. Chem.* **2013**, *78*, 10443–10451. [CrossRef]
42. Borgohain, H.; Devi, R.; Dheer, D.; Borah, B.J.; Shankar, R.; Das, S.K. Synthesis of Tetrahydroquinoline-Embedded Bridged Benzothiaoxazepine-1,1-dioxides. *Eur. J. Org. Chem.* **2017**, 6671–6679. [CrossRef]
43. Grosheva, D.S.; Rassadin, V.A.; Sokolov, V.V. A Route to Benzo-Annelated δ-Sultams through Michael Cyclization. *Eur. J. Org. Chem.* **2015**, *2015*, 1355–1363. [CrossRef]
44. Xu, Z.; Wang, Q.; Zhu, J. Total Syntheses of (−)-Mersicarpine, (−)-Scholarisine G, (+)-Melodinine E, (−)-Leuconoxine, (−)-Leuconolam, (−)-Leuconodine A, (+)-Leuconodine F, and (−)-Leuconodine C: Self-Induced Diastereomeric Anisochronism (SIDA) Phenomenon for Scholarisine G and Leuconodines A and C. *J. Am. Chem. Soc.* **2015**, *137*, 6712–6724. [PubMed]
45. Yang, Y.; Bai, Y.; Sun, S.; Dai, M. Biosynthetically Inspired Divergent Approach to Monoterpene Indole Alkaloids: Total Synthesis of Mersicarpine, Leuconodines B and D, Leuconoxine, Melodinine E, Leuconolam, and Rhazinilam. *Org. Lett.* **2014**, *16*, 6216–6219. [CrossRef] [PubMed]
46. Ma, X.; Gao, N.; Banwell, M.G.; Carr, P.D.; Willis, A.C. A Total Synthesis of (±)-3-O-Demethylmacronine through Rearrangement of a Precursor Embodying the Haemanthidine Alkaloid Framework. *J. Org. Chem.* **2017**, *82*, 4336–4341. [CrossRef] [PubMed]
47. Hassan, H.; Mohammed, S.; Robert, F.; Landais, Y. Total Synthesis of (±)-Eucophylline. A Free-Radical Approach to the Synthesis of the Azabicyclo[3.3.1]nonane Skeleton. *Org. Lett.* **2015**, *17*, 4518–4521. [PubMed]
48. Zhang, L.; Wang, Y.; Yao, Z.J.; Wang, S.; Yu, Z.X. Kinetic or Dynamic Control on a Bifurcating Potential Energy Surface? An Experimental and DFT Study of Gold-Catalyzed Ring Expansion and Spirocyclization of 2-Propargyl-β-tetrahydrocarbolines. *J. Am. Chem. Soc.* **2015**, *137*, 13290–13300. [CrossRef] [PubMed]
49. Pfaffenbach, M.; Roller, A.; Gaich, T. Synthesis of Indolophanes by Photochemical Macrocyclization. *Chem. Eur. J.* **2016**, *22*, 8444–8447. [CrossRef]
50. White, C.J.; Hickey, J.L.; Scully, C.C.G.; Yudin, A.K. Site-Specific Integration of Amino Acid Fragments into Cyclic Peptides. *J. Am. Chem. Soc.* **2014**, *136*, 3728–3731. [CrossRef]
51. Zaretsky, S.; Rai, V.; Gish, G.; Forbes, M.W.; Kofler, M.; Yu, J.C.Y.; Tan, J.; Hickey, J.L.; Pawson, T.; Yudin, A.K. Twisted amide electrophiles enable cyclic peptide sequencing. *Org. Biomol. Chem.* **2015**, *13*, 7384–7388. [CrossRef]
52. Chung, B.K.W.; White, C.J.; Scully, C.C.G.; Yudin, A.K. The reactivity and conformational control of cyclic tetrapeptides derived from aziridine-containing amino acids. *Chem. Sci.* **2016**, *7*, 6662–6668. [CrossRef]
53. Pandey, K.K. Theoretical insights into structure, bonding, reactivity and importance of ion-pair interactions in Kirby's tetrafluoroboric acid salts of twisted amides. *RSC Adv.* **2015**, *5*, 105668–105677. [CrossRef]
54. Pandey, K.K. Does hydrohalic acid HX (X = F, Cl) form true N-protonated twisted amide salts? Effects of anions on the ion-pair interactions and on the amide moiety in N-protonated tricyclic twisted amide salts. *New J. Chem.* **2016**, *40*, 7831–7839. [CrossRef]
55. Artamonov, O.S.; Slobodyanyuk, E.Y.; Volochnyuk, D.M.; Komarov, I.V.; Tolmachev, A.A.; Mykhailiuk, P.K. Synthesis of Trifluoromethyl-Substituted 3-Azabicyclo[n.1.0]alkanes: Advanced Building Blocks for Drug Discovery. *Eur. J. Org. Chem.* **2014**, *46*, 3592–3598. [CrossRef]
56. Amatov, T.; Jangra, H.; Pohl, R.; Cisařová, I.; Zipse, H.; Jahn, U. Unique Stereoselective Homolytic C-O Bond Activation in Diketopiperazine-Derived Alkoxyamines by Adjacent Amide Pyramidalization. *Chem. Eur. J.* **2018**, *24*, 15336–15345. [CrossRef]
57. Wang, S.; Taniguchi, T.; Monde, K.; Kawahata, M.; Yamaguchi, K.; Otani, Y.; Ohwada, T. Hydrogen bonding to carbonyl oxygen of nitrogen-pyramidalized amide—Detection of pyramidalization direction preference by vibrational circular dichroism spectroscopy. *Chem. Commun.* **2016**, *52*, 4018–4021. [CrossRef]
58. Mahesh, S.; Tang, K.C.; Raj, M. Amide Bond Activation of Biological Molecules. *Molecules* **2018**, *23*, 2615. [CrossRef] [PubMed]
59. Kovács, E.; Rózsa, B.; Csomos, A.; Csizmadia, I.G.; Mucsi, Z. Amide Activation in Ground and Excited States. *Molecules* **2018**, *23*, 2859. [CrossRef] [PubMed]

60. Glover, S.A.; Rosser, A.A. Heteroatom Substitution at Amide Nitrogen—Resonance Reduction and HERON Reactions of Anomeric Amides. *Molecules* **2018**, *23*, 2834. [CrossRef]
61. Morgan, K.M.; Ashline, D.J.; Morgan, J.P.; Greenberg, A. Electrospray Ionization (ESI) Fragmentations and Dimethyldioxirane Reactivities of Three Diverse Lactams Having Full, Half, and Zero Resonance Energies. *J. Org. Chem.* **2014**, *79*, 517–528. [CrossRef]
62. Glover, S.A.; Rosser, A.A. HERON reactions of anomeric amides: Understanding the driving force. *J. Phys. Org. Chem.* **2015**, *28*, 215–222. [CrossRef]
63. Szostak, R.; Shi, S.; Meng, G.; Lalancette, R.; Szostak, M. Ground-State Distortion in *N*-Acyl-*tert*-butyl-carbamates (Boc) and *N*-Acyl-tosylamides (Ts): Twisted Amides of Relevance to Amide N–C Cross-Coupling. *J. Org. Chem.* **2016**, *81*, 8091–8094. [CrossRef]
64. Meng, G.; Shi, S.; Lalancette, R.; Szostak, R.; Szostak, M. Reversible Twisting of Primary Amides via Ground State N–C(O) Destabilization: Highly Twisted Rotationally Inverted Acyclic Amides. *J. Am. Chem. Soc.* **2018**, *140*, 727–734. [CrossRef] [PubMed]
65. Szostak, R.; Szostak, M. N-Acyl-Glutarimides: Resonance and Proton Affinities of Rotationally-Inverted Twisted Amides Relevant to N−C(O) Cross-Coupling. *Org. Lett.* **2018**, *20*, 1342–1345. [CrossRef] [PubMed]
66. Liu, C.; Shi, S.; Liu, Y.; Liu, R.; Lalancette, R.; Szostak, R.; Szostak, M. The Most Twisted Acyclic Amides: Structures and Reactivity. *Org. Lett.* **2018**, *20*, 7771–7774. [CrossRef] [PubMed]

12

Computational Study of Mechanism and Thermodynamics of Ni/IPr-Catalyzed Amidation of Esters

Chong-Lei Ji †, Pei-Pei Xie † and Xin Hong *

Department of Chemistry, Zhejiang University, Hangzhou 310027, China; chjicl@zju.edu.cn (C.-L.J.); xppyyfd@zju.edu.cn (P.-P.X.)

* Correspondence: hxchem@zju.edu.cn

† These authors contributed equally to this work.

Abstract: Nickel catalysis has shown remarkable potential in amide C–N bond activation and functionalization. Particularly for the transformation between ester and amide, nickel catalysis has realized both the forward (ester to amide) and reverse (amide to ester) reactions, allowing a powerful approach for the ester and amide synthesis. Based on density functional theory (DFT) calculations, we explored the mechanism and thermodynamics of Ni/IPr-catalyzed amidation with both aromatic and aliphatic esters. The reaction follows the general cross-coupling mechanism, involving sequential oxidative addition, proton transfer, and reductive elimination. The calculations indicated the reversible nature of amidation, which highlights the importance of reaction thermodynamics in related reaction designs. To shed light on the control of thermodynamics, we also investigated the thermodynamic free energy changes of amidation with a series of esters and amides.

Keywords: amide C–N bond activation; nickel catalysis; amidation; DFT study; reaction thermodynamics

---

1. Introduction

Amide C–N bond activation and functionalization provide a powerful strategy to utilize amide as a central synthon [1–6]. Despite the remarkable value of this strategy, the development of amide C–N bond activation has progressed slowly due to the resonance nature of amide and the resulting difficulty for bond dissociation [7–16]. Recently, Garg and coworkers discovered the outstanding performance of nickel catalysis for amide C–N bond activation [17], leading to the development of a series of exciting Ni-catalyzed cross coupling reactions [18–26]. Szostak and coworkers developed a family of geometrically twisted amides which serves as a powerful synthetic platform [27–39] for both metal-catalyzed and metal-free transformations. Independent studies from Shi, Zou, Rueping, Maiti, Stanley, and Molander also contribute significantly to the synthetic advances involving amide C–N bond activation [40–52].

In addition, amide C–N bond activation stimulated the synthetic developments of amide C–N bond formation. Szostak and coworkers reported a series of transamidation reactions from twisted amides, which provides a distinctive approach for amide synthesis [39] (Scheme 1a). Garg, Houk, and coworkers discovered that Lewis acid plays an important role on the kinetics and thermodynamics of amidation of ester, which reverses the thermodynamic equilibrium of esterification of amide [53] (Scheme 1b). Newman and coworkers recently disclosed a Ni/IPr-catalyzed amidation of ester under Lewis acid-free conditions [54] (Scheme 1c). This transformation attracts our mechanistic interests [55–66]; particularly for the Lewis-acid free Ni-mediated C–O bond cleavage and C–N bond formation. Here we report a computational study on the mechanism of Ni/IPr-catalyzed amidation of ester and the thermodynamic equilibrium of amidation.

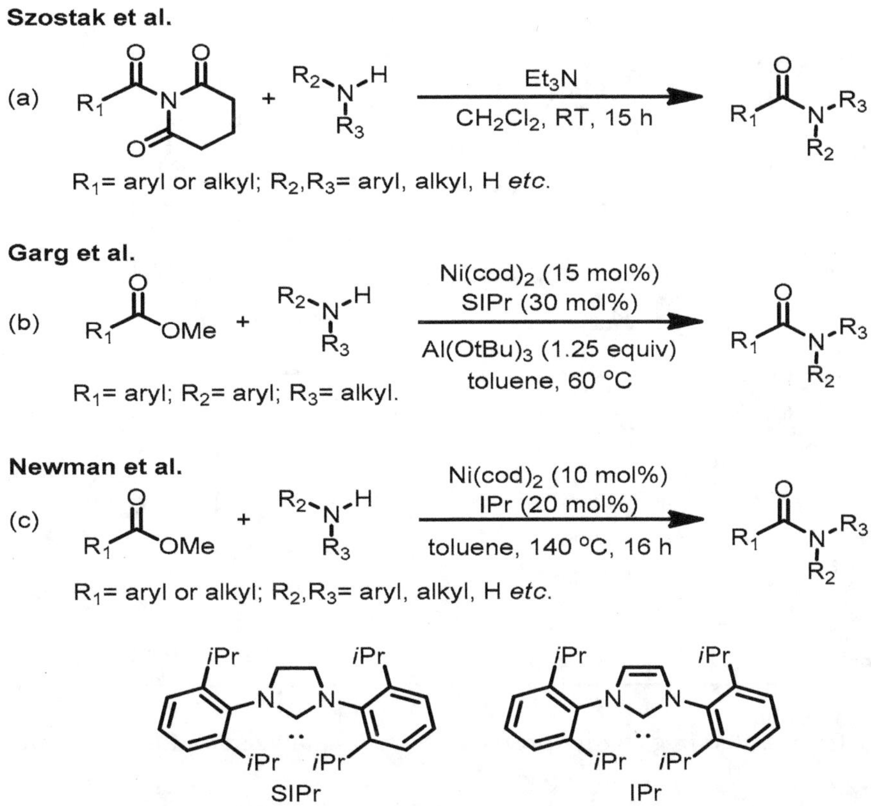

Scheme 1. (**a**) Transamidation of *N*-acyl-glutarimides with amines; (**b**) Ni/SIPr-catalyzed amidation of ester under Lewis acid conditions; and (**c**) Ni/IPr-catalyzed amidation of ester under Lewis acid-free conditions.

2. Results and Discussion

2.1. Reaction Mechanism of Aromatic Ester

Using methyl benzoate and morpholine as the model substrates, we first explored the free energy changes of Ni/IPr-catalyzed amidation with aromatic ester. The free energy profile is shown in Figure 1, and the optimized structures of selected species in the catalytic cycle are shown in Figure 2. From the Ni(IPr)(toluene) complex **1** [67], the initial ligand exchange with substrate is exergonic and leads to the substrate-coordinated complex **2**. Subsequent oxidative addition via TS3 requires a barrier of 20.4 kcal/mol, which generates the acylnickel species **4**. **5** then undergoes a facile inner-sphere proton transfer through TS6, and the dissociation of methanol leads to the LNi^{II}(acyl)(amino) intermediate **8**. From **8**, the C–N bond formation occurs via the reductive elimination transition state TS9, and the final exergonic product liberation produces the amidation product **11** as well as regenerates the catalytically active species **2**. We have confirmed the stability of wavefunctions for all Ni(II) species involved in the catalytic cycle. We were not able to locate the transition states for the ligand exchange steps (**4** to **5**, **7** to **8**) due to the flat energy surface. A series of constrained optimizations were performed to verify the flat energy surface of ligand exchange. Details are included in Supplementary Materials (Figure S1). Based on the DFT-computed free energy changes of the overall catalytic cycle, the on-cycle resting state is the substrate-coordinated complex **2**, and the rate-limiting step is the C–O bond activation step via TS3 with an overall barrier of 20.4 kcal/mol. Our computations also suggested that the overall amidation is reversible and almost thermodynamic neutral. The reaction barrier for the reversed transformation from **11** to **2** is only 20.3 kcal/mol. This corresponds well with the reversible nature of this amidation transformation, as reported by Newman [54]. It should be pointed out the generated methanol may leave the reaction system due to its volatility, which serves as an additional driving force for the amide formation.

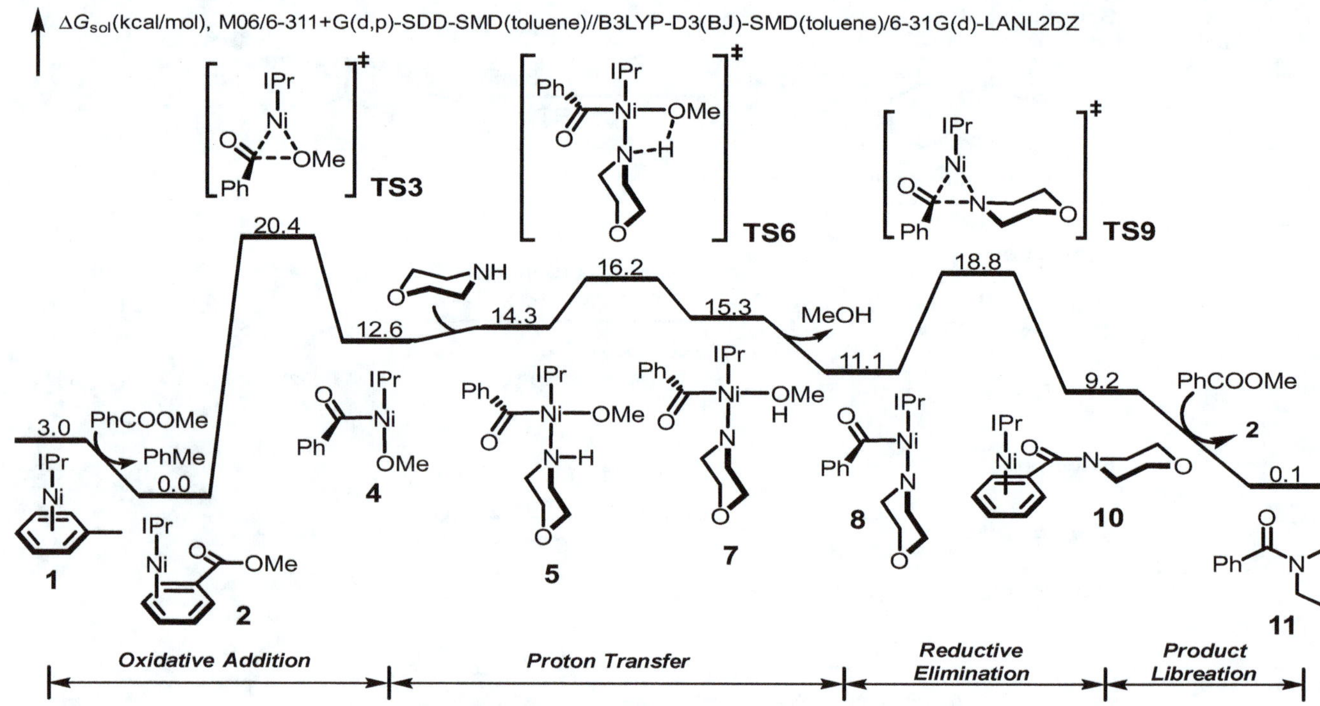

Figure 1. DFT-computed free energy profile of Ni/IPr-catalyzed amidation of aromatic ester.

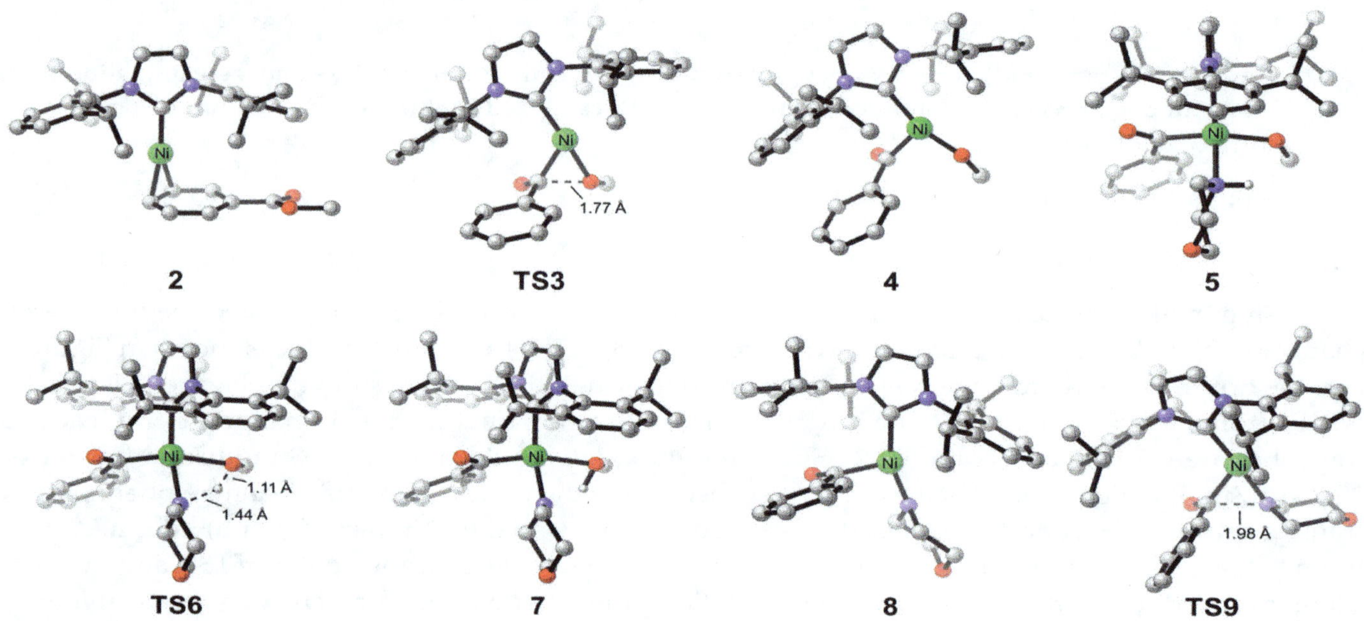

Figure 2. Optimized structures of selected intermediates and transition states in the catalytic cycle of Ni/IPr-catalyzed amidation of ester.

2.2. Reaction Mechanism of Aliphatic Ester

We next explored the Ni-catalyzed amidation with aliphatic ester, using methyl 3-phenylpropionate as the model substrate. The DFT-computed free energy profile is shown in Figure 3, and the optimized structures of selected species in the catalytic cycle are shown in Figure 4. The reaction mechanism of aliphatic ester is similar to that of aromatic ester, which involves sequential oxidative addition, proton transfer, reductive elimination and product liberation. The resting state is toluene-coordinated complex **1**, and the rate-limiting step is the proton transfer step with an overall barrier of 16.9 kcal/mol. The free energy change of the amidation with methyl 3-phenylpropionate is exergonic by 1.8 kcal/mol. Comparing the free energy profiles of aromatic and aliphatic esters, the major difference is the resting state and overall reaction barrier. Due to the lack of electron-deficient aromatic functionality in methyl 3-phenylpropionate, the substrate-coordinated complex **12** is 6.1 kcal/mol higher in free energy comparing with the toluene-coordinated species **1**. This leads to the low overall barrier of the amidation with aliphatic ester.

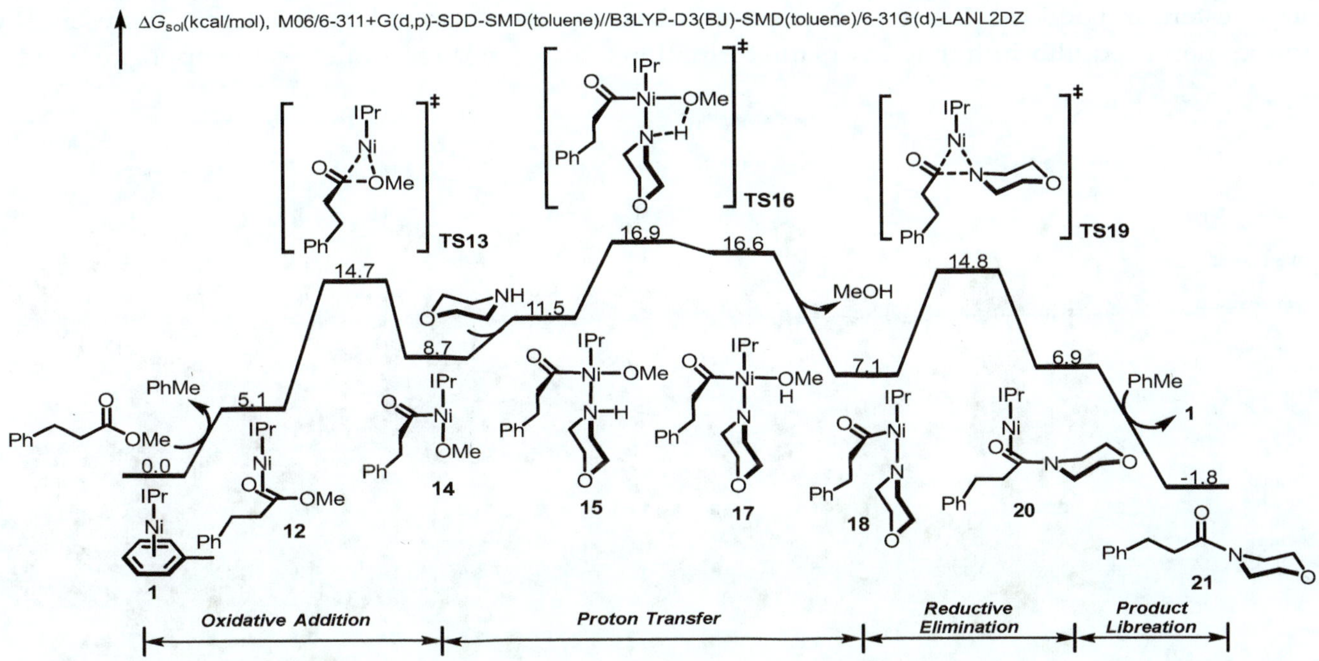

Figure 3. Density functional theory (DFT)-computed free energy profile of Ni/IPr-catalyzed amidation of aliphatic ester.

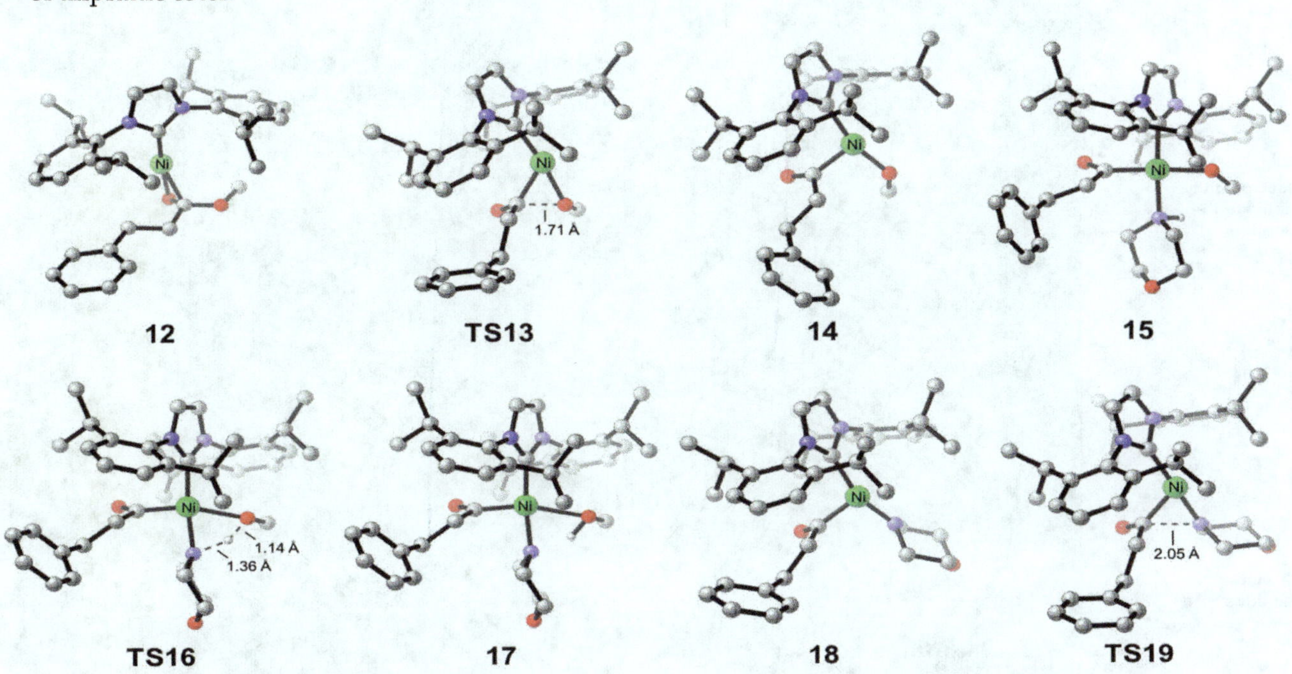

Figure 4. Optimized structures of selected intermediates and transition states in the catalytic cycle of Ni/IPr-catalyzed amidation of ester.

2.3. *Reaction Thermodynamics*

Based on the computed free energy profiles of Ni-catalyzed amidation of ester, we noticed that the major bottleneck for this type of transformation is the control of thermodynamics. The thermodynamic equilibrium has to be designed cautiously because the overall transformation does not have a strong force, unlike many other organic transformations. Therefore, the synthetic efforts could be unfruitful even the designed amidation or esterification has a surmountable reaction barrier. In order to understand the controlling factors of the thermodynamic equilibrium of amidation as well as provide a general thermodynamic guideline for related reaction designs, we computed the free energy changes with a series of esters and amides. The results are summarized as a heat map in Figure 5. Green represents the thermodynamic equilibrium that favors the formation of ester, and red indicates the opposite trend.

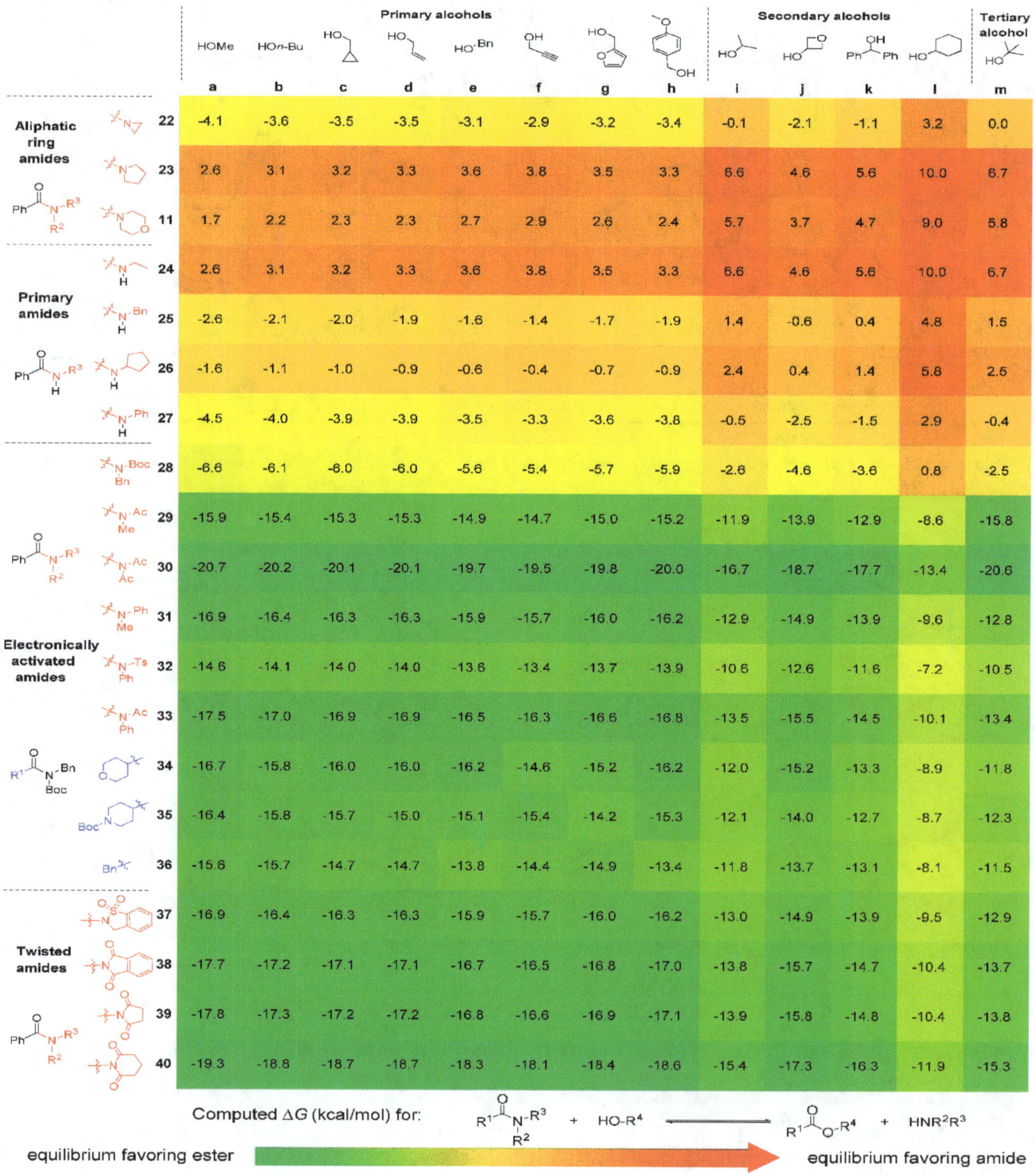

	a	b	c	d	e	f	g	h	i	j	k	l	m
22	-4.1	-3.6	-3.5	-3.5	-3.1	-2.9	-3.2	-3.4	-0.1	-2.1	-1.1	3.2	0.0
23	2.6	3.1	3.2	3.3	3.6	3.8	3.5	3.3	6.6	4.6	5.6	10.0	6.7
11	1.7	2.2	2.3	2.3	2.7	2.9	2.6	2.4	5.7	3.7	4.7	9.0	5.8
24	2.6	3.1	3.2	3.3	3.6	3.8	3.5	3.3	6.6	4.6	5.6	10.0	6.7
25	-2.6	-2.1	-2.0	-1.9	-1.6	-1.4	-1.7	-1.9	1.4	-0.6	0.4	4.8	1.5
26	-1.6	-1.1	-1.0	-0.9	-0.6	-0.4	-0.7	-0.9	2.4	0.4	1.4	5.8	2.5
27	-4.5	-4.0	-3.9	-3.9	-3.5	-3.3	-3.6	-3.8	-0.5	-2.5	-1.5	2.9	-0.4
28	-6.6	-6.1	-6.0	-6.0	-5.6	-5.4	-5.7	-5.9	-2.6	-4.6	-3.6	0.8	-2.5
29	-15.9	-15.4	-15.3	-15.3	-14.9	-14.7	-15.0	-15.2	-11.9	-13.9	-12.9	-8.6	-15.8
30	-20.7	-20.2	-20.1	-20.1	-19.7	-19.5	-19.8	-20.0	-16.7	-18.7	-17.7	-13.4	-20.6
31	-16.9	-16.4	-16.3	-16.3	-15.9	-15.7	-16.0	-16.2	-12.9	-14.9	-13.9	-9.6	-12.8
32	-14.6	-14.1	-14.0	-14.0	-13.6	-13.4	-13.7	-13.9	-10.6	-12.6	-11.6	-7.2	-10.5
33	-17.5	-17.0	-16.9	-16.9	-16.5	-16.3	-16.6	-16.8	-13.5	-15.5	-14.5	-10.1	-13.4
34	-16.7	-15.8	-16.0	-16.0	-16.2	-14.6	-15.2	-16.2	-12.0	-15.2	-13.3	-8.9	-11.8
35	-16.4	-15.8	-15.7	-15.0	-15.1	-15.4	-14.2	-15.3	-12.1	-14.0	-12.7	-8.7	-12.3
36	-15.6	-15.7	-14.7	-14.7	-13.8	-14.4	-14.9	-13.4	-11.8	-13.7	-13.1	-8.1	-11.5
37	-16.9	-16.4	-16.3	-16.3	-15.9	-15.7	-16.0	-16.2	-13.0	-14.9	-13.9	-9.5	-12.9
38	-17.7	-17.2	-17.1	-17.1	-16.7	-16.5	-16.8	-17.0	-13.8	-15.7	-14.7	-10.4	-13.7
39	-17.8	-17.3	-17.2	-17.2	-16.8	-16.6	-16.9	-17.1	-13.9	-15.8	-14.8	-10.4	-13.8
40	-19.3	-18.8	-18.7	-18.7	-18.3	-18.1	-18.4	-18.6	-15.4	-17.3	-16.3	-11.9	-15.3

Figure 5. DFT-computed thermodynamics of esterification of amides.

For the explored esters and amides, the alcohol component (derivatization of ester) have limited effects on the reaction thermodynamics. The effects from the change of primary alcohols are less than 1 kcal/mol for most amides (**a** to **h**). Comparing with the primary alcohols, the esters with secondary alcohols are less stable (**i** to **l**), especially for cyclohexanol. The thermodynamic equilibrium involving cyclohexanol favors the amide formation the most among all the studied alcohols. This suggest that the cyclohexanol derived esters can be useful substrates for amide synthesis. Interestingly, the tertiary alcohol is not the best alcohol for amide formation, potentially due to the steric repulsions in the corresponding esters.

Unlike the alcohols, the structure of amides has profound effects on the thermodynamic equilibrium of amidation. Without strong electron-activating groups or geometric twisting, amides are more stable than the corresponding esters regardless of the effects from alcohol (**23** and **24**). This follows the consensus that amide is generally more stable than ester due to the strong resonance nature. However, the electronically activated amides (**28–36**) and twisted amides (**37–40**) are significantly less stable in terms of thermodynamics. The equilibrium involving these amides all favor the ester formation, providing significant driving force for amide C–N bond activation and functionalization. Based on the computed thermodynamics, we found that the *N*-substitution has important effect on the thermodynamic stability of amides. This corroborates Szostak's design of twisted amides [27–39], and also highlights the importance of amide design for future synthetic transformations involving amides.

3. Computational Methods

All density functional theory (DFT) calculations were carried out using Gaussian16 software package [68]. Geometry optimizations and frequency calculations were conducted using B3LYP [69,70] functional with Grimme's D3(BJ) empirical dispersion correction [71], the LANL2DZ [72] basis set for nickel, and 6-31G(d) basis set for all the other elements. The solvation energy corrections were also included in the geometry optimizations and frequency calculations. Based on the solution phase-optimized geometries, single-point energies were calculated with the M06 [73] functional, the SDD [74,75] basis set for nickel, and 6-311G++(d,p) basis set for all other elements. Solvation energies were calculated using the SMD solvation model [76] with toluene as the solvent. No frequency scaling factor was used to calculate Gibbs energy corrections. The 3D diagrams of computed species were visualized using CYLview [77].

4. Conclusions

In summary, the mechanism and thermodynamics of Ni/IPr-catalyzed amidation of esters have been explored with DFT calculations. The reaction proceeds via the general cross-coupling mechanism. Nickel catalyst first cleaves the C–O bond of ester via oxidative addition, and subsequent proton transfer between LNi(acyl)(OR) species and amine leads to LNi(acyl)(amino) intermediate. This intermediate undergoes C–N reductive elimination to produce the amidation product. The studied aromatic and aliphatic esters both have fairly low reaction barriers, 20.4 kcal/mol for methyl benzoate and 16.9 kcal/mol for methyl 3-phenylpropionate. However, the overall amidation with morpholine is almost thermodynamic neutral. This suggests that the amidation only has limited thermodynamic driving force, and the evaporation of methanol could be critical for productivity. Therefore, the key to reaction success of Ni/NHC-catalyzed amidation of ester is the control of thermodynamic equilibrium.

For the thermodynamic free energy changes of the esterification of amide, the O-substitution of ester has limited effects, and cyclohexanol derivative is the most stable ester among all the studied cases. In contrast to the O-substitution of ester, the N-substitution of amide has profound effects on the thermodynamic stability of amides. Electron-activating groups and geometric twisting can significantly destabilize amides and provide the desired driving force towards the amide C–N bond activation. These computations highlight the design of amide for future reaction development involving amide C–N bond activation.

Author Contributions: X.H. conceived and designed the research. C.-L.J. and P.-P.X. performed the calculations. All authors analyzed the data and wrote the paper.

Acknowledgments: Calculations were performed on the high-performance computing system at the Department of Chemistry, Zhejiang University.

References

1. Meng, G.; Shi, S.; Szostak, M. Cross-Coupling of Amides by N–C Bond Activation. *Synlett* **2016**, *27*, 2530–2540.
2. Meng, G.; Szostak, M. Palladium-Catalyzed Suzuki–Miyaura Coupling of Amides by Carbon–Nitrogen Cleavage: General Strategy for Amide N–C Bond Activation. *Org. Biomol. Chem.* **2016**, *14*, 5690–5707. [CrossRef] [PubMed]
3. Liu, C.; Szostak, M. Twisted Amides: From Obscurity to Broadly Useful Transition-Metal-Catalyzed Reactions by N–C Amide Bond Activation. *Chem. Eur. J.* **2017**, *23*, 7157–7173. [CrossRef] [PubMed]
4. Dander, J.E.; Garg, N.K. Breaking Amides using Nickel Catalysis. *ACS Catal.* **2017**, *7*, 1413–1423. [CrossRef] [PubMed]
5. Takise, R.; Muto, K.; Yamaguchi, J. Cross-coupling of aromatic esters and amides. *Chem. Soc. Rev.* **2017**, *46*, 5864–5888. [CrossRef] [PubMed]
6. Gao, Y.; Ji, C.-L.; Hong, X. Ni-mediated C–N activation of amides and derived catalytic transformations. *Sci. China Chem.* **2017**, *60*, 1413–1424. [CrossRef]
7. Wiberg, K.B.; Breneman, C.M. Resonance interactions in acyclic systems. 3. Formamide internal rotation revisited. Charge and energy redistribution along the C–N bond rotational pathway. *J. Am. Chem. Soc.* **1992**, *114*, 831–840. [CrossRef]
8. Wiberg, K.B. The Interaction of Carbonyl Groups with Substituents. *Acc. Chem. Res.* **1999**, *32*, 922–929. [CrossRef]
9. Kemnitz, C.R.; Loewen, M.J. "Amide Resonance" Correlates with a Breadth of C–N Rotation Barriers. *J. Am. Chem. Soc.* **2007**, *129*, 2521–2528. [CrossRef] [PubMed]
10. Glover, S.A.; Rosser, A.A. Reliable Determination of Amidicity in Acyclic Amides and Lactams. *J. Org. Chem.* **2012**, *77*, 5492–5502. [CrossRef] [PubMed]
11. Szostak, R.; Aubé, J.; Szostak, M. Determination of Structures and Energetics of Small- and Medium-Sized One-Carbon Bridged Twisted Amides using ab Initio Molecular Orbital Methods. Implications for Amidic Resonance along the C–N Rotational Pathway. *J. Org. Chem.* **2015**, *80*, 7905–7927. [CrossRef] [PubMed]
12. Hu, F.; Lalancette, R.; Szostak, M. Structural Characterization of *N*–Alkylated Twisted Amides: Consequences for Amide Bond Resonance and N–C Cleavage. *Angew. Chem. Int. Ed.* **2016**, *55*, 5062–5066. [CrossRef] [PubMed]
13. Szostak, R.; Meng, G.; Szostak, M. Resonance Destabilization in *N*-Acylanilines (Anilides): Electronically-Activated Planar Amides of Relevance in N–C(O) Cross-Coupling. *J. Org. Chem.* **2017**, *82*, 6373–6378. [CrossRef] [PubMed]
14. Szostak, R.; Szostak, M. *N*-Acyl-Glutarimides: Resonance and Proton Affinities of Rotationally-Inverted Twisted Amides Relevant to N–C(O) Cross-Coupling. *Org. Lett.* **2018**, *20*, 1342–1345. [CrossRef] [PubMed]
15. Bisz, E.; Piontek, A.; Dziuk, B.; Szostak, R.; Szostak, M. Barriers to Rotation in ortho-Substituted Tertiary Aromatic Amides: Effect of Chloro-Substitution on Resonance and Distortion. *J. Org. Chem.* **2018**, *83*, 3159–3163. [CrossRef] [PubMed]
16. Meng, G.; Shi, S.; Lalancette, R.; Szostak, R.; Szostak, M. Reversible Twisting of Primary Amides via Ground State N–C(O) Destabilization: Highly Twisted Rotationally Inverted Acyclic Amides. *J. Am. Chem. Soc.* **2018**, *140*, 727–734. [CrossRef] [PubMed]
17. Hie, L.; Nathel, N.F.F.; Shah, T.K.; Baker, E.L.; Hong, X.; Yang, Y.-F.; Liu, P.; Houk, K.N.; Garg, N.K. Conversion of amides to esters by the nickel-catalysed activation of amide C–N bonds. *Nature* **2015**, *524*, 79–83. [CrossRef] [PubMed]
18. Weires, N.A.; Baker, E.L.; Garg, N.K. Nickel-Catalysed Suzuki-Miyaura Coupling of Amides. *Nat. Chem.* **2016**, *8*, 75–79. [CrossRef] [PubMed]

19. Simmons, B.J.; Weires, N.A.; Dander, J.E.; Garg, N.K. Nickel-Catalyzed Alkylation of Amide Derivatives. *ACS Catal.* **2016**, *6*, 3176–3179. [CrossRef]
20. Baker, E.L.; Yamano, M.M.; Zhou, Y.; Anthony, S.M.; Garg, N.K. A Two-Step Approach to Achieve Secondary Amide Transamidation Enabled by Nickel Catalysis. *Nat. Commun.* **2016**, *7*, 11554. [CrossRef] [PubMed]
21. Dander, J.E.; Weires, N.A.; Garg, N.K. Benchtop Delivery of $Ni(cod)_2$ using Paraffin Capsules. *Org. Lett.* **2016**, *18*, 3934–3936. [CrossRef] [PubMed]
22. Hie, L.; Baker, E.L.; Anthony, S.M.; Desrosier, J.; Senanayake, C.; Garg, N.K. Nickel-Catalyzed Esterification of Aliphatic Amides. *Angew. Chem. Int. Ed.* **2016**, *55*, 15129–15132. [CrossRef] [PubMed]
23. Medina, J.M.; Moreno, J.; Racine, S.; Du, S.; Garg, N.K. Mizoroki–Heck Cyclizations of Amide Derivatives for the Introduction of Quaternary Centers. *Angew. Chem. Int. Ed.* **2017**, *56*, 6567–6571. [CrossRef] [PubMed]
24. Weires, N.A.; Caspi, D.D.; Garg, N.K. Kinetic Modeling of the Nickel-Catalyzed Esterification of Amides. *ACS Catal.* **2017**, *7*, 4381–4385. [CrossRef] [PubMed]
25. Dander, J.E.; Baker, E.L.; Garg, N.K. Nickel-Catalyzed Transamidation of Aliphatic Amide Derivatives. *Chem. Sci.* **2017**, *8*, 6433–6438. [CrossRef] [PubMed]
26. Boit, T.B.; Weires, N.A.; Kim, J.; Garg, N.K. Nickel-Catalyzed Suzuki–Miyaura Coupling of Aliphatic Amides. *ACS Catal.* **2018**, *8*, 1003–1008. [CrossRef] [PubMed]
27. Meng, G.; Szostak, M. Sterically-Controlled Pd-Catalyzed Chemoselective Ketone Synthesis via N–C Cleavage in Twisted Amides. *Org. Lett.* **2015**, *17*, 4364–4367. [CrossRef] [PubMed]
28. Meng, G.; Szostak, M. General Olefin Synthesis by the Palladium-Catalyzed Heck Reaction of Amides: Sterically-Controlled Chemoselective N–C Activation. *Angew. Chem. Int. Ed.* **2015**, *54*, 14518–14522. [CrossRef] [PubMed]
29. Meng, G.; Szostak, M. Rhodium-Catalyzed C–H Bond Functionalization with Amides by Double C–H/C–N Bond Activation. *Org. Lett.* **2016**, *18*, 796–799. [CrossRef] [PubMed]
30. Shi, S.; Meng, G.; Szostak, M. Synthesis of Biaryls via Nickel Catalyzed Suzuki–Miyaura Coupling of Amides by Carbon–Nitrogen Cleavage. *Angew. Chem. Int. Ed.* **2016**, *55*, 6959–6963. [CrossRef] [PubMed]
31. Liu, Y.; Meng, G.; Liu, R.; Szostak, M. Sterically-Controlled Intermolecular Friedel-Crafts Acylation with Twisted Amides via Selective N–C Cleavage under Mild Conditions. *Chem. Commun.* **2016**, *52*, 6841–6844. [CrossRef] [PubMed]
32. Shi, S.; Szostak, M. Efficient Synthesis of Diaryl Ketones by Nickel-Catalyzed Negishi Cross-Coupling of Amides via Carbon–Nitrogen Bond Cleavage at Room Temperature Accelerated by Solvent Effect. *Chem. Eur. J.* **2016**, *22*, 10420–10424. [CrossRef] [PubMed]
33. Lei, P.; Meng, G.; Szostak, M. General Method for the Suzuki–Miyaura Cross-Coupling of Amides Using Commercially Available, Air- and Moisture-Stable Palladium/NHC (NHC = N-Heterocyclic Carbene) Complexes. *ACS Catal.* **2017**, *7*, 1960–1965. [CrossRef]
34. Shi, S.; Szostak, M. Decarbonylative Cyanation of Amides by Palladium Catalysis. *Org. Lett.* **2017**, *19*, 3095–3098. [CrossRef] [PubMed]
35. Shi, S.; Szostak, M. Nickel-Catalyzed Negishi Cross-Coupling of N-Acylsuccinimides: Stable, Amide-Based, Twist-Controlled Acyl-Transfer Reagents via N–C Activation. *Synthesis* **2017**, *49*, 3602–3608.
36. Lei, P.; Meng, G.; Ling, Y.; An, J.; Szostak, M. Pd-PEPPSI: Pd-NHC Precatalyst for Suzuki−Miyaura Cross-Coupling Reactions of Amides. *J. Org. Chem.* **2017**, *82*, 6638–6646. [CrossRef] [PubMed]
37. Liu, C.; Szostak, M. Decarbonylative Phosphorylation of Amides by Palladium and Nickel Catalysis: The Hirao Cross-Coupling of Amide Derivatives. *Angew. Chem. Int. Ed.* **2017**, *56*, 12718–12722. [CrossRef] [PubMed]
38. Osumi, Y.; Liu, C.; Szostak, M. *N*-Acylsuccinimides: Twist-Controlled, Acyl-Transfer Reagents in Suzuki–Miyaura Cross-Coupling by N–C Amide Bond Activation. *Org. Biomol. Chem.* **2017**, *15*, 8867–8871. [CrossRef] [PubMed]
39. Liu, Y.; Achtenhagen, M.; Liu, R.; Szostak, M. Transamidation of *N*-Acyl-Glutarimides with Amines. *Org. Biomol. Chem.* **2018**, *16*, 1322–1329. [CrossRef] [PubMed]
40. Hu, J.; Zhao, Y.; Liu, J.; Zhang, Y.; Shi, Z. Nickel-Catalyzed Decarbonylative Borylation of Amides: Evidence for Acyl C-N Bond Activation. *Angew. Chem. Int. Ed.* **2016**, *55*, 8718–8722. [CrossRef] [PubMed]
41. Hu, J.; Wang, M.; Pu, X.; Shi, Z. Nickel-Catalyzed Retro-Hydroamidocarbonylation of Aliphatic Amides to Olefins. *Nat. Commun.* **2017**, *8*, 14993. [CrossRef] [PubMed]

42. Li, X.; Zou, G. Acylative Suzuki coupling of amides: Acyl-nitrogen activation via synergy of independently modifiable activating groups. *Chem. Commun.* **2015**, *51*, 5089–5092. [CrossRef] [PubMed]
43. Yue, H.; Guo, L.; Lee, S.; Liu, X.; Rueping, M. Selective Reductive Removal of Ester and Amide Groups from Arenes and Heteroarenes through Nickel–Catalyzed C–O and C–N Bond Activation. *Angew. Chem. Int. Ed.* **2017**, *56*, 3972–3976. [CrossRef] [PubMed]
44. Srimontree, W.; Chatupheeraphat, A.; Liao, H.; Rueping, M. Amide to Alkyne Interconversion via a Nickel/Copper-Catalyzed Deamidative Cross-Coupling of Aryl and Alkenyl Amides. *Org. Lett.* **2017**, *19*, 3091–3094. [CrossRef] [PubMed]
45. Chatupheeraphat, A.; Liao, H.; Lee, S.; Rueping, M. Nickel-Catalyzed C–CN Bond Formation via Decarbonylative Cyanation of Esters, Amides, and Intramolecular Recombination Fragment Coupling of Acyl Cyanides. *Org. Lett.* **2017**, *19*, 4255–4258. [CrossRef] [PubMed]
46. Liu, X.; Yue, H.; Jia, J.; Guo, L.; Rueping, M. Synthesis of Amidines from Amides using a Nickel Catalyzed Decarbonylative Amination via CO Extrusion Intramolecular Recombination Fragment Coupling. *Chem. Eur. J.* **2017**, *23*, 11771–11775. [CrossRef] [PubMed]
47. Lee, S.; Guo, L.; Yue, H.; Liao, H.; Rueping, M. Nickel-Catalyzed Decarbonylative Silylation, Borylation, and Amination of Arylamides via a Deamidative Reaction Pathway. *Synlett* **2017**, *28*, 2594–2598.
48. Lee, S.; Liao, H.; Chatupheeraphat, A.; Rueping, M. Nickel-Catalyzed C–S Bond Formation via Decarbonylative Thioetherification of Esters, Amides and Intramolecular Recombination Fragment Coupling of Thioesters. *Chem. Eur. J.* **2018**, *24*, 3608–3612. [CrossRef] [PubMed]
49. Liu, X.; Hsiao, C.; Guo, L.; Rueping, M. Cross-Coupling of Amides with Alkylboranes via Nickel-Catalyzed C–N Bond Cleavage. *Org. Lett.* **2018**, *20*, 2976–2979. [CrossRef] [PubMed]
50. Dey, A.; Sasmal, S.; Seth, K.; Lahiri, G.K.; Maiti, D. Nickel-Catalyzed Deamidative Step-Down Reduction of Amides to Aromatic Hydrocarbons. *ACS Catal.* **2017**, *7*, 433–437. [CrossRef]
51. Walker, J.A.; Vickerman, K.L.; Humke, J.N.; Stanley, L.M. Ni-Catalyzed Alkene Carboacylation via Amide C–N Bond Activation. *J. Am. Chem. Soc.* **2017**, *30*, 10228–10231. [CrossRef] [PubMed]
52. Amani, J.; Alam, R.; Badir, S.; Molander, G.A. Synergistic Visible-Light Photoredox/Nickel-Catalyzed Synthesis of Aliphatic Ketones via N–C Cleavage of Imides. *Org. Lett.* **2017**, *19*, 2426–2429. [CrossRef] [PubMed]
53. Hie, L.; Nathel, N.F.F.; Hong, X.; Yang, Y.-F.; Houk, K.N.; Garg, N.K. Nickel-Catalyzed Activation of Acyl C–O Bonds of Methyl Esters. *Angew. Chem. Int. Ed.* **2016**, *55*, 2810–2814. [CrossRef] [PubMed]
54. Halima, T.B.; Masson-Makdissi, J.; Newman, S.G. Nickel-Catalyzed Amide Bond Formation from Methyl Esters. *Angew. Chem. Int. Ed.* **2018**, *57*, 12925–12929. [CrossRef] [PubMed]
55. Li, Z.; Zhang, S.; Fu, Y.; Guo, Q.; Liu, L. Mechanism of Ni-Catalyzed Selective C–O Bond Activation in Cross-Coupling of Aryl Esters. *J. Am. Chem. Soc.* **2009**, *131*, 8815–8823. [CrossRef] [PubMed]
56. Yu, H.-Z.; Fu, Y. Mechanistic Origin of Cross-Coupling Selectivity in Ni-Catalysed Tishchenko Reactions. *Chem. Eur. J.* **2012**, *18*, 16765–16773. [CrossRef] [PubMed]
57. Hong, X.; Liang, Y.; Houk, K.N. Mechanisms and Origins of Switchable Chemoselectivity of Ni-Catalyzed C(aryl)–O and C(acyl)–O Activation of Aryl Esters with Phosphine Ligands. *J. Am. Chem. Soc.* **2014**, *136*, 2017–2025. [CrossRef] [PubMed]
58. Lu, Q.; Yu, H.-Z.; Fu, Y. Mechanistic Study of Chemoselectivity in Ni-Catalyzed Coupling Reactions between Azoles and Aryl Carboxylates. *J. Am. Chem. Soc.* **2014**, *136*, 8252–8260. [CrossRef] [PubMed]
59. Xu, H.; Muto, K.; Yamaguchi, J.; Zhao, C.; Itami, K.; Musaev, D.G. Key Mechanistic Features of Ni-Catalyzed C–H/C–O Biaryl Coupling of Azoles and Naphthalen-2-yl Pivalates. *J. Am. Chem. Soc.* **2014**, *136*, 14834–14844. [CrossRef] [PubMed]
60. Muto, K.; Yamaguchi, J.; Musaev, D.G.; Itami, K. Decarbonylative organoboron cross-coupling of esters by nickel catalysis. *Nat. Commun.* **2015**, *6*, 7508. [CrossRef] [PubMed]
61. Li, Z.; Liu, L. Recent advances in mechanistic studies on Ni catalyzed cross-coupling reactions. *Chin. J. Catal.* **2015**, *36*, 3–14. [CrossRef]
62. Liu, L.; Chen, P.; Sun, Y.; Wu, Y.; Chen, S.; Zhu, J.; Zhao, Y. Mechanism of Nickel-Catalyzed Selective C–N Bond Activation in Suzuki–Miyaura Cross-Coupling of Amides: A Theoretical Investigation. *J. Org. Chem.* **2016**, *81*, 11686–11696. [CrossRef] [PubMed]
63. Xu, Z.-Y.; Yu, H.-Z.; Fu, Y. Mechanism of Nickel-Catalyzed Suzuki–Miyaura Coupling of Amides. *Chem. Asian J.* **2017** *12*, 1765–1772. [CrossRef] [PubMed]

64. Ji, C.-L.; Hong, X. Factors Controlling the Reactivity and Chemoselectivity of Resonance Destabilized Amides in Ni-Catalyzed Decarbonylative and Nondecarbonylative Suzuki–Miyaura Coupling. *J. Am. Chem. Soc.* **2017**, *139*, 15522–15529. [CrossRef] [PubMed]
65. Jiang, Y.-Y.; Liu, T.; Sun, X.; Xu, Z.; Fan, X.; Zhu, L.; Bi, S.-W. Computational study of the mechanism of amide bond formation via CS2-releasing 1,3-acyl transfer. *Org. Biomol. Chem.* **2018**, *16*, 5808–5815. [CrossRef] [PubMed]
66. Li, Y.; Zou, L.; Bai, R.; Lan, Y. Ni(I)–Ni(III) vs. Ni(II)–Ni(IV): Mechanistic study of Ni-catalyzed alkylation of benzamides with alkyl halides. *Org. Chem. Front.* **2018**, *5*, 615–622. [CrossRef]
67. Saper, N.I.; Hartwig, J.F. Mechanistic Investigations of the Hydrogenolysis of Diaryl Ethers Catalyzed by Nickel Complexes of N-Heterocyclic Carbene Ligands. *J. Am. Chem. Soc.* **2017**, *139*, 17667–17676. [CrossRef] [PubMed]
68. Frisch, M.J.; Trucks, G.W.; Schlegel, H.B.; Scuseria, G.E.; Robb, M.A.; Cheeseman, J.R.; Scalmani, G.; Barone, V.; Petersson, G.A.; Nakatsuji, H.; et al. *Gaussian 16*; Revision A.03; Gaussian, Inc.: Wallingford, CT, USA, 2016.
69. Head-Gordon, M.; Pople, J.A.; Frisch, M. MP2 energy evaluation by direct methods. *Chem. Phys. Lett.* **1988**, *153*, 503–506. [CrossRef]
70. Lee, C.; Yang, W.; Parr, R.G. Development of the Colle-Salvetti correlation-energy formula into a functional of the electron density. *Phys. Rev. B Condens. Matter Mater. Phys.* **1988**, *37*, 785–789. [CrossRef]
71. Grimme, S.; Antony, J.; Ehrlich, S.; Krieg, H. A consistent and accurate ab initio parametrization of density functional dispersion correction (DFT-D) for the 94 elements H-Pu. *J. Chem. Phys.* **2010**, *132*, 154104. [CrossRef] [PubMed]
72. Hay, P.J.; Wadt, W.R. Ab initio effective core potentials for molecular calculations. Potentials for K to Au including the outermost core orbitals. *J. Chem. Phys.* **1985**, *82*, 299. [CrossRef]
73. Zhao, Y.; Truhlar, D.G. The M06 suite of density functionals for main group thermochemistry, thermochemical kinetics, noncovalent interactions, excited states, and transition elements: Two new functionals and systematic testing of four M06-class functionals and 12 other functionals. *Theor. Chem. Acc.* **2008**, *120*, 215–241.
74. Häussermann, U.; Dolg, M.; Stoll, H.; Preuss, H.; Schwerdtfeger, P.; Pitzer, R.M. Accuracy of energy-adjusted quasirelativistic ab initio pseudopotentials. *Mol. Phys.* **1993**, *78*, 1211–1224. [CrossRef]
75. Küchle, W.; Dolg, M.; Stoll, H.; Preuss, H. Energy-adjusted pseudopotentials for the actinides. Parameter sets and test calculations for thorium and thorium monoxide. *J. Chem. Phys.* **1994**, *100*, 7535–7542. [CrossRef]
76. Marenich, A.V.; Cramer, C.J.; Truhlar, D.G. Universal Solvation Model Based on Solute Electron Density and on a Continuum Model of the Solvent Defined by the Bulk Dielectric Constant and Atomic Surface Tensions. *J. Phys. Chem. B* **2009**, *113*, 6378–6396. [CrossRef] [PubMed]
77. *CYLview, 1.0b*; CYLview 2.0; Universitéde Sherbrooke: Sherbrooke, QC, Canada, 2009.

13

Density Functional Studies on Secondary Amides: Role of Steric Factors in Cis/Trans Isomerization

Balmukund S. Thakkar *, John Sigurd M. Svendsen and Richard A. Engh *

Department of Chemistry, UiT The Arctic University of Norway, N-9037 Tromsø, Norway; john-sigurd.svendsen@uit.no

* Correspondence: balmukund.thakkar@uit.no (B.S.T.); Richard.Engh@uit.no (R.A.E.)

Abstract: Cis/trans isomerization of amide bonds is a key step in a wide range of biological and synthetic processes. Occurring through C-N amide bond rotation, it also coincides with the activation of amides in enzymatic hydrolysis. In recently described QM studies of cis/trans isomerization in secondary amides using density functional methods, we highlighted that a peptidic prototype, such as glycylglycine methyl ester, can suitably represent the isomerization and complexities arising out of a larger molecular backbone, and can serve as the primary scaffold for model structures with different substitution patterns in order to assess and compare the steric effect of the substitution patterns. Here, we describe our theoretical assessment of such steric effects using *tert*-butyl as a representative bulky substitution. We analyze the geometries and relative stabilities of both trans and cis isomers, and effects on the cis/trans isomerization barrier. We also use the additivity principle to calculate absolute steric effects with a gradual increase in bulk. The study establishes that bulky substitutions significantly destabilize cis isomers and also increases the isomerization barrier, thereby synergistically hindering the cis/trans isomerization of secondary amides. These results provide a basis for the rationalization of kinetic and thermodynamic properties of peptides with potential applications in synthetic and medicinal chemistry.

Keywords: density functional theory; cis/trans isomerization; secondary amides; dipeptides; steric effects; *tert*-butyl; additivity principle

---

1. Introduction

The chemistry of the amide bond has attracted the interest of chemists with diverse specializations. Its unique characteristics arise from the delocalization of electrons from nitrogen to the carbonyl group, which confers a partial double-bond character to the C-N bond and stabilizes a planar geometry with a relatively high energy rotational barrier that hinders the free rotation, giving rise to cis and trans isomers [1–3]. The resonance effect also protects the amide moiety against nucleophilic attacks at the carbonyl carbon (e.g., it is virtually immune to hydrolysis at ambient temperature and pH in non-enzymatic conditions); hence, it is a common practice to activate amides using Lewis acids for chemical transformation. However, studies on enzymatic hydrolysis of amides have revealed that distortion in the amide bond planarity via C-N bond rotation also results in amide bond activation, increasing susceptibility to nucleophilic attack [4–6]. Cis/trans isomerization is one phenomenon whereby the amide moiety loses its planarity, as significant geometric and hybridizational changes occur throughout C-N bond rotation [7]. Therefore, information regarding the stabilities of cis and trans forms of amides, C-N bond rotation in terms of cis/trans isomerization, and relevant energy barriers can be useful for understanding the activation by deformation for a variety of amides, especially peptides.

While 3°-amides (e.g., prolyl peptide bonds) have often been observed to undergo cis/trans isomerization due to small energy differences between cis and trans isomers [8–10], 2°-amides also undergo cis/trans isomerization via higher energy states in diverse important phenomena, such as chemo-mechanical cycling of motor proteins [11], the protein folding [12–14] and catalytic activity [15] of enzymes (such as cyclophilin A), cascade dissociation of peptide cation radicals for peptide sequencing [16], and cyclization reactions of peptides (e.g., as in the formation of piperazine-2,5-diones) [17].

With advances in computational capabilities since the 1990s, theoretical studies on trans and cis isomers of 2°-amides and their interconversion have revealed diverse phenomena, such as: effects of pyramidalization of the amide and geometries of transition states [18]; the role of conjugation [19]; simulated solvent effects with molecular dynamics [20]; comparison of theoretically obtained rotational barrier values with experimental values [21,22]; and the generation of ensembles of transition state geometries [23]. Recently, we have conducted theoretical studies [7,24] on secondary amides using density functional methods and molecular dynamics to provide a detailed account of geometry changes during cis/trans isomerization, as well as the effects of solvent models, using glycylglycine methyl ester (GGMe, Figure 1) as an example. We described that cis/trans isomerization can occur via either of the two paths: one via the *anti*-type transition state, and one via the *syn*-type transition state (Figure 2). We also showed that the salient features of the cis/trans isomerization remained consistent when the studies were extended from *N*-methylacetamide to the peptidic scaffolds of GGMe, thus serving as a simple peptide prototype to study conformational flexibilities and complexities relevant to larger molecular backbones. In the present work, we extend our studies to substituted derivatives of GGMe (Figure 1).

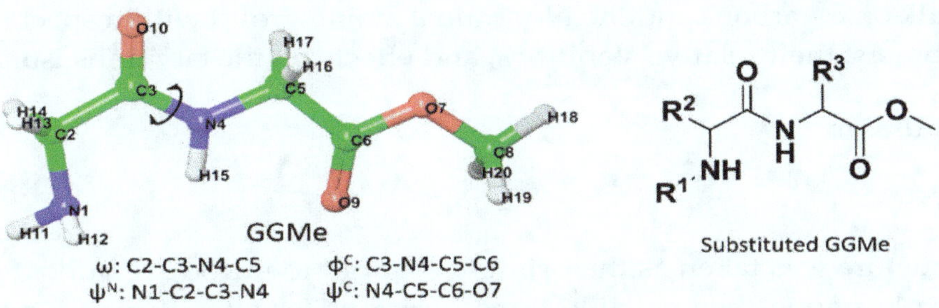

Figure 1. The structures of glycylglycine methyl ester (**left**) and the substituted GGMe scaffold (**right**). The relevant dihedral angles are defined by reference to the GGMe atom names (**left**).

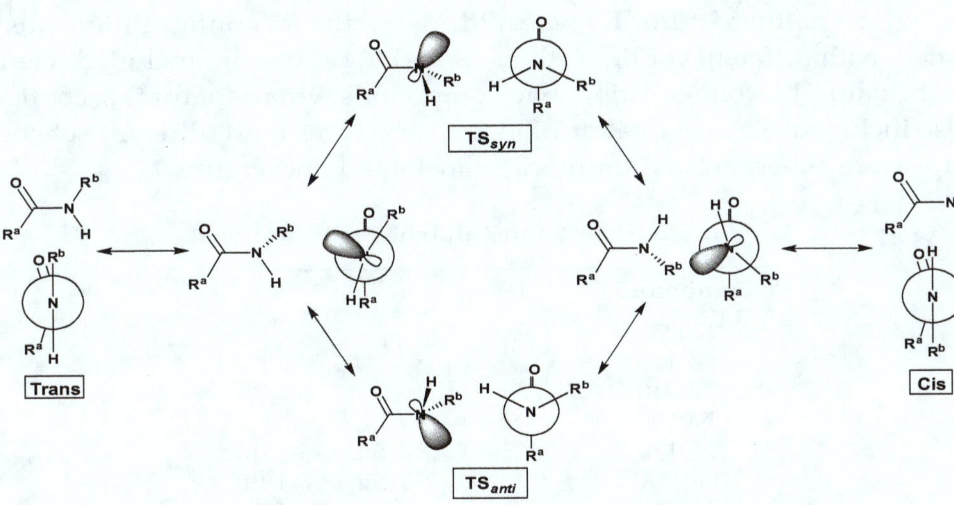

Figure 2. Two paths of cis/trans isomerization via the *syn* transition state (TS_{syn}) and *anti* transiton state (TS_{anti}), respectively. R^a and R^b represent substitutions attached to the carbonyl carbon and amide nitrogen, respectively. For simplicity, the rotation is shown only for positive ω values between 0° (cis) and 180° (trans). See [7] for more details.

It has been observed experimentally [25,26] that the steric bulk on both sides of the amide moiety affects the cis/trans isomerization barrier. This demonstrates the key role of amino-acid side-chains in peptides in the rate of isomerization and extent of equilibria between the two isomers. Corresponding attempts have been made to tailor the flexibility of the rotamers and overall conformations by introducing constraints, such as intramolecular hydrogen bonding and/or steric bulk around the amide bonds [27,28]. In addition to targeted flexibility design, the use of unnatural isomers (especially D-amino acids) and the introduction of *N*-alkylated chains or functionalities in peptide chemistry for the generation of novel peptidomimetics [29,30] provide further variation of the steric bulk around the peptide bond (as well as diversification towards enzymatic activation). Thus, a systematic theoretical study on the effect of steric bulk on the relative stabilities of trans and cis isomers and their interconversion via cis/trans isomerization is in order.

The diversity of side chains and substitutions, along with their differing extents of steric effects, greatly complicates their study. Bigger and bulkier side chains and substitutions introduce more potential interactions, asymmetry, flexibility, and resonance effects, and hence have prohibitively greater demands for CPU time for QM studies at higher levels of theory. Hence, a systematic study requires representative substitution group(s) that can serve to introduce simple "bulk", devoid of any resonance or hydrogen-bonding effects. With the A value > 4 [31,32] and υ_{ef} value = 1.2 [33,34], the *tert*-butyl group can therefore be an ideal choice for "bulk". In synthetic chemistry also, it is a common practice to use *tert*-butyl groups as bulky substitutions to analyze the effect of steric bulk [35–37].

With this background, we present our theoretical assessment of the impact of stereospecific patterns of steric bulk on α-carbons and the N-terminal amino group with respect to the geometries of the trans and cis isomers, their relative stabilities, and effects on the cis/trans isomerization barrier.

2. Results and Discussion

2.1. Model Structures

The GGMe structure was taken as the primary scaffold to introduce bulk at different positions. As shown in Figure 1, among the available three positions for substitutions, two are α-carbons of amino acids and are thus chiral centers. In line with the natural amino acids, the default configuration of substitution at α-carbons was kept as the *S*-configuration. However, when both chiral centers had substitutions (i.e., where neither R^2 nor R^3 was hydrogen), the R^2 configuration was kept constant (*S*), and both isomers with different configurations (*R* and *S*) of R^3 were included. Hence, along with compounds with the natural *S*-configuration, two compounds with unnatural *R*-configurations at the α-carbon were also included. Thus, a total of 10 model structures with different substitution patterns of *tert*-butyl groups were generated, as summarized in Table 1 and Figure 3.

Table 1. Model structures and their substitutions.

Compound [a]	R^1	R^2	R^3
N0000 (GGMe)	H	H	H
N0010	H	H	*S*-*t*-Bu
N0100	H	*S*-*t*-Bu	H
N0110	H	*S*-*t*-Bu	*S*-*t*-Bu
N011′0	H	*S*-*t*-Bu	R-t-Bu
N1000	t-Bu	H	H
N1010	*t*-Bu	H	*S*-*t*-Bu
N1100	*t*-Bu	*S*-*t*-Bu	H
N1110	*t*-Bu	*S*-*t*-Bu	*S*-*t*-Bu
N111′0	*t*-Bu	*S*-*t*-Bu	*R*-*t*-Bu

[a] The first, second, and third digits in the name of each compound represent the variants of R^1, R^2, and R^3, respectively. The fourth digit (0 in all) represents methyl ester and is specified in order to enable comparison to future variation at this site.

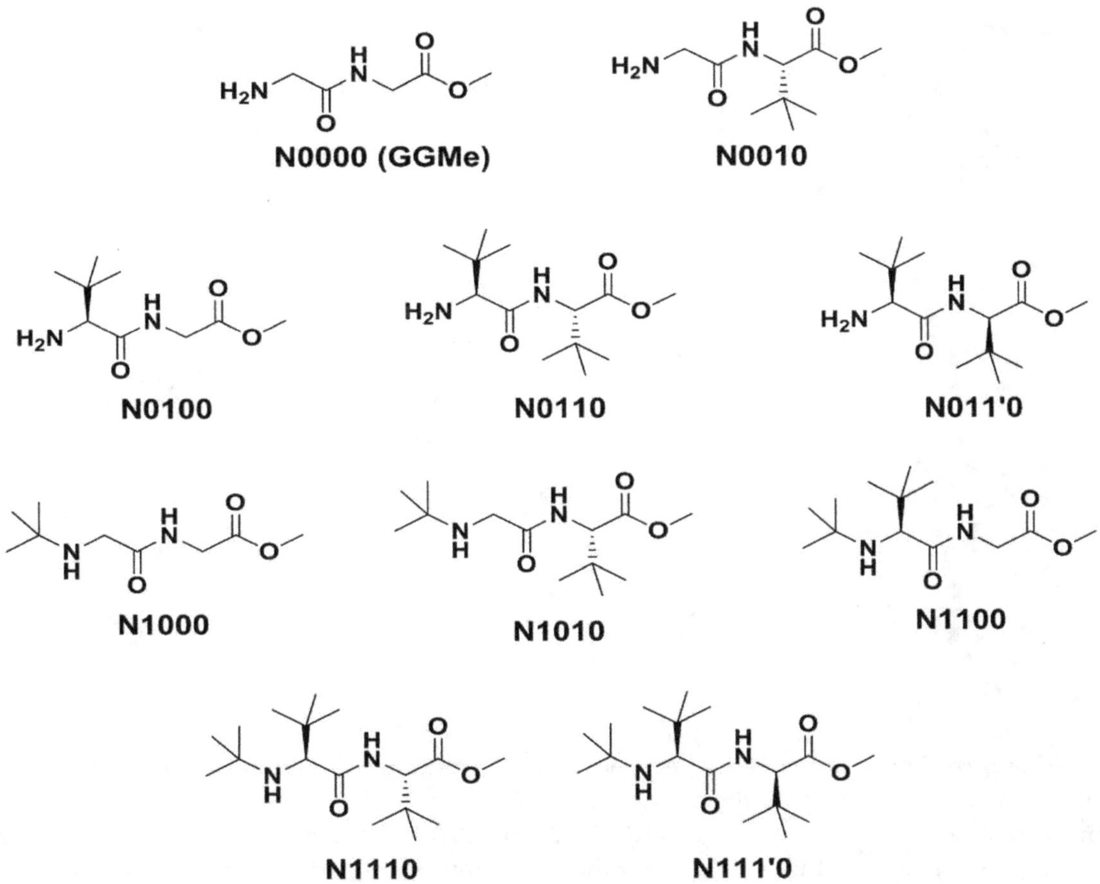

Figure 3. GGMe and its *tert*-butyl-substituted derivatives as model dipeptide esters.

2.2. Trans Geometries

In both gas and water phases (Figures 4 and 5), the optimized minimum energy geometries of trans isomers showed an interesting blend of peptidic features with many of the characteristics previously observed for GGMe [24]. For example, due to the absence of an amide group at the C-terminal, C_7-forms (γ-foldings) were not observed and the geometries with unsubstituted chiral carbons showed a preference for the extended planar C_5-form [38]. However, the substitutions at chiral carbons introduced peptidic folding. A majority of geometries also showed the presence of a hydrogen bond between the N-terminal amino group and amidic hydrogen [38].

The amide bond planarity is often described quantitatively in terms of the dihedral, ω. The minimum energy geometries showed that the presence of bulky groups on both α-carbons can cause distortion in the amide bond planarity beyond $\pm 5°$ of ideal perfect planarity (Table 2), and therefore can be used as an alternate approach towards acyclic twisted amides in combination with non-covalent bonding strategies.

In order to assess the magnitude of steric factors in terms of energy (${}^{st}E$), the energy values calculated based on the additivity principle [31] for *tert*-butyl-substituted structures (${}^{Add}E$) were compared with their direct energy estimates for their minimum energy geometries (${}^{opt}E$). As explained in Figure 6A, the energy difference between methane and neopentane was taken to represent the additive energy value of *tert*-butyl substitution on an sp^3-carbon ($E_{tb}{}^{C}$), while the energy difference between ammonia and *tert*-butylamine was taken to represent the additive energy value of *tert*-butyl substitution on pyramidal sp^3-nitrogen of the amino group ($E_{tb}{}^{N}$). Both values were calculated for both gas and water phases (Tables 3 and 4) each. Then, as shown in Figure 6B, the energy value for the optimized minimum energy geometry of N0000 (GGMe), or ${}^{opt}E_{N0000}$, was taken as the basis for the calculations of other compounds having *tert*-butyl substitutions on the α-carbon or amino nitrogen.

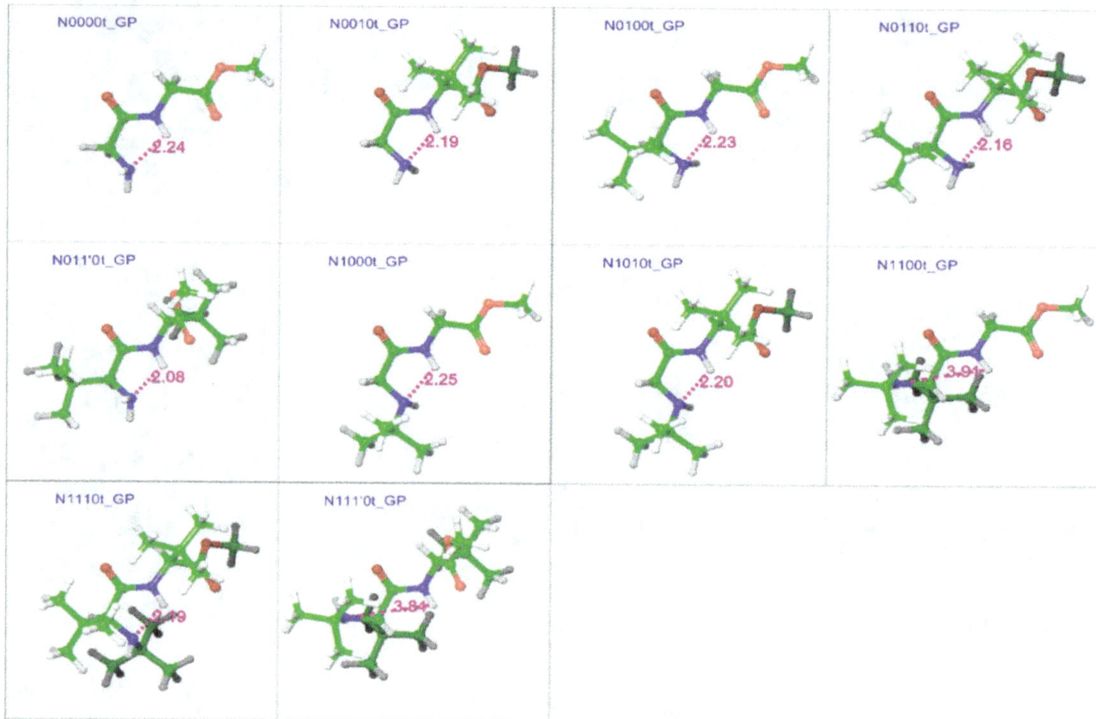

Figure 4. Optimized minimum energy geometries of substituted GGMe derivatives in the gas phase, where all geometries are aligned at the amide moiety. The distance between N-terminal amino nitrogen and amidic hydrogen, in magenta, highlights the presence (or absence when distance > 2.3 Å) of the H-bond between the two atoms. The "t" after the compound name indicates the "trans" isomer.

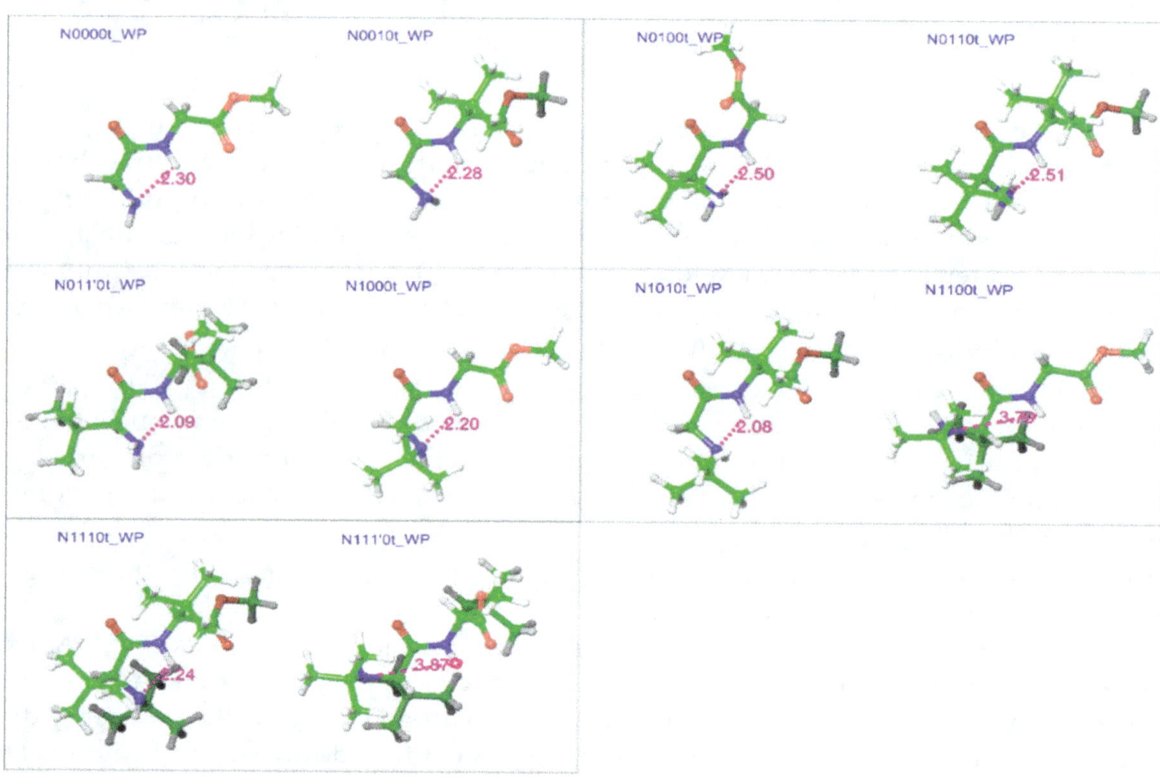

Figure 5. Optimized minimum energy geometries of substituted GGMe derivatives in the water phase, where all geometries are aligned at the amide moiety. The distance between N-terminal amino nitrogen and amidic hydrogen, in magenta, highlights the presence (or absence when distance > 2.3 Å) of the H-bond between the two atoms. The "t" after the compound name indicates the "trans" isomer.

For each compound, the additive energy value (^{Add}E) was determined by the addition of corresponding additive energy values of the *tert*-butyl group (E_{tb}), depending on the atom of attachment (carbon or nitrogen), number of *tert*-butyl groups, and the phase (gas or water). The steric effect (^{st}E) was calculated as the difference between ^{Add}E and the energy values for optimized minimum energy geometries of respective compounds (^{opt}E) in their respective phase—that is, the gas or water phases (Tables 5 and 6).

Table 2. Values (in degrees) of different dihedrals for the minimum energy geometries of different compounds in gas and water phases. The presence of bulky groups on both α-carbons can cause distortion in the planarity of amide moiety in terms of the ω dihedral, as highlighted in grey.

Compound	Gas Phase				Water Phase			
	ψ^N	ω	φ^C	ψ^C	ψ^N	ω	φ^C	ψ^C
N0000t	12.8	178.3	175.6	−179.3	15.2	179.2	178.5	−177.4
N0010t	−14.3	177.0	−122.7	141.7	−19.3	178.8	−124.4	142.6
N0100t	−22.8	−179.7	−173.1	178.8	−40.5	−177.9	66.3	−146.4
N0110t	−23.8	169.5	−114.3	141.2	−50.8	171.3	−135.3	152.0
N011′0t	14.3	−178.8	123.0	−140.9	13.4	178.5	127.3	−143.2
N1000t	−15.6	−178.0	−179.1	179.6	12.7	179.5	175.3	−177.3
N1010t	−17.9	176.6	−123.7	141.8	−12.2	178.7	−127.5	141.4
N1100t	140.7	177.5	−175.0	−179.9	134.6	174.6	−163.7	171.2
N1110t	−27.1	172.5	−117.0	140.5	−32.7	174.2	−123.1	141.1
N111′0t	143.0	−169.7	127.2	−145.3	148.2	−173.4	136.1	−147.7

The "t" after the compound name indicates the "trans" isomer.

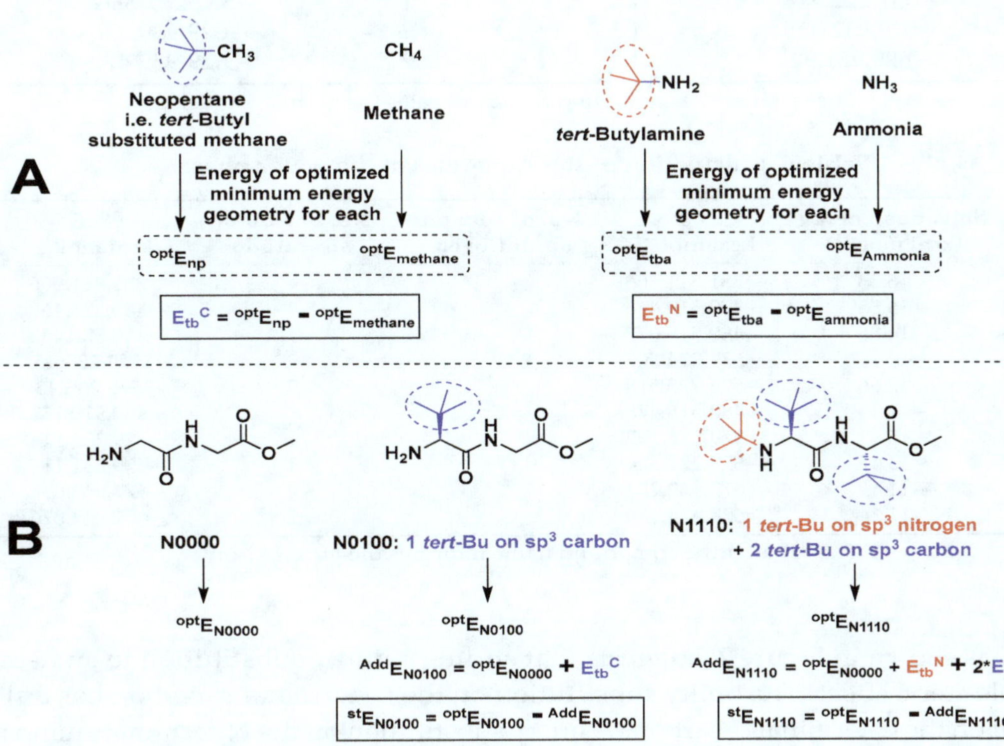

Figure 6. Using the additivity principle: (**A**) Calculation of the additive energy values of *tert*-butyl substitution (**B**) Calculation of the magnitude of the steric effect in terms of energy based on GGMe, explained with examples of N0100 and N1110.

Table 3. Calculation of *tert*-Bu additive energy in the gas phase.

Compound	^{opt}E	Attachment Point	*tert*-Bu Additive Energy kcal mol^{-1} (E_{tb})
NH_3 *tert*-Bu-NH_2	−35,508.975 −134,218.106	sp^3 nitrogen	E_{tb}^{N} = −98,709.131
CH_4 *tert*-Bu-CH_3	−25,437.531 −124,149.665	sp^3 carbon	E_{tb}^{C} = −98,712.134

Table 4. Calculation of *tert*-Bu additive energy in the water phase.

Compound	^{opt}E	Attachment Point	*tert*-Bu Additive Energy Kcal mol^{-1} (E_{tb})
NH_3 *tert*-Bu-NH_2	−35,514.577 −134,222.223	sp^3 nitrogen	E_{tb}^{N} = −98,707.646
CH_4 *tert*-Bu-CH_3	−25,436.175 −124,148.137	sp^3 carbon	E_{tb}^{C} = −98,711.962

Table 5. Calculation of steric energies in the gas phase.

Compound [a]	^{opt}E kcal mol^{-1}	No. of *t*-Bu on sp^3 Nitrogen	No. of *t*-Bu on sp^3 Carbon	^{Add}E kcal mol^{-1}	^{st}E kcal mol^{-1}
N0000t	−333,806.071			−333,806.071	0
N0010t	−432,516.46		1	−432,518.205	1.745
N0100t	−432,515.901		1	−432,518.205	2.304
N0110t	−531,225.709		2	−531,230.339	4.63
N011′0t	−531,225.494		2	−531,230.339	4.845
N1000t	−432,514.375	1		−432,515.202	0.827
N1010t	−531,224.575	1	1	−531,227.336	2.761
N1100t	−531,222.518	1	1	−531,227.336	4.818
N1110t	−629,932.315	1	2	−629,939.47	7.155
N111′0t	−629,931.92	1	2	−629,939.47	7.55

[a] The "t" after the compound name indicates the "trans" isomer.

Table 6. Calculation of steric energies in the water phase.

Compound [a]	Solvation Energy (kcal mol^{-1})	^{opt}E kcal mol^{-1}	No. of *t*-Bu on sp^3 nitrogen	No. of *t*-Bu on sp^3 carbon	^{Add}E kcal mol^{-1}	^{st}E kcal mol^{-1}
N0000t	−16.33	−381,811.506			−381,811.506	0
N0010t	−16.85	−480,519.643		1	−480,523.468	3.825
N0100t	−14.86	−480,518.077		1	−480,523.468	5.391
N0110t	−16.1	−579,224.507		2	−579,235.43	10.923
N011′0t	−12.66	−579,225.518		2	−579,235.43	9.912
N1000t	−11.16	−480,518.478	1		−480,519.152	0.674
N1010t	−15.97	−579,226.192	1	1	−579,231.114	4.922
N1100t	−13.47	−579,222.324	1	1	−579,231.114	8.79
N1110t	−11.49	−677,932.254	1	2	−677,943.076	10.822
N111′0t	−11.34	−677,929.486	1	2	−677,943.076	13.59

[a] The "t" after the compound name indicates the "trans" isomer.

The trend, as shown in Figure 7, suggests that adding a bulky substitution to an α-carbon results in a considerable steric effect. The bulky substitution on the N-terminal α-carbon has a slightly larger impact than that on the C-terminal α-carbon. A bulky substitution on the N-terminal amino nitrogen also leads to a steric effect that is consistently present across corresponding pairs, becoming stronger with bulk on α-carbons. This steric effect seen in terms of energies is also reflected in strained geometries, and is stronger when using the polarizable continuum model (the water phase). The comparison between different substitution patterns using the additivity principle provides interesting insights about the importance of the hydrogen bond between the N-terminal amino nitrogen and the amide hydrogen for conformational stability. For example, the difference between the stabilities of the minimum energy conformers of N0010 and N0100 in the gas phase is ~0.5 kcal mol^{-1}, but with *tert*-butyl substitution on the N-terminal nitrogen (N1010 and N1100), the same stability difference increases to ~2.0 kcal mol^{-1}, as N1100 lacks the hydrogen bond (Figure 4). Similarly, in the water phase, the minimum energy geometry of N011′0 is more stable (by ~1 kcal mol^{-1}) than that of N0110, while N1110 is more stable (by ~2.7 kcal mol^{-1}) than N111′0. Both of the more stable geometries (N011′0 and N1110) show the presence of the hydrogen bond, while both less-stable geometries lack it (Figure 5). Additionally, as is evident from Figure 8, the *tert*-butyl groups in the less-substituted N0110 and N011′0 can remain apart, but the same is not possible for the highly substituted geometries of N1110 and N111′0 with their additional bulk on the N-terminal amino nitrogen, ultimately resulting in the higher energy difference.

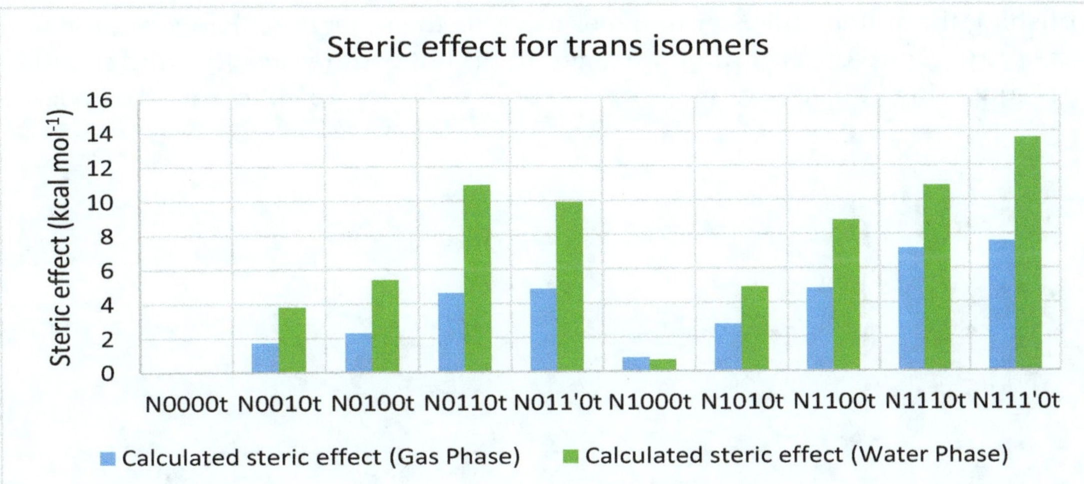

Figure 7. Calculated steric effect for trans-isomers in the gas and water phase, using the additivity principle. The "t" after the compound name indicates the "trans" isomer.

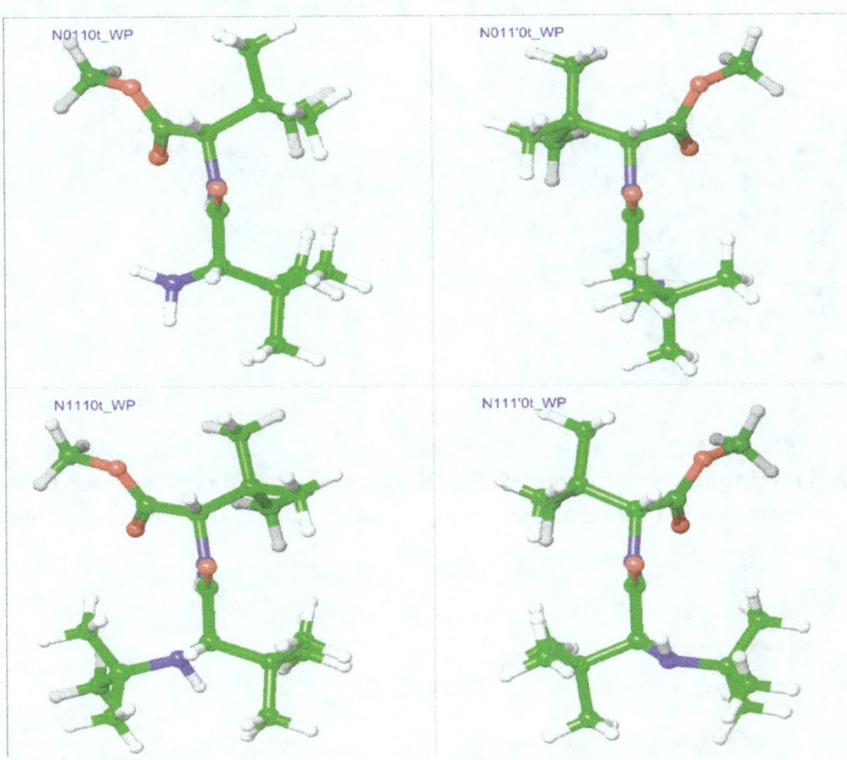

Figure 8. Comparison of minimum energy geometries of N0110, N011′0, N1110, and N111′0 in the water phase. The "t" after the compound name indicates the "trans" isomer.

2.3. *Cis Geometries*

In line with the trans isomers, the minimum energy geometries of the cis isomers of the compounds in both gas and water phases showed a preference for the extended form at unsubstituted α-carbons, but also showed folding at substituted α-carbons (Figures 9 and 10). With both α-carbons in proximity of each other in the cis-geometries, the impact of steric bulk was evident from the strained geometries. For example, compounds N0010, N0110, and N1110 in the gas phase show similar chain-folding, but the dihedral Φ for the C-terminal residue shows dramatic changes, while the *tert*-butyl groups on both α-carbons are also forced to come closer (Figure 11). With the *tert*-butyl group only on the C-terminal α-carbon, the dihedral Φ for N0010 stands at −115°, which narrows sharply to −81° with the addition of a *tert*-butyl group at the N-terminal α-carbon, as both *tert*-butyl groups try to stay apart (2.82 Å) by pushing the amino and ester moieties towards the other side. However, the repulsion due to another *tert*-butyl group on the amino nitrogen in N1110 counteracts this and broadens the Φ to −87°, in turn "squeezing" the *tert*-butyl groups on both α-carbons even closer (2.49 Å).

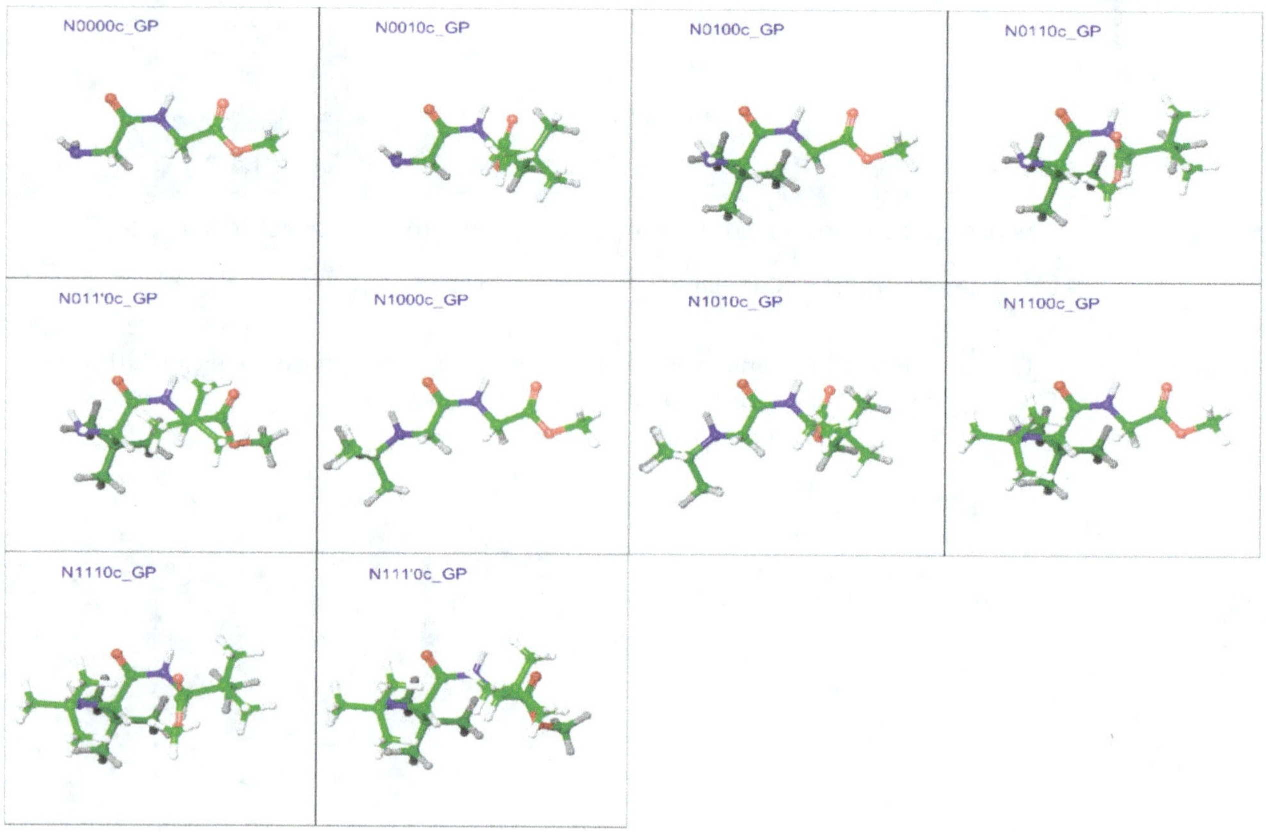

Figure 9. Optimized minimum energy geometries of the cis isomer of the substituted GGMe derivatives in the gas phase; all geometries are aligned at the amide moiety. The "c" after the compound name indicates the "cis" isomer.

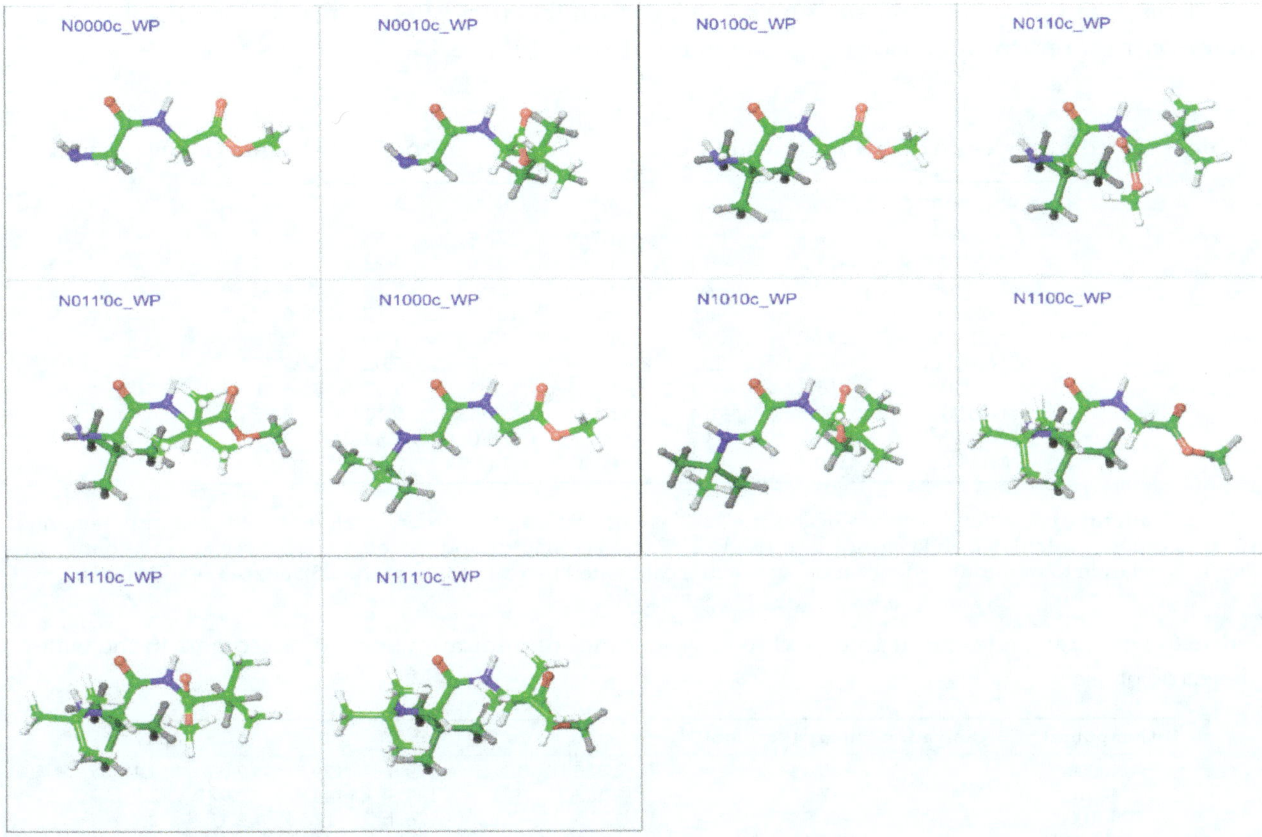

Figure 10. Optimized minimum energy geometries of the cis isomers of the substituted GGMe derivatives in the water dielectric phase; all geometries are aligned at the amide moiety. The "c" after the compound name indicates the "cis" isomer.

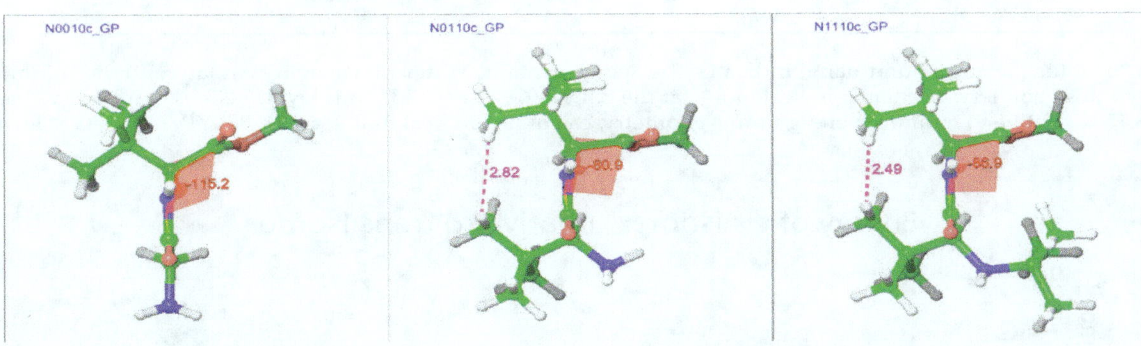

Figure 11. Comparison of cis geometries of N0010, N0110, and N1110 in the gas phase shows increasing strain in the geometries with the successive addition of *tert*-butyl groups. The "c" after the compound name indicates the "cis" isomer.

2.3.1. The Relative Stabilities of Cis vs. Trans Isomers

Tables 7 and 8 describe the relative energies of the minimum energy cis geometries in gas and water phases, respectively, along with the corresponding Gibbs free energy change estimates. Interestingly, the relative energy values of the minimum energy geometries of most cis isomers (compared to the minimum energy geometries of the trans isomers) were found to be higher in the gas phase than in the water phase (Figure 12). This may be attributed partially to the fact that the trans isomers in the water phase already have much higher energy values than the corresponding gas-phase structures, as revealed in steric-factor calculations using the additivity principle for the trans isomers. Furthermore, the solvation energy values for most cis isomers are higher (i.e., more negative) than corresponding trans isomers by 2–3 kcal mol^{-1}. This indicates better stabilization/solvation of the cis isomers than the trans isomers in the water phase, which may, in turn, be attributed to the higher dipole moment of cis isomers compared to trans isomers, as is evident from Figure 13.

Table 7. Dihedral angles (in degree) and relative energies of minimum energy cis isomers in the gas phase.

Compound [a]	ψ^N	ω	φ^C	ψ^C	ΔE [b]	ΔG [c]
N0000c	179.9	0.1	−179.9	179.8	4.63	4.92
N0010c	−176.7	−6.1	−115.2	132.8	5.23	5.14
N0100c	130.7	−2.4	−178.5	−179.3	5.12	4.92
N0110c	130.0	−8.5	−80.9	131.9	6.17	6.07
N011′0c	131.2	0.0	127.3	−136.5	4.96	5.18
N1000c	−162.2	2.1	−177.6	178.5	5.02	5.77
N1010c	−158.2	−5.1	−121.5	135.7	5.67	6.11
N1100c	137.5	−6.4	−175.1	−178.3	5.17	5.05
N1110c	138.9	−3.5	−86.9	132.0	7.52	7.71
N111′0c	140.4	6.9	104.4	−143.9	7.52	7.61

[a] The "c" after the compound name indicates the "cis" isomer. [b] Gas-phase energy relative to the minimum energy trans geometry, calculated in kcal mol^{-1} at the B3LYP/6-311++G(3df,3pd) level. [c] Gibbs free energy change at 298.15 K relative to minimum energy trans geometry, calculated in kcal mol^{-1} at the B3LYP/6-31++G** level.

Table 8. Dihedral angles (in degree) and relative energies of minimum energy cis-isomers in the water dielectric phase.

Compound [a]	Solvation Energy (kcal mol^{-1})	φ^N	ω	φ^C	ψ^C	ΔE [b]	ΔG [c]
N0000c	−19.43	179.5	0.1	−179.6	179.7	2.03	2.1
N0010c	−17.85	−163.8	−0.6	−129.8	144.4	2.97	2.9
N0100c	−16.88	127.5	−2.1	−176.4	−177.4	2.11	2.3
N0110c	−15.67	128.2	4.3	−80.8	143.6	1.38	1.7
N011′0c	−14.7	130.7	8.3	133.4	−141.9	2.80	2.4
N1000c	−19.86	−174.0	0.7	−167.2	178.3	2.79	2.9
N1010c	−18.63	−159.0	0.0	−128.3	143.1	3.34	3.1
N1100c	−14.35	140.1	0.7	113.8	−163.8	3.92	3.7
N1110c	−11.87	134.6	−5.4	−94.3	138.8	7.66	8.1
N111′0c	−11.76	141.2	9.5	109.1	−147.5	4.86	4.94

[a] The "c" after the compound name indicates the "cis" isomer. [b] Water phase energy relative to the minimum energy trans geometry, calculated in kcal mol^{-1} at the B3LYP/6-311++G(3df,3pd) level. [c] Gibbs free energy change at 298.15 K relative to minimum energy trans geometry, calculated in kcal mol^{-1} at the B3LYP/6-31++G ** level.

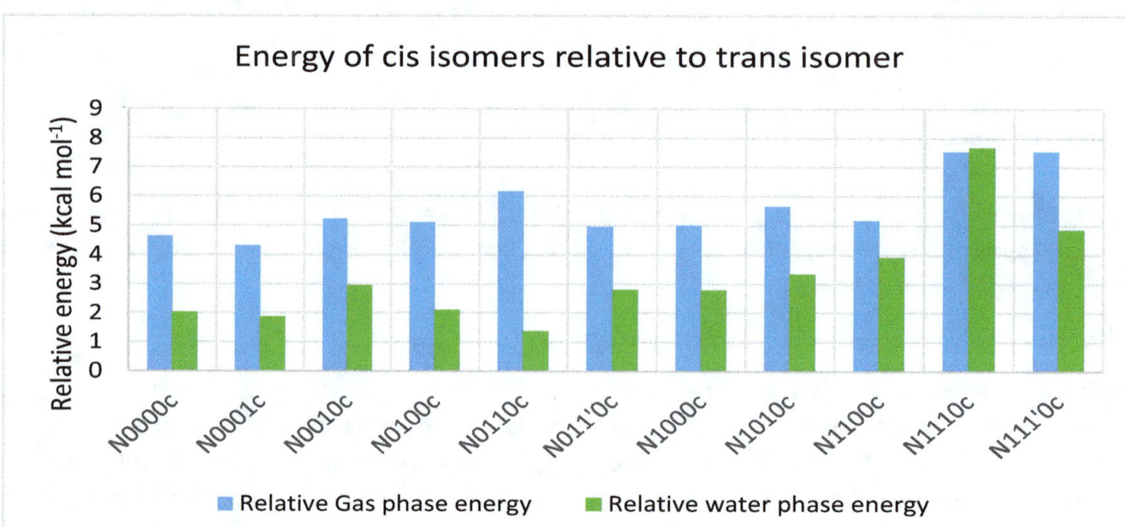

Figure 12. Energies of minimum-energy geometries of cis isomers relative to that of trans-isomers in the gas phase (blue bars) and the water phase (green bars). The "c" after the compound name indicates the "cis" isomer.

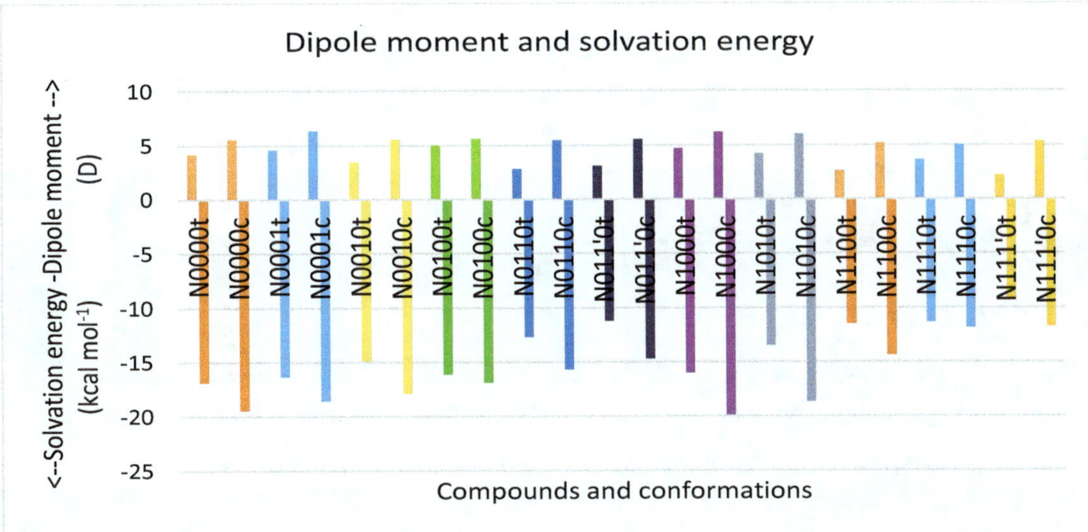

Figure 13. Comparison of the dipole moment and solvation energy for trans and cis isomers (denoted by the "t" and "c" after the compound names, respectively).

2.3.2. The Effect of Bulk on Cis Isomers

Tables 9 and 10 describe the calculation of steric effects on cis-isomers based on the additivity principle in the gas and water phases respectively, which show a trend similar to that for the trans-isomers (Figure 14), as discussed in Section 2.2.

Table 9. Steric-factor effect on the cis-isomer based on the additivity principle in the gas phase.

Compound [a]	^{opt}E kcal mol^{-1}	No. of *t*-Bu on sp^3 Nitrogen	No. of *t*-Bu on sp^3 Carbon	^{Add}E kcal mol^{-1}	^{st}E kcal mol^{-1}
N0000c	−333,801.441			−333,801.441	0
N0010c	−432,511.233		1	−432,513.575	2.342
N0100c	−432,510.785		1	−432,513.575	2.79
N0110c	−531,219.543		2	−531,225.709	6.166
N011′0c	−531,220.53		2	−531,225.709	5.179
N1000c	−432,509.352	1		−432,510.572	1.22
N1010c	−531,218.91	1	1	−531,222.706	3.796
N1100c	−531,217.348	1	1	−531,222.706	5.358
N1110c	−629,924.792	1	2	−629,934.84	10.048
N111′0c	−629,924.4	1	2	−629,934.84	10.44

[a] The "c" after the compound name indicates the "cis" isomer.

Table 10. Steric-factor effect on the cis-isomer based on the additivity principle in the water phase.

Compound [a]	^{opt}E kcal mol^{-1}	No. of *t*-Bu on sp^3 Nitrogen	No. of *t*-Bu on sp^3 Carbon	^{Add}E kcal mol^{-1}	^{st}E kcal mol^{-1}
N0000c	−333,819.556			−333,819.556	0
N0010c	−432,526.761		1	−432,531.518	4.757
N0100c	−432,526.055		1	−432,531.518	5.463
N0110c	−531,233.207		2	−531,243.48	10.273
N011′0c	−531,232.806		2	−531,243.48	10.674
N1000c	−432,525.775	1		−432,527.202	1.427
N1010c	−531,232.932	1	1	−531,239.164	6.232
N1100c	−531,228.49	1	1	−531,239.164	10.674
N1110c	−629,934.673	1	2	−629,951.126	16.453
N111′0c	−629,934.711	1	2	−629,951.126	16.415

[a] The "c" after the compound name indicates the "cis" isomer.

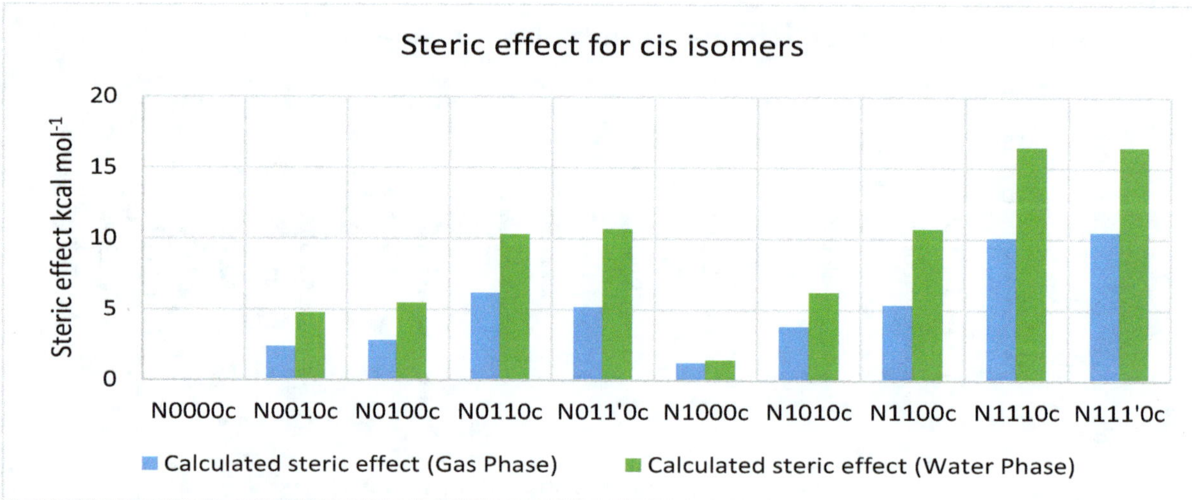

Figure 14. Steric effect for cis-isomers in the gas and water phases using the additivity principle. The "c" after the compound name indicates the "cis" isomer.

2.4. Cis/Trans Isomerization Barrier

The calculation of energy barriers for substituted derivatives of GGMe is a complex task. As we have previously reported [7], the existence of multiple rotamers makes it difficult to identify every possible transition state. In the case of GGMe, however, the lack of substitutions on α-carbons still significantly simplifies the calculation, as the isomerization study along the rotation coordinate ω between 180° and 0° suffices due to the symmetry. This becomes more complex with the introduction of substitutions and stereospecificity at α-carbons, which introduces asymmetry along the rotation coordinates, such as the ω-dihedral, resulting in different energy barriers for rotation along the positive vs. negative directions of rotation, with the simultaneous existence of multiple rotamers. Therefore, the number of possible transition-state conformations increases enormously.

In order to simplify this complex problem, it was necessary to use a specific method to enable comparison among same types of geometries for different compounds. Such a comparison can provide information about the overall effect of substitution patterns, if not for specific conformers. From the examples of *N*-methylacetamide and GGMe [7], it became evident that the energy-barrier geometries obtained with stepwise RCS would provide a reasonably accurate estimation of transition states and corresponding energies. Therefore, we decided to find energy-barrier geometries (EBGs) to estimate energy-barrier values. For compounds without substitutions on the α-carbon, the calculation in one direction (between 0° and 180°, or between 180° (=−180°) and 360° (=0°)) was sufficient. However, for other substituted compounds, calculation in both directions was carried out. Correspondingly, from the trans isomers, *syn* EBGs were obtained close to $\omega = \pm 60°$ and from the cis isomers, *anti* EBGs were obtained close to $\omega = \pm 120°$. Figures 15 and 16 represent the most stable energy barrier geometry for each compound in the gas phase and water phase, respectively.

Tables 11 and 12 describe energy-barrier values corresponding to *syn/anti* EBGs in the gas phase and water phase, respectively. It is evident that the energy-barrier value strongly depends on the direction of rotation for compounds with substitution on chiral α-carbons, and in such cases, the energy difference between the same types of energy-barrier geometries in two different directions of rotation can be significant, from 1 to 6 kcal mol^{-1}.

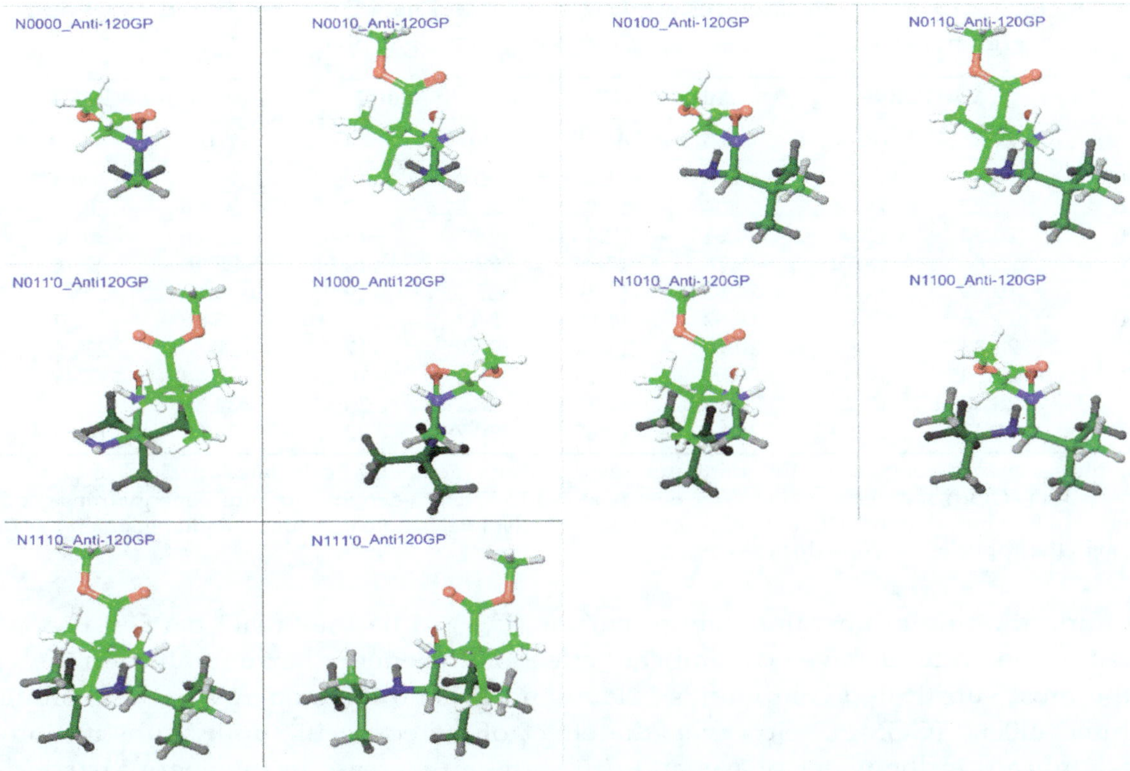

Figure 15. Cis/trans isomerization energy barrier geometries of substituted GGMe derivatives in the gas phase; all geometries are aligned at the amide moiety.

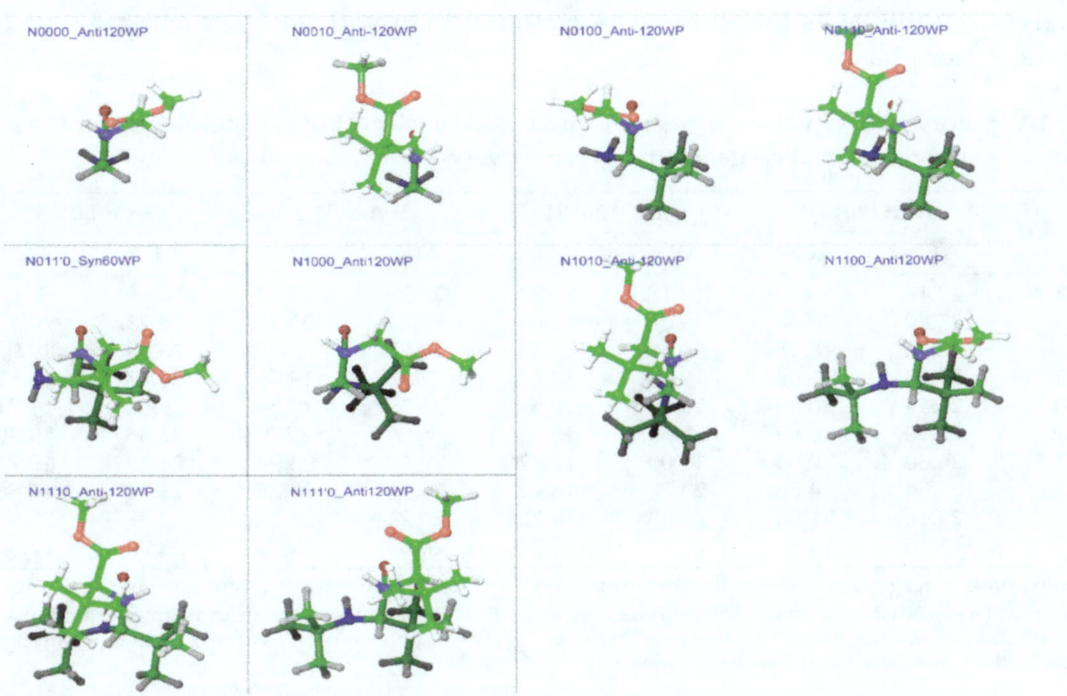

Figure 16. Cis/trans isomerization energy barrier geometries of substituted GGMe derivatives in the water phase; all geometries are aligned at the amide moiety.

Table 11. Energy-barrier values corresponding to *syn* and *anti* EBGs generated from optimized minimum energy trans and cis geometries, respectively, in the gas phase.

Compound	Anti120GP		Anti-120GP		Syn60GP		Syn-60GP		ΔG_{eff}
	ΔE [a]	ΔG [b]	ΔE [a]	ΔG [b]	ΔE [a]	ΔG [b]	ΔE [a]	ΔG [b]	
N0000	19.53	13.70	19.53	13.72	25.46	19.91	25.26	19.84	13.70
N0010	24.22	18.34	19.34	11.93	25.22	17.86	27.99	20.23	11.93
N0100	20.96	13.40	20.51	12.43	26.30	18.63	25.26	20.38	12.43
N0110	25.64	18.72	19.76	12.34	23.98	17.63	28.18	22.10	12.34
N011′0	19.99	12.91	24.92	17.54	27.33	19.06	29.92	22.79	12.91
N1000	19.55	11.03	19.82	11.09	ND [c]	ND	25.09	19.31	11.03
N1010	24.33	17.92	20.10	14.82	24.95	17.65	27.80	22.20	14.82
N1100	21.54	12.90	20.59	13.14	26.61	19.13	25.40	17.49	12.90
N1110	26.08	17.27	20.15	11.17	28.29	20.62	34.39	28.17	11.17
N111′0	21.43	12.59	26.18	17.45	30.79	22.72	23.68	14.51	12.59

[a] Gas-phase energy relative to the minimum-energy trans geometry, calculated in kcal mol^{-1} at the B3LYP/6-311++G(3df,3pd) level. [b] Gibbs free energy at 298.15 K relative to the minimum-energy trans geometry, calculated in kcal mol^{-1} at the B3LYP/6-31++G** level. [c] ND: Not determined, assumed similar geometry as for the rotation in the opposite direction due to symmetry.

Assuming the minimum-value energy barrier (G_{eff}) to be the "real" barrier, it appears that bulky substitution patterns have no significant effect on the energy barrier values in the gas phase, as even the most substituted compounds N1110 and N111′0 have energy barriers similar to the unsubstituted N0000 (GGMe). However, the effect of stereospecific bulky substitution patterns becomes significant in the water phase, and can show an increase in the energy barrier by up to 4–5 kcal mol^{-1}. Diastereomers N0110 and N011′0 (with S- and R- configuation on the C-terminal α-carbon) show a difference of ~5 kcal mol^{-1}. A similar difference is seen also in the case of the other pairs of diastereomers, N1110 and N111′0. Thus, the stereochemistry of substitutions has a profound effect on the energy-barrier heights and can provide an opportunity to control the flexibility selectively. Interestingly, a *syn* EBG was found to be more stable than both *anti* EBGs in the case of compound N011′0 in the water phase.

Table 12. Energy-barrier values corresponding to *syn* and *anti* EBGs generated from the optimized minimum-energy trans and cis geometries, respectively, in the water phase.

Compound	Anti120WP		Anti-120WP		Syn60WP		Syn-60WP		ΔG_{eff}
	ΔE [a]	ΔG [b]	ΔE [a]	ΔG [b]	ΔE [a]	ΔG [b]	ΔE [a]	ΔG [b]	
N0000	19.18	13.82	19.18	13.74	24.29	20.07	ND [c]	ND	13.74
N0010	25.40	20.48	19.03	13.67	29.42	23.23	24.23	20.47	13.67
N0100	22.45	18.28	20.66	16.24	23.31	18.6	26.09	19.91	16.24
N0110	25.13	21.06	19.32	13.46	22.23	19.06	20.47	16.61	13.46
N011′0	26.44	20.83	26.55	20.67	23.39	18.55	26.51	21.71	18.55
N1000	22.30	16.89	22.35	16.85	ND	ND	23.86	19.40	16.85
N1010	26.60	20.05	19.00	12.72	29.60	23.96	24.29	19.71	12.72
N1100	22.54	14.06	23.71	16.92	26.50	19.28	23.04	16.86	14.06
N1110	29.46	21.96	24.53	19.22	29.24	24.21	30.37	23.98	19.22
N111′0	23.15	14.50	27.71	21.34	26.65	19.74	23.78	17.50	14.50

[a] Water-phase energy relative to the minimum-energy trans geometry, calculated in kcal mol^{-1} at the B3LYP/6-311++G(3df,3pd) level. [b] Gibbs free energy at 298.15 K relative to the minimum energy trans geometry, calculated in kcal mol^{-1} at the B3LYP/6-31++G** level. [c] ND: Not determined, assumed similar geometry as for the rotation in the opposite direction due to symmetry.

Overall, bulky substitutions were found to affect cis/trans isomerization in two ways: by making the cis isomer more unstable and thereby shifting the equilibrium towards the trans isomer, and by increasing the reaction barrier height and thereby decreasing the reaction rate. As both effects discourage trans to cis isomerization, compounds with bulky substitutions are expected to take longer and/or require higher temperatures for the same effect. Moreover, as trans to cis isomerization is an essential step in the cyclization of dipeptides to piperazine-2,5-diones, it is certain that slow isomerization would also result in decelerated dipeptide cyclization.

3. Methods

3.1. Generation of Structures

The relevant structures were initially generated as 2D trans isomers using Marvin Sketch 15.6.8, 2015, (ChemAxon, Budapest, Hungary), and were imported to the Maestro [39] module of the Schrodinger suite (Schrödinger, LLC. New York, NY, USA), from which all further studies were performed. The cis geometries were generated by torsional adjustment of the ω dihedral.

3.2. Conformational Search

All trans and cis geometries were subjected to the MacroModel [40] conformational search algorithm, using the MMFFs force-field. From the conformational search output, the geometries within a potential energy of up to 21 kJ/mol over the minimum energy geometry were kept.

All further studies were performed using the Jaguar [41,42] program of the Schrodinger suite.

3.3. Geometric Optimization

The geometries kept after the conformational search were subjected to geometric optimization at the B3LYP/6-31++G** level with maximum grid density and the "accurate" accuracy level of SCF. Frequency calculation studies were carried out to confirm zero imaginary frequencies for an optimized geometry. For each compound, the most stable geometry (i.e., with the minimum energy) was chosen for further study. Calculations for the water dielectric continuum (water phase) were performed using the same parameters and the additional Poisson Boltzmann Finite (PBF) solvent model. The Cartesian coordinates of the optimized geometries are provided in the supplementary information.

3.4. Relaxed Coordinate Scan (RCS)

The optimized minimum energy cis and trans geometries of all compounds were subjected to relaxed coordinate scans (RCS), on the ω dihedral coordinate at the B3LYP/6-31++G** level with maximum grid density and the "accurate" accuracy level of SCF. The optimized geometry at each step was used to generate the starting geometry for the next step. For water-phase calculations, the Poisson Boltzmann Finite (PBF) solvent model was used. Each RCS was performed in three rounds, with a 15° step-size for the first round and a 2° step-size (near the rotation barriers) for the second round. The step-size for the third round (closest to the rotation barrier) was kept at 0.125° and 0.250° for the gas and water phase calculations, respectively. Except for compounds with only hydrogens on both chiral α-carbons, the relaxed coordinate scan was performed in both directions (positive and negative direction, clockwise and anticlockwise). The *syn* and *anti* energy-barrier geometries (EBG) were identified and used as plausible energy-barrier geometries. (Henceforth, EBGs refers to the energy-barrier geometries obtained from RCS). The Cartesian coordinates of the EBGs are provided in the supplementary information.

3.5. Single-Point Energy Calculations

Accurate single-point energies were calculated for all optimized and reaction-barrier geometries at the B3LYP/6-311++G(3df,3pd) level with maximum grid density and the "accurate" accuracy level of SCF. Vibrational analyses were carried out at the B3LYP/6-31++G** level, and the free energy values obtained for 298.15 K were used to calculate relative Gibbs free energies. For water phase calculations, the Poisson Boltzmann Finite (PBF) solvent model was used.

4. Conclusions

In this work, we conducted a systematic theoretical study of the effects of steric bulk on the relative stabilities with varied substitution patterns on a peptidic scaffold using density functional methods. We used model compounds based on the simple glycylglycine methyl ester peptidic scaffold with patterns of stereospecific substitutions of the *tert*-butyl group to represent steric bulk. The study establishes how bulky substitutions significantly destabilize the cis isomers and also increase the energy barrier for cis/trans isomerization (especially in the water phase). These effects synergistically discourage the cis/trans isomerization of secondary amides. Therefore, secondary amides with bulky and more substitutions are much less likely to undergo cis/trans isomerization than the unsubstituted or less substituted second amides. Because cis/trans isomerization is an important step in the synthesis of piperazine-2,5-diones via cyclization of dipeptide esters, the rate of reaction for highly substituted dipeptide esters can be expected to be much slower than those with less substitutions. Moreover, for medicinal chemistry purposes, such highly substituted peptidomimetics can be expected to possess a very rigid peptide backbone.

Author Contributions: Conceptualization, B.S.T.; Formal analysis, B.S.T. and R.A.E.; Funding acquisition, J.S.M.S.; Investigation, B.S.T.; Methodology, B.S.T.; Project administration, R.A.E.; Supervision, J.S.M.S. and R.A.E.; Visualization, B.S.T. and R.A.E.; Writing-original draft, B.S.T.; Writing-review & editing, J.S.M.S. and R.A.E.

Acknowledgments: The authors thank Department of Chemistry, UiT The Arctic University of Norway for supporting this research work. The publication charges for this article have been funded by a grant from the publication fund of UiT The Arctic University of Norway. This paper is taken in part from the Ph.D. thesis of B.S.T.

References

1. LaPlanche, L.A.; Rogers, M.T. Cis and trans Configurations of the Peptide Bond in *N*-Monosubstituted Amides by Nuclear Magnetic Resonance. *J. Am. Chem. Soc.* **1964**, *86*, 337–341. [CrossRef]
2. Ramachandran, G.N.; Mitra, A.K. An explanation for the rare occurrence of cis peptide units in proteins and polypeptides. *J. Mol. Biol.* **1976**, *107*, 85–92. [CrossRef]
3. Ramachandran, G.N.; Sasisekharan, V. Conformation of polypeptides and proteins. *Adv. Protein Chem.* **1968**, *23*, 283–438. [PubMed]
4. Liu, C.; Szostak, M. Twisted Amides: From Obscurity to Broadly Useful Transition-Metal-Catalyzed Reactions by N−C Amide Bond Activation. *Chem. Eur. J.* **2017**, *23*, 7157–7173. [CrossRef] [PubMed]
5. Mock, W.L. Torsional strain considerations in the mechanism of the proteolytic enzymes, with particular application to carboxypeptidase A. *Bioorg. Chem.* **1975**, *4*, 270–278. [CrossRef]
6. Mock, W.L. Torsional-strain considerations in enzymology: Some applications to proteases and ensuing mechanistic consequences. *Bioorg. Chem.* **1976**, *5*, 403–414. [CrossRef]
7. Thakkar, B.S.; Svendsen, J.-S.M.; Engh, R.A. Cis/Trans Isomerization in Secondary Amides: Reaction Paths, Nitrogen Inversion, and Relevance to Peptidic Systems. *J. Phys. Chem. A* **2017**, *121*, 6830–6837. [CrossRef] [PubMed]
8. Stewart, D.E.; Sarkar, A.; Wampler, J.E. Occurrence and role of cis peptide bonds in protein structures. *J. Mol. Biol.* **1990**, *214*, 253–260. [CrossRef]
9. Weiss, M.S.; Jabs, A.; Hilgenfeld, R. Peptide bonds revisited. *Nat. Struct. Mol. Biol.* **1998**, *5*, 676. [CrossRef] [PubMed]
10. Pal, D.; Chakrabarti, P. Cis peptide bonds in proteins: Residues involved, their conformations, interactions and locations. *J. Mol. Biol.* **1999**, *294*, 271–288. [CrossRef] [PubMed]
11. Tchaicheeyan, O. Is peptide bond cis/trans isomerization a key stage in the chemo-mechanical cycle of motor proteins? *FASEB J.* **2004**, *18*, 783–789. [CrossRef] [PubMed]
12. Odefey, C.; Mayr, L.M.; Schmid, F.X. Non-prolyl cis-trans peptide bond isomerization as a rate-determining step in protein unfolding and refolding. *J. Mol. Biol.* **1995**, *245*, 69–78. [CrossRef]
13. Vanhove, M.; Raquet, X.; Palzkill, T.; Pain, R.H.; Frère, J.-M. The rate-limiting step in the folding of the *cis*-Pro167Thr mutant of TEM-1 β-lactamase is the trans to cis isomerization of a non-proline peptide bond. *Proteins Struct. Funct. Bioinform.* **1996**, *25*, 104–111. [CrossRef]

14. Wheeler, K.A.; Hawkins, A.R.; Pain, R.; Virden, R. The slow step of folding of *Staphylococcus aureus* PC1 β-lactamase involves the collapse of a surface loop rate limited by the *Trans* to *Cis* isomerization of a non-proline peptide bond. *Proteins Struct. Funct. Bioinform.* **1998**, *33*, 550–557. [CrossRef]
15. Agarwal, P.K. Cis/trans isomerization in HIV-1 capsid protein catalyzed by cyclophilin A: Insights from computational and theoretical studies. *Proteins Struct. Funct. Bioinform.* **2004**, *56*, 449–463. [CrossRef] [PubMed]
16. Ledvina, A.R.; Chung, T.W.; Hui, R.; Coon, J.J.; Tureček, F. Cascade dissociations of peptide cation-radicals. Part 2. Infrared multiphoton dissociation and mechanistic studies of z-ions from pentapeptides. *J. Am. Soc. Mass Spectrom.* **2012**, *23*, 1351–1363. [CrossRef] [PubMed]
17. Baldoni, H.A.; Zamarbide, G.N.; Enriz, R.D.; Jauregui, E.A.; Farkas, Ö.; Perczel, A.; Salpietro, S.J.; Csizmadia, I.G. Peptide models XXIX. cis–trans Isomerism of peptide bonds: Ab initio study on small peptide model compound; the 3D-Ramachandran map of formylglycinamide. *J. Mol. Struct. THEOCHEM* **2000**, *500*, 97–111. [CrossRef]
18. Wiberg, K.B.; Rablen, P.R.; Rush, D.J.; Keith, T.A. Amides. 3. Experimental and Theoretical Studies of the Effect of the Medium on the Rotational Barriers for *N*,*N*-Dimethylformamide and *N*,*N*-Dimethylacetamide. *J. Am. Chem. Soc.* **1995**, *117*, 4261–4270. [CrossRef]
19. Lauvergnat, D.; Hiberty, P.C. Role of Conjugation in the Stabilities and Rotational Barriers of Formamide and Thioformamide. An ab Initio Valence-Bond Study. *J. Am. Chem. Soc.* **1997**, *119*, 9478–9482. [CrossRef]
20. Mantz, Y.A.; Gerard, H.; Iftimie, R.; Martyna, G.J. Ab Initio and Empirical Model MD Simulation Studies of Solvent Effects on the Properties of *N*-Methylacetamide along a cis−trans Isomerization Pathway. *J. Phys. Chem. B* **2006**, *110*, 13523–13538. [CrossRef] [PubMed]
21. Taha, A.N.; True, N.S. Experimental 1H NMR and Computational Studies of Internal Rotation of Solvated Formamide. *J. Phys. Chem. A* **2000**, *104*, 2985–2993. [CrossRef]
22. Kang, Y.K.; Park, H.S. Internal rotation about the C.–N. bond of amides. *J. Mol. Struct. THEOCHEM* **2004**, *676*, 171–176. [CrossRef]
23. Mantz, Y.A.; Branduardi, D.; Bussi, G.; Parrinello, M. Ensemble of Transition State Structures for the Cis−Trans Isomerization of *N*-Methylacetamide. *J. Phys. Chem. B* **2009**, *113*, 12521–12529. [CrossRef] [PubMed]
24. Thakkar, B.S.; Engh, R.A. Comparative conformational analyses and molecular dynamics studies of glycylglycine methyl ester and glycylglycine *N*-methylamide. *RSC Adv.* **2018**, *8*, 4445–4453. [CrossRef]
25. Yoder, C.H.; Gardner, R.D. Multiple-substituent parameter analysis of the effects of substituents at nitrogen on the barriers to rotation in amides. *J. Org. Chem.* **1981**, *46*, 64–66. [CrossRef]
26. Fischer, G. Chemical aspects of peptide bond isomerisation. *Chem. Soc. Rev.* **2000**, *29*, 119–127. [CrossRef]
27. Moure, A.; Sanclimens, G.; Bujons, J.; Masip, I.; Alvarez-Larena, A.; Pérez-Payá, E.; Alfonso, I.; Messeguer, A. Chemical modulation of peptoids: Synthesis and conformational studies on partially constrained derivatives. *Chem.-Eur. J.* **2011**, *17*, 7927–7939. [CrossRef] [PubMed]
28. Laursen, J.S.; Engel-Andreasen, J.; Fristrup, P.; Harris, P.; Olsen, C.A. Cis-trans amide bond rotamers in β-peptoids and peptoids: Evaluation of stereoelectronic effects in backbone and side chains. *J. Am. Chem. Soc.* **2013**, *135*, 2835–2844. [CrossRef] [PubMed]
29. Goodwin, D.; Simerska, P.; Toth, I. Peptides as therapeutics with enhanced bioactivity. *Curr. Med. Chem.* **2012**, *19*, 4451–4461. [CrossRef] [PubMed]
30. Avan, I.; Hall, C.D.; Katritzky, A.R. Peptidomimetics via modifications of amino acids and peptide bonds. *Chem. Soc. Rev.* **2014**, *43*, 3575–3594. [CrossRef] [PubMed]
31. Muller, P. Glossary of terms used in physical organic chemistry (IUPAC Recommendations 1994). *Pure Appl. Chem.* **2009**, *66*, 1077–1184. [CrossRef]
32. Hirsch, J.A. Table of Conformational Energies—1967. In *Topics in Stereochemistry*; Allinger, N.L., Eliel, E.L., Eds.; John Wiley & Sons, Inc.: Manhattan, NY, USA, 1967; pp. 199–222, ISBN 978-0-470-14710-8.
33. Charton, M. Steric effects. I. Esterification and acid-catalyzed hydrolysis of esters. *J. Am. Chem. Soc.* **1975**, *97*, 1552–1556. [CrossRef]
34. Charton, M. The upsilon steric parameter—Definition and determination. In *Steric Effects in Drug Design*; Topics in Current Chemistry; Springer: Berlin/Heidelberg, Germany, 1983; pp. 57–91, ISBN 978-3-540-12398-9.

35. Icli, S.; Kandil, K.A.; Thankachan, C.; Tidwell, T.T. Steric Crowding and the Reactivity of Substituted tert-Butyl Perbenzoates. *Can. J. Chem.* **1975**, *53*, 979–985. [CrossRef]
36. Marcelin, G.; Brooks, P.R. Steric hindrance in potassium atom-oriented molecule reactions. Methyl iodide and tert-butyl iodide. *J. Am. Chem. Soc.* **1975**, *97*, 1710–1715. [CrossRef]
37. Bartell, L.S.; Bradford, W.F. Molecular structures of neopentane and di-tert-butylmethane by vapor-phase electron diffraction. *J. Mol. Struct.* **1977**, *37*, 113–126. [CrossRef]
38. Toniolo, C. Intramolecularly hydrogen-bonded peptide conformations. *CRC Crit. Rev. Biochem.* **1980**, *9*, 1–44. [CrossRef] [PubMed]
39. *Schrödinger Release 2016-1: Maestro*; Schrödinger, LLC: New York, NY, USA, 2016.
40. *Schrödinger Release 2016-1: MacroModel*; Schrödinger, LLC: New York, NY, USA, 2016.
41. Bochevarov, A.D.; Harder, E.; Hughes, T.F.; Greenwood, J.R.; Braden, D.A.; Philipp, D.M.; Rinaldo, D.; Halls, M.D.; Zhang, J.; Friesner, R.A. Jaguar: A high-performance quantum chemistry software program with strengths in life and materials sciences. *Int. J. Quantum Chem.* **2013**, *113*, 2110–2142. [CrossRef]
42. *Schrödinger Release 2016-1: Jaguar*; Schrödinger, LLC: New York, NY, USA, 2016.

Permissions

All chapters in this book were first published in MDPI; hereby published with permission under the Creative Commons Attribution License or equivalent. Every chapter published in this book has been scrutinized by our experts. Their significance has been extensively debated. The topics covered herein carry significant findings which will fuel the growth of the discipline. They may even be implemented as practical applications or may be referred to as a beginning point for another development.

The contributors of this book come from diverse backgrounds, making this book a truly international effort. This book will bring forth new frontiers with its revolutionizing research information and detailed analysis of the nascent developments around the world.

We would like to thank all the contributing authors for lending their expertise to make the book truly unique. They have played a crucial role in the development of this book. Without their invaluable contributions this book wouldn't have been possible. They have made vital efforts to compile up to date information on the varied aspects of this subject to make this book a valuable addition to the collection of many professionals and students.

This book was conceptualized with the vision of imparting up-to-date information and advanced data in this field. To ensure the same, a matchless editorial board was set up. Every individual on the board went through rigorous rounds of assessment to prove their worth. After which they invested a large part of their time researching and compiling the most relevant data for our readers.

The editorial board has been involved in producing this book since its inception. They have spent rigorous hours researching and exploring the diverse topics which have resulted in the successful publishing of this book. They have passed on their knowledge of decades through this book. To expedite this challenging task, the publisher supported the team at every step. A small team of assistant editors was also appointed to further simplify the editing procedure and attain best results for the readers.

Apart from the editorial board, the designing team has also invested a significant amount of their time in understanding the subject and creating the most relevant covers. They scrutinized every image to scout for the most suitable representation of the subject and create an appropriate cover for the book.

The publishing team has been an ardent support to the editorial, designing and production team. Their endless efforts to recruit the best for this project, has resulted in the accomplishment of this book. They are a veteran in the field of academics and their pool of knowledge is as vast as their experience in printing. Their expertise and guidance has proved useful at every step. Their uncompromising quality standards have made this book an exceptional effort. Their encouragement from time to time has been an inspiration for everyone.

The publisher and the editorial board hope that this book will prove to be a valuable piece of knowledge for researchers, students, practitioners and scholars across the globe.

List of Contributors

Jiaxi Xu
State Key Laboratory of Chemical Resource Engineering, Department of Organic Chemistry, College of Chemistry, Beijing University of Chemical Technology, Beijing 100029, China

Lazaros Papadopoulos and Dimitrios N. Bikiaris
Laboratory of Polymer Chemistry and Technology, Department of Chemistry, Aristotle University of Thessaloniki, GR-541 24 Thessaloniki, Greece

Tobias Robert
Fraunhofer Institute for Wood Research—Wilhelm-Klauditz-Institut WKI, Bienroder Weg 54E, 38108 Braunschweig, Germany

Marcel Kluge
Fraunhofer Institute for Wood Research—Wilhelm-Klauditz-Institut WKI, Bienroder Weg 54E, 38108 Braunschweig, Germany
Institute for Chemistry of Renewable Resources, Department of Chemistry, University of Natural Resources and Life Sciences Vienna, Konrad-Lorenz-Straße 24, 3430 Tulln, Austria

Weijia Shi and Gang Zou
School of Chemistry & Molecular Engineering, East China University of Science & Technology, 130 Meilong Rd, Shanghai 200237, China

Zrinka Rajić, Maja Beus, Ivan Kosalec, Toma Keser and Branka Zorc
Faculty of Pharmacy and Biochemistry, University of Zagreb, A. Kovačića 1, 10000 Zagreb, Croatia

Hana Michnová and Josef Jampílek
Department of Pharmaceutical Chemistry, Faculty of Pharmacy, Comenius University, Odbojárov 10, 83232 Bratislava, Slovakia

Josipa Vlainić
Laboratory for Advanced Genomics, Division of Molecular Medicine, Rudjer Bošković Institute, Bijenička cesta 54, 10000 Zagreb, Croatia

Leentje Persoons and Dominique Schols
Laboratory of Virology and Chemotherapy, Rega Institute for Medical Research, KU Leuven, Herestraat 49, 3000 Leuven, Belgium

Igor V. Komarov
Institute of High Technologies, Taras Shevchenko National University of Kyiv, Vul. Volodymyrska 64/13, 01601 Kyiv, Ukraine

Aleksandr Yu. Ishchenko, Aleksandr Hovtvianitsa, Viacheslav Stepanenko and Serhii Kharchenko
Enamine Ltd., Vul. Chervonotkatska 78, 02094 Kyiv, Ukraine

Andrew D. Bond and Anthony J. Kirby
University Chemical Laboratory, Cambridge CB2 1EW, UK

Sriram Mahesh, Kuei-Chien Tang and Monika Raj
Department of Chemistry and Biochemistry, Auburn University, Auburn, AL 36849, USA

Jyunichi Terasawa, Yu Shibata, Miho Fukui and Ken Tanaka
Department of Chemical Science and Engineering, Tokyo Institute of Technology, O-okayama, Meguro-ku, Tokyo 152-8550, Japan

Ervin Kovács
Institute of Materials and Environmental Chemistry, Research Centre for Natural Sciences, Hungarian Academy of Sciences, Budapest H-1117, Hungary

Balázs Rózsa
Laboratory of 3D Functional Imaging of Neuronal Networks and Dendritic Integration, Institute of Experimental Medicine, Hungarian Academy of Sciences, Budapest H-1083, Hungary

Attila Csomos
Institute of Chemistry, Eötvös University, Budapest H-1117, Hungary

Imre G. Csizmadia
Department of Chemistry, University of Toronto, Toronto, ON M5S 3H6, Canada

Zoltán Mucsi
Department of Chemistry, University of Toronto, Toronto, ON M5S 3H6, Canada
Department of Chemistry, Femtonics Inc., Budapest H-1094, Hungary

Sungsoo Kim, Jieun Kim, Daeun Won, Suk-Kyu Chang and Sangdoo Ahn
Department of Chemistry, Chung-Ang University, Seoul 156-756, Korea

Jungyu Kim
Center for Molecular Spectroscopy and Dynamics, Institute for Basic Science (IBS), Seoul 02841, Korea

Kyungwon Kwak
Center for Molecular Spectroscopy and Dynamics, Institute for Basic Science (IBS), Seoul 02841, Korea
Department of Chemistry, Korea University, Seoul 02841, Korea

Wansik Cha
Nuclear Chemistry Research Division, Korea Atomic Energy Research Institute, Daejeon 34057, Korea

Keunhong Jeong
Department of Chemistry, Korea Military Academy, Seoul 01805, Korea

Hengzhao Li, Zemin Lai, Adila Adijiang, Hongye Zhao and Jie An
College of Science, China Agricultural University, No. 2 Yuanmingyuan West Road, Beijing 100193, China

Roman Szostak
Department of Chemistry, Wroclaw University, F. Joliot-Curie 14, 50-383 Wroclaw, Poland

Michal Szostak
Department of Chemistry, Rutgers University, 73 Warren Street, Newark, NJ 07102, USA

Chong-Lei Ji, Pei-Pei Xie and Xin Hong
Department of Chemistry, Zhejiang University, Hangzhou 310027, China

Balmukund S. Thakkar, John Sigurd M. Svendsen and Richard A. Engh
Department of Chemistry, UiT The Arctic University of Norway, N-9037 Tromsø, Norway

Index

www.ingramcontent.com/pod-product-compliance
Lightning Source LLC
LaVergne TN
LVHW081022210926
848485LV00004B/925